Polymer Characterization

Spectroscopic, Chromatographic, and Physical Instrumental Methods

Clara D. Craver, EDITOR

Chemir Laboratories

Based on a symposium sponsored

by the ACS Macromolecular Secretariat

at the 181st Meeting

of the American Chemical Society,

March 29–April 3, 1981,

Atlanta, Georgia

ADVANCES IN CHEMISTRY SERIES 203

AMERICAN CHEMICAL SOCIETY

WASHINGTON, D.C.   1983

Library of Congress Cataloging in Publication Data

Polymer characterization.

(Advances in chemistry series, ISSN 0065–2393; 203)

"Based on a symposium sponsored by the Macro-molecular Secretariat of the ACS at the 181st Meeting of the American Chemical Society, March 29–April 3, 1981, Atlanta, Georgia."

Includes bibliographies and index.

1. Polymers and polymerization—Congresses.
I. Craver, Clara D. II. American Chemical Society. Macromolecular Secretariat. III. American Chemical Society. National Meeting (181st: 1981: Atlanta, Ga.) IV. Series.

QD1.A355 [QD380] 540s [547.8'4] 82–24496
ISBN 0–8412–0700–3

QD
1
. A355
V. 203
1983

Advances in Chemistry Series

M. Joan Comstock, *Series Editor*

FOREWORD

Advances in Chemistry Series was founded in 1949 by the American Chemical Society as an outlet for symposia and collections of data in special areas of topical interest that could not be accommodated in the Society's journals. It provides a medium for symposia that would otherwise be fragmented, their papers, distributed among several journals or not published at all. Papers are reviewed critically according to ACS editorial standards and receive the careful attention and processing characteristic of ACS publications. Volumes in the Advances in Chemistry Series maintain the integrity of the symposia on which they are based; however, verbatim reproductions of previously published papers are not accepted. Papers may include reports of research as well as reviews since symposia may embrace both types of presentation.

ABOUT THE EDITOR

CLARA D. CRAVER, president and founder (1958) of Chemir Laboratories, received her B.S. degree in chemistry from Ohio State University in 1945 and was awarded the honorary degree of Doctor of Science by Fisk University in 1974. Her work at Esso Research Laboratories, 1945–1949, resulted in patents on characterization of complex hydrocarbon mixtures. She established an infrared laboratory at Battelle Memorial Institute in 1949, and, acting as group leader until 1958, conducted spectroscopic work for Battelle's research projects as well as original research work that resulted in scientific publications in the areas of drying oils, asphalts, paper, and resins. Her work in organic coatings won her the 1955 Carbide and Carbon Award of the Division of Organic Coatings and Plastics Chemistry of the ACS. In 1975 she became chairman of that Division. She initiated the Coblentz Society's spectral publication program and is editor of five books of Special Collections of IR Spectra. She served as consultant to ASTM for the Evaluated IR Spectral Publication Program supported by the National Bureau of Standards, and is chairman of the Joint Committee on Atomic and Molecular Physical Data. She is past-chairman of the ASTM Committee on Molecular Spectroscopy, and was named a Fellow of ASTM in 1982 when she received that organization's highest honor, the Award of Merit. She is a Fellow of the American Institute of Chemists and a Certified Professional Chemist.

CONTENTS

vii

PREFACE

ODAY'S POLYMER SCIENTIST makes use of a large array of characterization methods to develop data relating the performance of polymeric materials to chemical and physical structure. Polymer applications demand conflicting performance characteristics: strong adhesion with easy peel, hardness with resiliency, flexibility with dimensional stability, inertness with miscibility, smoothness with paintability, rigidity with lack of brittleness, and crack resistance. The desired properties are obtained by the selection of chemical composition, including residual or grafted functional groups. Control of properties can be exercised by alternating structures, by alternating clumps of functional groups or by synthesizing interpenetrating polymer networks. Some properties are controlled by molecular dimensions: linear, branched, cross-linked. Others are controlled by molecular size, by curing conditions and cross-linking agents, by blending and by additives including plasticizers, antioxidants, surfactants, stabilizers, and fillers.

The characterization and resultant understanding of polymeric systems require input from many fields: chromatography, spectroscopy, rheology, surface chemistry, electron microscopy, thermal analysis, and physical testing. Other highly specialized techniques involve light scattering, neutron scattering, excimer fluorescence, photoacoustic spectroscopy, and pyrolytic degradation, with exact identification of small molecular subunits by chromatography combined with IR and mass spectrometry.

In drawing upon so many diverse and specialized fields, polymer characterization has profited by the cross-fertilization of ideas that occur when specialists meet and communicate with one another. Serving this purpose, the American Chemical Society Divisions of Organic Coatings and Plastics Chemistry, Polymer Chemistry, Rubber Chemistry, Colloid Chemistry, and Cellulose and Paper Chemicals hold cosponsored symposia at the National ACS meetings. The Macromolecular Secretariat, a programming group composed of all five divisions, designated Polymer Characterization as its topic for 1981, and with the participation of the Division of Analytical Chemistry, programmed a full week's symposium organized and chaired by the editor. This book, covering significant advances in polymer characterization, is based on that symposium. It is

xi

organized around seven major methods with single chapters on a few new or highly specialized techniques.

Comprehensive coverage of dynamic mechanical methods is provided in nine chapters by experts selected by the 1978 Borden Award winner, John K. Gillham, of Princeton University. The latest developments in automation of equipment for dynamic mechanical spectroscopy, torsion pendulum and torsional braid analyses, and shear in polymer melts are reported. Applications include impregnated cloth analysis, mechanical analysis of organic coatings, and strength and viscoelastic characterization of polymeric solids.

The thermal characteristics of polymers bear important relationships to their performance. Thermal gravimetric analysis (TGA) may be used for identification or determination of purity of polymers. TGA can measure solvent retention and provide activation energies and heats of reactions. Bernhard Wunderlich, of Rensselaer Polytechnic Institute, describes sources of experimental errors in differential scanning calorimetry (DSC) and summarizes the state of the art in instrumentation. The treatment of kinetic data and kinetic modeling, including applications to polymer degradation, is reported by J. Flynn and B. Dickens of the National Bureau of Standards. Application to coatings is reported by investigators from Glidden Coatings and Resins Division of SCM Corporation.

Chromatographic methods of polymer fractionation and the resultant correlation of chromatograms with polymer composition and performance are among the most widely used methods of characterization. However, the data must be interpreted cautiously because many variables affect polymer distribution as reflected by chromatograms. Among these are nonuniformity of chemical composition of the various particle size ranges which in turn affect solubilities and detector response. The coverage of this subject as arranged by Theodore Provder, of SCM Corporation, includes the theory and understanding of calibration problems of polymer chromatography and techniques for minimizing them. Novel partition methods such as foam fractionation are also presented.

Studies of polymer surfaces by scanning electron microscopy alone and in combination with x-ray analysis and micro-Raman techniques is especially useful in organic coatings analysis and research. Results reported here include investigation of mildew attack on paint surfaces and corrosion of metal substrates in microscopic-sized blisters at the coating–metal interface. Raman microprobe, MOLE, permits chemical analysis of microareas. An example of its usefulness is the identification of "fisheyes" in a copolymer sheet as being homopolymer impurities. The capability of correlating metal analysis from x-ray maps of surfaces with organic analysis by Raman microprobe makes these combined techniques a powerful research tool. L. H. Princen, USDA, organized an application-oriented report on these significant instrumental advances.

Nearly half of the chapters in this book report on advances in the use of spectroscopy for polymer characterization. These tools are often combined, not only with each other, but with thermal and mechanical techniques to relate molecular structure to polymer performance and stability.

E. Brame, du Pont, has provided an overview of NMR applications. Advances in ^{13}C NMR, made possible by the increased sensitivity of pulsed Fourier transform instruments, provide improved determination of tacticity, comonomer sequence and branching of synthetic polymers. There are five full chapters reporting on advances in polymer structure determination by NMR including ^{1}H, ^{13}C, and ^{19}F. The chapter authored by Frank Bovey, of Bell Laboratories, is rich in interpretation of the significance of NMR data as applied to polymer synthesis, stability, and morphology. Practical use of sophisticated structural analysis arises from such findings as the detection of tertiary chloride atoms in poly(vinyl chlorine) which may contribute to its thermal instability.

NMR is usually carried out on polymers in solution or in the molten state. Only low resolution or broadline NMR has been attainable on solids. Recent advances in magic-angle spinning and dipolar decoupling along with multipulse NMR have made it possible to resolve individual resonances of solid state samples. Research directed toward describing the segmental motions in amorphous and semicrystalline polymers and understanding secondary transitions makes use of these NMR techniques applied to samples at varying temperatures.

IR spectroscopy has long been a valuable tool for chemical characterization of polymers and for following the chemical changes in curing or degradation reactions. The increased sensitivity now obtainable from computer averaging of data and the increased speed of determining a spectrum made possible by Fourier transform are expanding the applications of IR to minor composition differences and to conformation at surfaces and in bulk samples under stress. J. Koenig, Case Western Reserve University, arranged the coverage of this subject and included advances in the theory of IR intensities and developments in photoacoustic spectroscopy.

Advances in the techniques for observing microstructural changes by IR data simultaneously with stress–strain data are described by S. L. Hsu and D. J. Burchell, University of Massachusetts. Degradation studies of poly(ethylene terephthalate) and polyacrylonitrile polymers are reported by B. Bulkin, E. Pearce, and C. Chen, of the Polytechnic Institute of New York, and M. Coleman and G. Sivy of Pennsylvania State University, respectively. James Boerio, University of Cincinnati, describes reflection/absorption spectra showing polymer orientation at interfaces of ultrathin coatings on polished metal surfaces. Advances in the application of photoacoustic Fourier transform IR to polymer analysis is discussed by Warren Vidrine of Nicolet Instruments.

Major advances have been made in the last few years in the theory of IR and Raman intensities. A wealth of molecular structure information is available from absolute band intensities, but spectroscopists have lacked the theoretical tools to interpret the significance of these intensities. The manuscript contributed by Giuseppe Zerbi, of Istituto Chimica, Milan, Italy, describes advances in this field. This contribution provides examples that demonstrate the significance of the breakthrough in theory to the understanding of molecular structure of polymers. Other new areas reported by Dr. Zerbi deal with the calculation of vibrational frequencies of very large molecules with no symmetry and the determination of chain lengths from the Longitudinal Acoustic Mode (LAM).

Improvements in techniques for pyrolytic degradation of polymers into reproducible fragments that can be analyzed by gas chromatography and mass spectrometry permit atom-by-atom molecular arrangements to be identified. The coverage of analytical pyrolysis was organized by Shirley Liebman, of Chemical Data Systems, Inc. S. Tsuge, of Nagoya University, Japan, discusses the fundamental conditions for obtaining characteristic and reproducible high-resolution pyrograms and the application to sequence distribution investigation in copolymers. The specificity of laser sources and the usefulness of being able to vary power densities is demonstrated in the report by David Hercules et al., University of Pittsburgh, on laser microprobe mass analysis (LAMMA). These techniques are applicable to diverse macromolecules. Coal characterization is reported by K. Voorhees and coworkers, Colorado School of Mines. Forrest Bayer provides an overview of the specificity and sensitivity of these techniques to a wide range of biopolymers. E. Reiner and F. Moran, of Georgia Institute of Technology, characterize details of pathogenic microorganisms and mammalian cells including ones reflecting genetic defects which may produce disease such as cystic fibrosis.

Larry Brydia, of Union Carbide, represented the Analytical Division and served as chairman for specialized techniques. Two subjects from that symposium that are covered in single chapters are advances in scattering techniques by R. S. Stein, University of Massachusetts, and the use of excimer fluorescence as a probe of polymer structure by C. W. Frank of Stanford University.

I am indebted to the subject chairmen and the authors for providing well-balanced coverage in each field. They have included advances in theory along with experimental details essential for obtaining good data. The practical value of the results from these major methods of chemical and physical characterization of polymers has been interpreted in terms of stability and performance of the many polymers which were investigated. I am especially grateful to all of them for their cooperation and patience which made this book possible.

CLARA D. CRAVER
Chemir Laboratories
Glendale, MO 63122

November 1982

PHYSICAL PROPERTIES

Automated Dynamic Mechanical Testing

RAYMOND F. BOYER

Michigan Molecular Institute, Midland, MI 48640

During the past 10 years a number of fully automated instruments for the dynamic mechanical testing of polymers became commercially available. Our purpose is to provide an introductory survey that covers key references, the goals of such test methods, general but not specific differences between the several types of instruments, and factors to consider in choosing an instrument. These latter topics include fixed vs. variable (resonant) frequency, frequency range, resolving power, types of specimens that can be handled, and maintenance costs. We also emphasize that all suppliers and users of such instruments are striving to improve and/or extend the range of capabilities of their respective instruments. A special section discusses the problem of determining activation enthalpies for any loss process. Finally, we point out that parallel developments are occurring in dielectric testing via new techniques: automated self-balancing dielectric bridges, and new methods of treating dielectric data. Thermally stimulated creep and thermally stimulated current are relatively new methods. We consider these alternate techniques as both competitive with, and complementary to, dynamic mechanical methods. Possible artifacts introduced by the use of porous substrates to support fluid polymers are discussed and dismissed.

GENERAL ASPECTS OF DYNAMIC MECHANICAL and dielectric testing, automated or not, are discussed in this chapter. Background material is given in References 1–10. For many years I collected transition and relaxation temperature data from the literature, whether obtained by static methods such as thermal expansion or by dynamic methods including both mechanical and dielectric loss. This endeavor pro-

0065–2393/83/0203–0003$06.75/0
© 1983 American Chemical Society

vided an opportunity to compare numerous test methods and instruments while being relatively neutral as regards any one method (1 –9).

I have been closely associated with J. K. Gillham in the preparation of publications (see, for example, Reference 11) involving data obtained by the torsional braid method described in Reference 12. In 1979, we acquired a du Pont 981 dynamic mechanical analyzer (DMA) (13). Several joint publications with K. Varadarajan resulted from studies with this instrument (14 –16). These personal experiences will not detract from my goal of objectivity in preparation of this introductory chapter. I plan to emphasize certain principles and features of dynamic test methods that should be considered by those who are interested in acquiring an instrument for a specific research program.

I will comment on two of my previous publications (17, 18) to emphasize rapid developments in this field. In 1972, I compared general features of the Rheovibron and the torsion pendulum, both of which were commercial and neither of which was automated; and also the newly emerging torsional braid analysis (TBA) invented by Lewis and Gillham and perfected by Gillham (12). Reference 19 gives a historical review and comparison of both the torsion pendulum and the TBA (19).

In 1976 (18) I projected a complete swing to the automation of dynamic mechanical testing. This viewpoint hardly required a crystal ball. Gillham's apparatus now is perfected, much as he has summarized it (12). Kenyon et al. had already presented a lecture on their success in automating the Rheovibron (20). Automation of dynamic testing equipment indeed was at hand, evidenced by the Dynostat instrument developed by Sternstein (22); the mechanical spectrometer developed by Starita (23); and later the PL dynamic mechanical thermal analyzer developed by Wetton (24). Still another automated dynamic mechanical tester developed in connection with Hedvig is the Unirelax (Tetrahedron Associates). Although in commercial use, it temporarily was withdrawn from the market pending development of a new model. Hence, this instrument is not discussed in this volume.

This proliferation in instrumentation thus includes automation of older instruments such as the Rheovibron, the introduction of totally new types, and extensive use of computer data plotters to permit representation of any desired mechanical loss function in any desired format. [See also the recent paper by Ikeda and Starkweather (25).]

Two net consequences of the pattern just mentioned are the following: (1) A tremendous advance in the state of the art in acquiring precision test results on dynamic mechanical and related rheological data by a variety of sophisticated competitive methods; and (2) bewilderment and confusion for the research person faced with a choice of purchasing one of these new units to perform the tasks he has in mind, while generally working within a prescribed budget limitation.

Perhaps the most useful function I can perform is to suggest a series of considerations that might be helpful in selecting an instrument.

Meaning of Dynamic Mechanical and Dielectric Loss Peaks

When any material is subject to a small mechanical stress and hence low strain, the stress–strain curve is linear. If the substance is truly elastic, all of the stored elastic energy, namely the area under the stress–strain line, is recovered on removing the load, as at the left of Figure 1. However, most materials are not truly elastic and will exhibit a hysteresis loop, as at the right of Figure 1, on removing the load. Zener called this latter behavior anelastic. Some of the stored energy is dissipated in the form of heat. The area of the hysteresis loop, which is a measure of heat generation per deformation cycle, varies with temperature. Over a short temperature region near a molecular relaxation process, the loop may be exceedingly small at temperatures above and below the relaxation process with a maximum in area at the temperature of the relaxation. Over a broad temperature region in a polymer with multiple transitions, several maxima in the area of the hysteresis loops, one for each relaxation process, will be evident.

For a dielectric stress applied to a polar polymer, one or more hysteresis loops also arise as frequency or temperature is increased. Dielectric heating takes advantage of this fact. In dynamic mechanical testing, heating effects are usually small but nonetheless real. An example of such heating effects was given by Törmälä et al. (26). A cylinder of polycarbonate (PC) is sandwiched between two steel rods—an elastic input bar and an elastic output bar. Impacting the input bar with a striker generates a stress pulse that is transmitted through the specimen to the output bar. Heat generated by this dynamic loading of the polymer bar is measured with a thermocouple for a series of starting temperatures. These temperature increases can

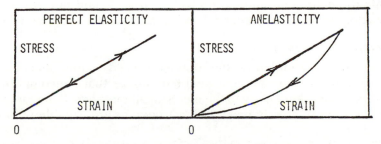

Figure 1. Schematic stress–strain plots: left, a perfectly elastic material; and right, a material that is not perfectly elastic. Area of hysteresis loop represents energy dissipated in the form of heat.

be converted to dynamic loss moduli by the following approximate equation:

$$G'' \equiv 4mC_p\Delta T/\pi e^2_{max}$$

where m is the mass of the sample, C_p is its specific heat, ΔT is the temperature rise, and e_{max} is the compressive strain. Loss peaks for T_γ (230 K), T_β (350 K), and T_g (above 430 K) are depicted. We have found (27) these three loss peaks (at lower temperatures because of lower frequencies) using the 981 DMA unit described in Reference 13. The existence of a true β-relaxation near 350 K has created some controversy, which was summarized recently (16). The temperature increase measured by Törmälä et al. (26) apprently offers additional evidence for a β-loss process in PC.

The various dynamic mechanical testing units currently available are simply very sensitive, sophisticated instruments used to measure energy loss per cycle (and hence heat generation) by a variety of nonthermal sensing devices, such as damping per cycle or phase angle. The various instruments measure and report the following quantities, depending on their mode(s) of operation: real, G', and imaginary, G'', shear moduli; the corresponding compliances, J' and J''; real, E', and imaginary, E'', tensile moduli. The loss angle, δ, usually is reported as tan $\delta = G''/G' = E''/E'$; the dynamic melt viscosity, η, is reported as $\eta = G''/\omega$ where ω represents the angular frequency, namely, $\omega = 2\Pi f$ where f is the frequency in hertz. Quantities proportional to some of these parameters also may be given. Instruments in which the frequency varies also may display the frequency directly or may provide information from which it can be calculated. Table I lists units and relations between units.

Although an applied alternating voltage stress gives rise to heat generation, especially at high frequencies and voltages, temperature rise is not used normally to measure dielectric loss (but may be as discussed in Reference 5). Dielectric quantities commonly reported include dielectric constant, ϵ'; dielectric loss, ϵ''; and tangent of the loss angle, tan $\delta = \epsilon''/\epsilon'$. Other related quantities may be given.

Competitive Techniques

Anyone contemplating the purchase of dynamic mechanical equipment must be aware of several techniques that are competitive with and/or complementary to the mechanical methods. Some of these techniques are relatively new; all have been or are capable of being automated.

Thermally Stimulated Creep. This mechanical technique was developed by Lacabanne et al. (27–29). The specimen is subjected to

Table I. Common Terms, Symbols, and Units for Dynamic Mechanical and Dynamic Melt Viscosity Data

Term	Symbol	Units — Old	Units — New
Shear modulus	G'	dynes/cm² kg/cm² lb/in.²	N/m²[a]
Shear loss	G''	lb/in.	N/m²
Tensile modulus	E'	lb/in.	N/m²
Tensile loss	E''	lb/in.	N/m²
Loss angle	δ	degrees	same
Loss tangent	tan δ[b]	—	—
Decrement	Δ[c]	—	—
Frequency		cycles/s	Hz
Angular frequency	ω	radians/s	same
Dynamic melt viscosity	η'[d]	Ps[e]	same

Note: Conversion factors are approximately as follows:
$$1 \text{ kg force/cm}^2 = 14.2 \text{ PSI}$$
$$= 0.968 \text{ atm}$$
$$= 0.981 \text{ bar}$$
$$= 9.81 \times 10^4 \text{ N/m}^2$$

(Because E' and G' may be of the order of 10^9 N/m² $= 10^9$ Pascals, the term gigapascals is being used with symbol gP.)

[a] Newtons per meter squared, also called Pascals.
[b] Tan $\delta = G''/G' = E''/E'$.
[c] $\Delta = \pi$ tan δ. Log Δ is plotted directly by some instruments.
[d] $\eta' = G''/\omega$.
[e] Pascal seconds.

a torsional load at an elevated temperature and cooled rapidly with the load still applied. On warming, both strain and its time derivative are recorded as a function of temperature. The method is capable of high resolving power for detecting multiple transitions. It uses a torsion pendulum with the specimen supported on a glass braid if the melt state is to be studied.

Dielectric Methods. These methods are of two types: (a) direct reading bridges that provide either recording or digital readout of dielectric constant, dielectric loss, tan δ, and other quantities as a function of temperature; some instruments provide a choice of several frequencies; and (b) thermally stimulated current discharge (TSC). This method is the electrical analog of thermally stimulated creep. A high voltage is applied to a polymer film at elevated temperature and the sample is cooled rapidly with the voltage on. After removal of the applied voltage, the specimen is heated at a slow rate while current flow through the specimen is measured. A current peak occurs at each transition (*see* References 8 and 9 for details). Figure 2 shows a current trace for atactic polymethyl methacrylate (PMMA) (*30*). An auto-

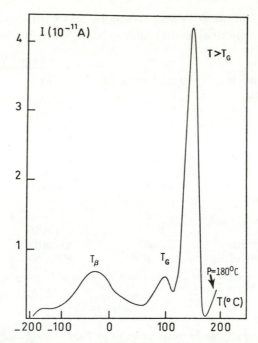

Figure 2. TSC trace on atactic polymethyl methacrylate that was polarized at 180 °C for 2 min by an applied field of 10^6 V/m, rapidly cooled under voltage, and heated at about 1 K/min. Effective frequency was $\sim 10^{-4}$ Hz. Vertical axis is electric current in amperes generated by the polarized specimen because of reorientation of dipoles and/or motion of charge carriers. See Ref. 28 for details.

mated instrument for obtaining such data is the Electret Thermal Analyzer (Atlas Electric Devices). Resolving power is very high and many small current peaks, each indicative of some transition or relaxation, may be seen in addition to major transitions observed in dynamic testing. TSC has several advantages besides high resolving power and simplicity. First, it can work in the molten state where polymers are mechanically weak, very sticky, and subject to viscous flow. Lacabanne et al. used TSC to observe the liquid–liquid (T_{ll}) transition in PS (31) and PMMA (30). Second, having scanned a specimen from low to high temperature and located all of the current peaks, one can do a new experiment combining selection of the polarizing temperature followed by short-circuiting at some lower temperature to isolate any desired transition region. Examples appear in Reference 31 for PS and Reference 30 for PMMA. Application of the TSC technique to other specific polymers may be cited as follows: several polystyrenes (32); polyacrylonitrile (33); polycarbonate (PC) (34); several crystalline polymers [polyethylene (PE), Teflon, polyvinylidene fluoride

(PVDF)] (35, 36); polyblends (37); epoxies (38); and nylons (39). Of course, numerous examples appear in References 8 and 9.

A (qualitative) comparison between dynamic mechanical and dielectric test results may be instructive. Except in special cases, relating a dynamic mechanical loss peak to a specific type of mechanical motion may be difficult. Of course, J. Heijboer's torsion pendulum study (40) of cyclohexyl and other cycloalkyl ring motions is a noted exception in which the type of motion is known in detail. If a polymer contains a polar group (including mildly oxidized PE) so that there is direct coupling between the dielectric field and that polar group, the nature of the motion may be known with precision. For a complicated polymer like PC, a combination of many techniques was needed to unravel the low temperature loss peaks: dynamic mechanical, dielectric, IR, TSC, and NMR. We reviewed the pertinent literature on this polymer (14, 16).

Dielectric testing always has faced a dilemma between direct current conduction loss at low frequency and poor resolving power at high frequency. A technique recently reported by Klaase and Van Turnhout (41) allows for circumvention of the problems associated with conduction currents at relatively low frequency. Traditionally, ϵ'' loss peaks may be buried completely by conduction currents, usually at high fluidity, whether the result of high temperature, the presence of diluents, or both. Classical methods can correct this problem. However, Reference 41 points out that the dielectric constant, ϵ', is not affected by conduction currents. Hence, the derivative, $d\epsilon'/dT$ or $d\epsilon'/d \log f$, will exhibit a maximum at an ϵ'' loss peak buried under the conduction current. This technique can greatly extend the utilization of dielectric testing. Application of this method to dielectric loss at 1 Hz as applied to PVDF has been demonstrated (41).

I predict a resurgence of interest in dielectric testing sparked by the electronic applications of electrets and other piezoelectric devices prepared from polymers and copolymers of vinylidene fluoride. This resurgence will be aided both by the self-balancing bridges and by the improved data treatment methods just mentioned. Moreover, a dielectric bridge is a convenient method for extending the frequency range of dynamic mechanical loss equipment, thus permitting calculation of more accurate activation energies. (*See* section entitled "Test Frequency.")

Hedvig (4) described the principles of various self-balancing and/or direct recording bridges. Yalof and Wrasidlo (42) discussed a specific direct balancing–recording unit now commercially available as Audrey (automatic dielectrometer) from Tetrahedron. This group (42) strongly recommends the combined use of dielectric testing with one other method, for example, differential scanning calorimetry (DSC), in production quality control.

We use a Hewlett-Packard automatic balancing bridge that provides digital readouts of capacitance, resistance, inductance, tan δ, and Q at any of three preselected frequencies, 10^2, 10^3, and 10^4 Hz (42). Designed for production line quality control testing of electrical components, it has proven to be a useful research tool. Its lower limit of tan δ is 0.001. This automatic balancing bridge can be interfaced with recording equipment. Its use in detecting T_{ll} in polychlorostyrene is discussed elsewhere (43). The liquid–liquid transition (T_{ll}) in polypropylene glycol 4000 also was detected with this same instrument (44).

Choosing an Instrument

This task is formidable, in part because of the variety of high caliber instruments available from which to make a choice; in part because usually more than one need and/or more than one group must be satisfied; and finally because there may be money restrictions. Some compromises frequently must be made in optimizing the choice in instrumentation. I attempt only to point out some factors for consideration.

I shall not attempt to make a point-to-point comparison of all commercial instruments. I do speak of specific brand names in relation to a situation or problem under discussion; however, the citing of a specific brand name should not be construed as endorsing that instrument over competitive brands. The main factors that I wish to cover as guidelines of choice are the following:

1. Defining the task(s) or problem(s) of the instrument being chosen. This includes immediate as well as possible future needs. This is an individual problem and can best be solved by considering the following specific areas.
2. Types of specimens, temperature range, frequency range, and deformation mode. (*See* Table II.)
3. Fixed vs. variable frequency. (*See* the section on "Test Frequency.")
4. Properties (dynamic loss, storage modulus, etc.) as a function of frequency vs. such properties as a function of temperature, or as a function of both.
5. Resolving power, accuracy, and reproducibility. (*See* the section on "Resolving Power.")
6. Total capability of a given instrument; a one-line characterization of any instrument can be misleading. (*See* the section on "Over Categorization.")

Table II. Types of Specimens and Temperature Ranges

Specimen Type	*Comment*
1. Fixed liquids such as plasticizers that may have transitions such as melting and T_g	A substrate to carry the specimen is needed with most instruments [a]
2. Elastomers	May require vulcanization and special sample clamping device
3. Amorphous thermoplastics measured up to or just above T_g	Usually no problem unless specimen is very brittle [b]
4. Same as 3 but measured above T_g	Mechanical support needed [a]
5. Semicrystalline polymers up to T_m	Usually no problems
6. Same as 5 but going above T_m	Mechanical support needed [a]
7. Thermosets (already cured) up to thermal decomposition temperatures	Usually no problems
8. Curable low mol. wt. liquids or solids that change to thermoplastics or thermosets on cure	Support system needed [a]
9. Composites consisting of a thermoplastic polymer blended with particulate or fibrous filler	Filler may provide internal support into the $T > T_g$ temperature region
10. Composite thermosets	Usually no problems

[a] Substrate may be a porous system impregnated with the polymer; or a solid such as a metal strip coated with the polymer. *See* section on "The Braid Controversy."
[b] Low molecular weight polymers tend to have low strength and may be quite brittle.

Temperature vs. Frequency as an Independent Plotting Variable. Dynamic mechanical functions such as E', G', E'', G'', and tan δ may be plotted as a function of temperature at a single fixed frequency (isochronal plot); against frequency, f, at a single fixed temperature (isothermal plot); combinations in which T or f is the independent variable and different isochrones or isotherms appear on the same plot as a function of T or f, respectively; or finally, on a reduced frequency plot against log fa_T where a_T is the reduction factor. Such plots combine different sets of temperature and frequency data into a single curve commonly given at a reference temperature of 25 °C, although any reference temperature may be selected. Examples of these various types of plots may be inspected in standard reference works $(1-5)$.

Some instruments may come with accessories that include a self-contained computer–plotter for plotting any dependent variable against one or more independent variables. More simply, data from the instrument may be tape punched and used in a standard computer-data plotting system. One sophisticated device of great versatility was described elsewhere (25).

Any dynamic mechanical unit capable of being preset to a fixed frequency (*see* the section entitled "Test Frequency") will have a relatively small number of these fixed frequencies to select as well as a small total range. Temperature increments at a fixed frequency can be as small as the experimenter desires, such as 1, 5, 10, and 20 °C, or, with some instruments, they can increase continuously. Plots against temperature show locations of transitions at temperatures that can be correlated with data from quasi-static devices such as DSC, thermal expansion, and other devices.

Finally, two practical suggestions are based on the assumption that an individual or group has arrived at a choice between two competitive instruments or even a single instrument. One suggestion is to talk to one or several individuals already using one or the other of the two instruments in question. Subtle points about mechanical dependability, speed of service, replacement costs, etc., may emerge. The other point is to request from the instrument manufacturer a test run on one or several typical specimens that are of current interest to the eventual user. There are, of course, many other practical considerations. For example, if the intended use of an instrument covers a study of elastomers, block polymers, or any system with T_g well below room temperature, one should compare alternate instruments in costs for liquid nitrogen and/or mechanical refrigeration systems.

Over Categorization

It is relatively easy to attach to any given instrument a label that implies limited use for that instrument. Such labels are restrictive and misleading. In an early review we stated that the Rheovibron was limited to the tension mode at temperatures up to T_g. We learned (*see* footnotes to Table I of Reference 17) that it had indeed been used in shear above T_g. Manufacturers are aware of the diverse, even though seldom used, capabilities of their own instruments as developed by themselves or described in the literature.

Gillham emphasizes that a TBA unit can be used as a torsion pendulum by substituting a polymer film for the loaded braid; and, conversely, that any torsional pendulum can function in the TBA mode by using a loaded braid in place of the film. The only substantive differences are that (a) the braid plus polymer system is a composite that does not yield an absolute value of G'; and (b) one can not operate appreciably above T_g or T_m in the pendulum mode.

We expect the trend to grow such that any given instrument will have one or several principal modes of operation but it need not be confined to those modes. Competition between suppliers of instruments and/or ingenuity on the part of users will lead to a constant stream of innovations.

The mode of operation of the du Pont DMA 981 unit has puzzled many persons. Lear (45) clarified this situation by showing that at a low L/t (length/thickness) ratio the deformation is primarily shear; at high L/t, the deformation is in bending, hence tension.

Of course, before the development of TBA, techniques had been developed and theory provided as early as 1951 for polymer-coated metal strips to be measured by the torsion pendulum method, thus permitting study of low molecular weight polymers and/or polymers above T_g (*see* footnote to Table I in Reference 17 and Appendix C).

Subsequently, numerous techniques have been described: use of porous supports for torsion pendulum (46) and Rheovibron (47); polymer-coated coiled springs in the Rheovibron, a method called dynamic spring analysis (DSA) (48); use of a glass monofilament coated with a uniform thickness of polymer as the specimen in a TBA unit (49). Although the specimen is a composite, simple geometry permits calculation of polymer constants. Problems concern initial uniformity of coating and vertical flow of polymer during the experiment. Obviously, the same experiment can be conducted in a conventional torsion pendulum.

Test Frequency

Available dynamic mechanical test equipment is of several types: (a) a small number of fixed, preselected frequencies, with temperature as the variable; (b) a frequency sweep against a linear or logarithmic frequency scale; and (c) resonant systems in which the frequency decreases with increasing temperature as the modulus of the specimen decreases. The torsion pendulum and DMA 981 units are of this type.

The intended end use for a dynamic mechanical tester will be a major factor in deciding which of three types of units to choose. One may wish to correlate with impact testing where the effective frequency is about 1000 Hz; or with acoustic energy absorption where a range from about 100 to about 10,000 Hz may be of interest. Alternatively, the investigator may wish to construct relaxation maps, namely, plots of log frequency against reciprocal temperature for each of the several loss peaks in a specimen, from which apparent enthalpies of activation, ΔH_a, can be calculated. In this case, a frequency range of at least two and preferably a minimum of three decades is desirable to get meaningful values of ΔH_a. If extreme precision in ΔH_a is desired, several dynamic mechanical test instruments may be needed. For example, one group advocates the use of five different instruments to cover the frequency range from 10^{-4} Hz (creep) to 10^6 Hz (50).

Conversely, low frequency mechanical loss results may be supplemented with dielectric measurements in which a single dielectric bridge may cover the range from 60 to 10^5 Hz. If, however, one's interest is primarily in locating mechanical loss peaks along the temperature scale as a function of chemical structure, polymer blends, block copolymers, percent crystallinity, cross-linking, additive content such as fillers or plasticizers, a resonant-type apparatus is generally adequate. The glass transition temperature for most common thermoplastics and thermosets is relatively invariant with frequency over a range of 100–1000 Hz.

With resonant systems, the initial frequency can be varied within limits by changing the geometry of the instrument and/or the specimen dimensions. A torsion pendulum may choose any of several inertia disks to alter initial frequency, although this approach is not common. The du Pont DMA 981 unit can employ specimens of different lengths and/or thicknesses to alter the starting frequency.

Our philosophy about frequency variation for calculating ΔH_a values is now best described as cautious though not callous. This attitude is based on the following observations and considerations:

1. Many relaxation maps have been determined and published, generally in convenient collections (1–7).

2. Information gained from these reports shows that ΔH_a for T_g increases monotonically (actually, exponentially) with increasing T_g so that knowing T_g by any low frequency method such as DSC or torsion pendulum allows a reasonably accurate estimate of ΔH_a for T_g to be made. (See Reference 10 and also Appendix B.)

3. In similar fashion, ΔH_a for the β-relaxation [$T_\beta(K) \simeq 0.75 T_g(K)$ for many polymers] increases linearly with $T_\beta(K)$ (51).

4. Knowing the several transitions in order of decreasing temperature at a low frequency such as about 1 Hz, namely, melting, α_c, T_g, T_β, T_γ, T_δ, and other transitions, the relaxation map most likely will consist of a series of straight lines drawn from these 1-Hz values and intersect at or near $10^{12}/T(K)$ on the vertical scale at a temperature very near to T_m (52).

5. Numerical analytical procedures for determining activa-

tion energies from a single frequency have been proposed (53). The treatment of resonant modes is a special
case.

6. Occasionally, we collected scattered mechanical loss
data from the literature and constructed a relaxation
map, usually of limited scope. In one instance, we
merged our low frequency DMA data on a β-loss peak
with higher frequency literature data to estimate a reliable value of ΔH_a (16).

These remarks should not be construed to imply a negative attitude on our part about relaxation maps and values of ΔH_a. Such numerical values and plots clearly assist in the overall understanding of
relaxation and transition phenomena. Rather, we are making two
points: (1) If an individual or group is forced, for whatever reason, to
purchase an instrument lacking controlled frequency settings, various
alternatives are available for estimating ΔH_a values, as just enumer-
ated; and (2) one may get carried away by the seeming luxury of
twisting knobs without realizing some of the attendant problems in
analyzing the data.

Resolving Power

Resolving power denotes herein the ability of a technique or an
instrument to separate two adjacent loss peaks. This ability will depend on several factors: (1) actual separation of the transitions in Kelvins for a low frequency test such as thermal expansion or DSC; (2)
strengths of the two transitions; (3) frequency of the test method; (4)
inherent sensitivity of the test instrument; (5) activation enthalpies for
the two loss processes; and (6) mode of deformation, namely shear,
tension, or bending.

A good polymer with which to check resolving power is PVDF at
low temperature. According to unpublished torsion pendulum data of
Heijboer at 1 Hz, a T_g loss peak at -40 °C and a still lower temperature
peak at -100 °C are well resolved. Our experience in examining literature data from different researchers on different specimens, but all
using a Rheovibron, has been that these two peaks are not resolvable
at a frequency of 110 Hz, are just resolved at about 35 Hz, and are
resolved clearly at 11 Hz and lower.

To check this observation further, we conducted experiments on a
single specimen of PVDF using the du Pont 981 DMA unit at three
starting frequencies determined by varying the sample length and
thickness. The results at about 10 Hz are shown in Figures 3 and 4.
The pattern was much the same: not resolved at 50 Hz; partly resolved

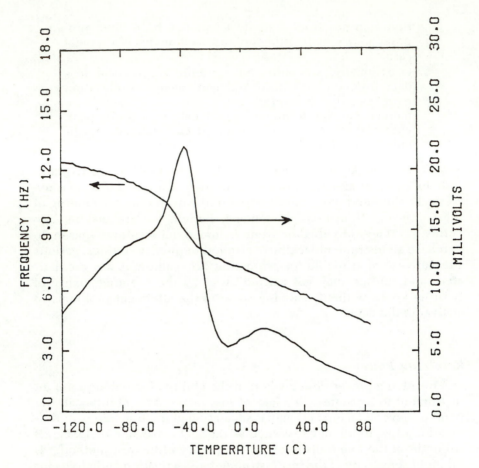

Figure 3. DMA 981 dynamic mechanical loss data in millivolts for PVDF, showing the resolved β-loss peak at −84 °C for f = 10 Hz, as well as the lower (L) and upper (U) glass transition loss peaks, $T_g(L)$ = −38 °C, and $T_g(U)$ = 14 °C.

at 25 Hz; and well resolved at 10 Hz. Various techniques such as use of a curve resolver or computer simulation can help in doubtful cases, but use of frequencies below 20–25 Hz are generally most reliable. Data on other pairs of transitions and/or other polymers may give different results. Also, the automated Rheovibron is said to have better resolving power than the older models (20).

The basic problem at issue is shown schematically in Figure 5, which represents a relaxation map for a polymer having two low temperature transitions with frequency dependence as indicated. The two lines ultimately converge at higher frequencies. Resolution of the two processes is obviously best at lower frequencies and will ultimately fail at the higher frequencies, regardless of the capability of a given instrument.

Even at low frequency (~1 Hz) with the most sensitive of the dynamic mechanical instruments, there are still loss peaks whose details cannot be resolved. Yet other methods can achieve a factor of 3 to 10 in greater detail. A case involving TSC (*see* section entitled "Competitive Techniques") and the γ-peak in PC recently was reviewed (*14, 16*).

Another striking case is afforded by a study (*29*) on atactic PMMA using both thermally stimulated current and thermally stimulated creep whose results were combined with literature data on phos-

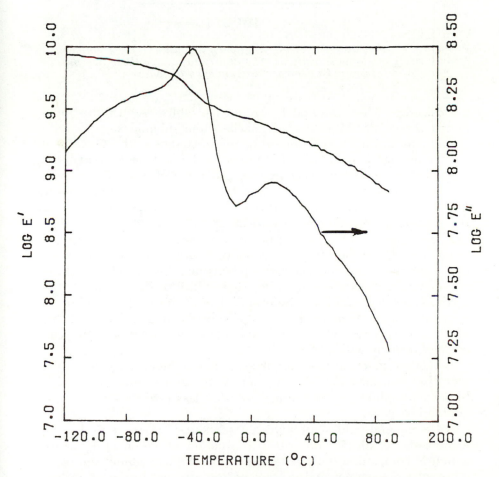

Figure 4. Data of Figure 3 converted to E′ and E″, both in units of Pascals. Logarithmic scale for E″ enhances apparent strength of β-peak compared to T_g(L) peak. T(°C): T_g(U), 14; T_g(L), −38; and T_β, −84.

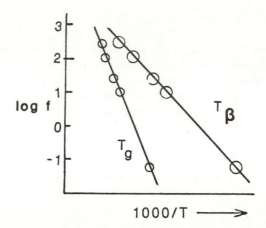

Figure 5. Schematic plot of log frequency against 1/T for a polymer with two close lying transitions. The ability to resolve these two processes must decrease as frequency increases.

phorescence. These several techniques established that the long-known β-peak of PMMA contains at least four submodes.

Atactic PMMA was observed in the bulk state with ^{13}C NMR at room temperature (p. 103ff of Reference 3). The motion of each of the several types of carbon atoms present, $-CH_2-$, CH_3-, $C=O$, etc., was resolved and their respective motional frequencies were determined. Possibly, one or more of these types of motion may contribute to the several sub-β processes in PMMA.

The question of resolving power may seem to be primarily of academic interest; however, it has many practical aspects. For example, absorbed water can affect the mechanical loss spectra of many polymers such as the nylons (*see* Figure 12.9 of Reference 1); the methacrylates (Figure 8.9 of Reference 1); the epoxies (*54*), and other systems in some manner that can be subtle or major. Usually, this effect is most prevalent in polymers containing polar groups that can interact strongly with water.

Another practical case involves polymer blends, for example, blends of two incompatible polymers with similar values of T_g: the blend will exhibit either a broad, single glass peak or a resolvable double loss peak.

The general field of additives presents many possible problems. Armeniades et al. (*55*) described a situation in which mineral oil, which is compatible with general purpose PS at ambient temperatures, phase separates and gives a narrow distinct loss peak at low temperature.

Concerning all three problems that were lumped together (resolving power, reproducibility, and accuracy), we think the ultimate answer lies in having duplicate (possibly multiple) runs on a specific material on the instrument(s) being considered for acquisition.

The Braid Controversy

Three papers appeared recently using rheological data and/or rheological arguments to claim that a transition (relaxation) observed by TBA above T_g in PS is an artifact caused by an interaction between glass braid and molten polymer (56–58). Gillham and I assembled various facts to counter these claims (11, 59, 60). We use both rheological data obtained by methods not employing a braid and non-rheological types of data in which no net transport of molten polymer occurs such as thermal expansion, specific heat, and several types of spectroscopic evidence. The interested reader can consult the three references just cited, a paper by Cowie (61), and the study of Nguyen and Maxwell on melt-elasticity of PS (62).

A specific problem involves the TBA braid and T_{ll}. Gillham and Schwartz reported (63) that the T_{ll} of PS depends on the braid material. They ascribed this phenomenon to an interfacial tension problem. However, Hedvat subsequently demonstrated that these effects arise because of the composite system, such that moduli of braid and polymer play a role both at T_g and T_{ll} (64, 65).

A more important area concerns the use of TBA to follow the curing of thermosets (66). Here, Gillham observes a pregelation loss peak. My belief, based on the studies just noted in the T_{ll} region, is that this and other aspects of the curing process observed by TBA are genuine.

Actually, the basic problem goes beyond that of a specific technique, namely, TBA. Any dynamic mechanical technique involving a support system for a polymer that is weak or fluid may raise the same issues. Other supports reportedly used include blotting paper, filter paper, metal wire mesh, glass cloth, incompatible polymers, compatible polymers, metal strips, etc. Varadarajan and I (67) reviewed this subject elsewhere. Koleske and Faucher also reviewed the use of porous substrates (68).

One may ignore any general and/or theoretical arguments as to whether the braid in TBA (or, for that matter, the porous substrate in any dynamic mechanical test) introduces one or more false transitions, i.e., artifacts. The real issue is whether or not loss peaks observed by any of these porous support methods correlate in a meaningful way with end-use properties.

Summary

This chapter presented (1) a tutorial introduction for individuals new to the field but interested in the nature and potential of dynamic mechanical testing and (2) an overview of the types of automated equipment currently available. Factors to consider before purchasing an instrument were listed. At the same time, it is emphasized that various new techniques such as thermally stimulated current and thermally stimulated creep, as well as improvements in dielectric loss measurements, can yield information that may be superior to, or more generally complementary to, that obtained by dynamic mechanical testing.

Resolving power, or the ability to observe detail between and/or in mechanical loss peaks, appears to have two aspects:

1. Resolution of the β-peak (at $0.75\,T_g$) from T_g with modern instruments appears to require use of a frequency in the range of $1-10$ Hz, as is illustrated with PVDF (*see* Figures 3 and 4).

2. Fine structure within the β-peak of PMMA or within the γ-peak (circa -120 °C) of PC requires the use of techniques such as thermally stimulated current or thermally stimulated creep, with effective frequencies in the range of 10^{-4} Hz, and hence much higher resolving power.

Finally, the use of porous substrates (braids, filter paper, and wire mesh) as supports for fluid systems (oligomers, curable monomers, or molten polymers) was discussed from the viewpoint of possible artifacts arising from the composite nature of the specimen. I believe that, because transitions observed with supported systems are generally observable by a variety of methods not involving supports, the artifact problem has been exaggerated.

Appendix A. Additional Automated Instruments

Subsequent to the preparation of the foregoing material, we learned of two additional automated instruments that recently became commercially available. One such instrument is the Torsionautomat supplied by Brabender. As the name implies, this instrument is a fully automated torsional pendulum. A four-color plotter is employed to record quantities such as G', G'', log decrement, and frequency as a function of temperature or time.

A dynamic viscosimeter was offered by Rodem. It consists of oscillating concentric cylinders driven mechanically with continuously variable frequency over a range of 100 to 5×10^{-5} Hz. The parameters G', G'', and loss angle can be obtained on solutions, gels, and molten polymers. Digital readout is standard for a computer with data handling being optional.

Most likely, other automated instruments are under development or are even already announced; a survey was not conducted.

Appendix B.   Activation Energies for T_g, T_β, and T_{ll}

As mentioned earlier, ΔH_a values tend to increase as the transition temperature in question increases. Figure 6 is a semilogarithmic plot of ΔH_a for T_g as a function of T_g.

Figure 7 is a similar plot for T_β. We find the latter plot to be slightly more linear than the linear fit proposed by Heijboer (32). Values of ΔH_a on which Figures 6 and 7 were based can be found in Tables III and IV of Reference 10.

Plots of $\log f - 1/T$ for T_{ll} are curved decidedly, being concave downward. They do not follow an Arrhenius equation but rather a Vogel–WLF type of relationship, with $\exp - [B/(T - T_0)]$ where T_0 is a reference temperature, and B is a constant. Examples of curved plots for T_{ll} may be found in References 30 and 31 and in Figures 20 and 22 of Reference 60. Hence, ΔH_a for T_{ll} decreases with increasing frequency. Cowie (61) suggests that ΔH_a for T_{ll} is about half of that for T_g, but this statement must refer to frequencies of the order of 100 Hz.

Appendix C.   Polymer-Coated Metal Strips Used with DMA

Polymer-coated metal strips have been used as specimens in the torsion pendulum for many years (69–71), but this practice is not widespread. Starkweather and Giri (72) developed a variation of this

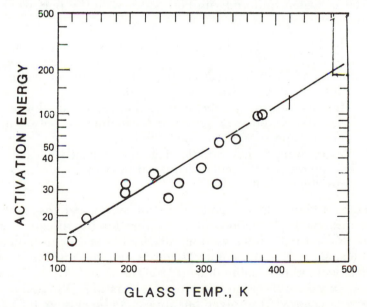

Figure 6. Apparent activation energy, $\triangle H_a$, in kilocalories per mole, as a function of quasistatic value of T_g. The rectangle in upper right corner indicates literature range of T_g and $\triangle H_a$ for poly(2,6-dimethylphenylene oxide).

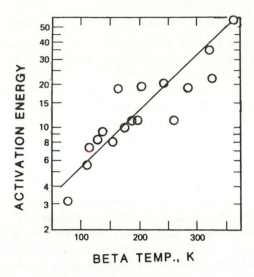

Figure 7. Same as Figure 6 but for β-transition ($T_\beta < T_g$). (See Ref. 29.)

old concept in which a low mechanical loss metal strip (e.g., brass shim stock) is used as a support for polymers to be studied in the du Pont 981 DMA unit. Several advantages are found, especially when the metal is coated uniformly on both sides with the polymer to be studied.

1. Absolute modulus of the polymer in bending can be calculated.
2. Loss peaks can be located although absolute values of the loss may not always be observed.
3. Measurement of the dynamic properties is possible at least 100 K above T_g provided that adhesion between polymer and support is maintained.
4. Resonant frequency variation is reduced greatly because the modulus of the metal changes little over the temperature range of interest.

Figure 8 shows loss expressed in millivolts and resonant frequency in hertz as a function of temperature for 10-mil copper shim stock coated on both sides with about 10 mils of a high molecular weight (ca. $M_n = 5 \times 10^5$, where M_n represents the number-average molecular weight) polyisobutylene (PIB). The T_g, $T > T_g$, or T_{ll} transitions and a weak β-relaxation are evident. The ratio $T_{ll}(K)/T_g(K)$ is 1.22. The T_β peak is not well resolved but appears to be close to 0.75 $T_g(K)$ = −107 °C. Morgan et al. (73) were unable to locate a β-peak in PIB using the vibrating reed technique but a β-peak was observed by Wei (74) in the region of −90 °C on six different specimens of PIB varying in $\bar{M}_n$ from 760−8.6 × 10⁵, using TBA. Resolution was best in the

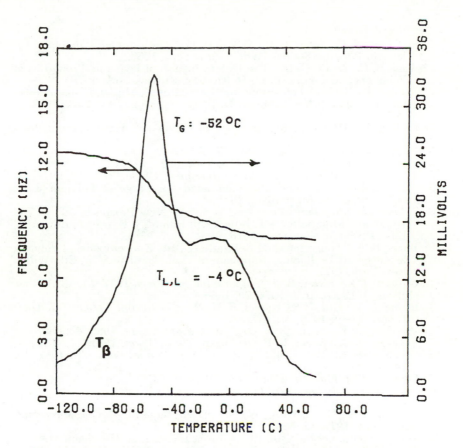

Figure 8. DMA 981 loss curve for PIB coated on both sides of 10-mil copper shim stock, which permits loss measurements to be extended to about 100 K above T_g at a relatively constant frequency.

specimen of lowest $\bar{M}_n$ but not appreciably better than that shown in Figure 8.

Acknowledgments

I am indebted to K. Varadarajan, a former postdoctoral fellow at Michigan Molecular Institute, for the DMA results presented in Figures 3, 4, and 8. I also am indebted to Howard Starkweather of E. I. du Pont de Nemours & Co., Inc. Experimental Station, for discussing the advantages of polymer-coated shim stock in DMA prior to publication. C. Lacabanne provided us, prior to publication, with the manuscript from which Figure 2 was prepared. Helpful comments were received from John Gillham, Princeton University and Philip S. Gill, du Pont Instrument Products Department, Wilmington, DE.

This chapter is dedicated to Dale J. Meier, Michigan Molecular Institute, on the occasion of his 60th birthday.

Literature Cited

1. McCrum, N. G.; Read, B. E.; Williams, G. "Anelastic and Dielectric Effects in Polymer Solids"; Wiley: New York, 1967.
2. Read, B. E.; Dean, G. D. "The Determination of Dynamic Properties of Polymers and Composites"; Halsted Press: New York, 1979.
3. "Molecular Basis of Transitions and Relaxations", Midland Macromolecular Monograph No. 4; Meier, D. J., Ed.; Gordon: New York, 1978. 1978.
4. Hedvig, P. "Dielectric Spectroscopy of Polymers"; Halsted Press: New York, 1977.
5. Hedvig, P. *Macromol. Rev.* **1980,** *15,* 375.
6. McCall, D. W. *NBS Spec. Publ. U.S.* **1967,** *301.*
7. Törmälä, P. *J. Macromol. Sci. Rev. Macromol. Chem.* **1979,** *17,* 297.
8. Van Turnhout, J. "Thermally Stimulated Discharge of Polymer Electrets"; Elsevier: Amsterdam, 1975.
9. "Thermally Stimulated Processes in Solids: New Prospects", Proc. Int. Workshop, Montpelier, VT, June 1976; Fillard, J. P.; van Turnhout, J., Eds.; Elsevier: New York, 1978.
10. Boyer, R. F. "Transitions and Relaxations"; In "Encyclopedia of Polymer Science and Technology Supplement" Vol. 2; Bikales, N., Ed.; Wiley: New York, 1977.
11. Gillham, J. K.; Boyer, R. F. *J. Macromol. Sci. Phys.* **1977,** *13,* 497.
12. Gillham, J. K. *AIChE J.* **1974,** *20,* 1066.
13. Blaine, R. L.; Gill, P. S.; Hassel, R. L.; Woo, L. *J. Appl. Polym. Sci. Appl. Polym. Symp.* **1978,** *34,* 157.
14. Varadarajan, K.; Boyer, R. F. *Org. Coat. Plast. Chem.* **1980,** *42,* 689.
15. Varadarajan, K.; Boyer, R. F. *Org. Coat. Plast. Chem.* **1981,** *44,* 409.
16. Varadarajan, K.; Boyer, R. F. *J. Polym. Sci., Polym. Phys. Ed.* **1981,** *19,* 141.
17. Boyer, R. F. *J. Macromol. Sci. Phys.* **1974,** *9,* 187.
18. Boyer, R. F. *Polymer* **1976,** *17,* 996.
19. Boyer, R. F. (Guest Ed.) *Polym. Eng. Sci.* **1979,** *19,* 661−759.
20. Kenyon, A. S.; Grote, W. A.; Wallace, D. A.; Rayford, M. C. *J. Macromol. Sci. Phys.* **1977,** *13,* 553.
21. Technical product bulletin, Rheovibron, Imass, Inc., Hingham, MA.
22. Technical product bulletin, Dynastat, Imass, Inc., Hingham, MA.
23. Technical bulletin, The Mechanical Spectrometer, Rheometrics, Inc., Union, NJ.
24. Technical bulletin, PL Dynamic Mechanical Analyzer, Stow, OH.
25. Ikeda, R. M.; Starkweather, H. W. *Polym. Eng. Sci.* **1980,** *20,* 322.
26. Törmälä, P.; Paakonen, E.; Kemppainen, P. *J. Mater. Sci.* **1981,** *16,* 275.
27. Chatain, D.; Lacabanne, C.; Monpagens, J.-C. *Makromol. Chem.* **1977,** *178,* 583.
28. Lacabanne, C. et al. *Org. Coat. Plast. Chem.* **1980,** *43,* 550.
29. Lamarre, L.; Schreiber, H. P.; Wertheimer, M. R.; Chatain, D.; Lacabanne, C. *J. Macromol. Sci., Phys.* **1980,** *18,* 195.
30. Lacabanne, C.; Chatain, D. *J. Polym. Sci., Polym. Phys. Ed.* **1973,** *11,* 2315.
31. Lacabanne, C.; Goyaud, P.; Boyer, R. F. *J. Polym. Sci., Polym. Phys. Ed.* **1980,** *18,* 277.
32. Marchal, E.; Benoit, H.; Vogl, O. *J. Polym. Sci., Polym. Phys. Ed.* **1978,** *16,* 949.
33. Stupp, S. I.; Carr, S. H. *J. Polym. Sci., Polym. Phys. Ed.* **1978,** *16,* 13.
34. Aoki, Y.; Brittain, J. O. *J. Polym. Sci., Polym. Phys. Ed.* **1977,** *15,* 199.
35. Takamatsu, T.; Fukada, E. *Polym. J.* **1970,** *1,* 101.
36. Fukada, E.; Sakurai, T. *Polym. J.* **1971,** *2,* 656.
37. Carr, S. H. *Proc. 10th North Am. Therm. Anal. Soc.*, Boston, MA, Oct. 26−29, 1980, 1−4.

38. *Ibid.*, 161–166.
39. Chatain, D.; Gautier, P.; Lacabanne, C. *J. Polym. Sci., Polym. Phys. Ed.* **1973**, *11*, 1631.
40. Heijboer, J., Ph.D. Thesis, Leiden Univ., Delft, Netherlands, 1972.
41. Klaase, P. T. A.; van Turnhout, J. *IEE Conf. Publ. 177* **1979**, Delft, Netherlands.
42. Yalof, S.; Wrasidlo, W. *J. Appl. Polym. Sci.* **1972**, *16*, 2159.
43. Varadarajan, K.; Boyer, R. F. *Org. Coat. Plast. Chem.* **1981**, *44*, 402.
44. Varadarajan, K.; Boyer, R. F. *Polymer* **1982**, *23*, 314.
45. Personal communication, Lear, J., E. I. du Pont de Nemours & Co., Inc., Instrument Products, Wilmington, DE.
46. Boyer, R. F. (Guest Ed.) *Polym. Eng. Sci.* **1979**, *19*, 716.
47. Boyer, R. F. (Guest Ed.) *Polym. Eng. Sci.* **1979**, *19*, 709.
48. Senich, G. A.; MacKnight, W. J. *J. Appl. Polym. Sci.* **1978**, *22*, 2633.
49. Gillham, J. K.; Manzione, L. T.; Tu, C. F.; Poek, V. C. *Org. Coat. Plast. Chem.* **1979**, *41*, 357.
50. Struik, L. C. E.; Bree, H. W.; Tak, A. G. M., private publ., 75–61, Central Laboratory, TNO Delft, Netherlands.
51. Heijboer, J. *Int. J. Polym. Mater.* **1977**, *6*, 11.
52. Starkweather, H. W., Jr. *J. Macromol. Sci., Phys.* **1968**, *2*, 781.
53. Schatzki, T. In "Molecular Basis of Transitions and Relaxations", Midland Macromolecular Monographs No. 4; Meier, D. J., Ed.; Gordon: New York, 1978; 263–75.
54. Gillham, J. K.; Glandt, C.; McPherson, C. A. "Chemistry and Properties of Cross-linked Polymers"; Labana, S. S., Ed.; Academic: New York, 1977.
55. Armendiades, C. D.; Baer, E.; Rieke, J. K. *J. Appl. Polym. Sci.* **1970**, *14*, 2635.
56. Nielsen, L. E. *Polym. Eng. Sci.* **1977**, *17*, 713.
57. Neumann, R. M.; Senich, G. A.; MacKnight, W. *J. Polym. Eng. Sci.* **1978**, *18*, 624.
58. Heijboer, J. *Polym. Eng. Sci.* **1979**, *19*, 670.
59. Boyer, R. F. *Polym. Eng. Sci.* **1979**, *19*, 732.
60. Boyer, R. F. *J. Macromol. Sci., Phys.* **1980**, *18*, 461.
61. Cowie, J. M. G. *Polym. Eng. Sci.* **1979**, *19*, 709.
62. Maxwell, B.; Nguyen, M. *Polym. Eng. Sci.* **1979**, *19*, 1140.
63. Schwartz, B. G.; Gillham, J. K. *Org. Coat. Plast. Chem.* **1979**, *41*, 442, 447.
64. Hedvat, S. *Org. Coat. Plast. Chem.* **1980**, *42*, 667.
65. *Polym. Eng. Sci.* **1981**, *21*, 129.
66. Gillham, J. K. *Polym. Eng. Sci.* **1979**, *19*, 676.
67. Varadarajan, K.; Boyer, R. F. *Org. Coat. Plast. Chem.* **1981**, *44*, 409.
68. Koleske, J. V.; Faucher, J. A. *Polym. Eng. Sci.* **1979**, *19*, 716.
69. Oberst, H.; Frankenveld, K. *Acoustica* **1951**, *2*, 4.
70. Van Oort, W. P. *Mikroteknik* **1952**, *7*, 246.
71. Schwarzl, F. *Acoustica* **1958**, *8*, 164.
72. Personal communication, Starkweather, H. W.; Giri, M. R., E. I. du Pont de Nemours & Co., Inc., du Pont Experimental Station, Wilmington, DE; *see J. Appl. Polym. Sci.* **1982**, *27*, 1243.
73. Morgan, R. J.; Nielsen, L. E.; Buchdahl, R. *J. Appl. Phys.* **1971**, *42*, 4683.
74. Wei, L. M., B.S. Thesis, Princeton Univ., Princeton, NJ, 1979.

RECEIVED for review October 14, 1981. ACCEPTED February 12, 1982.

2

Torsional Braid Analysis
Time–Temperature–Transformation Cure Diagrams of Thermosetting Epoxy/Amine Systems

JOHN B. ENNS and JOHN K. GILLHAM

Princeton University, Department of Chemical Engineering, Polymer Materials Program, Princeton, NJ 08544

The conversion of liquid resin to solid thermoset during the process of cure can be monitored using a substrate coated with the reactive system as the specimen in a torsion pendulum experiment [torsional braid analysis (TBA)]. Measurement of times to gelation and vitrification at a series of temperatures results in an isothermal time–temperature–transformation (TTT) cure diagram that can be used for comparing different systems. In this chapter, molecular and macroscopic TTT cure diagrams are considered: TBA is demonstrated to be a convenient method for generating macroscopic TTT cure diagrams. The influence of chemical structure on the kinetics and macroscopic properties was investigated by comparing aliphatic and aromatic tetrafunctional amines with diepoxides of varying molecular weights. Automation of the TBA instrument using a desktop calculator also is described.

T HE TORSION PENDULUM has proven to be an important and versatile tool in the study of dynamic mechanical properties of materials. We have applied this technique primarily to polymers, although elsewhere it has been applied to a wide variety of materials, ranging from liquids to metals and ceramics. The basis of its wide appeal lies in its fundamental simplicity: information about the complex modulus of the material under investigation is obtained by simply observing the decaying oscillations of the pendulum. After the pendulum is set in motion, it is permitted to oscillate freely at its resonant frequency while the amplitude of the oscillatory wave decays. In an unautomated system, a relatively simple but tedious task is to calculate the

0065–2393/83/0203–0027$10.25/0

shear modulus (or rigidity) and the loss modulus from the period of the oscillation, its logarithmic decrement, and the geometric constants of the system. The independent variable in the investigation of dynamic mechanical properties of a material often is temperature, but it also can be time, as for chemically reactive or physically aging systems.

The technique is made more versatile by using a supported specimen as in torsional braid analysis (TBA). Impregnating a glass braid with a solution (or melt) facilitates fabrication of a mounted specimen, permits use of small quantities of sample (25 mg), and allows study of the specimen in its solid and liquid states, as well as its change in state from liquid to solid and vice versa.

In this chapter, an automated TBA technique (1–3), controlled by a digital desktop computer, is discussed, and its utility in the characterization of thermosetting resins is illustrated (4).

Instrumentation

A schematic diagram of the torsion pendulum (Plastics Analysis Instruments, Inc.) is shown in Figure 1. The pendulum is intermittently set into motion to generate a series of damped oscillations while the material behavior of the specimen changes with temperature and/or time. The nondriven, free oscillations are initiated by the step-displacement of an upper gear. The natural frequency range of the vibrations is 0.05–5 Hz. Conversion of the damped oscillations to electrical analog signals is accomplished using an optical transducer.

A key factor in the instrumentation was the development of a nondrag, optical transducer that produces an electrical response that is a linear function of the angular deformation of the pendulum. A polarizing disk is employed as the inertial member of the pendulum, and a stationary second polarizer is positioned in front of a photocell whose response is a linear function of intensity. The intensity of light transmission through two polarizers is a cosine squared function of their angular displacement. Over a useful range symmetrical between the crossed and parallel positions of the pair of polarizers, the transmission function approaches linearity. As the properties of the specimen change, twisting of the specimen may cause the transducer to drift out of the linear range. The automated system was designed to compensate for this behavior.

The specimen is supported in the cylindrical vertical shaft in a copper block around which are band heaters and cooling coils (for liquid nitrogen). The apparatus operates over a temperature range of $-190-400$ °C with a temperature spread of <1 °C over a 2-in. specimen. A temperature programmer/controller system permits experiments to be performed with linearly increasing, linearly decreasing, and isothermal (±0.1 °C) temperature modes. Measurements have been

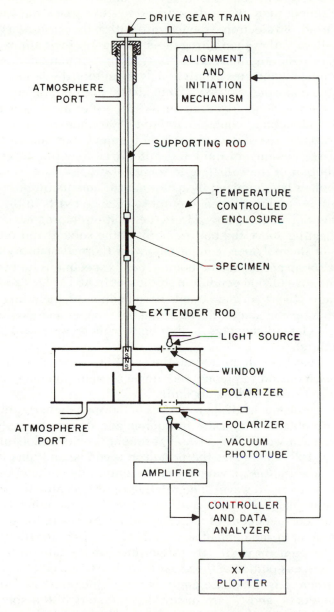

Figure 1. Scheme of the automated torsion pendulum. An analog electrical signal results from using a light beam passing through a pair of polarizers, one of which oscillates with the pendulum. The pendulum is aligned for linear response and initiated by a computer that also processes the damped waves to provide the elastic modulus and mechanical damping data, which are plotted on an XYY plotter vs. temperature or time.

made from 4 K and to 700 °C in other apparatus. The atmosphere is tightly controlled: inert, water-doped, and reactive gases and vacuum have been used. No electronic devices are within the specimen chamber. Dry helium, rather than nitrogen, is used as an inert atmosphere because of its higher thermal conductivity at low temperatures. An on-line electronic hygrometer continuously monitors the water vapor content of the atmospheres from <20 to 20,000 parts per million H_2O.

The substrate generally used in a TBA experiment is a loose, heat-cleaned glass braid containing about 3600 filaments (Figure 2). The large surface area of the assembly of filaments permits pickup of relatively large amounts of fluid and minimizes flow due to gravity. The contribution of the substrate to torsional properties of the composite specimen is minimized by using multifilaments (rather than a rod). A braid is employed in an attempt to balance twists in the component yarns. Solvent is removed from the solution-impregnated braid in situ by heating above the boiling point of the solvent and into the fluid state of the polymer (compatible with thermal stability). The apparatus also can accommodate homogeneous specimens for conventional quantitative torsion pendulum studies [as in ASTM D−2236 (5)].

Substrates other than glass braids have been used, including glass, carbon, cellulose, and aramid fibers; metal foil (copper and aluminum); paper; single glass filament (as in fiber optics); metal wire; and plastic film (polyimide).

Material has been deposited on the substrate from solution, melt, emulsion, suspension (aqueous and nonaqueous), and powder (by heating the substrate with an air gun).

Two dynamic mechanical properties of the specimen, rigidity and damping, are obtained from the frequency and decay constants that characterize each wave. The TBA experiment provides plots of relative rigidity ($1/P^2$, where P is the period in seconds) and logarithmic decrement [$\Delta = \ln(\theta_i/\theta_{i+1})$, where θ_i is the amplitude of the ith oscillation of a freely damped wave]. The relative rigidity is directly proportional to the in-phase or elastic portion of the shear modulus (G'); for example, for rod specimens of radius r and length L and for an oscillating system with moment of inertia I, $G' \cong (1/P^2)(8\pi IL/r^4)$. The logarithmic decrement is directly proportional to the ratio of the out-of-phase or viscous portion of the shear modulus (G'') to G' ($\Delta \cong \pi G''/G'$ $= \pi\tan \delta$, where δ is the phase angle between the stress and strain). The parameters G' and G'' are material parameters of the specimen and they characterize the storage and loss of mechanical energy on cyclic deformation, respectively. The energy stored during deformation is $\frac{1}{2}G'\epsilon^2$ at strain ϵ, and $\pi G''\epsilon_{max}^2$ is an approximation for the energy dissipated per cycle, where ϵ_{max} is the peak strain.

Figure 2. Automated torsion pendulum: photograph of apparatus (see text).

A photograph of a pendulum/computer system that has been used since 1979 is illustrated in Figure 2. It shows the pendulum (enclosed in a cabinet, top left; cabinet door open, bottom left) and the major components of the assembly. The temperature controller, digital voltmeter, scanner, and computer are in the rack at the right. The printer is to the left of the rack; the plotter is to its right. The analog temperature and wave signals are monitored continuously on a two-pen, strip-chart recorder. An atmosphere control panel and liquid nitrogen container are shown in the background.

Film and unimpregnated braid specimens are shown at the bottom right of Figure 2. A film is shown assembled with upper and lower extension rods ready for lowering into the apparatus and then coupling with the polarizer disk of the transducer (shown bottom right). The polarizer disk contains a magnet at its center that couples to the end of the lower extension rod. Dimensions of specimens are selected to provide periods of oscillation in the range 0.2–20 s.

The system schematic for interfacing the TBA unit to a desktop computer (HP–9825B) that can synchronize the control and data processing of the torsion pendulum to produce results in real time is shown in Figure 3. Earlier experience (6) with a hierarchical digital computer system aided the development of the presently used data processing schemes (7).

For each damped wave, the computer goes through a control sequence, which is represented schematically in Figure 4. An alignment motor rotates the pendulum (or the stationary polarizer) to the same reference position at the start of each control sequence. To initiate the oscillations, a second motor then rotates the pendulum a specified angular displacement against the tension of a spring. The pendulum is held in this cocked position until any oscillations set up by the alignment and cocking have subsided, at which time the clutch of the second motor is disengaged and the inertial mass swings back so as to oscillate about the reference position. The temperature (or time, for isothermal experiments) then is measured and the oscillation data are collected and processed. After plotting the reduced data, the oscillation is monitored until it decays to within specified limits and then the control cycle repeats.

The response function of the torsion pendulum to mechanical excitation is described to good approximation by

$$\theta(t) = \theta_0 \exp(-\alpha t) \cos(\omega t + \phi)$$

which is a solution to the equation of motion with form

$$\frac{d^2\theta}{dt^2} + A_1 \frac{d\theta}{dt} + A_2\theta = 0$$

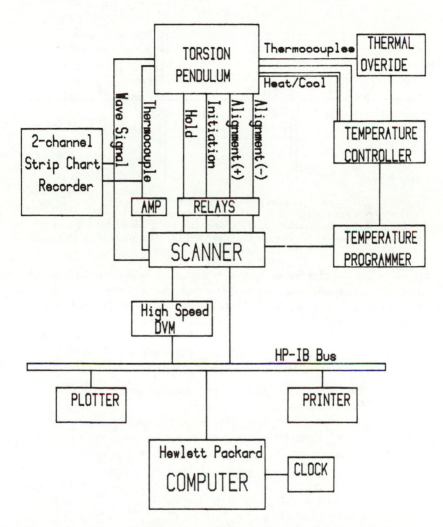

Figure 3. Automated torsion pendulum: system schematic for inter-facing with a digital computer. The motors that align the specimen and initiate the waves are under computer control. The wave and amplified analog thermocouple signals reach the computer digitized via a digital voltmeter (HP−3437A). The scanner (HP−3495A) supervises the input/output activity. Upon receiving the digitized raw data, the computer calculates the frequency and damping parameters, and plots the dynamic mechanical properties of the specimen as a function of tem-perature and time.

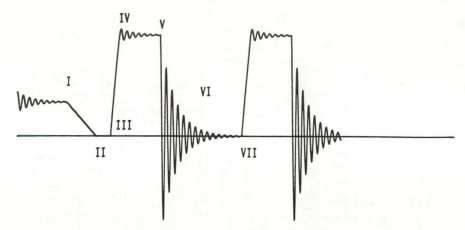

Figure 4. Automated torsion pendulum: control sequence. Key: I, previous wave decays, drift detected, and correction begins; II, reference level of polarizer pair reached; III, wave-initiating sequence begins; IV, decay of transients; V, free oscillations begin; VI, data collected; and VII, control sequence repeated.

where ω is the angular frequency, ϕ is a phase angle depending on the timing of data acquisition, and θ_0, A_1 ($= 2\alpha$), and A_2 ($= \alpha^2 + \omega^2$) are constants where α is the exponential decay constant. In complex format, the equation of motion is

$$I\frac{d^2\theta}{dt^2} + C(G' + iG'')\theta = 0$$

where G' and G'' are the in-phase and out-of-phase shear moduli, respectively, and C is a geometric constant.

Because an experiment may run over the course of several days and generate more than 1000 damped waves, manual techniques for reducing the experimental analog waves are slow and tedious. These techniques traditionally involved measuring the decay of successive peak amplitudes to provide the logarithmic decrement and the period $P(= 2\pi/\omega)$ for each wave. The data for torsion pendulum studies usually are presented as G' and Δ, G' and G'' ($\cong G'\Delta/\pi$), or as G' and tan δ ($= G''/G'$). For TBA experiments, because of the small size and the dependence of G' on the inverse fourth power of the radius, the irregular geometry, and the composite nature of the specimens, simplified parameters have been used. These parameters, which refer to the composite specimen, are relative rigidity ($1/P^2$) and the logarithmic decrement (Δ).

Time–Temperature–Transformation Cure Diagrams

The cure of thermosetting systems can be considered on a molecular as well as a macroscopic level. On the macroscopic time–temperature–transformation (TTT) cure diagram (Figure 5) the cure can be characterized by the resin rheology as a function of time and isothermal temperature. The boundaries between the regions delineated in Figure 5 are not as sharp as indicated, but rather diffuse; the time/temperature at which a transition occurs depends on the frequency of the test method and the dynamic mechanical property used to monitor it. ("Transition" is used in this chapter to denote a change from one state to another.) Between $_{gel}T_g'$ and $T_{g\infty}$ (defined later), the viscous liquid changes to a viscoelastic fluid, then to a rubber, and finally to a glass. The S-shaped curve indicates the onset of vitrification, which represents the time at which the glass transition temperature of the reacting system has reached the isothermal temperature of cure. The parameter $T_{g\infty}$ is the glass transition temperature of the fully cured resin, above which the system is a rubber in its fully cured state. As indicated in Figure 5, the time to vitrification passes through a minimum between $_{gel}T_g'$ and $T_{g\infty}$. This behavior reflects the competition between the increased rate constant for reaction and the increasing chemical conversion required to achieve vitrification as the temperature is increased. Also, as indicated in Figure 5, the time to vitrification passes through a maximum between $_{resin}T_g$ and $_{gel}T_g'$. This latter behavior reflects the competition between the temperature and time-dependences of viscosity of the reacting system. A liquid-to-rubber transition curve indicates the onset of the rubbery plateau. At lower isothermal temperatures (approaching $_{resin}T_g$) the state of the resin changes directly from a liquid to a glass without going through an intermediate rubbery state because of its low molecular weight. The parameter $_{gel}T_g'$ represents the lowest isothermal cure temperature that will result in a material with elastomeric properties; below $_{gel}T_g'$ the solid formed isothermally will be an ungelled glass, which on heating becomes a viscous liquid (before further reaction occurs). Below $_{resin}T_g$ the resin remains an ungelled glass with essentially no chemical reactivity.

The remaining curve in Figure 5 is considered to represent an isoviscosity contour. A rheological parameter often measured in thermosetting resins is the time to reach a particular viscosity (usually a function of the measurement technique), which often is identified as the gel point.

At the molecular level, isothermal cure of thermosetting systems is characterized by (molecular) gelation at a particular chemical conversion (8, 9) (depending on the functionality of the reactants) and by a continual reduction of molecular mobility and chemical reactivity caused by increasing molecular weight and cross-linking. Extent of conversion contours (one of which is the gelation curve) on an isothermal molecular TTT cure diagram (Figure 6) indicate that the reaction becomes diffusion controlled and eventually quenched, re-

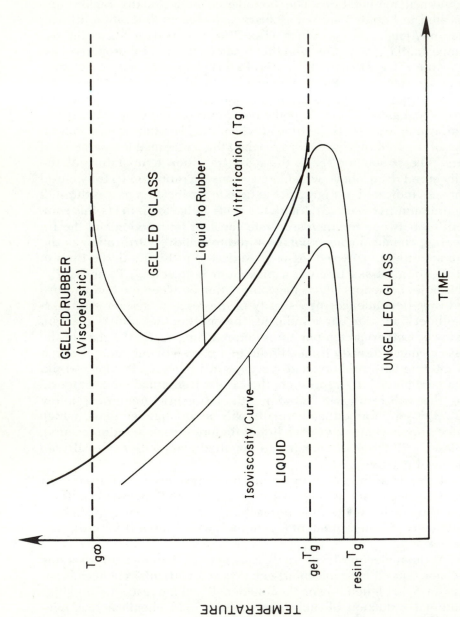

Figure 5. Macroscopic TTT cure diagram.

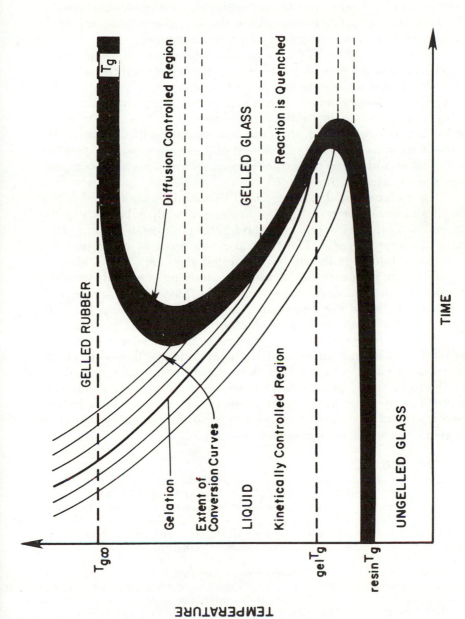

Figure 6. Molecular TTT cure diagram.

gardless of whether or not gelation has occurred. The temperature at which the time to gel equals the time to vitrify is defined as $_{gel}T_g$; the resin will not gel below this temperature. This parameter is related to the temperature, $_{gel}T_g'$, below which the liquid-to-rubber transition disappears in the macroscopic TTT cure diagram.

Each point on the TTT cure diagrams represents a unique state, independent of the path by which it was attained, only if the system is homogeneous and the cure is limited to a single reaction path. If the system is nonhomogeneous, phase separation (morphology and solubility) may be a function of the cure temperature. In a system where competing reactions exist, the cure temperature will determine the favored reaction path. In the absence of these complications, an isocure path would follow the extent of conversion curve if the temperature were to change (10).

The molecular and macroscopic TTT diagrams are related in that the liquid-to-rubber transition is associated with gelation and the molecular T_g (quenching of reaction curve) is associated with vitrification (macroscopic T_g). The thermosetting process (cure) involves four states: liquid, gelled rubber, ungelled glass, and gelled glass.

TTT cure diagrams can be used for studying chemical structure/physical property relationships because each system is unique with respect to both the kinetics of cure and the physical properties. To study structure/property relationships of thermosetting systems, the systems must be cured above $T_{g\infty}$ to ensure that they have been fully cured. If they are not fully cured, their properties will depend on the extent of cure, as well as on their chemical structure and morphology. The purpose of the present work is to use isothermal TTT cure diagrams and dynamic mechanical temperature scans of both incompletely and fully cured systems as a basis for investigating structure/property relationships in epoxy/amine systems.

Experimental

Each of a series of difunctional epoxies of the diglycidyl ether of bisphenol A type (Shell Epon 825, Epon 828, and Epon 834) was reacted with stoichiometric quantities (one epoxy/one amine hydrogen) of diaminodiphenyl sulfone (DDS, Aldrich). The Epon 828 also was reacted separately with 4,4'-methylenedianiline (MDA, Aldrich), bis(p-aminocyclohexyl)methane (PACM−20, du Pont), and trimethylene glycol di-p-aminobenzoate (TMAB, Polaroid).

Macroscopic TTT cure diagrams were generated from TBA experiments in which the relative rigidity ($\propto G'$) and logarithmic decrement (Δ) were obtained from the damped oscillations of the freely decaying torsion pendulum (~1 Hz). The supported specimens were made by dipping a glass braid in a solution of the epoxy and amine in methyl ethyl ketone, except for Epon 828/PACM−20, which was applied neat because it is a liquid at room temperature. The specimens were then mounted in the TBA apparatus at the temper-

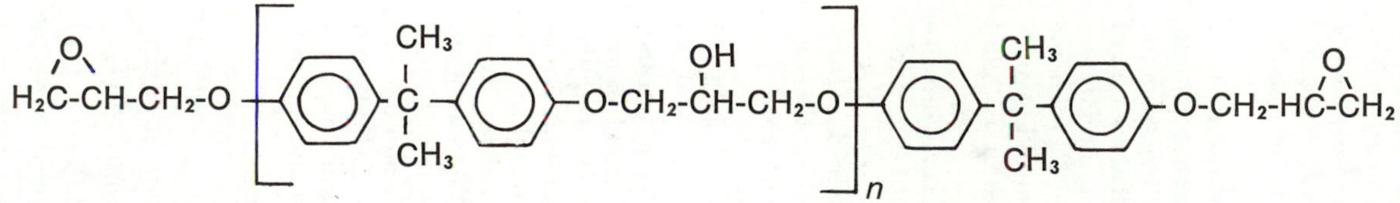

Epon 825: $n = 0$
Epon 828: $n = 0.15$
Epon 834: $n = 0.6$

Amine curing agents

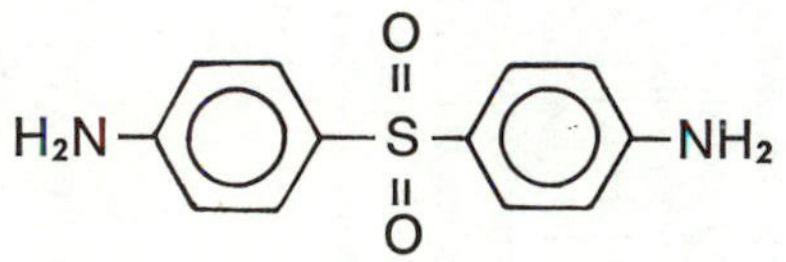

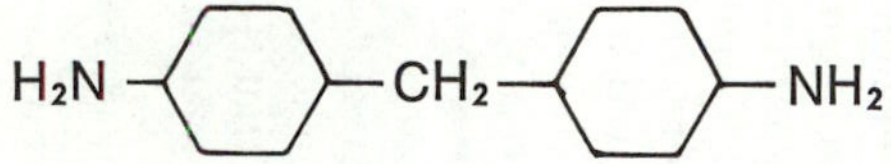

Bis(*p*-aminocyclohexyl)methane
(PACM-20)

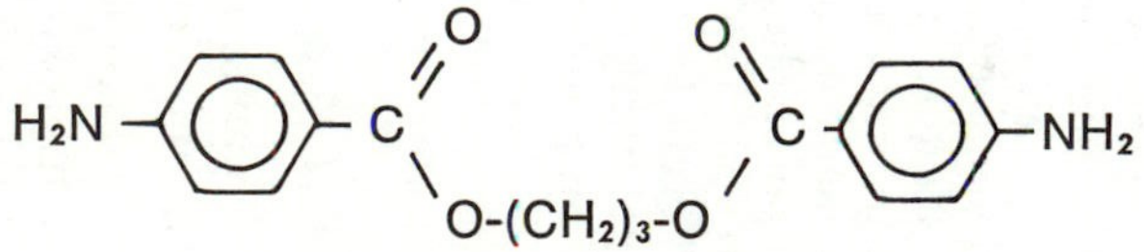

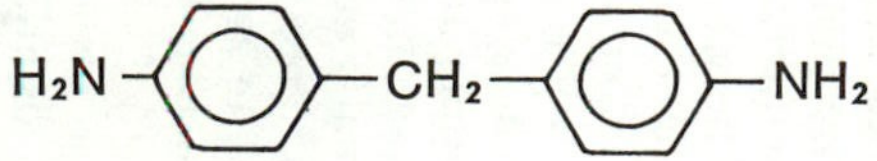

4,4'-Methylenedianiline
(MDA)

4,4'-Diaminodiphenyl Sulfone
(DDS)

Trimethylene glycol
di-*p*-aminobenzoate (TMAB)

ature of cure in a helium atmosphere, and their moduli were monitored as a function of time. The transitions that occur during cure were identified by the maxima in the logarithmic decrement curves of the dynamic mechanical plots. Having cured the specimens until the reaction was quenched (as indicated by the leveling off of the modulus), thermomechanical spectra were obtained at 1.5 °C/min by first cooling the specimen to −190 °C and then heating it to 250 °C to obtain the spectrum of the partially cured resin. The specimens were then cycled between 250 and −190 °C until no further change in T_g occurred, indicating that the cure was complete. In this way, $T_{g\infty}$, as well as the spectra of the fully cured resins, was obtained for each cure temperature.

A typical isothermal plot (11) shown in Figure 7 displays three distinct processes. The elapsed time to each of these events is plotted vs. the isothermal cure temperature, and in this way each of the curves on the macroscopic TTT cure diagram is generated. Examples of the thermomechanical spectra of the isothermally cured resin and the subsequent fully cured resin are shown in Figure 8. Note that both the glassy state relaxation, T_{sec}, and T_g increase with increasing extent of cure, and that the rigidity of the undercured resin is greater than that of the fully cured resin in the region between T_{sec} and T_g (12, 13).

Dynamic viscosities G' and G'' for Epon 828/PACM−20 also were obtained from oscillatory shear measurements at 1.6 Hz in a nitrogen atmosphere using a Rheometrics dynamic spectrometer. Samples were placed between parallel plates 25 mm in diameter, 0.5 mm apart, and cured isothermally. The maxima in G'' and the preceding shoulder, as well as the two corresponding stages in G' (Figure 9), provide another set of rheological data that also can be plotted on the TTT diagram.

Extent of conversion data for the molecular TTT diagram were obtained by IR techniques [Figure 10 (11, 14)] and, in principle, also can be obtained by differential scanning calorimetry (DSC) or titration. Theoretically, molecular gelation occurs at a specific extent of conversion. The gel point at any given temperature can be estimated from a plot of gel fraction vs. time, by noting the time at which the resin begins to be insoluble. The gel fractions of the isothermally curing resins were measured periodically by weighing the resin fraction insoluble in methyl ethyl ketone (dichloromethane for Epon 828/PACM−20). The gel fraction data for the Epon 828/PACM−20 system are shown in Figure 11: these data occur between 70 and 80% conversion (Figure 10). (The gel fraction experiments for the Epon 828/PACM−20 systems were performed on samples cured in a helium atmosphere. For Epon 825/DDS and Epon 834/DDS, gel fraction experiments were performed on samples cured in air. For Epon 828/DDS, in which cure was conducted both in air and helium, the gelation times were the same.)

Results and Discussion

The dynamic mechanical (TBA) plots of the fully cured resins (Figure 12) show that whereas the glass transition temperature ($T_{g\infty}$) varies for each system, the broad secondary relaxation ($T_{sec\infty} < T_{g\infty}$) remains at approximately −40 °C, indicating that it is probably due to localized backbone motion in the epoxy moiety, which is common to all of these systems (12, 15, 16). The $T_{g\infty}$ of each resin was obtained by cycling between −190 and 250 °C until no further change in T_g was

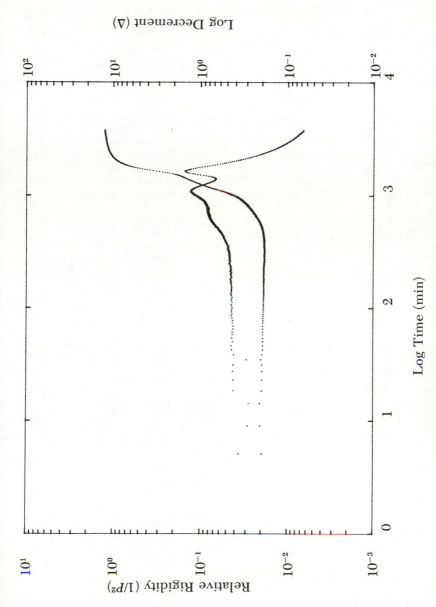

Figure 7. Isothermal cure of a thermosetting epoxy (TBA): Epon 828/TMAB at 80 °C. In general, three events are discerned, the last being vitrification and the second, gelation. Times to each of these events are measured for different isothermal temperatures as a basis for generating a TTT cure diagram for the particular thermosetting system (see Figures 19–24).

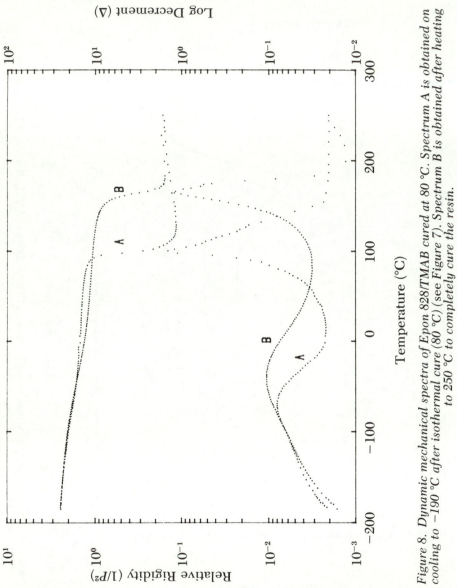

Figure 8. Dynamic mechanical spectra of Epon 828/TMAB cured at 80 °C. Spectrum A is obtained on cooling to −190 °C after isothermal cure (80 °C) (see Figure 7). Spectrum B is obtained after heating to 250 °C to completely cure the resin.

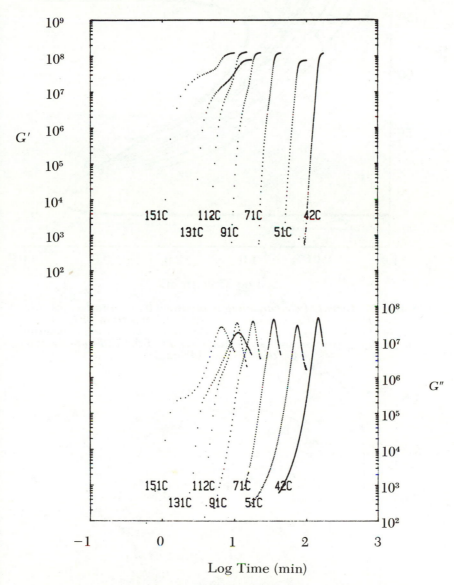

Figure 9. Dynamic mechanical data (Rheometrics), G' and G", obtained by isothermally curing Epon 828/PACM−20 at several temperatures.

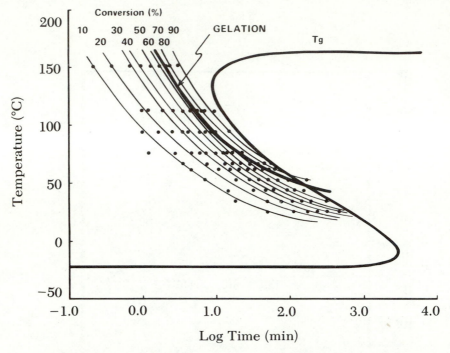

Figure 10. *Extent of conversion data obtained by monitoring the ratio of the 1184- and 915-cm⁻¹ peaks in the FTIR spectrum of Epon 828/ PACM−20 as a function of time at different temperatures, superimposed on the gelation and vitrification curves of the TTT cure diagram for that system (Figure 23).*

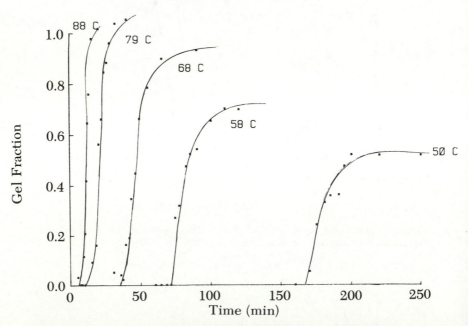

Figure 11. *Gel fraction data for Epon 828/PACM−20. Gel points were obtained by monitoring the gel fraction of the resin during cure in a helium atmosphere.*

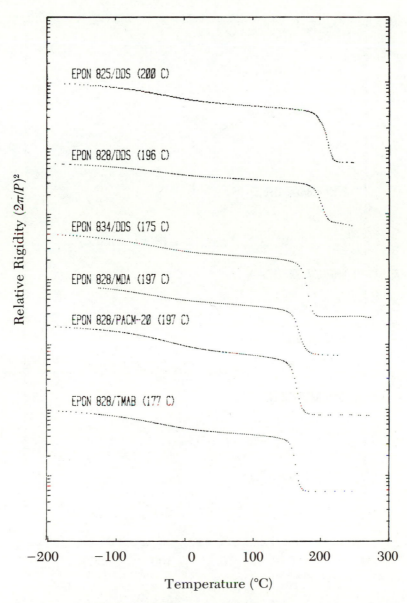

Figure 12a. TBA spectra of fully cured epoxy systems: relative rigidity. Specimens were cured isothermally (at the designated temperature) and post-cured by cycling to 250 °C.

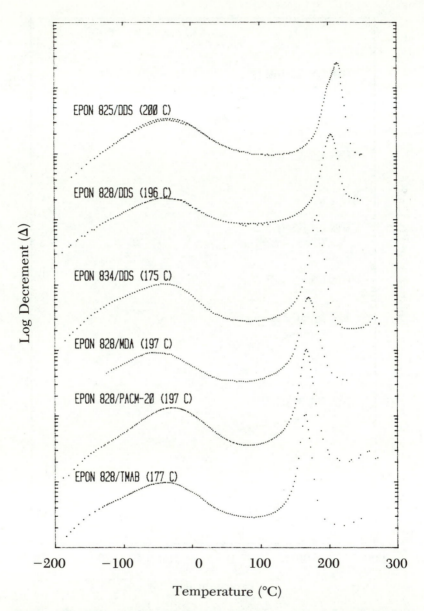

*Figure 12b. TBA spectra of fully cured epoxy systems: logarithmic dec-
rement. (See Figure 12a for details.)*

observed. Figures 13–18 show T_{sec} and T_g for each of the resins cured isothermally until the modulus leveled off, as well as $T_{sec\infty}$ and $T_{g\infty}$ of the fully cured resins, as a function of cure temperature. The T_g of the resin as identified by the maximum in the logarithmic decrement curve is higher than the cure temperature (as long as T_{cure} is below $T_{g\infty}$). At vitrification, the rate of diffusion begins to dominate the kinetics of the reaction, but the reaction is not quenched until the glass transition has risen well above the cure temperature. This difference between the glass transition temperature and the temperature at which molecular motion is quenched is evident in phenomena such as sub-T_g annealing and is predicted by theoretical expressions such as the WLF equation (*17*). The difference $T_g - T_{cure}$ varies with each resin system, and is tabulated in Table I along with values of $T_{sec\infty}$ and $T_{g\infty}$.

Macroscopic TTT cure diagrams were obtained from TBA experiments for each of the six epoxy/amine systems. Rheological data

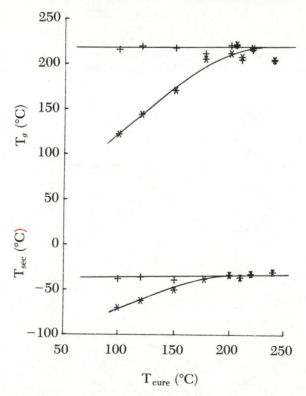

Figure 13. The parameters T_{sec} and T_g as a function of cure temperature for Epon 825/DDS. Key: *, as cured; +, post-cured.

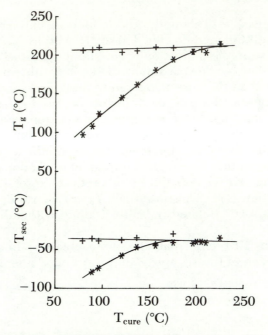

Figure 14. The parameters T_{sec} and T_g as a function of cure temperature for Epon 828/DDS. Key: *, as cured; +, post-cured.

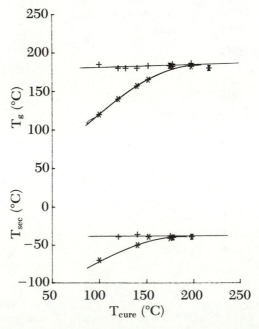

Figure 15. The parameters T_{sec} and T_g as a function of cure temperature for Epon 834/DDS. Key: *, as cured; +, post-cured.

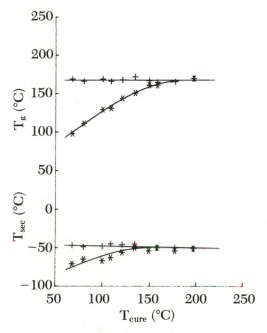

Figure 16. *The parameters T_{sec} and T_g as a function of cure temperature for Epon 828/MDA. Key: *, as cured; +, post-cured.*

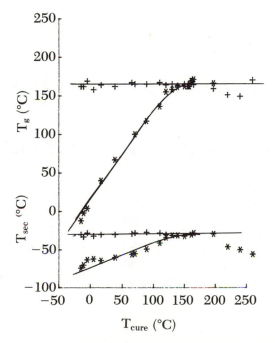

Figure 17. *The parameters T_{sec} and T_g as a function of cure temperature for Epon 828/PACM−20. Key: *, as cured; +, post-cured.*

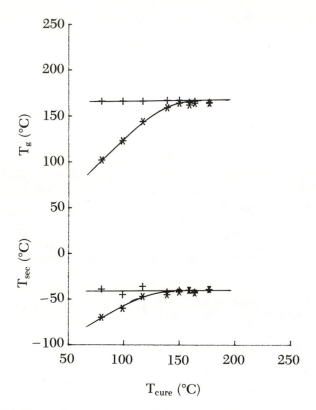

Figure 18. The parameters T_{sec} and T_g as a function of cure temperature for Epon 828/TMAB. Key: *, as cured; +, post-cured.

Table I. Cured Resins (Figures 13–18): Characteristic Parameters

Epoxy	$T_{sec\infty}$	$T_{g\infty}$	$T_g - T_{cure}$
Epon 825/DDS	−35	220	24
Epon 828/DDS	−35	210	25
Epon 834/DDS	−37	185	20
Epon 828/MDA	−50	168	30
Epon 828/PACM−20	−33	165	30
Epon 828/TMAB	−40	166	24

NOTE: All values are in degrees Centigrade.

using the Rheometrics dynamic spectrometer were obtained only for Epon 828/PACM−20, and gel fraction data were obtained for all except those cured with MDA and TMAB. Composite TTT diagrams are presented in Figures 19−24 consisting of the TBA, Rheometrics, and gel fraction data. Activation energies, obtained by plotting the data in an Arrhenius fashion, are listed in Table II. With the exception of Epon 828/PACM−20, which is the only aliphatic amine cured system, the activation energies of the epoxy systems studied in this work were similar. Characteristic parameters of the TTT cure diagrams are listed in Table III; $_{gel}T_g$ is the lowest temperature at which the resin gels, and $_{min}T_g$ is the cure temperature at which the time to vitrify is a minimum.

   The results fit the pattern of the macroscopic TTT cure diagram (Figure 5). In general, the TBA results show three events, except for the Epon 828/DDS system in which the first event is missing.

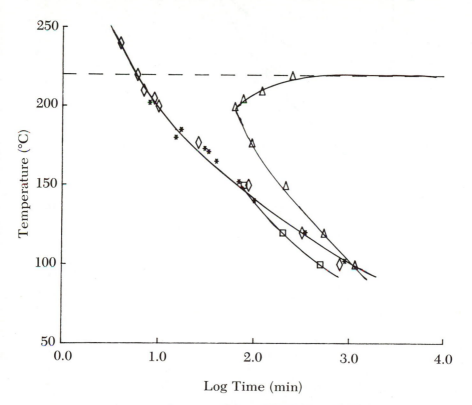

*Figure 19. TTT cure diagram of Epon 825/DDS, including TBA and gel fraction data. Key: □, isoviscous (TBA); ◇, liquid-to-rubber (TBA); △, vitrification (TBA); and *, gelation (gel fraction).*

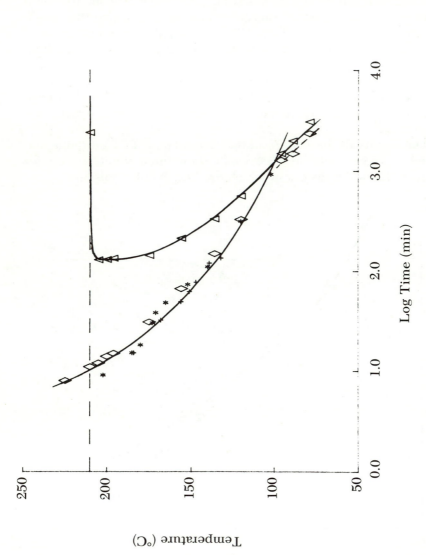

*Figure 20. TTT cure diagram of Epon 828/DDS, including TBA and gel fraction data. Key: ◇, liquid-to-rubber (TBA); △, vitrification (TBA); *, gelation (gel fraction air); and +, gelation (gel fraction helium).*

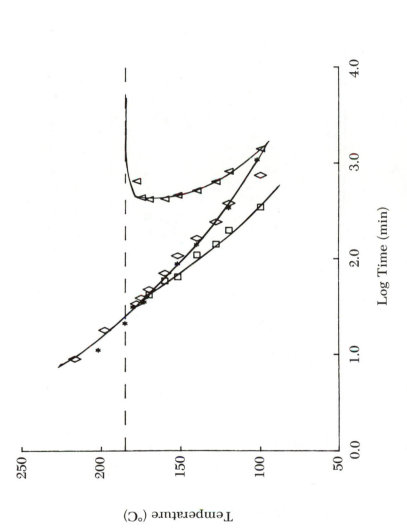

*Figure 21. TTT cure diagram of Epon 834/DDS, including TBA and gel fraction data. Key: □, isovis-cous (TBA); ◇, liquid-to-rubber (TBA); △, vitrification (TBA); and *, gelation (gel fraction).*

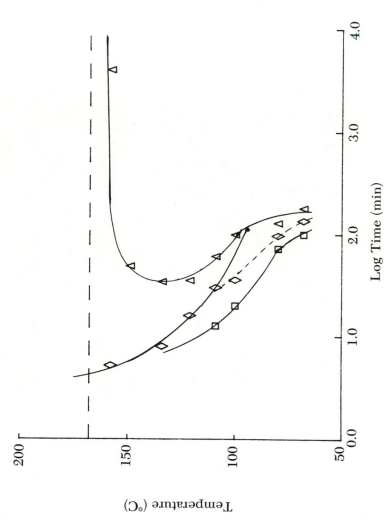

Figure 22. TTT cure diagram of Epon 828/MDA, TBA data. Key: □, *isoviscous (TBA);* ◇, *liquid-to-rubber (TBA); and* △, *vitrification (TBA).*

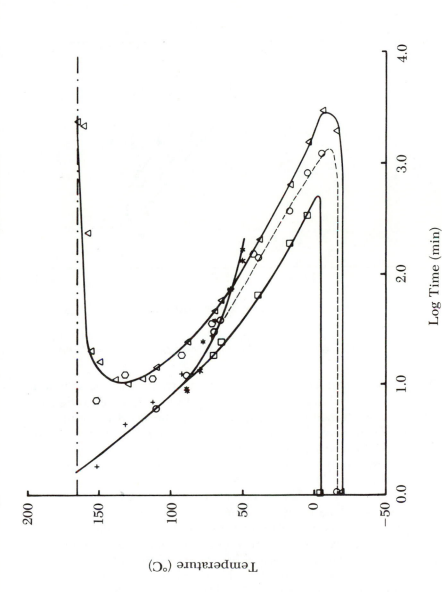

*Figure 23. TTT cure diagram of Epon 828/PACM −20, including TBA, Rheometrics, and gel fraction data. Key: □, isoviscous (TBA); ○, liquid-to-rubber (TBA); △, vitrification (TBA); * gelation (gel fraction); +, liquid-to-rubber (Rheometrics); and ○, T_g (Rheometrics).*

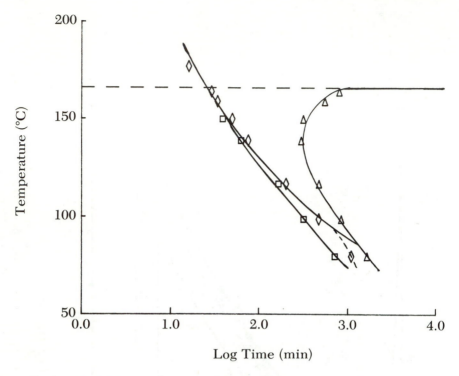

Figure 24. TTT cure diagram of Epon 828/TMAB, TBA data. Key: □, *isoviscous (TBA);* ◇, *liquid-to-rubber (TBA); and* △, *vitrification (TBA).*

The third event is associated with vitrification, as can be seen, for example, from the relative rigidity curve of Figure 7, which indicates a rubber-to-glass transition corresponding to the third maximum in logarithmic decrement. The entire S-shaped vitrification curve was obtained only for the Epon 828/PACM−20 system (Figure 23) because it was the only one in which a solvent was not required in sample preparation. The other systems all show the characteristic upper portion of the S-shaped curve. The G' data obtained from the Rheometrics dynamic spectrometer parallel the TBA relative rigidity data in the common temperature range that data were obtained. The lack of exact coincidence may be due to the transducer in the dynamic spectrometer approaching its upper limit.

The second event (Figure 7) can be associated with the liquid-to-rubber transition curve above $_{gel}T_g'$. The modulus shows a corresponding rise from a liquid-to-rubbery state (Figures 7 and 9), and the Rheometrics data (Figure 9) show a shoulder in G'' that corresponds to the second event (TBA) in the Epon 828/PACM−20 system (Figure 7). Below $_{gel}T_g'$, the shoulder in the Rheometrics dynamic spectrometer

Table II. Activation Energies (Figures 19–24)

Epoxy	E_a (isoviscous)	E_a (liquid-to-rubber)	E_a^b (gel)	E_a^c
Epon 825/DDS	11.5	14.8	16.6	—
Epon 828/DDS	—	14.3	16.3	—
Epon 834/DDS	10.0	14.0	15.7	—
Epon 828/MDA	13.6	15.4	—	—
Epon 828/PACM−20	8.5	10.7	16.8	12.6
Epon 828/TMAB	12.2	13.7	—	—

NOTE: All values are in kilocalories per mole.
[a] TBA.
[b] Gel fraction.
[c] IR (11).

Table III. TTT Cure Diagrams (Figures 19–24): Characteristic Parameters

Epoxy	$_{gel}T_g$ (time)	$_{min}T_g$ (time)	$T_{g\infty} - {_{min}}T_g$
Epon 825/DDS	100 (1000)	200 (63)	20
Epon 828/DDS	100 (1000)	205 (30)	5
Epon 834/DDS	100 (1000)	170 (425)	15
Epon 828/MDA	—	130 (35)	38
Epon 828/PACM–20	50 (165)	130 (10)	35
Epon 828/TMAB	—	140 (300)	27

NOTE: All values are in degrees Centigrade and minutes.

G'' data is not evident, and G' shows only a liquid-to-glass transition. The TBA data exhibit the transition prior to vitrification over the entire temperature range; above $_{gel}T_g'$ the shoulder corresponds to the liquid-to-rubber transition, but between $_{resin}T_g$ and $_{gel}T_g'$ it may or may not be attributed to the composite nature of the specimen. A similar effect called the liquid–liquid transition $(T_{\ell\ell})$ is observed in TBA measurements of amorphous thermoplastics (18) in which, at high molecular weights $(\bar{M}_n > M_c$, where $\bar{M}_n$ is the number-average molecular weight, and M_c is the critical molecular weight for chain entanglement), this transition occurs at the onset of the rubbery plateau. It persists even at low molecular weight just as it does in the thermosetting resins.

The first event may be associated with an isoviscosity level. The interactions of the viscous liquid and the braid in a TBA experiment give rise to a maximum in logarithmic decrement, as the viscosity increases (19, 20, 21). Other methods of monitoring the cure of thermosets (rising bubble, stirring motor, etc.) identify (macroscopic) gelation with different viscosity levels.

Complementary continuous heating transformation (CHT) cure diagrams were obtained by monitoring the dynamic mechanical properties of a thermosetting resin by TBA while raising the temperature at a series of constant rates. A temperature scan of a solid, initially unreacted, resin would be expected to reveal in sequence: $_{resin}T_g$, $T_{\ell\ell}$, an isoviscous event, an isoviscous event on reaction, the liquid-to-rubber (or $T_{\ell\ell}$) transition, vitrification, devitrification, and degradation. The isoviscous transitions would be observed only when the viscosity of the resin is below the isoviscous level. The scans in Figure 25 clearly show $_{resin}T_g$, $T_{\ell\ell}$, the liquid-to-rubber transition, and vitrification, in addition to another glass transition (devitrification), and a high temperature event corresponding to an intramolecular rearrangement reaction (11). The number of events observed depends on the heating rate; if the heating rate is too fast, the vitrified state will be missed. The CHT diagram is generated by plotting the temperature

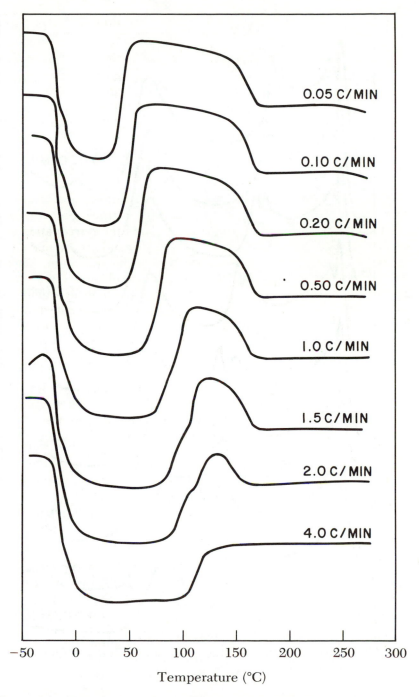

Figure 25a. Temperature scans of Epon 828/PACM−20 from −50 to 275 °C at various scanning rates: relative rigidity.

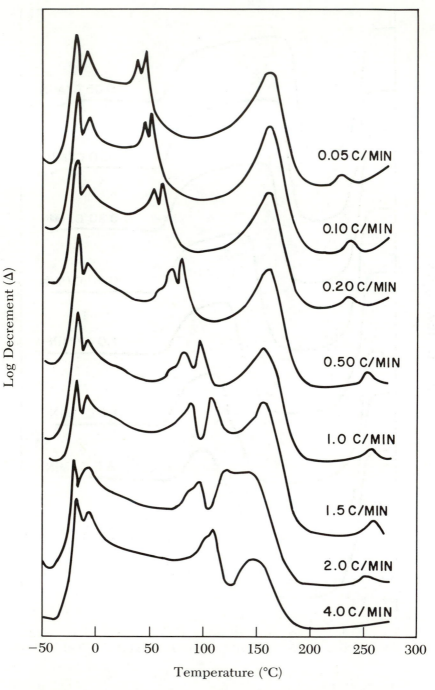

Figure 25b. Temperature scans of Epon 828/PACM−20 from −50 to 275 °C at various scanning rates: logarithmic decrement.

vs. the time at which the events were observed. As one example, the CHT diagram for Epon 828/PACM−20 is shown in Figure 26.

Although the CHT diagram has the same features as the TTT cure diagram, albeit in a somewhat skewed form, it has a practical value. Most of the features of the CHT diagram, such as $_{resin}T_g$, $_{gel}T_g'$, and $T_{g\infty}$ can be obtained from three to four temperature scans, whereas many more are required for the TTT cure diagram. Furthermore, the choice of cure temperatures is more critical for the TTT diagrams, especially when the characteristic parameters of a system are unknown.

The results also fit the pattern of the molecular TTT cure diagram (Figure 6). The extent of conversion data obtained isothermally by Fourier transform IR (Figure 10) for the Epon 828/PACM−20 system follow the trend indicated in Figure 6. As the extent of conversion curves approach the S-shaped glass transition curve, they become horizontal lines as the reaction becomes diffusion controlled and ul-timately is quenched. The gelation curve (determined from gel frac-tion data) lies parallel to the extent of conversion curves; for Epon

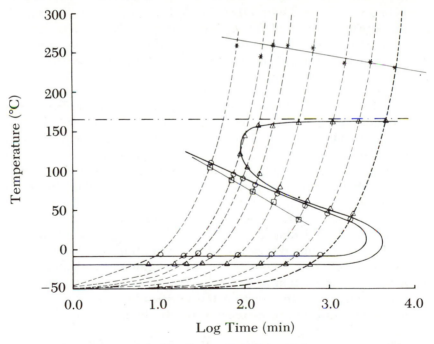

*Figure 26. CHT diagram of Epon 828/PACM−20: temperature scans were performed at 0.05, 0.1, 0.2, 0.5, 1.0, 1.5, 2.0, and 4.0 °C/min from −50 to 275 °C. Key: △, vitrification and devitrification; ○, liquid-to-rubber transition; □, isoviscous phenomenon; *, intramolecular rear-rangement. Heating rates greater than 2 °C/min are required to cure the resin without the occurrence of vitrification.*

828/PACM−20 it lies between 70 and 80% extent of conversion. In the other systems for which gel fraction data were obtained, the gelation curves follow the same form. For Epon 828/PACM−20, the lowest temperature at which gel fractions were obtained was 50 °C, whereas for the epoxy resins cured with DDS, the lowest temperature was 102 °C. The TTT cure diagrams show that the extrapolated gelation curve crosses the vitrification curve just below this temperature (50 °C for PACM−20, 102 °C for DDS), which corresponds to $_{gel}T_g$ in Figure 6.

The TTT cure diagrams indicate a correspondence between gelation and the liquid-to-rubber transition, as well as between $_{gel}T_g'$ and $_{gel}T_g$. Although the activation energies for the liquid-to-rubber transition are slightly less than those obtained for gelation (Table II), a close correlation between the two events exists. In the one case where an activation energy was obtained also by IR measurements, which might be considered to yield a more accurate value, the liquid-to-rubber transition provides a lower estimate, whereas gelation (gel fraction) provides a higher estimate. A comparison of $T_{g\infty}$ for the series of epoxies cured with DDS (Table I) indicates that $T_{g\infty}$ is quite sensitive to the length of the epoxy molecule, that is, the distance between cross-links. As the molecular weight of the epoxy monomer increases (as n increases from 0 to 0.6), $T_{g\infty}$ changes from 220 to 185 °C. On the other hand, $T_{g\infty}$ is rather insensitive to the flexibility of the curing agent. One would expect the flexibility to increase in the order: MDA < PACM−20 < TMAB, but $T_{g\infty}$ is approximately the same (165 °C) for all of them. The difference in $T_{g\infty}$ for Epon 828/DDS and Epon 828/MDA, Epon 828/PACM−20, or Epon 828/TMAB can be attributed to the polarity of the DDS moiety and its capability of being a site for hydrogen bonding, which could increase the apparent cross-link density thereby giving rise to the higher glass transition temperature.

The difference in the reactivity of the curing agents also is demonstrated in the TTT diagrams. Although the energy of activation, E_a, (Table II) as determined from liquid-to-rubber transition data, is practically the same for all but one of the curing agents investigated in this work, the rates of reaction vary widely. This behavior can be seen from the horizontal shift in the gelation and vitrification curves. The curing agents ranked in order of increasing reactivity are DDS < TMAB < MDA < PACM−20.

The chemical reactivity, as well as the value of $T_{g\infty}$, influences the sharpness of the upper portion of the TTT diagram, of which $T_{g\infty} - _{min}T_g$ (Table III) is a measure.

Conclusion

The curing process can be understood, in terms of the molecular and macroscopic TTT cure diagrams, as the complex interaction of chemical reaction and changing physical properties. No single technique exists with which one could extract both macroscopic and molecular level information from a thermosetting system. However, a series of isothermal TBA experiments can provide a qualitative description of a system's TTT cure diagram, including $_{resin}T_g$, $T_g\infty$, the S-shaped vitrification curve, and, through the correlation of the liquid-to-rubber transition with gelation, $_{gel}T_g$. This procedure was used to compare various amine-cured epoxy systems from the viewpoint of structure/property relationships.

Acknowledgment

Partial support was provided by the Chemistry Branch of the Office of Naval Research.

Literature Cited

1. Gillham, J. K. *AIChE J.* **1974**, *20*, 1066.
2. Gillham, J. K. *Polym. Eng. Sci.* **1979**, *19*, 676.
3. Gillham, J. K. In "Developments in Polymer Characterisation—3" Dawkins, J. V., Ed.; Applied Science Publishers: London, 1982, p. 5.
4. "Organic Matrix Structural Composites: Quality Assurance and Reproducibility"; National Materials Advisory Board Publication 356, 1981; Chap. 3.
5. American Society for Testing and Materials, Philadelphia, PA.
6. Gillham, J. K.; Stadnicki, S. J.; Hazony, Y. *J. Appl. Polym. Sci.* **1977**, *21*, 40.
7. Enns, J. B.; Gillham, J. K. In "Computer Applications in Applied Polymer Science," Provder, Theodore, Ed.; ACS SYMPOSIUM SERIES No. 197, ACS: Washington, D.C., 1982; p. 329.
8. Flory, P. J. "Principles of Polymer Chemistry"; Cornell University Press: Ithaca, NY, 1953.
9. Lunak, S.; Vladyka, J.; Dusek, K. *Polymer* **1978**, *19*, 931.
10. Lee, C. Y.-C.; Goldfarb, I. *J. Polym. Eng. Sci.* **1981**, *21*, 951.
11. Enns, J. B., Ph.D. Thesis, Princeton Univ., Princeton, NJ, 1982.
12. Hartman, B.; Lee, G. F. *J. Appl. Polym. Sci.* **1977**, *21*, 1341.
13. Aherne, J. P.; Enns, J. B.; Doyle, M. J.; Gillham, J. K. *Coat. Plast. Prepr. Pap. Meet.* (*Am. Chem. Soc., Div. Org. Coat. Plast. Chem.*) **1982**, *46*, 574.
14. Enns, J. B.; Gillham, J. K.; Small, R. D. *Polym. Prepr., Am. Chem. Soc., Div. Polym. Chem.* **1981**, *22*, 123.
15. Arridge, R. G. C.; Speake, J. H. *Polymer* **1972**, *13*, 450.
16. Pogany, G. A. *Polymer* **1970**, *11*, 66.
17. Billmeyer, F. R. "Textbook of Polymer Science"; Wiley: New York, 1971.
18. Gillham, J. K. *Polym. Eng. Sci.* **1979**, *19*, 749.
19. Gillham, J. K.; Boyer, R. F. *Polym. Prepr., Am. Chem. Soc., Div. Polym. Chem.* **1976**, *17*, 171.
20. Nielsen, L. E. *Polym. Eng. Sci.* **1977**, *17*, 713.
21. Neumann, B. M.; Senich, G. A.; MacKnight, W. J. *Polym. Eng. Sci.* **1978**, *18*, 624.

RECEIVED for review December 7, 1981. ACCEPTED May 4, 1982.

Application of Torsion Impregnated Cloth Analysis (TICA) to Study Resin Cure

C. Y-C. LEE and I. J. GOLDFARB

Wright-Patterson Air Force Base, Air Force Wright Aeronautical
Laboratories/MLBP, Dayton, OH 45433

Torsion impregnated cloth analysis (TICA) is a forced torsion technique used to study resin mechanical behavior while the resin is being supported by glass cloth. A constant frequency or a multifrequency scan can be employed. The transition peak temperatures of the neat materials and the TICA specimens agree reasonably well. The isothermal curing TICA results of an epoxy are also compared with neat resin results. The gel peak observed in TICA measurements is concluded to be the result of resin—substrate interaction. The TICA technique has been able to distinguish between loss peaks due to molecular transitions, and those from kinetic effects based on their frequency dependency. Isocure state curves have been constructed using a two-step curing method, and a calibration method to interpolate partially cured T_g has been demonstrated with the technique.

VAST DIFFERENCES IN STRESS LEVELS between a liquid polymer and a solid polymer necessitate separate experimental techniques for their mechanical characterization (1). To characterize fully a polymer through its various states of matter, at least two different experiments have to be performed. The data can then be reconstructed to cover the entire region of interest. Similar problems are encountered in studying cure rheology of thermosetting systems. Although it is rather straightforward to study the curing in the liquid state, it is almost impossible to follow the vitrification process through conventional techniques.

0065−2393/83/0203−0065$06.00/0

Mechanical properties of thermosetting resins during cure through the various states of matter have been studied with different techniques (2, 3) that all had one feature in common: the resins to be studied were supported by inert substrates. Torsion impregnated cloth analysis (TICA) is a variation of such techniques (4). The resin being studied is impregnated onto a glass cloth that is subsequently folded into the shape of a rectangular bar. The specimen is mounted on the rheometrics mechanical spectrometer (RMS) and is subjected to a forced torsion analysis that measures directly the in-phase and out-of-phase components of the specimen's response.

The TICA sample preparation is very similar to that of torsion braid analysis (TBA) (2). The sample arrangement has the advantage of allowing a continuing study of a resin with the same specimen through its various physical states: liquid, rubber, and glassy. Also, this arrangement shares the advantage of the relatively small sample size required, compared with most neat-resin measurements. Because of the presence of the resin–substrate interaction, liquid-state results have to be treated with caution. Only the relative changes in the resin responses are obtained by this technique. Absolute modulus values cannot be extracted from the measurements without further developmental studies of the technique.

Unlike TBA, which is a free oscillation technique, TICA utilizes the torsion feature of the RMS. Instead of relying on the frequency and the decay of the oscillation amplitude to yield the rigidity and loss tan δ information, TICA measures directly the in-phase and out-of-phase components of the specimen's response that correspond to the storage and loss modulus. Because of the forced oscillation mode, a constant frequency can be used throughout the entire TICA experiment, thus avoiding the complication of the changing frequency effect on the observed response. More importantly, a multifrequency scan can be employed so the frequency effect of the response can be studied as well.

This chapter reviews the application of the TICA technique in our laboratory to the problem of studying the mechanical properties of resins.

Experimental

The sample preparation procedure had been described in detail elsewhere (5). The resin to be studied was dissolved with appropriate solvents and a fiber glass cloth about 10 cm wide was wetted with the solution. After hanging in an exhaust hood overnight, the cloth was evacuated for solvent extraction at room temperature for 1 week before use. The cloth was cut into rectangular patches of 10 × 7.5 cm. These were folded into strips 10 cm long and 1.25 cm wide, with the cut edges folded inside the strips. Three strips were stacked together to compose one TICA specimen.

Fixtures were fabricated to hold the TICA specimen in the sample chamber of the RMS. The fixtures were stainless steel plates measuring 17.8 × 12.6 × 2.3 mm. The cloth ends were sandwiched by two fixture plates held together by a screw at the center. Aluminum foil was used to separate the fixture from the cloth so that the cured resin would bond to the foil instead of the fixtures. The resulting specimens had dimensions at both ends similar to those of torsion bar specimens of the RMS, and were mounted on the sample-holding chucks of the RMS the same way as the torsion bars. The specimens were held at both ends between the servomotor and transducer of the RMS in an environmental chamber. A sinusoidal strain was induced by the oscillatory motion of the servomotor. The resultant sinusoidal stress was detected with the stress gauge, whose output voltage was tapped into a frequency analyzer. By comparing stress and strain waveforms, the in-phase and out-of-phase responses of the specimen were calculated. The environmental chamber was capable of a temperature range from −150 to 400 °C.

Comparison with Thermoplastics Neat Properties (4)

Polyphenylsulfone (Radel, Union Carbide Corporation) had been characterized mechanically in neat form in our laboratory (6) using an RMS. The region below and through T_g was characterized in the torsional bar mode, and above the T_g region was measured in the parallel plates mode. The polymer was also characterized with TICA from −150 to 380 °C. The TICA result is shown in Figure 1, and the peak maximum temperatures, together with those from the neat resin measurements, are listed in Table I. The loss peaks at 170 °C are due to kinetic effects (*see* later discussions) and are not included in the table.

Because of the presence of the glass cloth in TICA, all loss peaks due to molecular motion transitions were smaller in amplitude, in comparison to the neat-form results. The temperature and frequency dependence of the transition agreed well in the glassy and glass transition regions. Low-frequency TICA measurements gave very noisy loss components, so the TICA sub-T_g loss transitions in the low-frequency (<1-rad/s) range were not well characterized. In the region above T_g, the G'' maxima associated with flow shifted to higher temperatures when the polymer was measured with the TICA technique. Tan δ peaks not observed in the parallel plates measurements were presented in the TICA results. There are indications of the presence of the resin−substrate interactions, so TICA results in the liquid region should be used with care.

Isothermal Curing of Thermosetting Systems (4)

An isothermal curing experiment of an uncured resin usually begins with the resin in the liquid state. As the cure is advanced, the resin will gel into a rubbery state and then vitrify into the glassy state. In following such a cure of an epoxy resin with the TICA technique, the a (in-phase) component of the specimen increases several orders

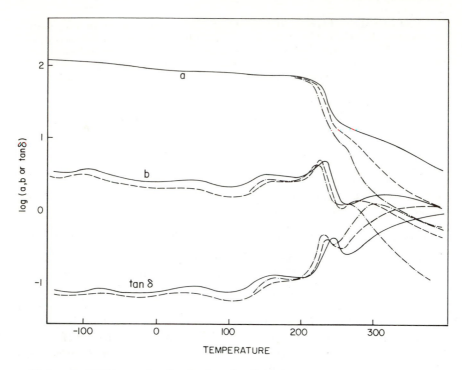

Figure 1. TICA result of polyphenylsulfone from −150 to 380 °C. Key:
—, 100 rad/s; ---, 10 rad/s; and −·−, 1 rad/s. (Reproduced with per-
mission from Ref. 4. Copyright 1981, Society of Plastics Engineers, Inc.)

Table I. Transition Temperatures of Polyphenylsulfone

Transition	Frequency (rad/s)	Neat Resin[a]	TICA
$T > T_g$	1	247	265
	10	260	280
	100	278	325
T_g	1	223 (228)	222 (228)
	10	226 (231)	225 (232)
	100	230 (237)	230 (237)
β	1	15	—
	10	32	44
	100	70	58
γ	1	−120	—
	10	−105	−112
	100	− 95	− 98

Note: All temperature values are in degrees Centigrade; all values refer to the loss modulus maxima except those in parentheses, which denote the values at the tan δ maxima.

[a] *See* Reference 6.

of magnitude as the resin changes from a liquid to a glass. There are also two tan δ maxima, followed by a prominent b (out-of-phase) component peak (Figure 2). The two tan δ peaks should correspond to the gel and vitrification peaks commonly observed in similar TBA experiments. The b maximum is also a vitrification peak, and, as expected, occurs later than the corresponding tan δ peak. This result is just the opposite of what one normally observes at the glass transition region. In the glass transition region, the low-frequency response is always leading the trend of changes during an increasing temperature sweep. For the isothermal curing, again the reverse is true; the high-frequency response is leading the trend.

Isothermal results from several curing temperatures can be used to compose a time−temperature−transformation (TTT) diagram (2). TBA results on the same system are available for comparison. Both sets of data are shown in Figure 3. The TICA results and the TBA results agree only qualitatively, not quantitatively. Both sets of data

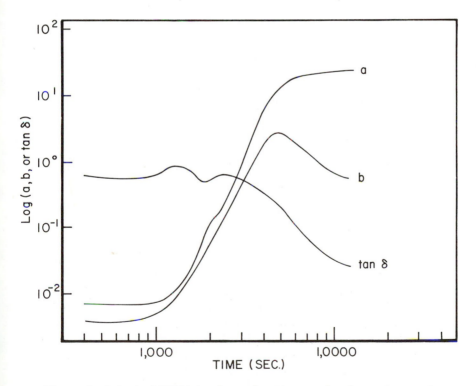

Figure 2. A typical TICA isothermal curing result of an epoxy resin. (Reproduced with permission from Ref. 4. Copyright 1981, Society of Plastics Engineers, Inc.)

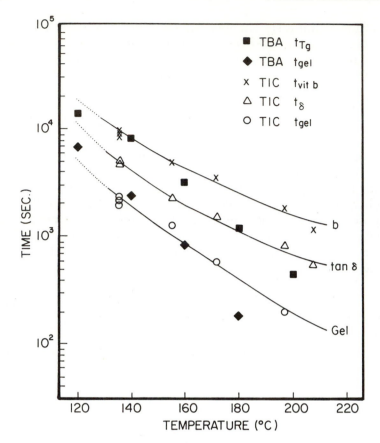

Figure 3. Time−temperature−transformation diagram comparing TICA and TBA results. (Reproduced with permission from Ref. 4. Copyright 1981, Society of Plastics Engineers, Inc.)

indicated that at the experimental temperature range, the system was between T_{gg} and T_g (∞) as defined by Gillham (2).

Data at the lower temperature regime agreed reasonably well in light of the difference between the two experiments in frequency effect. In the higher temperature range, the TICA times to peak maxima were consistently longer. These times are attributable to the different heat-up procedures employed by the two techniques. In both measurements, the samples were mounted onto the instrument at room temperature. The sample chamber temperature was then raised to the experimental value. For TBA, the temperature was raised at a rate of about 12 °C/min, and time zero of the experiment was when the final temperature was reached. In TICA, the experiments started at 0 °C, and the temperature was raised at a rate of 120 °C/min. Time zero was when the temperature increase was initiated.

Systematic errors are present in both procedures, but they are negligible when the reaction time is long in comparison to the heat-up time. When the condition is not satisfied, as in the high temperature region, then the TBA data will be consistently lower, and the TICA data consistently higher, than the real values. The isothermal curing results show a long time separation between the tan δ vitrification peak and the corresponding b maximum vitrification peak. By isothermally curing to the peak maximum at low temperature, the subsequent thermal scan will yield the corresponding glass transition peak at the same temperature if there is no additional curing during the temperature scan. These results indicate that the observed vitrification peaks truly correspond to the glass transition peaks. The long time separation between the tan δ and b maxima is due to the quenching of the reaction rate at high viscosity level.

Following the convention used for similar TBA experiments, the first tan δ peak in TICA isothermal experiments is referred to as a gel peak. The peak is not related to the conventional theory of gelation. The conventional theory of gelation is defined by a critical conversion condition and should be independent of the measurement temperatures. However, the TICA gel peak can be shifted to higher temperature by partial curing. A parallel plates study of the resin under isothermal conditions failed to reveal similar peaks in either the G'' or the tan δ component. We believe that the gel peak observed in TICA experiments is actually a result of the resin–substrate interaction.

Loss Peaks Due to Kinetic Effects (7)

Because of the multifrequency scan capability of TICA, loss peaks due to kinetic effects can be distinguished from those arising from molecular motion transitions. Because of transition energy barrier consideration, loss peaks due to molecular motion transitions should have a frequency dependence, with lower frequency peaks occurring at lower temperatures. For the loss peaks due to kinetic effects that are the results of a change in the glass transition temperature of the testing specimen during the experiment, there will be no frequency dependency. Such effects have been observed in thermoplastics, where the removal of residual solvents during the experiment causes a change in T_g, and also in partially cured thermosetting systems where the additional cure raises the glass transition temperature.

Figure 4 shows an example of a kinetic effect loss peak. The peak maxima temperature of the peaks at 180 °C are independent of frequency. Another important characteristic of kinetic effect loss peaks is the reversal of frequency trend. At the lower temperature regime of the peak, the low-frequency response leads the trend of changes in the a, b, and tan δ components. After the peak maximum, the high-fre-

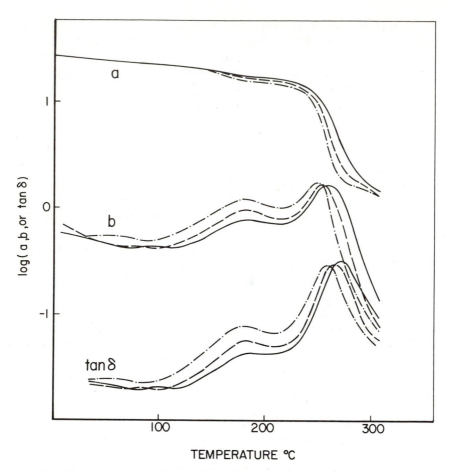

Figure 4. Loss peaks due to kinetic effects. Key: —, 100 rad/s; ---, 10 rad/s; and −·−, 1 rad/s. (Reproduced with permission from Ref. 4. Copyright 1981, Society of Plastics Engineers, Inc.)

quency response is leading the trend. On rescanning, the kinetic loss peaks should disappear.

The results from such experiments can be interpreted by describing the experiment as a function of time. With proper assumptions, the responses of the specimen can be expressed as a function of a reduced parameter $(T-T_g)$. The changes in the observed response are then a result of the changes in the reduced parameter which is controlled by the relative magnitudes of the thermal changes, dT/dt, and the kinetic changes, dT_g/dt. With the introduction of a rate conversion point, when $dT/dt = dT_g/dt$, and a four-stage scheme to distinguish various conditions encountered during the experiment, seemingly complex experimental results can be understood qualitatively.

Iso-cure State Curves (8)

The TTT diagram separates a time–temperature curing plot into liquid, rubbery, and glassy regions. The cure phase diagram concept is extended further by the introduction of an iso-cure state curve that joins all the points that represent the same cure state on the cure phase diagram (8). The superposition of the curve on the TTT diagram will allow reconstruction of the thermal properties of a partially cured state that may not be obtainable by direct thermoscan experiments because of the complications of additional cure during the measurement process.

An iso-cure state curve has been constructed with the TICA technique utilizing a two-step curing procedure. A batch of TICA specimens of the same partially cured state were prepared by subjecting them to an identical cure history. They were then post-cured at different temperatures to measure the time required to cure further to the state of the b maximum. By comparing these post-cure results with the isothermal curing results of virgin specimens one can obtain the time required to cure to the partially cured state of interest at various temperatures.

Figure 5 is the superposition of an iso-cure state curve and the TTT diagram of an epoxy system. The cure state represents a cure history of 5 h at 135 °C. A thermoscan of such a specimen yielded a kinetic effect loss peak at about 168°, and a glass transition at about 270 °C, which is the same as the T_g of the completely cured system. However, Figure 5 indicates that the cure state in question should have a T_g of 158 °C.

Iso-cure state curves can be a very useful tool to visualize thermal property changes as a function of cure. If the TTT diagram is accompanied by physical property measurements, e.g., volume and dielectric constant, a family of iso-cure state curves, as a function of cure time and temperature, will allow the extraction of information as to the manner in which the physical properties change over a certain temperature range as the cure is advanced. Of course, the concept of the iso-cure state curve on the TTT diagram requires certain assumptions (8). The choice of parameters to identify the cure state and the validity of the concept will have to be considered for each individual polymeric system, as well as for each individual physical property of interest.

T_g Calibration Method (8)

Because TICA specimens with known T_g values can be prepared by isothermally curing the specimen at the temperature of interest,

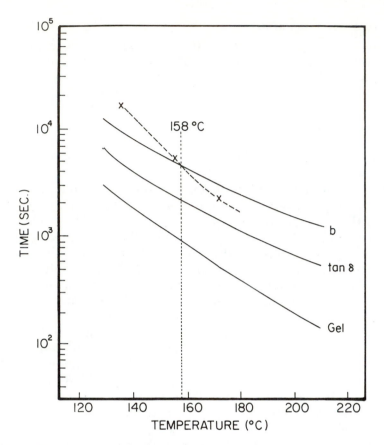

Figure 5. Superposition of iso-cured state curve (---) on time-temper-
ature-transformation diagram. (Reproduced with permission from
Ref. 8. Copyright 1981, Society of Plastics Engineers, Inc.)

and terminating the cure when the vitrification b maximum is reached, physical properties vs. T_g calibration curves can be constructed. When the T_g information of a partially cured specimen is needed, the physical properties of the unknown sample can be measured, and the T_g value can be interpolated through the calibration curve.

Such a concept has been demonstrated with the TICA technique by measuring the post-cure properties of the specimens. The time to b maximum, or a softening parameter, can be used as the calibration index. Samples of known T_g were prepared as described, and they were subjected to a high temperature post-cure with the same experimental procedure as in isothermal curing experiments. The times to

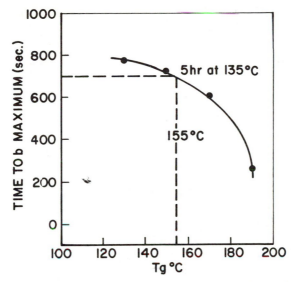

Figure 6. Time to b maximum calibration curve. (Reproduced with permission from Ref. 8. Copyright 1981, Society of Plastics Engineers, Inc.)

reach the b maximum were then used to correlate with the T_g. Figure 6 shows the result of using the time to b maximum as the calibration index. The same epoxy system in the iso-cure state example given above was used. For the state representing 5 h cure at 135 °C, the method showed a T_g of 155 °C, agreeing reasonably well with the iso-cure state example.

The softening parameter was also extracted from the post-cure experiment. Typical post-cure behavior is shown in Figure 7. The parameter R is defined as h_2/h_1. The amount of softening at the initial heat-up as indicated by h_1, and h_2 is the hardening due to additional curing. The ratio of the two values is needed to make consistent comparisons between specimens. The ratio is like normalizing the observed changes with an internal standard. When the additional curing rate is fast and the resolution of the time to b maximum is poor, the softening parameter is an attractive alternative to yield the partially cured T_g information. This method had been used to extract information of T_g advancement as a function of cure time, which is unobtainable through conventional thermoscan experiments (9). Again, the method suffers the same limitations as the iso-cure state method because of the necessity to make certain assumptions about the cure state, and its applicability should be considered for each individual system.

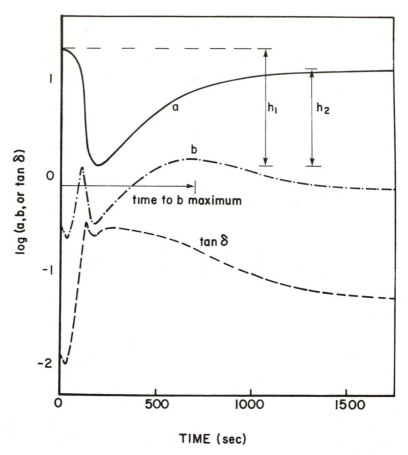

Figure 7. Post-cure curves of a partially cured specimen. Either the time to b maximum or the ratio h_2/h_1 can be used as calibration index. (Reproduced with permission from Ref. 8. Copyright 1981, Society of Plastics Engineers, Inc.)

Literature Cited

1. Ferry, J. D. "Visoelastic Properties of Polymer"; Wiley: New York, 1970.
2. Gilham, J. K. *Polym. Eng. Sci.* **1979**, *19*, 676.
3. Senich, G. A.; MacKnight, W. J. *J. Appl. Polym. Sci.* **1978**, *22*, 2633.
4. Lee, C. Y-C.; Goldfarb, I. J. *Polym. Eng. Sci.* **1981**, *21*, 390.
5. Lee, C. Y-C.; Goldfarb, I. J. AFWAL−TR−80−4159.
6. Lee, C. Y-C.; Henes, J. D.; Grossman, T. E. AFML−TR−79−4062.
7. Lee, C. Y-C.; Goldfarb, I. J. *Polym. Eng. Sci.* **1981**, *21*, 951.
8. Lee, C. Y-C.; Goldfarb, I. J. *Polym. Eng. Sci.* **1981**, *21*, 787.
9. Hedberg, F. L.; Lee, C. Y-C.; Goldfarb, I. J. AFWAL−TR−80−4179.

RECEIVED for review October 14, 1981. ACCEPTED December 24, 1981.

Dynamic Mechanical Analyzer for Thermal Mechanical Property Characterization of Organic Coatings

THEODORE PROVDER, RICHARD M. HOLSWORTH, and THOMAS H. GRENTZER

Glidden Coatings and Resins, Division of SCM Corporation, Strongsville, OH 44136

Dynamic mechanical analysis was used to study polymer/end-use property relationships for chemical coatings systems. The du Pont 981 Mechanical Analyzer (DMA) and a torsion pendulum gave comparable results for exterior acrylic latex and styrene−acrylic−acrylonitrile terpolymer free films. DMA damping profiles of can coatings correlated with the coatings performance during can manufacture. Gel coat DMA modulus and damping profiles correlated with the tendency toward environmental stress cracking. Combined DMA/DSC (differential scanning calorimetry) techniques were used to analyze a high solids coating foam entrapment problem. Cure kinetics methodology using a single dynamic DMA temperature scan was developed to obtain kinetics parameters. Comparative reaction kinetics were obtained by DMA/DSC techniques for the curing of EPON 825 with bis(4-aminophenyl)methane.

THE USE OF TORSION PENDULUM ANALYSIS (TPA) for evaluating organic coatings has been well documented (1−3). The first TPA in use at this laboratory was capable of obtaining modulus and loss profiles of thin (1−3 mil; 25−75 μm) organic coatings as well as thicker (10−20 mil; 250−500 μm) gel coat resin systems. This instrument is shown in Figures 1 and 2. The TPA can be used as a torsion braid analyzer by replacing the free film with a glass mat or braid to support coatings systems. This use has been demonstrated for a thermosetting acrylic powder coatings system as shown in Figure 3 (4). The use of the TPA was time consuming and required constant operator attention. An

Figure 1. Torsion pendulum designed for free paint films. (Reproduced with permission from Ref. 1. Copyright 1966, Journal of Coatings Technology.)

evaluation of the du Pont 981 dynamic mechanical analyzer (DMA) showed that the results were comparable to those obtained on the TPA for the same systems. The operation of the DMA did not require constant operator attention, thereby making it an excellent instrument for routine thermal mechanical property characterization of organic coatings.

Instrumentation and Methodology

Description of DMA Instrumentation. Dynamic mechanical analysis measures the deformation response of a sample which has been subjected to oscillatory forces. The du Pont 981 DMA follows the resonant frequency (or relative modulus) and energy dissipation of or-

Figure 2. Lower sample holder and phonograph cartridge torque sensing unit. (Reproduced with permission from Ref. 1. Copyright 1966, Journal of Coatings Technology.)

ganic coating systems as a dynamic function of time or temperature. From these measurements the modulus and tan δ can be calculated in the temperature range from −150 to 500 °C.

Figure 4 shows a scheme of the du Pont 981 DMA. The sample arms are fixed to the rigid block via low-friction flexure pivots. A compound resonance system is formed by clamping the sample between the arms. An electromechanical transducer is used to drive the active arm while the counterweighted passive arm is used for physical sup-

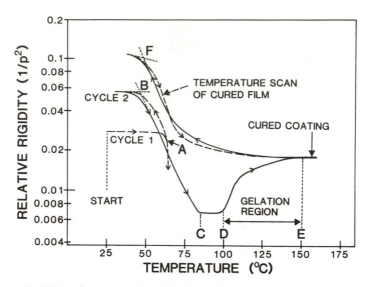

Figure 3. TBA relative rigidity as a function of temperature for a ther-
mosetting acrylic powder coating. Key: A, powder fusion; B, T_g of un-
cured film; C, end of melting; D, onset of gel; E, cure effected; and F, T_g
of cured film. (Reproduced with permission from Ref. 4. Copyright
1974, Journal of Coatings Technology.)

port. Horizontal angular displacement of the active arm about the pivot
flexure produces a few tenths of a millimeter deflection at the sample.
Thus, the sample is placed in flexural stress. After removing the dis-
placing force, the sample goes into resonant oscillation. The oscillation amplitude and frequency are sensed by a linear variable differential transformer (LVDT). The sample's natural resonant frequency
of oscillation is digitally displayed on the front panel of the 981 DMA
and is also plotted as a dynamic function of time or temperature. Normally, when a sample is deformed and released, the sample will oscillate at its resonant frequency with a decreasing amplitude of oscillation. The DMA sends the signal from the LVDT to the DMA driver
circuitry, which feeds the signal back to the electromechanical transducer to maintain a fixed amplitude of oscillation. The power required
to maintain this fixed amplitude of oscillation is related to the damping capacity of the material, and is plotted along with the resonant
frequency as a dynamic function of time and temperature (5, 6).

Experimental Conditions for Dynamic Mechanical Analysis. The
du Pont 981 DMA was used in conjunction with the 990 thermal analyzer programmer/recorder to obtain the experimental results. The
mechanical response of all samples was recorded as a dynamic func-

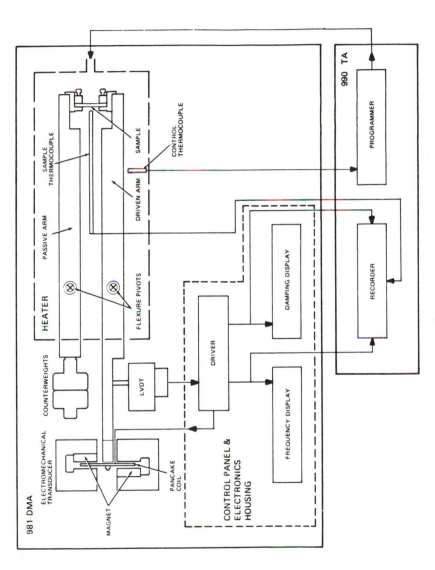

Figure 4. Scheme of du Pont 981 DMA. (Reproduced with permission from Ref. 5. Copyright 1979, E. I. du Pont de Nemours & Co., Inc.)

tion of temperature at a heating rate of 5 °C/min. The samples were cooled to subambient temperatures by purging the sample chamber with liquid nitrogen. After the sample temperature reached approximately −100 °C, the liquid flow rate was reduced and a dry nitrogen purge of 5 L/min was delivered to the sample compartment environment simultaneously. The sample preparation procedure and clamping arrangement were dependent on the type of coating investigated, as well as the information desired from the analysis. Therefore, this information will be discussed later for the specific coating system investigated.

Experimental Conditions for Differential Scanning Calorimetry (DSC). The du Pont 990 thermal analyzer programmer/recorder and the 910 DSC cell module were used to obtain the experimental results. The sample environment was either nitrogen or compressed air at a flow rate of 50 mL/min. The sample weight was kept in the 0.5-mg range. Hermetically sealed sample pans were used for reactions exhibiting a weight loss.

Reaction Kinetics

In previous studies (7–9), a method was developed for determining quantitative reaction kinetics by DSC with the use of a single dynamic temperature scan (one thermogram). In this chapter the mathematical approach was extended to obtain quantitative reaction kinetics of cure by dynamic mechanical analysis by the use of a single dynamic temperature scan.

The du Pont DMA can be used to follow the mechanical response of a material before, during, and after cure. However, in most cases, the thermosetting system must be supported by an inert material (i.e., a woven fiberglass braid) to follow the buildup of mechanical properties during cure. During a dynamic temperature scan, the DMA frequency profile (relative modulus) will increase in the curing temperature region for a thermoset system. This increase in relative modulus, which can be related to the fractional extent or degree of cure, $F(t,T)$, is defined by the following equation and schematically illustrated in Figure 5:

$$F(t,T) = \frac{G(t,T) - G_1}{G_2 - G_1} \tag{1}$$

where G_1, G, and G_2 are the relative modulus readings at the onset of cure, at a given time and temperature during the curing process, and after the curing process has ceased, respectively. Assuming that the functional form of the cure curve follows the general nth order rate expression:

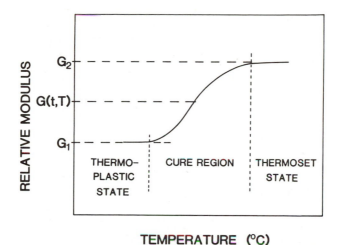

TEMPERATURE (°C)

Figure 5. DMA relative modulus profile as a function of temperature during cure.

$$\frac{dF(t,T)}{dt} = k(T)\big[1 - F(t,T)\big]^n \qquad (2)$$

where n is the order of reaction, then the rate constant $k(T)$ in reciprocal seconds takes on the following form:

$$k(T) = \frac{1}{\phi}\frac{dF(t,T)}{dT}\big[1 - F(t,T)\big]^{-n} \qquad (3)$$

where ϕ is the heating rate (dT/dt in degrees Centigrade per second) and $[dF(t,T)]/dT$ is the rate of change of degree of cure with temperature. The relationship between $F(t,T)$ and $dF(t,T)/dT$ is shown in Figure 6.

All of the quantities on the right side of Equation 3 are observable parameters except the reaction order n. Substituting the Arrhenius expression shown in Equation 4 into Equation 3 yields the working Equation 5.

$$\ln k(T) = \ln A - \frac{E}{RT} \qquad (4)$$

where A is the Arrhenius frequency factor (reciprocal seconds), E is the activation energy (kilojoules per mole), and R is the gas constant (kilojoules per mole degree).

$$\ln A - \frac{E}{RT} = \ln\frac{1}{\phi}\frac{dF(t,T)}{dT}\big[1 - F(t,T)\big]^{-n} \qquad (5)$$

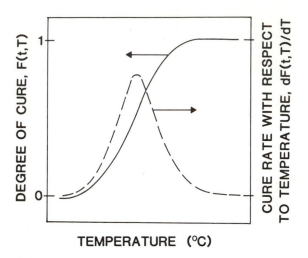

Figure 6. Relationship between degree of cure, F(t,T), *and temperature rate of cure,* dF(t,T)/dT.

The order of reaction is selected by evaluating the linearity of the Arrhenius plot. A series of Arrhenius plots is generated by varying the order of reaction in Equation 5 from 0.2 to 3.0. The best value of the reaction order is obtained by selecting the order of reaction giving the highest correlation coefficient of the linear least squares fit of the ln $k(T)$ vs. $1/T$ curve. The value for the reaction order is then substituted into Equation 5 to obtain the activation energy and Arrhenius frequency factor. The isothermal degree of cure as a function of time, t, or temperature, T, is obtained by integrating the rate expression to yield the following expressions:

$$F(t,T) = 1 - [(n-1)k(T)t + 1]^{1/1-n} \qquad n \neq 1 \qquad (6)$$

$$F(t,T) = 1 - e^{-k(T)t} \qquad n = 1 \qquad (7)$$

Equations 5–7 are quite similar to those obtained for the single dynamic temperature scan DSC method (7, 8).

Results and Discussion

Comparison of the DMA with the TPA. Dynamic mechanical property information on free films can be obtained with the DMA. In developing experimental methodology for the DMA, free film samples were analyzed by both DMA and TPA to compare and correlate the results. For the DMA studies the free film dimensions were 1.25 cm × 0.63 cm × 0.5 mm and the films were mounted horizontally in

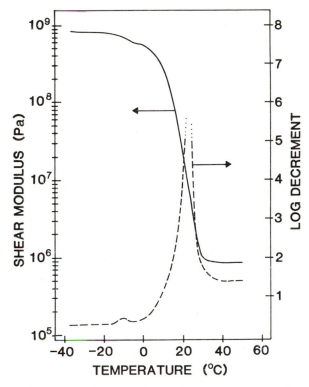

Figure 7. Shear modulus and logarithmic decrement of an exterior latex as a function of temperature. (Reproduced with permission from Ref. 1. Copyright 1966, Journal of Coatings Technology.)

the DMA clamps. The sample preparation procedure for TPA has been described previously (1, 2).

Figures 7 and 8 show the TPA and DMA results for an exterior acrylic latex, respectively. A rapid change in shear modulus and logarithmic decrement around 20 °C was observed by the TPA. Similar results were observed with the DMA. Figures 9 and 10 show the TPA and DMA results for a styrene–acrylic–acrylonitrile terpolymer. The point of maximum damping for the TPA curve lies between 10 and 15 °C. The DMA energy dissipation peak maximum is at 12 °C. Although the curve shapes have observable differences, in part due to the difference in experimental technique between TPA and DMA, the onset of the glass transition region for the relative modulus curve is within ±5 °C from both techniques. Thus, DMA results are readily comparable to TPA results for dynamic mechanical property analysis.

Can Coatings.   Stringent performance properties are required for coatings used in the beer and beverage can industry. A coating is

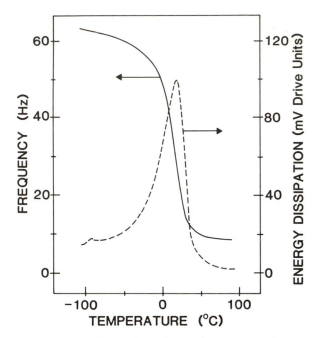

Figure 8. Dynamic mechanical analysis of an exterior latex as a function of temperature.

applied to the can metal stock and cured. The cans can be fabricated in a three- or two-piece process. The main body of the can is formed by iterative punching of the metal sheet by the appropriate tooling (draw–redraw) or by punching the metal sheet into a relatively flat, high diameter cup and then forcing the cup through a series of rings to iron out the can to full size. The can ends are stamped out from the coated metal sheets at very high rates. Coatings with tailor made mechanical properties are required for these processes.

Because the can coatings are deformed at high rates of elongation and deformation, the possibility of crazing or fracturing of the coating is increased. Dynamic mechanical analysis provides information on the structure–property performance characteristics of these coatings. The DMA can analyze the damping response of the coating on the metal can stock.

Figure 11 shows the DMA damping profiles for three can coating–steel substrate composites and the uncoated steel substrate. The coatings were applied to the steel sheets with a draw down bar. Samples of 2.54 × 1.25 cm were cut from the coated steel sheets and clamped vertically in the DMA. Except for Sample C, the samples were baked for 30 s at a peak metal temperature of 232 °C. Sample C was baked for 10 min at 204 °C. The coatings have a thickness of approximately 0.3 mil (7.6 μm), while the metal substrate had a thickness of approximately 30 mil (760 μm). The three coating samples are water-borne epoxy acrylic graft copolymers. Sample A has a broader damping peak than Sample B as shown in Figure 11. The damping curve for Sample A also showed signs of a doublet in the temperature region between 125 and 250 °C. Sample A showed good can fabrication performance, while Sample B fractured during the can fabrication process. Sample A differs from Sample B in the chemical composition of the acrylic portion of the graft copolymer. By analyzing damping profiles (along with a knowledge of the chemical composition), it was deduced that Sample A is more prone to microscopic aggregation due to increased differences in the solubility parameters between the graft and backbone polymer chains. Sample A has a more heterogeneous microscopic structure than Sample B. This increased heterogeneity is believed to result in an increase in impact strength during the can

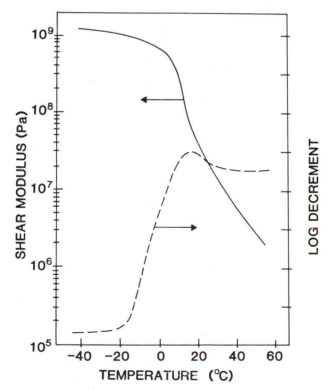

Figure 9. Shear modulus and logarithmic decrement of a styrene – acrylic – acrylonitrile terpolymer as a function of temperature.

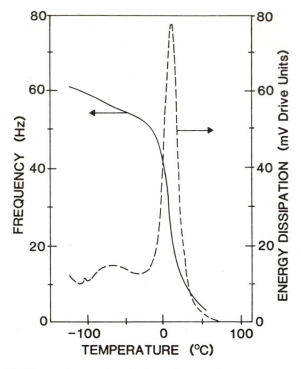

Figure 10. Dynamic mechanical analysis of a styrene−acrylic−acrylonitrile terpolymer as a function of temperature.

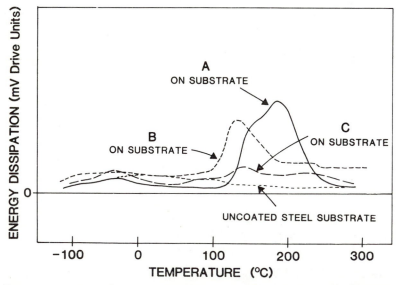

Figure 11. DMA damping profiles as a function of temperature for three can coating−steel substrate composites and the uncoated steel substrate.

fabrication process. Sample C shows the effect of overbaking a can coating system. The overbaked coating shows very little damping capacity by DMA analysis in Figure 11. This bake schedule also produced a can coating which failed by loss of adhesion as a result of overbaking.

Gel Coat System. Gel coats are protective/decorative coatings used in sanitary and marine applications. The gel coat is an in-mold integral part of the product. The gel coat system, along with catalyst and promoter, is sprayed into a mold. The gel coat undergoes free radical cure at ambient temperature until a tack-free surface is obtained. Then a fiberglass mat is placed on the gel coat and the system is reinforced by backup resin. If the fiberglass mat is applied before the gel coat reaches a tack-free state, the gel coat could pull away from the mold and produce an irregular surface.

The DMA can be used to correlate the end-use properties of gel coat resins to fundamental dynamic mechanical behavior and to the chemistry of the curing system. Three polyester gel coat formulations were analyzed as cured free films by the DMA and compared to environmental stress cracking results obtained after 1 year. The samples were clamped vertically in the DMA and had dimensions of 1.9 cm × 1.25 cm × 0.076 mm. Sample A is an orthophthalic–isophthalic acid based resin blend. Sample B is a neopentyl glycol-modified gel coat resin system, and Sample C is an isophthalic acid-based system and has more unsaturation than the other two systems. These systems were designed for use in the manufacture of fiberglass boats. Sample A showed cracking after 1 year of field service use. Sample C was the most crack-resistant formulation, and Sample B gave intermediate stress crack resistance. These gel coat samples also were cured in the laboratory and analyzed on the DMA. The modulus and tan δ curves are shown in Figures 12 and 13, respectively. Sample A has a higher modulus below its glass transition temperature than do Samples B or C. The glass transition region for Sample A covers a smaller temperature span than for Sample B or C. The damping peak for Sample A occurs at a lower temperature and has a narrower breadth of distribution than the other two samples.

Both the modulus and the damping curves for Sample B indicate a higher T_g material due to cross-linking of the additional unsaturation. Sample C, which had the best end-use performance, has a broad and intense damping peak. Therefore, Sample C has a greater potential for dissipating the energy of crack propagation from the polyester backup resin and performs well during service use. These results show the qualitative relationship between tan δ, modulus, and the gel coats' future tendency toward environmental stress cracking.

High Solids Coatings. A polyester–melamine coating exhibited a film defect (foam entrapment) upon curing. The foam entrapment problem was alleviated in the formulation by adding additional

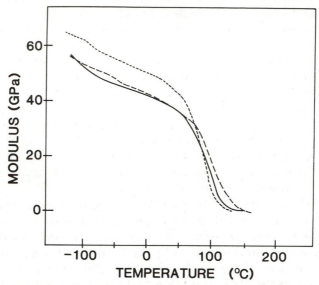

Figure 12. DMA modulus curves as a function of temperature for en-
vironmentally stress-cracked gel coat systems. Key: ---, severe cracking;
− − −, intermediate; and ——, crack-resistant.

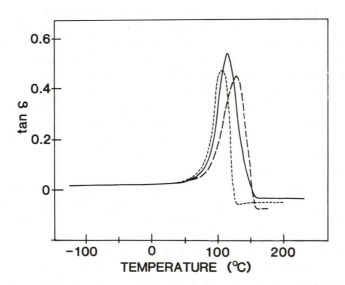

Figure 13. DMA tan δ curves as a function of temperature for environ-
mentally stress-cracked gel coat systems. Key is the same as in Figure 12.

cross-linker and decreasing the catalyst level. It was believed that this change in formulation produced differences in the chemical reactivity of the curing system.

A kinetics analysis using the single dynamic DSC temperature scan method (3–5) indicated that the two formulations had the same chemical reactivity. Therefore, for this particular system, the foam entrapment problem was not caused by differences in chemical reactivity of the curing system.

For the DMA studies, fiberglass braids, 2.54 × 1.25 cm were used as inert substrates for the uncured high solid resins. After air drying for 24 h, the braids were mounted horizontally in the DMA clamps. DMA cure studies also showed no discernable differences in cure rate between the two formulations. This result can be seen by examining the curing region in the relative modulus profiles for both formulations in Figure 14. This finding agrees with the DSC results. However, the relative modulus profiles for the cured coatings in Figure 15 show that the coating exhibiting foam entrapment has an onset temperature for the glass transition region of 55 °C compared to 32 °C for the coating that did not exhibit foam entrapment. Also, the glass transition region of the latter sample covers a broader temperature range, 124 °C vs. 62 °C. These results indicate that the reformulated coating,

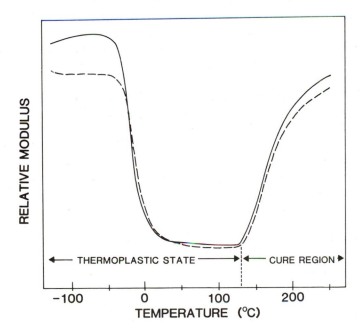

Figure 14. DMA relative modulus curves as a function of temperature for high solids coatings during the curing process. Key: – – –, with foam entrapment; and ——, good film.

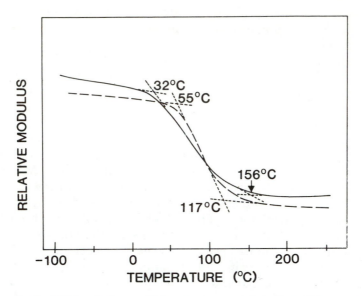

Figure 15. DMA relative modulus curves as a function of temperature for high solids coatings after cure. Key is the same as in Figure 14.

which does not exhibit defects due to foam entrapment, is more flexible with a higher molecular weight between cross-links. This would be expected to occur upon decreasing the catalyst level. The decrease in catalyst level may account for the elimination of foam entrapment by reducing the level of volatile evolution occurring during the melamine cure.

DMA/DSC Cure Kinetics.   The DSC monitors the heat flow into and out of the sample during the curing process. On the other hand, the DMA monitors rigidity buildup during cure. The DMA kinetics results are sensitive to volatilization, molecular weight between cross-links, backbone rigidity, viscosity, and functional group reactivity. Therefore, by using kinetics results from different techniques it should be possible to gain insight into the chemical and physical factors affecting the curing process.

The DMA experiment required the use of a fiberglass braid support for the epoxy cure reaction. Reaction kinetic parameters for the curing of a narrow molecular weight distribution epoxy, EPON 825, with bis(4-aminophenyl)methane by the single dynamic scan DSC and DMA methods are reported in Table I. The reaction kinetics results do not agree, as one might expect, because each method is sensitive to different physical phenomena. The differences between the fractional conversion (by DSC analysis) and the degree of cure (by DMA analysis) are shown in Figure 16. The two curves in Figure 16

Table I. Reaction Kinetics for the Curing of Epon 825 with Bis(4-aminophenyl)methane

Method	Activation Energy (kJ/mol)	Reaction Order	Arrhenius Frequency Factor (s^{-1})
DSC	58.6	1.05	8.9×10^{4}
DMA	109.7	1.10	7.8×10^{12}

were obtained by using Equation 6 along with the DSC and DMA kinetic parameters in Table I. The calculated degree of cure profile by DMA analysis indicates a faster reaction taking place at 150 °C than is indicated by DSC analysis.

The discrepancy between the DSC and DMA kinetic results may be due to the vitrification of the sample before complete chemical reaction has been achieved. Also, it is possible that this difference is due to the DMA's sensitivity to the rigidity of the amine molecule bis(4-aminophenyl)methane. Further work in determining isothermal vitrification points is planned.

Conclusions

Dynamic mechanical analysis is a valuable technique for solving production problems and aiding coatings failure analysis. This tech-

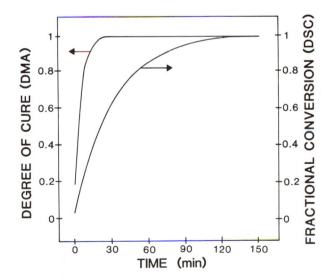

Figure 16. Comparison of DMA degree of cure curve and DSC fractional conversion curve for an epoxy–amine reaction at an isothermal temperature of 150 °C.

nique also is useful for establishing relationships between end-use performance properties and fundamental thermal–mechanical parameters.

The use of dynamic mechanical analysis cure kinetics studies can provide insight into the chemical and physical factors affecting the curing process. The kinetic parameters can be used to simulate oven bake conditions. This will allow coatings chemists to better design and formulate coatings to achieve desired performance properties.

Literature Cited

1. Pierce, P. E.; Holsworth, R. M. *J. Paint Technol.* **1966**, *38* *(496)*, 263.
2. Pierce, P. E.; Holsworth, R. M. *J. Paint Technol.* **1966**, *38* *(501)*, 585.
3. Bender, H. S. *J. Paint Technol.* **1969**, *41* *(535)*, 445.
4. Zicherman, J. B.; Holsworth, R. M. *J. Paint Technol.* **1974**, *46* *(591)*, 55.
5. E. I. du Pont de Nemours & Co., Inc., Scientific and Process Instruments Division, "Instruction Manual—981 Dynamic Mechanical Analyzer"; Wilmington, DE, 1979.
6. Murayama, T. In "Dynamic Mechanical Analysis of Polymeric Material"; Elsevier: New York, 1978; pp. 51, 52.
7. Grentzer, T. H.; Holsworth, R. M.; Provder, T.; Kline, S. "Proceedings of the Tenth North American Thermal Analysis Society Conference," Oct. 26–29, 1980, Boston, MA; p. 269.
8. Grentzer, T. H.; Holsworth, R. M.; Provder, T. *Org. Coat. Plast. Chem.* **1981**, *44*, 673.
9. Kah, A. F.; Koehler, M. E.; Grentzer, T. H.; Niemann, T. F.; Provder, T. *Org. Coat. Plast. Chem.* **1981**, *45*, 480.

RECEIVED for review October 14, 1981. ACCEPTED March 1, 1982.

The PL-Dynamic Mechanical Thermal Analyzer and Its Application to the Study of Polymer Transitions

R. E. WETTON
Loughborough University, Department of Chemistry,
Loughborough, United Kingdom

T. G. CROUCHER and J. W. M. FURSDON
Polymer Laboratories, Ltd., Church Stretton, Salop, United Kingdom

A new instrument for small strain dynamic mechanical measurements of polymers is reported. Sample temperature can be programmed from −150 to 300 °C (500 °C Special Head arrangement). Eight logarithmically spaced drive frequencies from 0.33 to 90 Hz can be selected on the front panel. Clamping geometrics for bending beam, shear, and elongational deformation are described. Applications using both single-frequency thermal scans and computer multiplexing of several frequencies are given. Examples of application to homopolymers, copolymers, and composites are presented and discussed.

MEASUREMENTS OF DYNAMIC MECHANICAL PROPERTIES of polymeric materials, which were previously laborious, can now be performed automatically with microprocessor-controlled instrumentation. The PL-dynamic mechanical thermal analyzer (PL-DMTA) uses the well-established phase and amplitude technique with digital measurement of phase to give tan δ resolutions of 0.0001. Measurements can be performed at fixed frequency over a three and one-half decade range. A dynamic stiffness range of over four decades allows measurement of all transitions with one sample. This, together with precise temperature programming, in the range from −150 to 300 °C, opens up new possibilities for research, analytical, and quality control measure-

0065−2393/83/0203−0095$06.00/0
© 1983 American Chemical Society

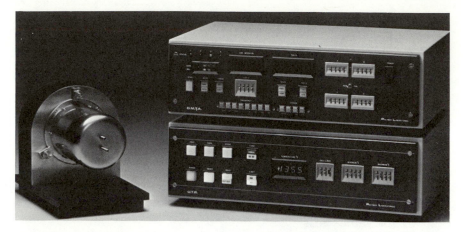

Figure 1. The PL-dynamic mechanical thermal analyzer. Key: the mechanical measuring head, left; the dynamic analyzer unit, top right; and the temperature programmer, bottom right.

ments. The instrument, as well as measurements on a range of polymer systems, is described.

Instrumentation

The PL-DMTA comprises three bench top units as shown in Figure 1. The mechanical head contains the sample in a temperature enclosure. The main deformation modes are bending beam (either double or single cantilever) and shear. The sample clamping details in these two arrangements are shown in Figures 2a and b. The sample disks for shear sandwich operation are mounted normally on the frame external to the instrument, and the complete assembly is then located on the instruments. This method of mounting is particularly useful when measuring difficult samples, such as adhesives. Results quoted in this chapter were taken in the bending beam mode (typical sample size 10 × 5 × 2 mm) unless otherwise stated. An alternative head assembly for the tensile deformation of films and fibers at constant tension is currently being made available.

The oscillating stress experienced by the specimen is proportional to the drive current from the oscillator. The strain produced is converted to a proportional voltage by a nonloading transducer. The dynamic analyzer unit compares the phase and amplitude of stress and strain and computes absolute values of log modulus and tan δ. The instrument is easily calibrated by an added mass technique that then makes the instrument absolute when the geometry constant for the sample under measurement is dialed in on the front panel. Calibration needs checking only periodically.

Frequency and strain amplitude are selected by push button with eight logarithmically spaced-frequencies from 0.03 to 90 Hz and three controlled strain levels covering a ×16 range. Temperature is sensed by a platinum resistance thermometer lying immediately behind the sample. A temperature programmer controls the balanced heat oven to produce uniform heating rates from 1°/10 min (or less) to 20°/min (nominal). The temperature may be controlled isothermally at any point, or series of points, with external computer control.

Accuracy depends on the optimization of a number of parameters. Temperature uniformity in thick specimens can only be achieved at a heating rate of a few degrees per minute. In thin specimens temperature definition is easier, but errors from dimensional measurements may become significant in modulus determinations. Tan δ is resolved to 0.0001 for values less than 1.0 and, because it is determined as kE''/kE', errors in the geometry term (k) do not produce errors in this term. The high resolution now achieved produces sufficiently fine steps in output vs. temperature such that essentially smooth curves are produced from traces on a normal two-pen recorder.

External computer control of the main functions f, T, strain, and heating rate is available through the IEEE (Institute of Electrical Engineers Standard 488) interface, and in this mode, results are displayed both in real time and stored for future manipulation. An example of the power of external computer control is shown in Figure 3 in which three frequencies are multiplexed during a single thermal scan at 2 °C/min. The sample is an extremely high modulus carbon-filled epoxy.

a   b

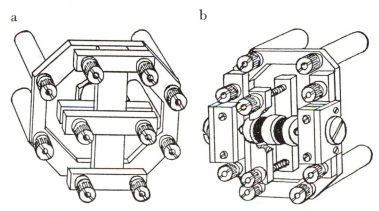

Figure 2. (a) Sample clamping arrangement with a dual cantilever. Drive clamp vibrates perpendicular to sample, which can be mounted on one side only to accommodate expansion or contraction in length. Clamp frames are interchangeable to give variable length; and (b) shear sandwich geometry for rubbers and gels. The black discs represent the two samples, and the central plate is oscillated by the drive clamp.

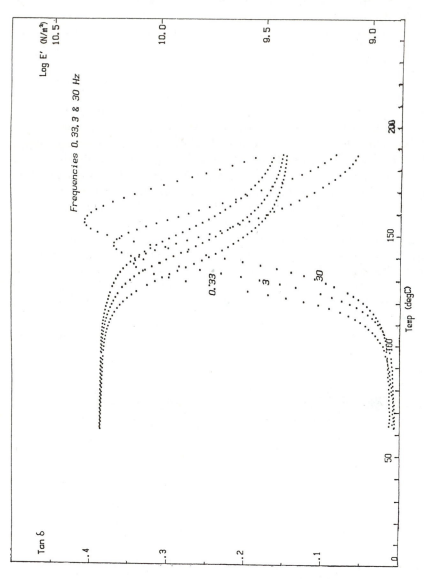

Figure 3. Control by an external computer allows collection of data at a number of frequencies during a single thermal scan. Data collected for a carbon filled epoxy through the main T_g region.

One advantage of this method is that any temperature errors in the sample are identical for each frequency.

Results and Discussion

Dynamic mechanical thermal analysis (DMTA) senses any change in molecular mobility in the sample as the temperature is raised or lowered. The time scale required for the molecular motion to manifest itself is determined by the frequency, f, of the impressed sinusoidal stress. A progressive change in storage modulus and peaking in tan δ (energy lost/energy stored per cycle) occurs when the average molecular relaxation time, τ, is $\frac{1}{2} \pi f$. DMTA is not dependent on temperature scanning rate in the same way that DSC-type measurements are. This can be demonstrated more acutely in the ability to obtain isothermal data by scanning through the motional time scale by changing the frequency. This information yields then the relaxation time spectrum ($1, 2$).

More than one transition will normally be encountered in the temperature plane. Figure 4 shows an example using a prequenched poly(ethylene terephthalate). The main transition (α_a) at 85 °C, 1 Hz is due to the onset of micro-Brownian motion of the main chains about T_g.

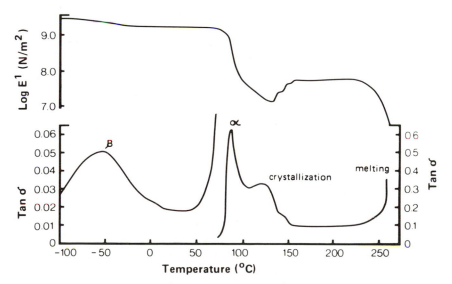

Figure 4. PL-DMTA (1 Hz) scan for a quenched poly(ethylene terephthalate) sample heated 5 °C/min. The scan shows α and β relaxation, crystallization, and melting.

This transition occurs in the amorphous phase, and has a large magnitude because of the high volume fraction of amorphous phase present in the quenched sample. The tan δ magnitude, 0.6, and half-width (ignoring recrystallization onset) at 13 °C, compare to quenched polystyrene values of 1.32 and 13 °C at the same scanning rate. The reduction in loss peak magnitude without peak broadening shows some crystallinity in the quenched sample. Quantitative information of this type has not been available previously as all experimental parameters must be controlled.

At low temperatures the broad β relaxation process is observed at -60 °C/Hz, and is due to localized molecular motion (local mode) of the methylene and carboxyl groups. It comprises at least two overlapping relaxations (3) and is relatively insensitive to crystallinity. When main chain relaxations are involved in the β process, good impact strength normally results.

As the temperature is raised above T_g, the molecular freedom acquired allows rapid crystallization at this high degree of supercooling. The modulus thus rises as crystallization proceeds in the range $110-150$ °C. Finally, at high temperatures catastrophic melting occurs at 200 °C, and clamping is lost. In principle, tan δ should go through a step at the melting point and increase without limit as the temperature is raised further.

Thermal scanning at reasonably high rates (5 °C/min) yields information that would not be obtained in very slow scanning or isothermal experiments. In isothermal experiments annealing processes may occur and the sample would crystallize at a significantly lower temperature.

Data through the melting region are difficult to obtain with relatively low molecular weight polymers that have high melting points, such as polyesters and nylons. When such materials melt the process is catastrophic because the equilibrium liquid has a low viscosity. Thus, clamping integrity is lost as soon as the crystalline interactions are lost. It is, however, relatively easy to obtain data through the melting range of a high molecular weight, low melting polymer, such as polyethylene oxide. Data in Figures 5 and 6 are for polyethylene oxide of a molecular weight of approximately one million (Polyox grade). A relatively short, thick sample with aspect ratio (length/thickness) of three was used to obtain reasonable sample stiffness on the high temperature side of the melting region. Some sacrifice of modulus accuracy due to unknown end corrections is made, particularly where modulus levels are high. Figure 5 shows results for a sample cooled quickly from the melt to room temperature, where it was held for 2 h before scanning at 4 °C/min and 1 Hz. Two peaks are exhibited in tan δ in sympathy with the decreases in modulus. The

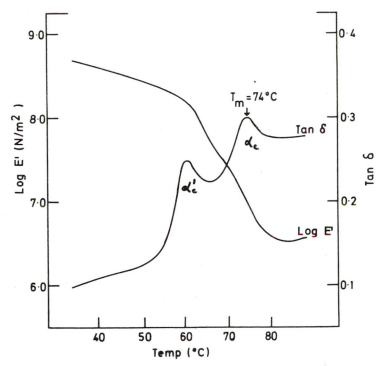

Figure 5. *High molecular weight polyethylene oxide. PL-DMTA scan (1 Hz) through the melting region for a sample rapidly cooled from melt to 20 °C.*

higher peak (α_c) coincides with the observed melting point, but the lower temperature peak (α_c') does not agree with any major thermal event seen by DSC-type techniques. Different thermal treatment alters the position and magnitude of the α_c' in Figure 6. The sample was crystallized at 60 °C for 5 h to allow complete primary crystallization, and the α_c' process occurs at a higher temperature. The α_c' process apparently depends on the perfection, or size of lamellae, and there is a strong parallel with similar relaxations in different types of polyethylene (4).

An inherent consequence of the dynamic mechanical method's response to motional time scale is that measurement at higher frequencies will cause the loss peak to be observed at higher temperatures. In amorphous homopolymers the loss peak shifts with minor shape changes in the temperature plane. The shift factor is given by the semiempirical Williams, Landel, and Ferry (WLF) equation (5), which approximates to a simple Arrhenius relation at temperatures far

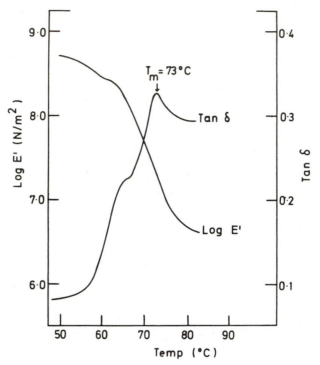

Figure 6. High molecular weight polyethylene oxide. PL-DMTA scan through the melting region for sample crystallized at 60 °C.

above T_g. However, many commercial polymers are partially crystalline, and copolymers are increasingly common. In such cases the loss curves will usually change shape radically with change of measurement frequency. Figure 7 shows this for an ethylene–vinyl acetate copolymer (24% vinyl acetate) where the sharpening of the loss curve with increasing temperature results from two effects. First there is a sharpening of the relaxation time spectrum that is inherently broad for random copolymers, and second there is an improvement in segment compatibility as the temperature is raised, leading to a more homogeneous material on the molecular scale. The loss peak half-width at 30 Hz is still 18 °C, and is much broader than observed for homopolymers in the same temperature range.

The PL-DMTA can be used as an analytical instrument in the same sense as IR is used to determine unknowns. The theory is less far developed, however, and the technique is in some cases empirical.

The loss peak location for a random copolymer moves normally between the location for the parent homopolymers in a similar way to T_g itself. However, in ethylene–vinyl acetate copolymers crystallinity level changes of the polyethylene sequences disturb the average amorphous phase composition, and, consequently, the α_a loss peak position changes only slightly (see Figure 8), but its magnitude decreases in sympathy with the amorphous phase content. The peak half-width in the temperature plane (30 Hz) increases with ethylene content until it is too broad to measure sensibly at 7.5% vinyl acetate composition. As more and more ethylene sequences crystallize, with increasing ethylene content, the amorphous phase becomes relatively richer in vinyl acetate composition, and this fact causes the lack of shift in peak position. Clearly, after proper thermal history control, a DMTA scan can provide a good indication of the composition, or alternatively, if composition is known, information is obtained on the nature of the amorphous phase and crystallite interaction/cross-linking.

The extreme cases of heterogeneity in copolymers is provided by block and graft copolymers particularly as exemplified by thermoplastic elastomers. In these materials physical phase separation of glassy or crystalline domains is necessary to provide cross-linking. Good clear cut phase separation, with each phase exhibiting a loss

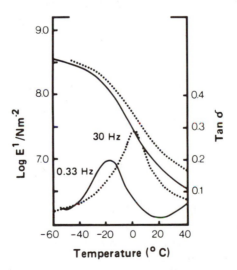

Figure 7. *Effect of frequency of measurement on loss peak position and shape for ethylene–vinyl acetate copolymer. Heating rate was 5 °C/min.*

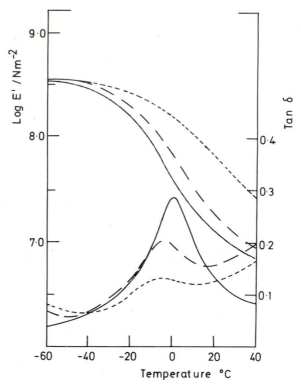

Figure 8. PL-DMTA scan (30 Hz) of ethylene–vinyl acetate copolymers with varying acetate contents. Crystallinity perturbs the amorphous phase composition. Key: —, 25% vinyl acetate; — —, 15% vinyl acetate; and ---, 7.5% vinyl acetate.

peak close to its homopolymer position and not significantly broadened, is the requirement. DMTA characterizes further the rubber and glassy state modulus levels. These features are not attainable by any other single technique. Phase studies of this type can be extended to polyurethane systems, which are essentially multi-block copolymers with poor chemical and physical definition. Figure 9 shows data for two types of polyurethane. Sample A is a normal formulation based on polyester, somewhat greater than bifunctional in−OH groups with butanediol/methylene diphenyl isocyanate as chain extenders. Phase separation is clear cut with the polyester α_a loss peak at −8 °C and 1 Hz, and some structural disordering at 70 °C. Structural integrity of the network persists up to 180 °C. In contrast, microphase separation with some phase mixing is achieved intentionally in Sample B to produce an optically clear product. The extent of

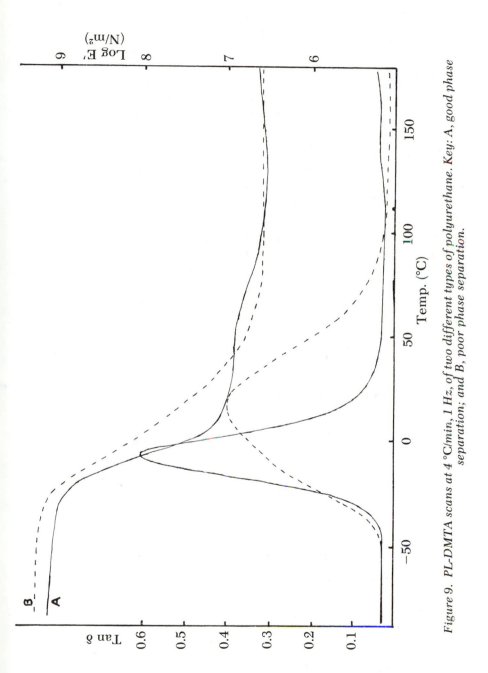

Figure 9. PL-DMTA scans at 4 °C/min, 1 Hz, of two different types of polyurethane. Key: A, good phase separation; and B, poor phase separation.

phase mixing is assessed by the DMTA trace. Again, cross-linking integrity is preserved up to 180 °C.

Alternative clamping geometry allows measurement of soft samples ($<10^8$ N/m²) in shear, but the main interest in these cases is for engineering data on rubbers that can be generated up to a 10% strain level. Measurements of this type are not relevant to the main theme of transition measurements, but this mode of deformation is useful for studying transition phenomena in adhesives and for studying the early part of cure in cross-linking resins.

Figure 10 shows an alternative geometry that allows tensile modulus measurement of thin polymer films, typically below 0.05 mm in thickness. The geometry constant in this mode is shown in the figure and is merely the normal tensile value modified by $\cos^2\theta$ (the angle between drive direction and film). The tension is monitored by the displacement produced in the drive spring and is seen directly as a proportional change in the transducer off-set voltage. After tension adjustments have been made, the sample is clamped by the central clamp with narrow clamp edges and the transducer is adjusted manually to its normal working range. Good data can be generated on films in regions of minor changes in tension through relaxation effects, but it is not practical if stress relaxation occurs strongly. The new tension head is designed to remedy this problem. Figure 11 shows data on rubber toughened polypropylene films mounted as shown in Figure 10. The region of ethylene–propylene rubber relaxation can be seen clearly in Samples A and B. The rubber segments have been incorporated differently, and this is reflected in the shape of the loss process. The DMTA technique is one of the few ways of studying fine differ-

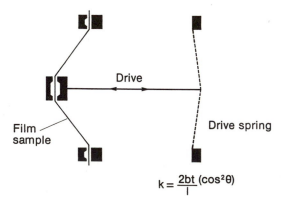

$$k = \frac{2bt}{l}(\cos^2\theta)$$

Figure 10. Arrangement for tensile dynamic modulus measurements in thin films. The angle θ is between film and drive directions, and b, t, and l are width, thickness, and length, respectively.

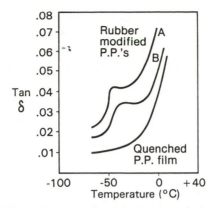

Figure 11. Rubber-modified polypropylene films measured through the ethylene–propylene rubber loss region. Geometry used is shown in Figure 10.

ences in film structure of this type. Sample A has a wider range of molecular heterogeneity associated with the rubber segments than Sample B.

Comparison of moduli obtained from bending of short film samples with those obtained from tensile measurements provides a method for looking at skin effects in films. Inner and outer layers are weighted equally in the tensile case, but the outer layers contribute more to the bending modulus.

High accuracy modulus and damping characteristics measured on the PL-DMTA have proved useful in studying the aging characteristics of polymer glasses. In both, cross-linked epoxy systems (6) and uncross-linked polystyrene (7, 8) the basic effects seem the same. Compared to the corresponding annealed (aged) glass a quenched sample has a lower glassy state modulus, but surprisingly a higher damping peak, when scanned in the temperature plane. The modulus change over a period of 1 week, annealed at 20 °C below T_g, is in excess of 100% and the damping level in the glass, at 30 °C below T_g, decreases by a factor of five in the case of annealed polystyrene. The low temperature (β) loss process in the epoxy systems shows a sharpening with annealing, but no significant shift in central position.

Conclusions

The PL-DMTA is able to give accurate data on the widest possible range of polymeric materials. Thermal scanning at constant frequency typically can be achieved in 1 h, comparable to DSC techniques. The technique is complementary to DSC in that motional transitions are observed, although thermodynamic transitions are always detected by the structural changes they produce. Secondary transitions are not observable by the DSC method nor can small reinforcing phases be studied. The PL-DMTA method detects all these transitions and allows formulation of definitive conclusions concerning sample morphology and phase composition.

External computer control allows the full capabilities of the instrument to be exploited and it is shown that several different measurement frequencies can be multiplexed during a single temperature scan. As data are stored in the computer any required data manipulation can be achieved subsequently.

Literature Cited

1. Ferry, J. D. In "Viscoelastic Properties of Polymers"; Wiley: New York, 1961; p. 63.
2. Schwarzl, F.; Stavermann, A. *J. Appl. Sci. Res.* **1953**, *A4*, 127.
3. Illers, K. H.; Breuer, H. *J. Colloid Sci.* **1963**, *18*, 1.
4. Mills, P. J.; Hay, J. N.; Cox, C., First PL-DMTA Users Meeting, Univ. of Warwick, 1981.
5. Ferry, J. D. In "Viscoelastic Properties of Polymers"; Wiley: New York, 1961; p. 203.
6. Richardson, M. J., First PL-DMTA Users Meeting, Univ. of Warwick, 1981; p. 9.
7. Wetton, R. E.; Woo, Y. H. *Polymer*, in press.
8. Wetton, R. E. *Chemical Soc., Analytical Proc.* Sept/Oct 1981.

RECEIVED for review October 30, 1981. ACCEPTED October 11, 1982.

Dynamic Mechanical Spectroscopy Using the Autovibron DDV-III-C

S. M. WEBLER, J. A. MANSON,[1] and R. W. LANG

Lehigh University, Materials Research Center, Bethlehem, PA 18015

The Autovibron DDV-III-C is a forced vibration unit capable of operating at several constant frequencies for the determination of the dynamic mechanical response of a system. The automation provides a programmed heating rate, continuous sample tensioning, and acquisition and reduction of data. Results obtained at several frequencies are reported for plain poly(vinyl chloride) (PVC), PVC modified with a methacrylate−butadiene−styrene terpolymer (MBS), and a commercially available mineral-reinforced polyamide. While problems have been encountered with sample alignment, tension adjustment, and measurement at low tan δ values, it is concluded that this instrument has good potential for the convenient determination of dynamic spectra of polymers and their composites.

IN DETERMINING THE DYNAMIC MECHANICAL RESPONSE of a system, it is often desirable to work with a forced-vibration instrument at a constant frequency. One of the instruments most commonly used for this purpose has been the direct-reading viscoelastometer originally developed by Takayanagi (1, 2)—the Rheovibron. A common model has been the model DDV-II (load capacity, 0.1 kgf) and in recent years a 5-kgf capacity model, the DDV-III-C, was introduced (3, 4).

While valuable research has been based on results obtained using such units, several problems have been recognized. Limitations include difficulty in working at $T > T_g$, and undesirably low ranges in tan δ and frequency. In addition, maintenance of proper tension on the specimen is often far from easy. Therefore, the operator must give constant attention to the instrument over a period of 4 h or more. More detailed discussions are available in the literature (5, 6).

---

[1] To whom correspondence should be addressed.

0065−2393/83/0203−0109$06.00/0

To remedy or alleviate some of these problems, the Rheovibron model DDV-II-B was modified and improved (5) by providing closed-loop control, and by improving and simplifying the technique used to determine the loss tangent and the storage modulus. Gains in accuracy, simplicity of operation, and adaptability to digital processing of the data were reported. The Rheovibron itself has been automated by the manufacturer (3) and the Rheovibron DDV-II was also automated (6) to provide automatic control of tension, increased sensitivity, and calculation and printout of E', E'', and tan δ. The latter unit has been commercialized (Imass, Inc.) (4) as the Autovibron, model DDV-II-C. Recently a generally similar adaptation was introduced based on the hydraulically operated Rheovibron DDV-III-C. Automation of a torsional pendulum (7) and a different constant-frequency instrument (8) have also been described.

Although a full critical analysis of the operation of the Autovibron DDV-III-C has not yet been possible, it is appropriate to describe our experience with this new instrument, and to make preliminary recommendations with respect to operation and future improvement. Because the instrument is the first of its type, the observations reported should be helpful to other investigators. Results obtained in our laboratory using an automated DDV-II unit are also described for comparison.

Instrumentation

As mentioned previously, the model DDV-III-C Rheovibron (Toyo Baldwin Co.) has been combined with an automation package (Imass, Inc.). The instrument maintains the essential characteristics of the Rheovibron DDV-III-C, utilizing the original sample bench, hydraulic system, load cell, and basic electronics. Four fixed frequencies of 3.5, 11, 35, and 110 Hz are available. Sample sizes up to 7 cm × 1 cm × 5 mm can be handled, with a claimed range for complex Young's modulus between 1 MPa and 100 GPa (1 GPa = 1 GN/m² = 10^{10} dynes/cm²). A low-temperature chamber allows measurements to be taken from about -140 to 175 °C with a programmed rate of temperature increase of ~ 1 °C/min. A second chamber is provided for temperatures up to 300 °C. The automation package is responsible for sample tensioning, phase angle measurements, temperature control, data acquisition, and data reduction. The key components of this package are a lock-in analyzer (Princeton Applied Research model 5204), a programmable calculator (Hewlett Packard model 9825A), a multiprogrammer (Hewlett Packard model 6940B), and an optional plotter (Hewlett Packard model 9872B). The automation package can also be interfaced readily with a Rheovibron model DDV-II. The essential

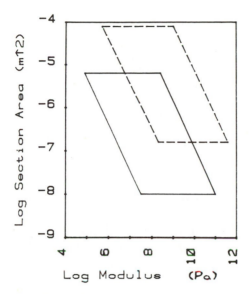

Figure 1. Comparison of the correct sample size vs. modulus for the DDV-II (—) and DDV-III (---) Rheovibrons.

differences in the two units are the driving unit and the load capacity. The hydraulic driving system of the DDV-III is replaced by an electromechanical driver in the smaller unit. The DDV-II is capable of handling sample sizes up to 5 × 0.05 × 0.4 cm (Figure 1 shows a comparison of sample sizes for the two models) with a maximum load capacity of 0.1 kgf and modulus range of 100 kPa– 100 GPa. A schema is given in Figure 2.

Temperature programming is effected through the calculator in conjunction with a platinum resistance thermometer. From − 140 to −45 °C the temperature is allowed to increase without regulation at a rate of 1 °C/min. At −45 °C power is supplied to the heaters, and the temperature is controlled by programming the application of power. The temperature rise can also be controlled at rates other than 1 °C/min through changes in the operating program. For temperatures above 225 °C, the high-temperature chamber must be used.

Phase-angle measurements using the lock-in analyzer were incorporated to simplify automation of the measurements, improve resolution of small angles, and increase the range of tan δ measurements (4). The calculator alternately switches the load (P) and displacement (X) signal through the multiprogrammer to the lock-in analyzer. After a programmed delay for setting of the signal, the in-phase and quadrature components of the respective signals are measured with respect to a reference signal from the Autovibron (Figure 3). The complex Young's modulus, E^*, is calculated using Equation 1.

$$E^* = \frac{R\,(P_I^2 + P_Q^2)^{\frac{1}{2}}\,L^2}{(X_I^2 + X_Q^2)^{\frac{1}{2}}\,V} \tag{1}$$

The subscripts I and Q designate the in-phase and quadrature components of the respective signals, R is a ranging and scaling factor, L is the sample length, and V is the original sample volume. The phase difference, δ is calculated using Equation 2.

$$\delta = \phi_1 - \phi_2 = \arctan\frac{P_Q}{P_I} - \arctan\frac{X_Q}{X_I} \tag{2}$$

With E^* and δ from Equations 1 and 2, respectively, the storage modulus, E', and loss modulus, E'', can be calculated.

$$E' = |E^*|\cos\delta \tag{3}$$

$$E'' = |E^*|\sin\delta \tag{4}$$

$$\tan\delta = E''/E' \tag{5}$$

The acquired data are displayed while the program is running and stored on magnetic-tape cartridges for further reduction. Results can be printed or plotted during the run with appropriate programming; programs for data reduction and plotting from tape are available.

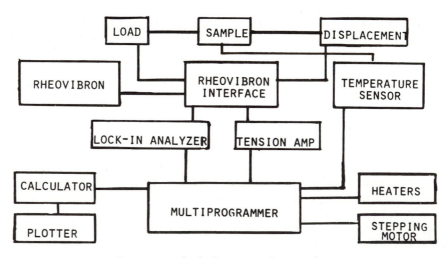

Figure 2. Block diagram of Autovibron.

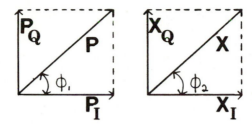

Figure 3. In-phase and quadrature components of the load (P) and displacement (X) signals. Subscripts I and Q refer to in-phase and out-of-phase (quadrature) components; ϕ_1 and ϕ_2 are phase angles.

Experimental

Problems and Their Correction. During start-up and subsequent trials using the DDV-III-C, several significant problems were encountered. Much of our work has been conducted using an automated DDV-II because of problems with the DDV-III. Preliminary work with the large load capacity unit has shown problems in sample tensioning, load control, measurement of small phase angles, and programming. These difficulties are discussed later.

Sample tension is controlled by the calculator through the multiprogrammer and a stepping-motor that moves the load arm. The original stepping-motor assembly used springs to control tension and resulted in enough lateral motion to preclude maintenance of alignment. Such alignment is of critical importance, otherwise serious errors in modulus and damping can result. A combination of shims to align the center of the load-arm with the center of the driver and a new screw-driven stepping-motor assembly (redesigned by Imass, Inc.) have minimized lateral motion. However, even with these modifications, great care must still be taken with clamping and alignment of the specimen. Also, bowing in the sample can be introduced by nonuniform tightening of the grips. Some scatter may be due to an inherent design problem; the manufacturer is currently revising the signal reading section. As was the case with the Autovibron DDV-II (6), sample mounting and alignment are two major flaws of the instrument. Reproducible methods for sample mounting have been reported for the DDV-II (9), and similar modifications should be included in further redesign of the DDV-III system.

The original software calculated the DC voltage of the load signal by sampling the sine wave, calculating the amplitude, and dividing by two to obtain the DC bias. Because of problems in measuring the load, the program was modified to measure the DC voltage bias directly, by momentarily switching off the sine wave. Load corrections are then made by the stepping-motor to maintain a preset limit. This new load-control program functions acceptably through a programmed temperature run except in the region around the glass transition temperature (T_g). Immediately after the transition, the sample is often put into compression. So far, reliable measurements of rubbery moduli on the order of 10 MPa have been obtained only occasionally.

While monitoring the load signal (P) with an oscilloscope, a problem was evident in the switching of the signal. The value of the load was intermittently recorded as zero. Because tension is a function of load, when zero loads are recorded the instrument reacts by making drastic changes in sample length, resulting in incorrect modulus and phase angle measurements. This problem appears to have been corrected by the replacement of a relay-readback board in the multiprogrammer.

Another problem has been the determination of optimum settings of the phase controls on the lock-in analyzer. After some experimentation it was determined that the in-phase and quadrature readings should be set approximately equal in magnitude and with the same sign (positive or negative) using the reference angle potentiometer and quadrant selector of the lock-in analyzer. Use of these settings appears to reduce the time required for the signal to stabilize, and facilitates ranging of the signals. The ranging subroutine has also been rewritten by Imass to alleviate a problem with signal saturation that occurred when the in-phase and quadrature components of the signal became unbalanced.

Performance.   Figures 4 and 5 compare data at 110 Hz from two samples of the same methacrylate–butadiene–styrene (MBS)-modified PVC in the as-received condition (PVC 132-3). These samples were analyzed by using the DDV-III-C Autovibron before manufacturer revisions. Table I shows PVC sample designations, weight-average molecular weight, M_w, and rubber contents. Previous literature contains detailed characterization of the PVC (10). The samples were of similar cross-sectional area and length (*see* Table I). The tests were run using an oscillating displacement, ΔL, of 2.5×10^{-3} cm, corresponding to an oscillating strain, $\Delta\epsilon$, of 0.05%. Great care was taken with sample mounting and alignment. Over the temperature range from -115 to 100 °C, values of the storage modulus (E') obtained in the two tests agreed within less than 5%. Below -75 °C (corresponding to tan δ ≤0.02) significant scatter was evident in the loss modulus (E'') and tan δ, and the slopes differed considerably, so that the value of tan δ at -100 °C is ~40% less in Figure 5 than in Figure 4. (The shapes of the E'' and tan δ curves in Figure 5 are in fact atypical.) Considerable scatter has also been seen at low values of tan δ with the Autovibron DDV-II-C (11). However, the peak for the MBS phase is clearly evident at about -60 °C. While data for Specimen A (Figure 4) could be taken up to 140 °C, it was not possible to exceed 100 °C with Specimen B

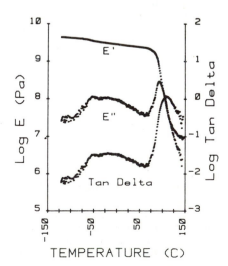

Figure 4. Dynamic mechanic spectra (110 Hz) of MBS-modified PVC (Sample 132−3A) using a DDV-III-C Autovibron.

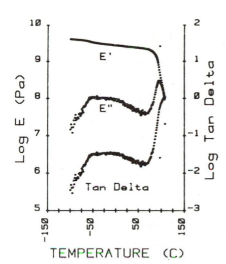

Figure 5. Dynamic mechanical spectra (110 Hz) of MBS-modified PVC (Sample 132–3B) using a DDV-III-C Autovibron.

which, for unknown reasons, deformed excessively. Values of T_g and complex modulus (E^*) are reported in Table II.

Figure 6 shows data obtained for a PVC 132-4 sample using the DDV-III-C unit after replacement of the pump and modification of the electronics by the manufacturer. Sample characteristics are given in Tables I and II. The test was run using an oscillating displacement, ΔL, of 2×10^{-3} cm (i.e., at $\Delta \epsilon \approx 0.05\%$). Tan δ and E'' values show considerably less scatter than before the instrument was modified (compare data in Figures 4, 5, and 6). In fact, E' results can now be reproduced at 110 Hz to within $\pm 2\%$ when using similar sample sizes. Nevertheless, the problem of excessive sample deformation at $T > T_g$ still exists.

Figure 7 illustrates data obtained with an automated Rheovibron DDV-II for two different specimens of the same material. The two samples had almost identical length, width, and thickness (*see* Table II). Both samples were machined in the same manner and run by the same operator under the same operating conditions. The initial oscillating displacement, ΔL, was 7.9×10^{-4} cm (i.e., $\Delta \epsilon \approx 0.01\%$) in both cases. Values of E^* agree to within 9% at $-100°C$

Table I. PVC Characterization

PVC Sample designation	Matrix[a] $M_w \times 10^{-5}$	phr MBS
131-1	0.67	0
131-4	0.67	14
132-1	0.95	0
132-3	0.95	10
132-4	0.95	14
135-1	2.08	0
135-4	2.08	14

[a] Values of M_w are from reference (*10*).

Table II. Comparison of Dynamic Data on a Standard and Modified PVC Using Autovibron and Rheovibron Model DDV-II

Sample	Rheovibron Model	Frequency (Hz)	T_g (°C)	Complex Modulus ($E^* \times 10^{-9}$ Pa)			L^a (cm)	L/W^b (cm^{-1})	L/A^c (cm^{-2})
				at −100 °C	at 0 °C	at 40 °C			
132-3A	III	110	94	4.240	2.867	2.501	4.942	12.7	59.5
132-3B	III	110	95	4.125	2.790	2.464	4.956	13.0	62.1
132-4	III	110	92	2.657	1.824	1.632	4.136	10.3	47.4
135-4	II	110	83	2.047	1.521	1.350	6.48	30.6	650
135-4	II	110	83	2.240	1.639	1.434	6.45	29.7	660
135-1	III	110	93	3.794	2.889	2.625	3.766	9.0	38
135-1	III	110	93	2.727	2.124	1.912	4.142	9.1	38
135-1	II	110	82	2.107	1.731	1.506	5.38	22.2	494
135-4	III	110	94	2.690	1.836	1.644	4.106	10.4	46.7
135-4-Q^a	III	110	93	2.674	1.796	1.526	4.048	9.8	44.3
135-4-7^e	III	110	92	2.638	1.774	1.555	3.990	10.1	47.6
131-4-Q^a	II	35	76	2.389	1.599	1.335	6.31	30.5	677

[a] L = sample length
[b] L/W = length ÷ width
[c] L/A = length ÷ area
[d] Q = quenched sample
[e] Annealed 7 d at 65 °C

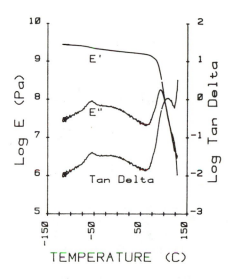

Figure 6. Dynamic mechanical spectra of MBS-modified PVC (Sample 132−4, $M_w = 0.95 \times 10^5$; 14 phr MBS) using DDV-III-C Autovibron after revisions (110 Hz).

and differ by less than 6% at 40 °C. In the T_g region E', E'', and tan δ are virtually identical. The two samples do not however, have the same E'' and tan δ in the region from −100 to 50 °C. For example, tan δ readings of 0.033 and 0.056 were taken at −25 °C for the two samples. These differences are as yet unexplained and suggest caution should be employed when analyzing data.

Figures 8 and 9 show data for PVC 135-1 using the DDV-III unit (before and after revisions) and the automated DDV-II. Tan δ values are comparable from −150 to 50 °C. The T_g measured using the DDV-III-C is 10°C higher than the T_g obtained with the DDV-II. This difference probably reflects an effect of the larger sample size in the DDV-III. Most likely the average sample temperature lags behind the furnace temperature; this temperature lag results in the apparent increase in T_g. A comparison of complex modulus (E^*) data at 0

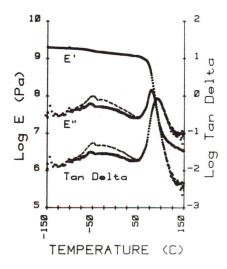

Figure 7. Dynamic mechanical spectra of two replicate MBS-modified PVC samples (Sample 135−4, $M_w = 2 \times 10^5$; 14 phr MBS) using an automated DDV-II Autovibron.

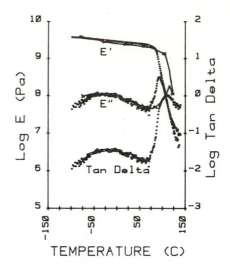

Figure 8. Comparison of dynamic mechanical spectra of PVC (Sample 135−1, $M_w = 2 \times 10^5$) using model DDV-III-C (···) and manual DDV-II (∗∗∗) Autovibron.

°C reveals an apparent 35% decrease after modification of the DDV-III, while E^* obtained at 0 °C using the DDV-II is 18% lower than the new DDV-III value. These unexplained differences suggest caution should be employed when comparing data obtained from different instruments and operators.

Error analysis for the manual DDV-II was addressed (9), and the analysis should be extended to the automated unit. Errors of up to 50% have been reported (9) and were attributed to the instrument compliance, sample yielding and slipping in the clamps, sample alignment, the instrument's inertia, variable sample sizes, and structural changes in the sample during testing. Each problem needs to be addressed before a thorough understanding of the automated unit will be possible, and true material properties can be measured with full confidence.

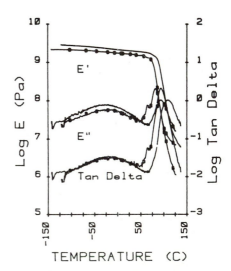

Figure 9. Comparison of dynamic mechanical spectra of PVC (Sample 135−1, $M_w = 2 \times 10^5$) using automated DDV-II (○) and DDV-III-C Autovibrons after manufacturer's revisions (—) at 110 Hz.

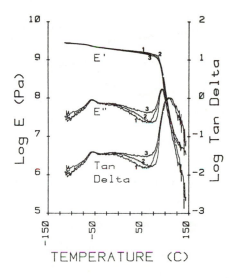

Figure 10. Dynamic mechanical spectra (110 Hz) of MBS-modified PVC (Sample 135–4, $M_w = 2 \times 10^5$; 14 phr MBS) as received (1), after 7 d at 65 °C (2), quenched from 110 °C (3). These spectra were obtained by using the DDV-III-C Autivibron after revisions.

Studies of the effect of thermal history and frequency of PVC and MBS-modified PVC are under way. Preliminary results obtained using the DDV-III-C on quenched PVC (quenched from 110 °C in ice) are shown in Figure 10 with respect to an as-received sample. The results show an increase in damping between T_β and T_g in the quenched sample similar to the results of Struik (*12*). Figure 10 also displays data for a sample that was quenched in ice from 110 °C and then annealed at 65 °C for a period of 7 d. The damping between 0 and 50 °C clearly had been affected by the aging process in a manner similar to data presented by Struik (*12*).

The manual Rheovibron is not used easily at frequencies <110 Hz, but limited results have been obtained using the automated DDV-II at 35 and 11 Hz. Data in Figure 11 show the effect of vibrations and resonances on data

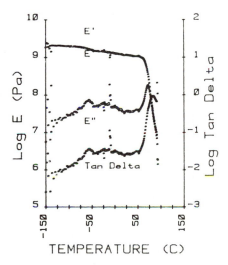

Figure 11. Dynamic mechanical spectra (110 Hz) of a quenched MBS-modified PVC (Sample 131–4–Q, $M_w = 7 \times 10^4$; 14 phr MBS) run at 35 Hz using an automated DDV-II Autovibron.

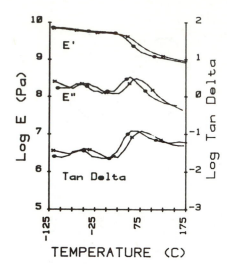

Figure 12. Effect of frequency on dynamic mechanical spectra (14) of Minlon 12T using DDV-III-C Autovibron (before revision). Key: ×, 110 Hz; and ●, 11 Hz.

obtained at 35 Hz. At $T = -10\ °C$, the scatter in E'' and tan δ was caused by lateral vibration of the sample, clamps, and rods. This problem has been seen and discussed by others (9). Results obtained using lower frequencies on the DDV-III have been less acceptable. Differences of up to 50% have been seen in E'' using the revised DDV-III at a test frequency of 35 Hz. The manufacturer is now considering the problem of low-frequency response for the DDV-III.

Studies of the effects of frequency and water content on the dynamic spectra and fatigue of various reinforced polymers (13) are also in progress.

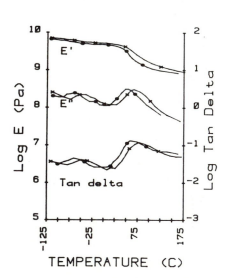

Figure 13. Effect of frequency on dynamic mechanical spectra (14) of Minlon 12T using DDV-III-C Autovibron (before revision). Key: ×, 110 Hz; and ●, 3.5 Hz.

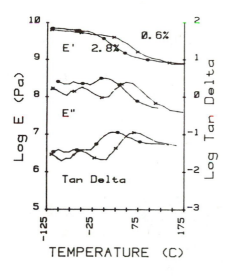

Figure 14. Effect of water on dynamic mechanical spectra (110 Hz) (14) of Minlon 12T using DDV-III-C Autovibron. Key: ×, 0.6% H_2O; and ●, 2.8% H_2O.

Typical best fits for the dynamic spectra for specimens of a mineral-reinforced nylon (Minlon 12T, du Pont) are shown in Figures 12–15. Three frequencies (3.5, 11, and 110 Hz) and five water contents (dry, 0.6, 1.3, 2.8, and 4.8%) were studied; the data show the trends expected with respect to frequency and water content. Apparent activation energies of the principal relaxation processes have been estimated to be: 68 kJ/mol (16 kcal/mol) for the β relaxation, and 160 kJ/mol (39 kcal/mol) for the α transition. The curves of the frequencies of the maxima in E″ are almost coincidental with those presented previously (14) for unmodified nylon 66.

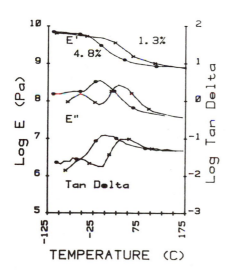

Figure 15. Effect of water on dynamic mechanical spectra (110 Hz) (14) of Minlon 12T using DDV-III-C. Autovibron. Key: ×, 1.3% H_2O; and ●, 4.8% H_2O.

Conclusions and Recommendations

Several conclusions and recommendations are in order and are mentioned here.

1. Although many problems have been encountered, the Autovibron DDV-III-C shows promise for the convenient determination of dynamic spectra of a variety of standard and multiphase polymer systems at 110 Hz.
2. Further work is needed on methods for clamping and alignment, the adjustment of tension at $T > T_g$ and for use at low frequencies.
3. A thorough analysis of errors resulting from instrument compliance, sample yielding and slipping in the clamps, variable sample sizes, and instrument inertia is needed in light of the recent work with the manual Rheovibron (9).
4. Further documentation from the manufacturer is in order.

Acknowledgments

The authors wish to acknowledge partial financial support by the Polymers Program, National Science Foundation, Grant No. DMR77-10063, and by the Office of Naval Research. Suggestions from E. Kozma (Lehigh University) and the cooperation of E. J. Tolle (Imass, Inc.) were also appreciated.

Literature Cited

1. Takayanagi, M. *Proc. Polym. Phys. (Japan)* **1962–1965.**
2. Rheovibron Instruction Manual 17, Toyo Baldwin Co., Ltd., Tokyo, August 1969.
3. Technical literature, Toyo Baldwin Co., Ltd., Tokyo.
4. Technical literature, Imass, Inc., Accord, MA.
5. Yee, A. F.; Takemori, M. T. *J. Appl. Polym. Sci.* **1977,** *21*, 2597.
6. Kenyon, A. S.; Grote, W. A.; Wallace, D. A.; Rayford, McC. *J. Macromol. Sci., Phys.* **1977,** *B13*(4), 553.
7. Ikeda, R. M.; Starkweather, H. W., Jr. *Polym. Eng. Sci.* **1980,** *20*, 321.
8. Wetton, R. E.; Croucher, T. G.; Fursdon, J. W. M. *Org. Coat. Plastics Chem.* **1981,** *44*, 520.
9. Wedgewood, A. R.; Seferis, J. C. *Polymer* **1981,** *22*, 966.
10. Skibo, M. D.; Manson, J. A.; Webler, S. M.; Hertzberg, R. W.; Collins, E. A. In "Durability of Macromolecular Materials," Eby, R. K., Ed.; ACS SYMPOSIUM SERIES No. 95, ACS: Washington, D.C., 1979; pp. 311-29.
11. Sikka, S.; Goldfarb, I. J. *Org. Coat. Plast. Chem.* **1980,** *40*(2), 1.
12. Struik, L. C. In "Physical Aging in Amorphous Polymers and Other Materials"; Elsevier: New York, 1978.
13. Lang, R. W.; Ph.D. Thesis, Lehigh Univ., Bethlehem, PA, 1982.
14. McCrum, N. G.; Read, B. E.; Williams, G. In "Anelastic and Dielectric Effects in Polymeric Solids"; Wiley: New York, 1967.

RECEIVED for review October 14, 1981. ACCEPTED April 5, 1982.

Transient and Dynamic Characterization of Viscoelastic Solids

S. S. STERNSTEIN

Rensselaer Polytechnic Institute, Materials Engineering Department, Troy, NY 12181

This chapter reviews the rudimentary aspects of linear viscoelasticity theory as applied to the transient and dynamic mechanical characterization of solids. Various sources of experimental error are discussed, and a detailed derivation of machine and load cell compliance corrections as applied to dynamic moduli data is given. The Dynastat system and associated computerized data acquisition and processing equipment are described. This instrument provides closed-loop control of either load or displacement covering a frequency range of DC to 200 Hz. A transient stress relaxation curve on a glassy polymer is given and illustrates the ability of the Dynastat to apply a displacement rapidly (15 ms) without overshoot or ringing. Dynamic data on a very stiff carbon–epoxy laminate (stiffness of 1000 N/mm) are presented versus both frequency and temperature. The effects of the matrix glass transition on the storage and loss stiffnesses of the laminate are illustrated.

APPLICATIONS OF LINEAR VISCOELASTIC test methods and data to the study of polymeric solids, melts, and solutions are well documented (1). The three-dimensional theory of viscoelasticity is presented with mathematical rigor by Christensen (2) and some examples of nonlinear viscoelastic behavior and theory are also given elsewhere (3, 4). Some examples of complexities introduced by the simultaneous existence of both volume and shear linear viscoelastic processes in polymeric solids are found in the literature (5, 6). The added complexities associated with three-dimensional effects (volume and shear processes), anisotropy, and nonlinearity will be ignored in this chapter. It

0065–2393/83/0203–0123$07.25/0
© 1983 American Chemical Society

is also assumed in this chapter that the material is subjected to a spatially uniform stress and strain.

In its simplest form, the one-dimensional theory of linear viscoelasticity states that the stress σ at the present time t is given by

$$\sigma(t) = \int_{-\infty}^{t} E_r(t - u)\frac{d\epsilon}{du} du \tag{1}$$

where E_r is the stress relaxation modulus, a function of time, and $\epsilon(u)$ is the strain history over all past time $u \leqslant t$.

An alternative formulation of the theory gives the strain at the present time $\epsilon(t)$ in terms of an arbitrary load history $\sigma(u)$

$$\epsilon(t) = \int_{-\infty}^{t} D_c(t - u)\frac{d\sigma}{du} du \tag{2}$$

where D_c is the creep compliance. Evaluating Equation 1 for a step function in strain occurring at time zero, that is $\epsilon(u) = 0$ for $u < 0$ and $\epsilon(u) = \epsilon_0$ for $u > 0$, one obtains

$$\frac{\sigma(t)}{\epsilon_0} = E_r(t) \tag{3}$$

This experiment defines an ideal stress relaxation test.

Similarly, a step function in stress may be applied, namely, $\sigma(u) = 0$ for $u < 0$ and $\sigma(u) = \sigma_0$ for $u > 0$, and Equation 2 may be integrated to obtain

$$\frac{\epsilon(t)}{\sigma_0} = D_c(t) \tag{4}$$

This experiment defines an ideal creep test. Clearly the histories that define E_r and D_c are different; however, these functions are related by the simultaneous solution of Equations 1 and 2 to obtain

$$\int_{0}^{t} E_r(t - u)\, D_c(u)\, du = t \tag{5}$$

Only for an elastic solid where both E and D are not time dependent does one obtain $ED = 1$.

The constant strain or stress histories used to define the material response functions $E_r(t)$ and $D_c(t)$ are but two of the many possible histories of deformation. Another history commonly used to define material response is a sinusoidal strain (or stress) history. This history is conveniently represented using phasor notation as in AC electrical networks, that is

$$\epsilon(u) = \epsilon_0 \, e^{i\omega u} \tag{6}$$

where ϵ_0 is the strain amplitude, ω is the angular frequency of the sine wave (radians per second), and u is historical time $u \leqslant t$. By using Equation 6 in Equation 1, the steady state ratio of instantaneous stress to instantaneous strain is seen to consist of both an in-phase and an out-of-phase component. Thus, the ratio can be expressed as a complex or dynamic modulus E^*

$$\frac{\sigma(t)}{\epsilon(t)} = E^* = E' + iE'' \tag{7}$$

where E' represents the ratio of in-phase stress to strain and E'' represents the ratio of out-of-phase stress to strain. The out-of-phase stress leads the strain by 90° and is therefore represented in Equation 7 as the imaginary part of E^*.

An alternative procedure for developing Equation 7 is given as follows. Consider that the material may be modeled as a parallel combination of a spring and a dashpot (the so-called Kelvin–Voight body). For the spring the stress developed is proportional to the strain (i.e., Hookean elastic) and can be written as

$$\sigma(\text{elastic}) = k\epsilon \tag{8}$$

where k is a spring constant. The dashpot represents a Newtonian viscous response and gives rise to a stress proportional to strain rate, namely

$$\sigma(\text{viscous}) = \eta\dot{\epsilon} \tag{9}$$

Where $\dot{\epsilon}$ is the strain rate and η is a viscosity constant. If the strain is sinusoidal, then

$$\epsilon = \epsilon_0 \sin \omega t \tag{10a}$$

$$\dot{\epsilon} = \epsilon_0 \, \omega \cos \omega t \tag{10b}$$

and the viscous stress component leads the elastic stress component by 90° as shown in Figure 1. Comparison of Equations 7–10 shows that, in this case, the in-phase modulus is given by $E' = k$ and the out-of-phase modulus by $E'' = \eta\omega$.

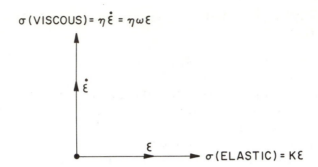

Figure 1. Phasor representation of strain (ϵ) and strain rate ($\dot{\epsilon}$) for a dynamic sinusoidal experiment and the resultant elastic and viscous material stresses.

The sample loss factor

$$\tan \delta = E''/E' = \eta \omega/k \tag{11}$$

is equal to the ratio of energy dissipated (viscous work) to energy storage (elastic work) per quarter cycle of deformation.

An alternative representation of Equation 7 is often used in the study of polymer melts and solutions. Such materials are more fluid than solid and it is convenient to divide the stress by the strain rate and not strain. The complex dynamic viscosity η^* is defined as

$$\frac{\sigma(t)}{\dot{\epsilon}(t)} = \eta^* = \eta' - i\,\eta'' \tag{12}$$

From Equation 6, one obtains $\dot{\epsilon} = i\omega\epsilon$, and comparing Equations 7 and 12, we find that

$$\eta' = E''/\omega \text{ and } \eta'' = E'/\omega \tag{13}$$

Thus, the viscosity η of Equation 9 is actually the in-phase dynamic viscosity η'.

Thus far, two basic test methods have been considered; namely, transient methods as defined by the ideal stress relaxation experiment, Equation 3, and the ideal creep experiment, Equation 4, and the dynamic method as defined by Equations 6–13. These two methods have different limitations and experimental difficulties as discussed later. It is appropriate, however, to consider first the theoretical (ideal) relationship between transient and dynamic data.

Substituting Equation 6 for a dynamic strain history in Equation 1 gives Equation 7, where the steady state E' and E'' are given by

$$E'(\omega) = E_e + \omega \int_0^\infty [E_r(t) - E_e]\sin \omega t \, dt \qquad (14)$$

$$E''(\omega) = \omega \int_0^\infty [E_r(t) - E_e]\cos \omega t \, dt \qquad (15)$$

where $E_e = E_r(\infty)$, the equilibrium value of the stress relaxation modulus. These equations may be inverted to obtain the stress relaxation modulus from either E' or E'' as follows:

$$E_r(t) = E_e + \frac{2}{\pi} \int_0^\infty [E'(\omega) - E_e]\frac{\sin \omega t}{\omega} d\omega \qquad (16)$$

$$E_r(t) = E_e + \frac{2}{\pi} \int_0^\infty E''(\omega)\frac{\cos \omega t}{\omega} d\omega \qquad (17)$$

Given the stress relaxation modulus $E_r(t)$, Equations 14 and 15 may be used to compute the components of the complex modulus at any desired frequency ω. Clearly, relaxation data at all times t from zero to infinity are required to evaluate these integrals. Conversely, it is clear from Equations 16 and 17 that dynamic data at all frequencies from zero to infinity are required to predict the stress relaxation modulus at any time t.

In theory, any one of the functions $[E_r(t), D_c(t), E'(\omega), E''(\omega)]$ contains all of the necessary information required to compute the other functions. The difficulty in implementing this statement rests with the experimental impossibility of obtaining meaningful stress relaxation data or creep data over all times, or dynamic data over all frequencies. The sources of error associated with each mode of testing are distinct and require different considerations. Before examining some of these however, it is appropriate to consider a specific example of Equations 3, 14, and 15 and introduce the concept of a material relaxation time that is quite useful in the experimental context but unneccessary in the theoretical treatment.

Let us suppose that a stress relaxation experiment is performed and that the observed response is that of a simple exponential decay of stress, namely

$$E_r(t) = A \, e^{-t/\tau} + E_e \qquad (18)$$

The constant τ has dimensions of time and is referred to as a relaxation time. Note that $E_r(0) = A + E_e$ and $E_r(\infty) = E_e$. Thus, A is the strength of the relaxation and τ is its time constant. Materials generally require a distribution of relaxation times to describe their linear viscoelastic response. Nevertheless, it is useful to consider an idealized situation of a single relaxation time.

Combining Equations 14, 15, and 18 and performing the required integrations give for the components of the complex modulus

$$\frac{E'(\omega) - E_e}{A} = \frac{\omega^2\tau^2}{1 + \omega^2\tau^2} \tag{19a}$$

$$\frac{E''(\omega)}{A} = \frac{\omega\tau}{1 + \omega^2\tau^2} \tag{19b}$$

These equations are plotted vs. the dimensionless variable $\omega\tau$ in Figure 2. Note that the real (or in-phase) component is sigmoidally shaped and that the frequency at which the inflection point occurs is given by $\omega\tau = 1$ or $\omega = 1/\tau$. At this same frequency, the imaginary (or out-of-phase) modulus goes through a maximum. Both E' and E'' are relatively flat with frequency when $\omega\tau < 0.1$ and $\omega\tau > 10$. It follows that a

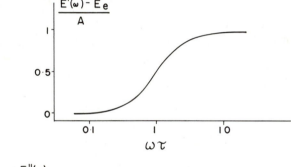

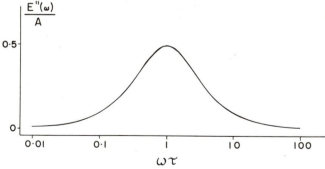

Figure 2. The storage modulus (E') component of dynamic modulus for a single relaxation time process as a function of reduced frequency $\omega\tau$, where ω is frequency (radians per second) and τ is the relaxation time (top); and the loss modulus (E") component of dynamic modulus for a single relaxation time process (bottom).

material having a significant stress relaxation component occurring in a timescale τ_1 (seconds) will have a corresponding dynamic activity at a frequency of ω_1 (radians per second) or $\omega/2\pi$ Hz, where $\omega_1 = 1/\tau_1$. For a distribution of relaxation times Equations 19a and 19b are readily generalized (*see* Reference 1). In theory, then, the same information is contained in transient and dynamic data. The necessity of doing both transient and dynamic experiments is due to experimental limitations and sources of error.

Machine Compliance

An example of experimental error that affects both transient and dynamic data is that of machine compliance. Often it is necessary to measure the displacement of one end of a sample relative to a supposedly fixed other end of a sample to determine the sample strain. Many sample geometries and materials do not lend themselves to true strain measurements obtained directly on the specimen, e.g., by strain gauges. In fact, specimen mounted transducers may give rise to as many problems as they solve under certain circumstances.

In this section, the necessary mathematics are developed for the correction of dynamic data due to finite machine compliance. Consider a specimen that is directly in series with an elastic spring as shown in Figure 3. The spring compliance (displacement/load) S is intended to represent the combined compliance of the load cell, machine frame, jaws, and extension rods. In a perfect system S would be zero, and a measurement of displacement (X in Figure 3) would be the true displacement of the sample. Unfortunately, S is never zero; at the very least the load cell always exhibits some displacement. To what extent, then, does a nonzero S affect dynamic measurements of E^*, especially for stiff solids?

A reasonable assumption for the apparatus described in this chapter is that the machine frame and load cell are purely elastic. It follows that S is a real number in the following derivation. Sample inertia is not considered either, which is tantamount to assuming that all positions in the sample are subjected to the same stress at the same time. At sufficiently high frequencies, depending on the sample dimensions, density, and modulus, wave propagation effects will dominate and the sample will not be uniformly stressed at each instant of time. Because the apparatus described here is designed for uniform stress measurements, wave propagation effects will not be considered.

With reference to Figure 3, the total (measured) displacement is X, the machine displacement due to the nonzero compliance S is Z, and the true sample displacement is given by Y, where

$$Y = X - Z = X - SF \tag{20}$$

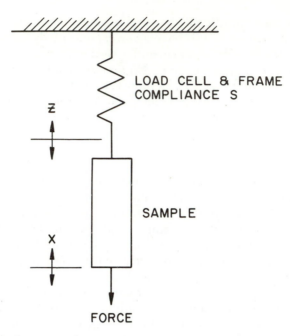

Figure 3. Schematic of a test frame showing a finite machine compliance S. Key: **X**, overall displacement; and **Z**, machine and load cell displacement only.

and **X**, **Y**, and **Z** are complex numbers. Note that **Y** and **X** are generally not in phase because the machine displacement **Z** is in phase with the load **F** and not the total displacement **X**. The apparent dynamic stiffness R^* is defined by the relation

$$\mathbf{F} = R^* \mathbf{X} = (R' + iR'') \mathbf{X} \tag{21}$$

and represents the amplitude and phase relationship of **F** to **X**.

The true (or desired) dynamic stiffness of the sample is denoted by M^* and defined by the amplitude and phase relationship between **F** and **Y**, namely

$$\mathbf{F} = M^* \mathbf{Y} = (M' + iM'') \mathbf{Y} \tag{22}$$

The sample dynamic modulus E^* (*see* Equation 7) is related to the dynamic stiffness M^* by a form factor B that involves only the sample geometry factors and test mode (e.g., shear or tension), that is $E^* = BM^*$. For the present purpose the form factor simply represents a numerical scaling factor. Thus, it is necessary to obtain M^* in terms of the measured R^*.

Combining Equations 20–22 gives

$$\mathbf{F} = M^* \, \mathbf{Y} = M^* \, (\mathbf{X} - S\mathbf{F}) = R^* \, \mathbf{X} \tag{23}$$

which rearranges to give

$$(M^* - R^*) \, \mathbf{X} = M^* \, S\mathbf{F} = M^* S R^* \mathbf{X}$$

Solving for M^* gives

$$M^* = \frac{R^*}{1 - SR^*} \tag{24}$$

The components of M^* are given by

$$M' = \frac{R' \, (1 - SR') - S(R'')^2}{D} \tag{25a}$$

$$M'' = \frac{R''}{D} \tag{25b}$$

where

$$D = (1 - SR')^2 + (SR'')^2 \tag{25c}$$

After some algebra, it can be shown that the magnitude of M^* is given by

$$|M^*| = \frac{|R^*|}{\sqrt{D}} \tag{26}$$

The loss factor of the sample is given by $M''/M' = \tan \delta$ and is obtained directly from Equations 25a and 25b. After some algebraic manipulation, one obtains

$$\tan \delta = \frac{M''}{M'} = \frac{\tan \beta}{1 - SR' \, (1 + \tan^2 \beta)} \tag{27}$$

where $\tan \beta = R''/R'$, the apparent (measured) value of loss factor without compliance correction. The correction is nonlinear with respect to $\tan \beta$.

For samples that are highly elastic (low loss), i.e., with $\tan \beta \ll 1$, Equation 27 can be approximated by

$$\tan \delta \doteq \frac{\tan \beta}{1 - SR'} \tag{28}$$

Clearly, even for this case the compliance correction can be large if the sample stiffness R' times the machine compliance S is significantly large.

Finally, the true sample displacement amplitude relative to the overall displacement amplitude is given by

$$\frac{|\mathbf{Y}|}{|\mathbf{X}|} = \sqrt{D} \tag{29}$$

The vector relationships among $\mathbf{X}$, $\mathbf{Y}$, $\mathbf{Z}$, $\tan \beta$, and $\tan \delta$ are shown in Figure 4.

The machine compliance S may be obtained either by replacing the sample with an exceedingly stiff (rigid) steel specimen and measuring displacement vs. force in a static test or by performing a

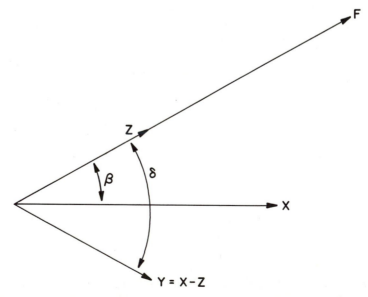

Figure 4. Phasor representation of the force (**F**), machine displacement (**Z**), overall displacement (**X**), corrected sample displacement (**Y**), measured viscoelastic loss factor (tan β), and true sample loss factor (tan δ).

dynamic test on the rigid specimen. To the degree of approximation that the sample is rigid, then $Y = 0$ and from Equations 29 and 25c, one obtains $R'' = 0$ and $S = 1/R'$.

Compliance tests on the Dynastat system described under "Instrumentation" give the value of S as typically 0.002 mm/kg. Corrections of less than 5% (Equation 29) are typical for rigid glassy polymers such as polymethyl methacrylate (PMMA) having a diameter of 5 mm and a gauge length of 25–50 mm. Corrections of less than 1% are typical for rubbery samples.

The Dynastat system automatically performs the corrections given in Equations 24–29 using digital computations supplied as part of the software associated with the Dynalyzer and scaling unit. Thus, meaningful dynamic data can be obtained on very rigid specimens such as carbon–epoxy laminates (discussed later) even though the compliance correction is as large as 40%. Typically, dynamic moduli on such stiff specimens are obtained at displacement levels as low as 10–50 μm.

In addition to the machine compliance factor, the effect of load cell assembly resonance must be considered as it relates to the measured vs. actual sample force amplitudes. This effect has been discussed in detail elsewhere (7). Suffice it to say that this inertial correction is performed automatically on dynamic data processed by the Datalyzer.

Instrumentation

The transient and dynamic testing apparatus described here was developed in this laboratory initially for the viscoelastic characterization of glassy polymers. It is now available commercially as the Dynastat. A complete system consisting of the Dynastat, Dynalyzer, scaling unit, and Dynatherm is shown in Figure 5. Additional instrumentation not shown is also described.

The test frame of the Dynastat contains the load cell with adjustable sample gauge length provisions, a linear motor consisting of a permanent magnet and spider suspended coil wound on a lightweight coil form, a lower push–pull rod rigidly coupled to the motor coil and mounted in an air bearing for transverse stiffness, two displacement transducers connected to the lower push rod, a temperature chamber, and associated control features.

The load cell is a dual action (tension–compression) beam type having very low compliance and a range of ±50 kg. The load range of the Dynastat is ±10 kg full scale with the associated electronic gain set up to give a full scale output of ±10 V. This configuration effectively decreases the load cell compliance by a factor of 5 relative to the use of a 10-kg load cell.

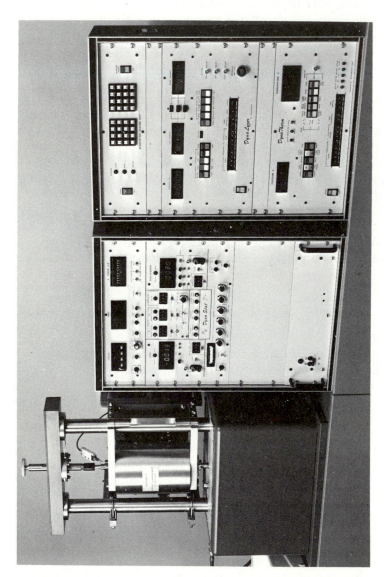

Figure 5. The Dynastat system consisting of the Dynastat, Dynalyzer, Dynatherm, scaling unit, and environmental chamber.

One of the two displacement transducers has a range of ±0.5 mm for a full scale output of ±10 V and is mounted in a precision zeroing device that may be set at any position in the stroke range of the instrument. The second displacement transducer has a range of ±5 mm full scale and is fixed in position. This transducer monitors the full scale displacement capabilities of the Dynastat. Dual electronics are provided which enable either displacement transducer to be used for closed-loop operation, but allow both transducers to be used for output of data. This arrangement is particularly useful when performing dynamic tests on compliant materials such as rubber where the 5-mm transducer is used to control the static and dynamic displacement but the 0.5-mm transducer can be positioned to account for the static displacement while it is used to monitor the dynamic signal. This provides for a 10:1 improvement in the dynamic signal-to-noise ratio. Similarly, in closed-loop load control, the ability to position the 0.5-mm transducer allows one to eliminate the offset displacement due to a static preload.

The main control unit of the Dynastat contains the transducer conditioning circuits, program and servo controls, logic circuits, and calibration and protection circuits. The load circuit and two displacement transducer circuits contain dynamic filters that are phase matched to within 1% at 2000 Hz, thereby ensuring negligible phase shift at or below 200 Hz. The program control allows for two static commands to be preset in level and sign (tension or compression) and for control of a dynamic (or external) program signal. All program switching is logic controlled, and switch closures are debounced. This is especially important in transient tests at very short times. External control of the program functions is provided for automated transient and dynamic tests (discussed later).

The servo system provides for closed-loop control using load or either displacement transducer for feedback. All feedback circuits including the power amplifier are DC coupled to provide for long-term static control (e.g., in creep or stress relaxation experiments). The servo compensation circuits are operator adjustable from the front panel and can be optimized for any sample size or geometry, and the type of test (transient or dynamic). A long-term control stability of 1 part in 10,000 full scale in a transient test is commonly achieved. Dynamic tests can be performed at frequencies to 200 Hz.

Several features of the Dynastat are patented (8) and among these is a risetime control circuit that allows step functions in load or displacement to be imposed on the sample without overshoot but with the risetime of an underdamped servo. Figure 6 gives an example that shows a stress relaxation experiment on a glassy PMMA sample. The strain transient is 90% complete in about 15 ms, but there is no overshoot. A conventional servo compensation scheme would give overshoot if the same risetime was used or would have a much slower risetime if no overshoot was called for. Interestingly, about 17% stress relaxation occurs in the glassy PMMA within 1 s of straining. Dynamic data on PMMA correspondingly show a peak in E'' at 1−10 Hz, and this is referred to as the β-relaxation.

Temperature programming and control is accomplished using a Dynatherm that employs a microprocessor-based digital servo system that can control one to five heating and cooling relays as well as a low voltage servo follower for solid-state power amplifiers. Temperature programming is flexible and can be set between −200 and 750 °C using from 1 to 99 temperature increments. Increment steps can be controlled by time alone or by a temperature−time criterion. The latter case is most useful when rheological data on isothermal samples is desired. A soak time of any value between 0.1 and 99.9 min can be set to commence when the temperature chamber is within a temperature window set by the operator. At the completion of each soak cycle, the Dynatherm issues a start signal for the automatic transient or dynamic test to begin and will not change temperature until the test is complete.

The servo settings on the Dynatherm are adjustable over a wide range. For example, the thermal time constant of the chamber to be controlled can be set between 1 min and 88 h using logarithmically spaced values. In practice, temperature control to ±0.1 °C can be achieved without offset error because the Dynatherm can be used as a totally integral controller (no proportional control). Conventional proportional control with integral reset is also provided. The damping factor for the system may be set to provide no overshoot after program temperature changes. This feature is useful when temperature overshoot may result in an unwanted sample reaction, e.g., during thermoset cure studies.

When used with the Dynastat temperature chamber, the operating range is −150−225 °C. Below ambient temperature, a liquid nitrogen supply is used to provide cooling. Continuous and intermittent supply solenoids and a heater are controlled by the Dynatherm. The servo system controls the duty cycles of the heater and intermittent flow solenoid. Above ambient temperature the system automatically switches to a heated air system with servo-controlled heat. This system provides programmed cooling at elevated temperatures at nearly the same rate as heating. Temperature sensing is by a platinum resistance thermometer adjacent to the sample. The temperature chamber contains a unique flow distribution baffle that effectively eliminates gradients. Temperature control is typically ±0.1 °C for indefinite time.

Dynamic modulus measurements are made with the aid of the Dynalyzer, also a microprocessor-based instrument. This unit generates the reference sine wave that is used as the external program input to the Dynastat. The sine wave is generated digitally using a 1024-word, 12-bit digital-to-analog conversion. The load and displacement output signals from the Dynastat are used as inputs to the Dynalyzer and are sampled using a 14-bit analog-to-digital (A/D) converter. Prior to the A/D conversion, a digital feedback scheme is used to remove the DC component of the input signal and autorange the dynamic signal for each input. Gains of $\times 1$ (± 10 V), $\times 10$ (± 1 V) and $\times 100$ (± 100 mV) are provided. This scheme allows a 50-mV dynamic signal riding on a 9.9-V static signal to be processed using the $\times 100$ range for optimum resolution.

The frequency range of the Dynalyzer is $0.1-199.9$ Hz, or $0.01-99.99$ Hz optionally. Frequency sweeps may be performed linearly using up to 99 frequency values between the start and stop values, or logarithmically using either 5 or 10 points per decade. Serial ASCII coded data outputs (RS-232) of frequency, temperature, load amplitude, displacement amplitude, and complex modulus (E^*, tan δ, E', and E'') are provided. The Dynalyzer handshakes with the Dynatherm for providing automatic temperature and frequency sweeps.

The complex modulus values are obtained using a Fourier-type correlation between the reference signal and each of the inputs (load and displacement). The fundamentals of each signal that are in phase and 90° out of phase with the reference are obtained. Complex algebra is then used to eliminate the phase shift between the reference signal and machine response, leaving only the relative in-phase and out-of-phase components of load with respect to displacement. The scheme provides a high degree of noise and harmonic rejection. The number of cycles sampled can be set between 1 and 15, and a delay feature is provided to eliminate transient (unsteady state) effects that the testing machine may exhibit following each frequency jump. This delay may be set at $1-9$ cycles for frequencies below 1 Hz or $1-9$ s above 1 Hz.

The Dynalyzer is available with a transient option which enables it to be used as a two-channel transient recorder having a logarithmic time sampling rate covering up to six decades and beginning with 10 ms. This feature is particularly useful with the Dynastat when creep and stress relaxation tests are to be performed. An auto-timer sequencer interfaces with the transient Dynalyzer and Dynatherm to provide automatic temperature sweeps and programmed multiple cycle creep and recovery, two-level creep, stress relaxation and recovery, and two-level stress relaxation tests.

A microprocessor-based scaling unit can manipulate the digital data from the Dynalyzer prior to printing. Sample geometry factors, load and displacement scales (e.g., 1 kg/V), and several correction factors such as machine compliance (*see* Equations 24–29) are entered on a keyboard. Nine algorithms for sample test mode are provided, including tension and compression of a rectangular or circular bar, double-shear specimen, three-point flexure, torsion of a cylinder, and cone and plate. (The Dynastat operates only in a linear motion configuration and does not perform rotary tests.) The results are printed out in terms of stress, strain, and moduli in either English or SI units.

A new accessory, the Datalyzer, is not shown in Figure 5. This subsystem provides all the features of the scaling unit but with several useful additions. Data entry is by means of a CRT keyboard, and a floppy disk and eight-color digital plotter are provided. The Datalyzer provides all necessary software to store and manipulate data and to plot the results of both dynamic and transient tests. Fifteen different ordinates may be plotted vs. frequency, time, or temperature, and full control of axes scaling is provided (both numerically and log or linear). In addition, the Datalyzer can compute and plot time–temperature or frequency–temperature superposition graphs.

Results

An example of a transient experiment, namely stress relaxation for a glassy PMMA sample, is given in Figure 6. The discussion following Equations 18 and 19 shows that relaxation processes that are characterized by relaxation times longer than about 50 ms may be studied using a transient test method such as indicated by Figure 6. However, relaxation times shorter than 50 ms pose a problem as far as the transient mode is concerned. Referring again to Figure 6, below about 20 ms, the strain (displacement) history is not at all constant. Thus, even though the risetime capability, without overshoot, of the Dynastat is quite small, it is not small enough to approximate a step function insofar as a 20-ms relaxation time is concerned. Faster risetimes are possible but result in added complications. For example, overshoot is almost certain to occur due to inertial and time lag effects on the servo system.

Another perhaps more subtle effect is that a very rapid risetime results in a material response function that is closer to adiabatic than to isothermal. Thus, a dynamic measurement is more appropriate to the study of short relaxation time phenomena, namely, shorter than about 50 ms, which corresponds to a frequency of 20 rad/s or 3.2 Hz. Dynamic results taken at a frequency of about 1 Hz are well within the transient capability of the Dynastat.

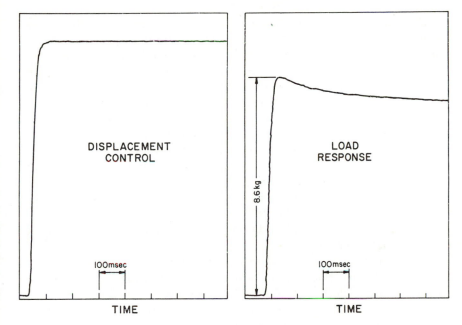

Figure 6. A stress relaxation experiment on glassy PMMA at 30 °C. The servo-controlled step function in displacement having a risetime of 15 ms without overshoot, and the resultant stress relaxation are shown. The data were obtained digitally with a high speed transient recorder and then replayed from memory using a strip-chart recorder in slow time.

Dynamic experiments pose a different set of problems, many of which are readily handled by the Dynastat. The dynamic correction for machine stiffness described in Equations 20–29 can be quite significant for stiff samples that exhibit a low loss modulus (E''). Fortunately, this correction may be performed automatically in the Dynastat system along with the inertial correction for load cell assembly resonance.

A not so easily corrected and potentially large error in dynamic measurements especially for high loss modulus materials is due to mechanical energy dissipation and the subsequent internal heating of the sample. In this regard, a transient test may prove to be more appropriate. For example, either a relaxation experiment or a dynamic experiment may be used to investigate a relaxation process of 1-s time constant. However, the dynamic test continually produces sample heat generation, whereas a transient test produces minimal heating. It is always prudent to test for sample self heating in a dynamic test by looking for time drift of the dynamic moduli.

Dynamic experiments may be performed in the Dynastat using either load or displacement control. Load control is particularly useful when the sample geometry requires that a fixed value of static to dynamic ratio be used, as when testing a thin film that buckles when the load goes compressive. In this case the static (tensile) load must be at least equal to the dynamic load amplitude and maintained at all temperatures. Displacement control is useful when testing elastomers because often the dynamic moduli are a direct function of the static strain level about which the dynamic test is performed. This is a result of certain nonlinear terms that are generated in the strain tensor at finite deformation. Consequently, the static strain, and not the static stress, is the source of nonlinearity. Both load- or displacement-controlled dynamic tests should be conducted at a fixed amplitude level and with precisely defined test frequency at each test point as in the case of the Dynastat.

A dynamic test conducted in load control is shown in Figure 7. The sample is a carbon–epoxy laminate and is being tested using a centro symmetric deformation (CSD) test described by Sternstein and Yang (9). Basically, the CSD geometry is a circular symmetry bending test that is particularly useful for stiff samples. As Figure 7 shows, the force amplitude is constant as the temperature of the sample is swept through the glass transition of the matrix. For three fixed frequencies, 0.1, 1, and 10 Hz, the magnitude of the dynamic modulus shows a pronounced reduction in value as the matrix softens. The corresponding displacement amplitude is also shown and is clearly the mirror image of the modulus curves because a constant force amplitude is being maintained.

The storage (M') and loss (M'') stiffness components of the dynamic stiffness are shown in Figure 8 for the same sample as Figure 7, but at a single frequency of 6.5 Hz. The glass transition region of the matrix is seen readily. The sample stiffness prior to the glass transition is on the order of 1000 N/mm and the dynamic displacements at which these data were taken are of the order of 20–50 μm. The results in Figures 7 and 8 were plotted automatically using the Datalyzer and represent only two of many possible graphs that can be plotted. These data are compliance corrected.

The storage component of stiffness (M') is shown vs. frequency at a number of temperatures in Figure 9 and the corresponding loss component (M'') in Figure 10. These data may be superposed using time–temperature equivalence to obtain master curves as shown in Figures 11 and 12. The details of the dynamic study of carbon–epoxy laminates using the Dynastat system and the CSD test method are given elsewhere (9).

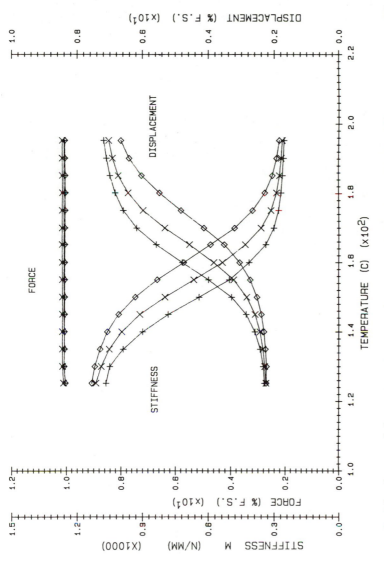

Figure 7. A load-control dynamic test on a carbon–epoxy laminate using the CSD test geometry. The constant force amplitude gives rise to a displacement amplitude that is the mirror image of the dynamic stiffness |M*|. The large change in dynamic stiffness is due to the glass transition of the matrix. Key to frequency: +, 0.1 Hz; ×, 1.0 Hz; and ◇, 10 Hz.

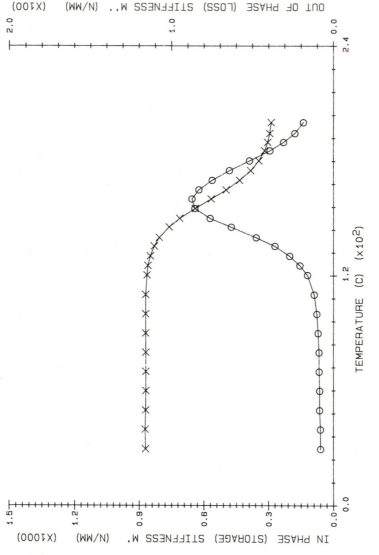

Figure 8. The in-phase (storage) (M') and out-of-phase (loss) (M") components of dynamic stiffness at 6.5 Hz vs. temperature for the sample used in Figure 7 are shown. Dynamic displacements are from ~10 to 50 μm. Key: ×, in phase; and ⊙, out of phase.

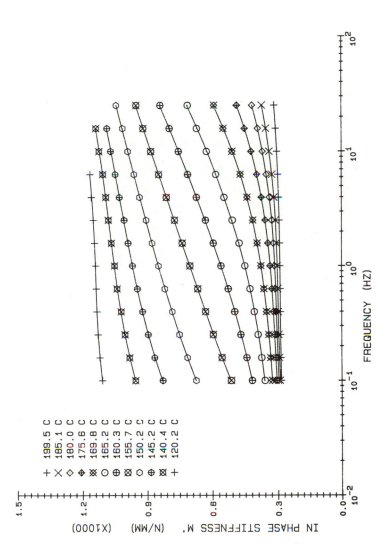

Figure 9. In-phase stiffness vs. frequency at various temperatures for the sample used in Figure 7.

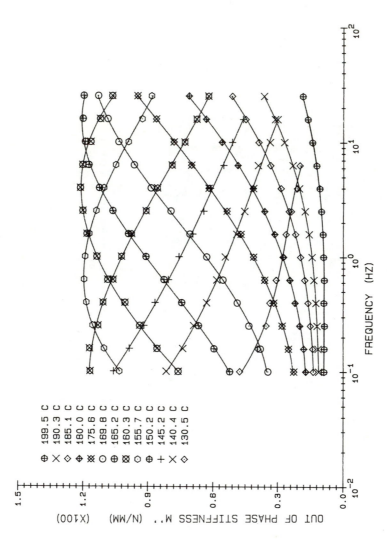

Figure 10. Out-of-phase stiffness vs. frequency at various temperatures for the sample used in Figure 7.

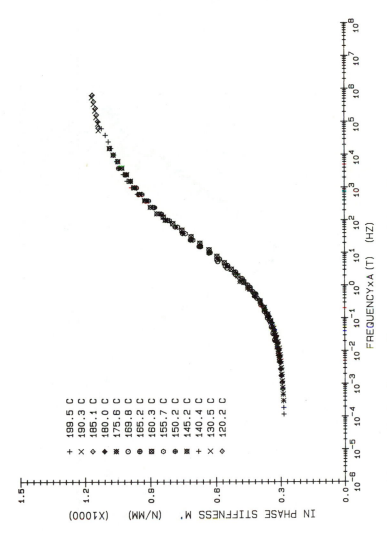

Figure 11. In-phase stiffness master curve for the data of Figure 9.

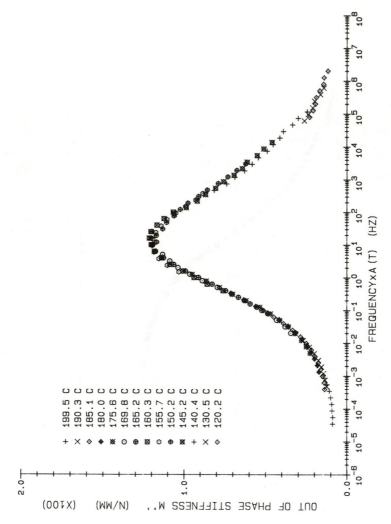

Figure 12. Out-of-phase stiffness master curve for the data of Figure 10.

Literature Cited

1. Ferry, J. D. "Viscoelastic Properties of Polymers"; Wiley: New York, 1970.
2. Christensen, R. M. "Theory of Viscoelasticity—An Introduction"; Academic: New York, 1971.
3. Ward, J. M. "Mechanical Properties of Solid Polymers"; Wiley: New York, 1971.
4. Kinder, D. F.; Sternstein, S. S. *Trans. Soc. Rheol.* **1976**, *20* (*1*), 119–40.
5. Sternstein, S. S.; Ho, T. C. *J. Appl. Phys.* **1972**, *43* (*11*), 4370–83.
6. Sternstein, S. S. In "Properties of Solid Polymeric Materials"; Schultz, J., Ed.; Academic: New York, 1977; pp. 541–98.
7. Smith, T.; Oppermann, W. "Dynastat Viscoelastic Analyzer for Studies of Polymer Properties", reprint of a paper presented at the ANTEC, May 1981, Report No. RJ3053 (37948) 2/10/81, IBM Research Lab, San Jose, CA.
8. Sternstein, S. S., U.S. Patent 4 096 741, 1978.
9. Sternstein, S. S.; Yang, P. "Characterization of the Matrix Glass Transition in Carbon-Epoxy Laminates Using the CSD Test Geometry", paper presented at USA–Italy Joint Symposium on Composite Materials, Capri, Italy, June 15–19, 1981, in press.

RECEIVED for review January 4, 1982. ACCEPTED August 13, 1982.

The Dynamic Response of Polymer Melts Subjected to Shear from the Quiescent State

BRYCE MAXWELL

Princeton University, Department of Chemical Engineering, Polymer Materials Program, Princeton, NJ 08544

Polymer melts were subjected to an applied shear starting from the quiescent state. The properties measured include the yield stress, the steady state stress, and the time-dependent recoverable strain. These properties indicate a transition, T_{ll}, in polystyrene and polymethyl methacrylate that increases with increasing molecular weight. The yield stress and recoverable strain in low density polyethylene was found to be strongly dependent on the high molecular weight end of the molecular weight distribution. Blends of polystyrene and polymethyl methacrylate exhibit double yield stresses and much greater recoverable strains than either of the components. This method of studying the behavior of polymer melts provides information that may be of value in practical processing situations.

Viscoelastic properties of polymer melts can be studied by the popular method of subjecting the specimen to an oscillatory sinusoidal stress or strain of small magnitude and known frequency and then measuring the response of the polymer in terms of the components of the complex modulus or the complex viscosity. Such tests are useful and informative, particularly in the area of structure–property relationships. But other dynamic methods and properties exist that can shed additional knowledge of the response of macromolecules to stress and strain. The purpose of this chapter is to present one such method and to present some results therefrom.

The dictionary defines dynamic as "dealing with forces and their relationship to motion." Therefore, the word is much broader than just oscillatory behavior. In an oscillatory experiment, the internal structural response mechanisms of the material equilibrate with the imposed oscillatory stress or strain and the response measured is that of the structure that has resulted from the applied forcing function.

Experimental

The properties of melts to be measured include the following: the shear modulus of elasticity, the yield stress or stress overshoot, the steady state shear stress or viscosity, and the time-dependent strain recovery. These properties are all a function of temperature, applied shear rate, and applied shear magnitude.

Figure 1 is a schematic illustration of the apparatus that was described in detail previously (1). The shear field is essentially that of a Couette viscometer consisting of a cup, a central cylindrical member coaxially aligned with the cup and the melt specimen in the annular space between the central cylindrical member, and the inside of the cup.

The experimental procedure consists of shearing the specimen at the desired rate, $\dot{\gamma}$, to the desired shear magnitude, γ_a, by rotating the cup while the central cylindrical member is prevented from rotating by a restraining arm acting against a force transducer. The output of the transducer is fed to a strip-chart recorder that plots the stress vs. strain curve. The initial slope gives the shear modulus of elasticity. As the test proceeds, the melt exhibits a yielding or stress overshoot, and then the stress decreases to the steady state value associated with the melt's viscosity.

When the desired magnitude of strain has been applied, the rotation of the cup is stopped and the central cylindrical member is released by removing the force transducer. The rotation of the central member as it turns in accordance with the elastic recoil of the melt is recorded photographically to give the recoverable strain, γ_r, as a function of time. The resulting data were reproducible within ±3%.

Results

Polystyrene.   Figure 2 shows the stress vs. strain characteristics of a series of polystyrenes at 170 °C. Weight average molecular weights ($\bar{M}_w$) ranged from 75,000 to 400,000 and the weight average to number average was approximately 1.7.

As the quiescent melt is sheared, the stress rises until it reaches a value high enough to overcome the interchain interactions [such as entanglements (2)], then the melt yields and the stress decreases to the stress associated with steady state flow. The higher the molecular weight, the higher the yield stress and, in general, the higher the steady state stress. These results are to be expected if one considers the quiescent melt to have a structure of interchain interactions that must be overcome by the applied shear before a steady flow consisting of an equilibrium of the formation of interchain interactions and the breaking of these interactions by the imposed shear can be achieved. As the molecular weight increases, more stress is required to disassociate the interactions that existed in the quiescent state. The magnitude of the yield stress has technological implications in such fab-

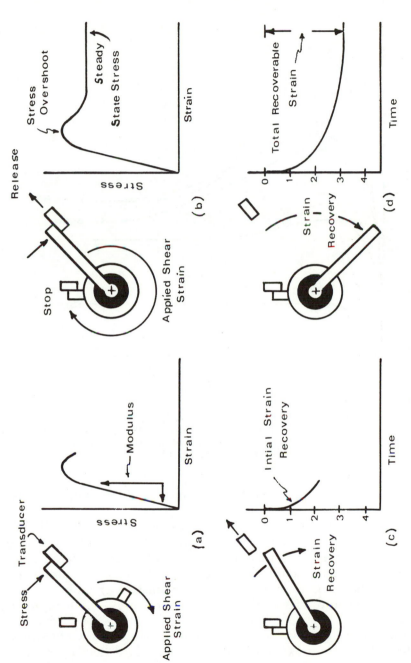

Figure 1. Schematic drawing of test apparatus.

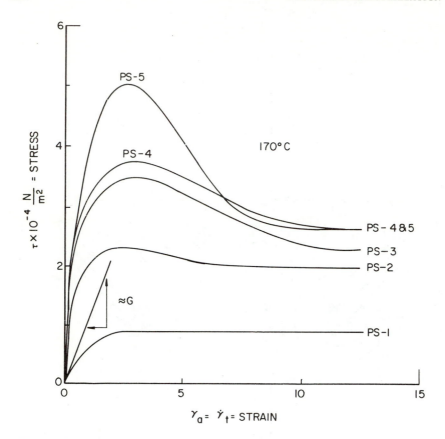

Figure 2. Melt stress vs. strain curves for polystyrene (PS) of various molecular weights: PS-1 = 75,000, PS-2 = 140,000, PS-3 = 310,000, PS-4 = 380,000, and PS-5 = 400,000 M_w. (Reproduced with permission from Ref. 1. Copyright 1979.)

rication operations as vacuum forming, blow molding, and velocity profiles in extrusion.

Figure 3 shows the stress vs. strain characteristics of the 310,000 $\bar{M}_w$ polystyrene at various melt temperatures. The higher the temperature, the lower the yield stress and the lower the steady state stress. This behavior again would be expected on the basis of interchain interactions being disassociated by the combination of applied shear and thermal energy. The greater the thermal energy, the less stress required.

From Figure 3, the temperature dependence of the yield stress and the steady state stress may be determined. These data are plotted in Figure 4. Apparently, an abrupt change in the temperature dependence of both these stresses occurs at 190 °C. This temperature is 90

°C above the glass transition and may be associated with the liquid–liquid transition, T_{ll}, proposed by Boyer (3). This point will be
discussed more extensively later in this chapter after further data have
been presented.

The results of recoverable strain experiments are shown in Figure
5 for polystyrenes of various molecular weights at a melt temperature
of 170 °C. The recoverable strain is plotted downward, indicating
recovery from a previously applied strain vs. linear time in seconds.
The first observation is that the amount of recoverable strain is large,
and in some cases the strain recovery continues for a long period of
time, greater than 15 min. The second observation is that a distinct
trend is not found with molecular weight.

To explore this latter point further, the recoverable strain for the
310,000 $\bar{M}_w$ polystyrene as a function of temperature is shown in Fig-

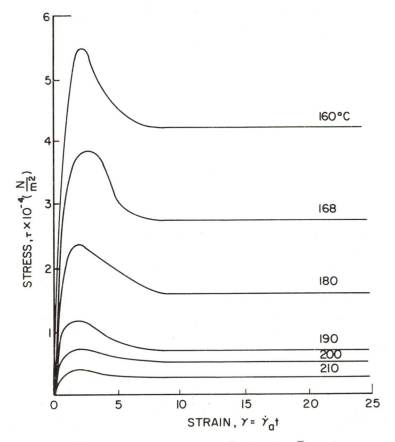

Figure 3. *Melt stress vs. strain curves for 310,000 $\bar{M}_w$ polystyrene at various T. (Reproduced with permission from Ref. 1. Copyright 1979.)*

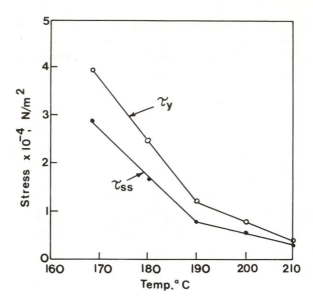

Figure 4. Melt yield stress and steady state stress as a function of T for 310,000 $\overline{M}_w$ polystyrene.

ure 6. Again, no distinct trend with temperature is observed. The temperature-dependent recoverable strain curves cross each other several times, but careful examination indicates that the recoverable strain curves fall distinctly into two categories. The 170 °C curve shows time-dependent recoverable strain that goes on slowly for a long period of time; this will be called Type I recovery. The 220 °C curve shows a very rapid strain recovery that is finished in a rather short time; this will be called Type II recovery.

The factor that most clearly differentiates Type I recovery from Type II recovery is the length of time required for the recovery to be finished. The time for the strain recovery to be finished is plotted in Figure 7 vs. temperature for three different molecular weight polystyrenes. Apparently, a rather sharp transition temperature is evident at which the response changes from Type I to Type II, and this transition temperature is higher for higher molecular weights. This result is consistent with Boyer's T_{ll}. At temperatures below the transition, the melt is capable of storing a large amount of elastic strain energy, but the recovery of this stored elastic strain energy is retarded causing the recoverable strain to require long times to finish the recoiling process. At temperatures above the transition, the recoiling process is not retarded severely and the recovery process is finished quickly.

Recognizing that a transition in the response of melts above the glass transition temperature exists is of assistance in understanding

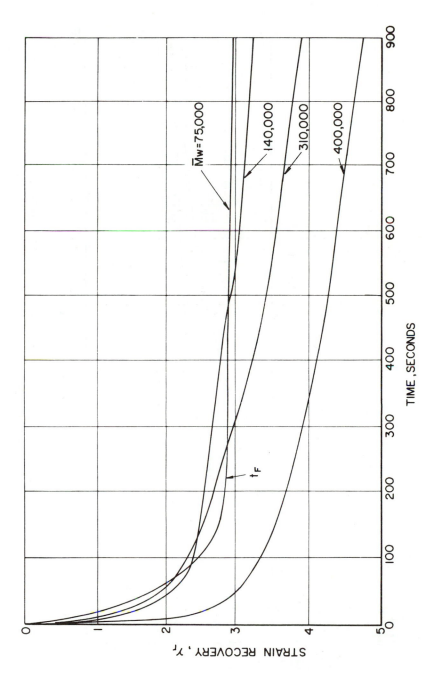

Figure 5. Recoverable strain as a function of time for various molecular weight polystyrenes at 170 °C.

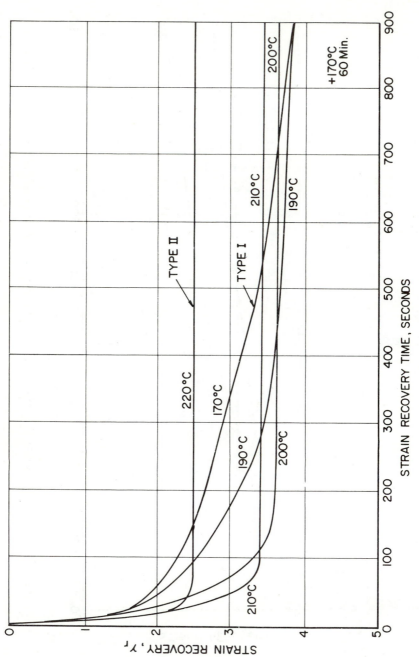

Figure 6. Recoverable strain as a function of T for 310,000 $\bar{M}_w$ polystyrene.

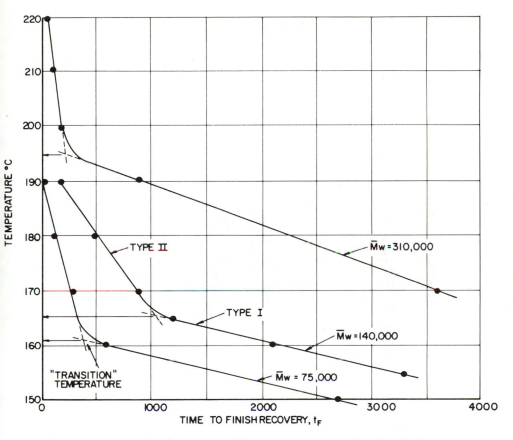

Figure 7. Time for the recoverable strain process to be finished as a function of T for polystyrene.

the recoverable strain as a function of molecular weight shown in Figure 5. The 310,000 and the 400,000 $\bar{M}_w$ materials exhibit Type I recovery curves because at the test temperature, 170 °C, they are below their respective transition temperatures. The 140,000 $\bar{M}_w$ material is just about at its transition at the test temperature, and the 75,000 $\bar{M}_w$ material exhibits a Type II response because at 170 °C it is well above its transition temperature.

Polyethylene. Four samples of long chain branched polyethylenes of various molecular weight distributions also were investigated using these techniques (4).

Figure 8 shows the melt stress vs. strain curves at 120 °C with the relative gel permeation chromatography (GPC) curves superimposed. Two materials, PE-1 and -6 (where PE represents polyethylene), are of narrow molecular weight distribution. The other two, PE-3 and -4,

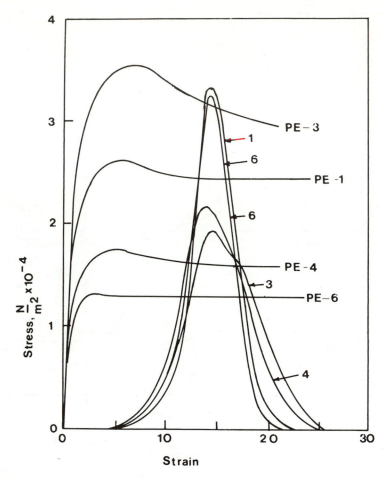

Figure 8. Melt stress vs. strain curves for low density polyethylenes of various molecular weight distributions.

have broad molecular weight distributions, and sample PE-3 has a distinct shoulder on the high molecular weight end of the distribution. Of the two narrow distribution materials, PE-1 shows a yield stress and a steady state stress that is approximately twice as large as those exhibited by PE-6. This behavior is attributed to the larger high molecular weight component in PE-1. Of the two broad molecular weight distribution materials, PE-3 has a much larger yield stress than PE-4. The steady state stress for PE-4 is low compared to PE-6, which does not quite reach a steady state stress even at large magnitudes of applied strain. Clearly, the high molecular weight shoulder on the distribution dominates these properties.

Figure 9 shows the recoverable strain characteristics for these four materials. PE-3 with the high molecular weight shoulder is the most elastic. PE-4, the other broad distribution material, is highly elastic but exhibits only about one-half as much recoverable strain as PE-3. The narrow distribution materials, PE-1 and PE-6, have a considerably smaller high molecular weight component and exhibit less melt elasticity. PE-1 with a molecular weight distribution shifted slightly to the high side has twice as much recoverable strain in comparison to the other narrow distribution material, PE-6. This finding shows again how the high molecular weight end of the distribution dominates the elastic behavior.

Polymethyl Methacrylate.   Investigations of the response of melts of polymethyl methacrylate (5) when sheared from the quiescent state showed similar behavior to that of polystyrene. Figure 10 presents the yield stress and steady state stress as a function of temperature. Again, a sharp transition is observed in both properties at approximately 205 °C (100 °C above the glass transition temperature).

Figure 11 shows the initial strain recovery during the first second

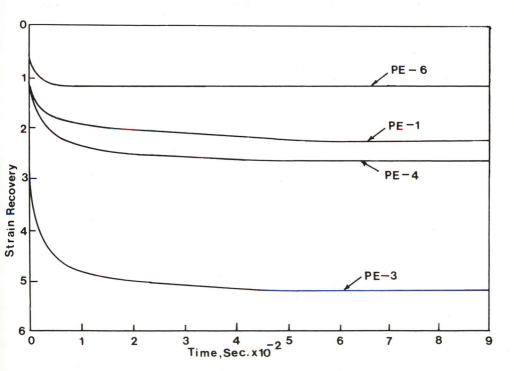

Figure 9. Recoverable strain as a function of time for low density polyethylenes of various molecular weight distributions.

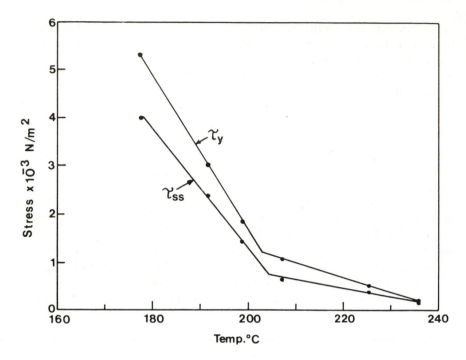

Figure 10. Melt yield stress and steady state stress as a function of T
for polymethyl methacrylate.

of the elastic recoil as a function of temperature. In general, a monotonic trend of decreasing recovery is observed as the temperature is increased. The amount of strain recovered during the first 0.5 and first 0.8 s of recoil is shown in Figure 12 as a function of temperature. Again, the transition is clearly evident at about 201 °C by this test.

Polymer Blends.   Blends of polystyrene and polymethyl methacrylate also were studied by these techniques (6). Figure 13 shows the stress vs. strain curves for pure polystyrene, pure polymethyl methacrylate, and a 40–60 wt% blend of the two at 200 °C. Two yield points are exhibited by the blend, indicating the possibility of two interpenetrating phases in the blend.

Figure 14 shows the recoverable strain behavior as a function of composition. The total amount of recoverable strain for the two pure components is relatively small. As one component is added to the other by extrusion melt blending, the recoverable strain increases to a maximum at 40% polystyrene–60% polymethyl methacrylate, and the recovery changes from Type II for the individual components to Type I for the blend. Corresponding to the maximum in recoverable strain, the time for the strain recovery to be finished goes through a maximum that is more than ten times the time for the strain recovery of the individual components.

The melt tension in fiber spinning these blends also goes through a maximum at the same composition.

Discussion

When a polymer melt is subjected to an applied shear rate from the quiescent state, the stress increases until a yield stress is reached and then decreases to some steady state stress associated with the flow process. This observation indicates that in the quiescent melt a structure is built up that resists flow, and this structure must be disassociated before a steady state flow can be achieved. This behavior is

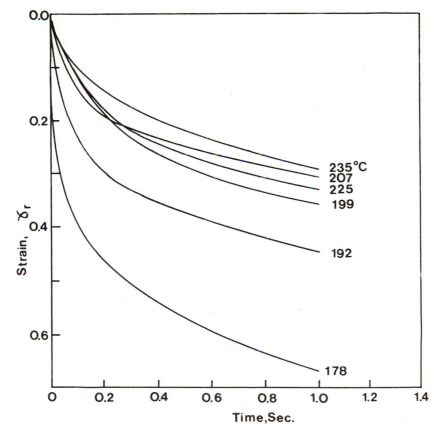

Figure 11. *Recoverable strain during the first second of recovery for polymethyl methacrylate at various T.*

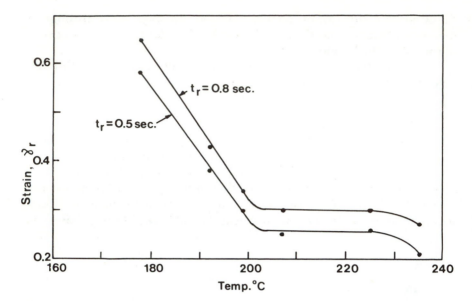

Figure 12. Recoverable strain at recovery times of 0.5 and 0.8 s as a function of T for polymethyl methacrylate.

shown in the various stress vs. strain curves to be a fairly general phenomenon.

In some cases (polystyrene and polymethyl methacrylate) the temperature dependence of the yield stress changes abruptly at a specific temperature, T_u. Below this temperature, the yield stress is high and is also strongly dependent on temperature. Above this temperature, the yield stress is low and only slightly temperature dependent. This result indicates that below the transition a structure exists in the melt that is strong but rapidly decreases in strength as temperature is increased.

Recoverable strain also shows this transition in polystyrene melts. Below the transition, the amount of recoverable strain is large and requires a long time to finish the recovery process (Type I). Above the transition, the amount of recoverable strain is smaller and the recovery process is finished in a much shorter time (Type II). This finding indicates, in confirmation with the yield stress results, that below the transition a strong interacting structure exists that permits a high elastic energy storage. Above the transition, this structure seems to disappear. The transition temperature increases with increasing molecular weight, indicating that the interactions in the melt structure increase with increasing molecular length.

The results on low density polyethylene show that both the yield stress and the recoverable strain are strongly dependent on the high molecular weight end of the molecular weight distribution. This finding again confirms that a stronger melt structure is built up with larger molecules.

The results on the blends of the two incompatible polymers, polystyrene and polymethyl methacrylate, demonstrate that the transition temperature, T_u, of a blend can be higher than either of its components. Both components of the blend were above their respective transitions at the test temperature, but when they are blended together the response changes from Type II to Type I. This result indicates the formation of a strong interacting structure in the melt.

Although these structures now are not identifiable in the melt, some practical conclusions can be drawn from these results. For example, if a high orientation is desired in blow molding, vacuum forming, or film formation then the stronger melt structure should be used. This task can be accomplished by using a processing temperature below the transition, T_u, or by increasing the high molecular weight end of the distribution, or by blending.

Conclusion

The author hopes that the investigation presented demonstrates that the molecule's behavior can demonstrate many interesting things in response to an applied shear starting from the quiescent state. Such studies are supplementary and complementary to more conventional dynamic testing.

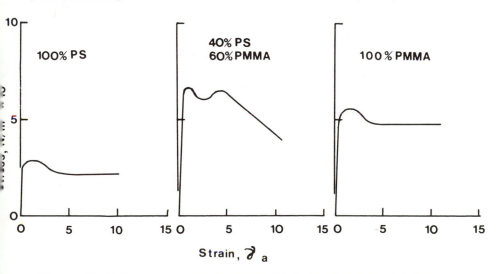

Figure 13. *Melt stress vs. strain curves for pure polystyrene (a), pure polymethyl methacrylate (b), and a 40–60% blend (c).*

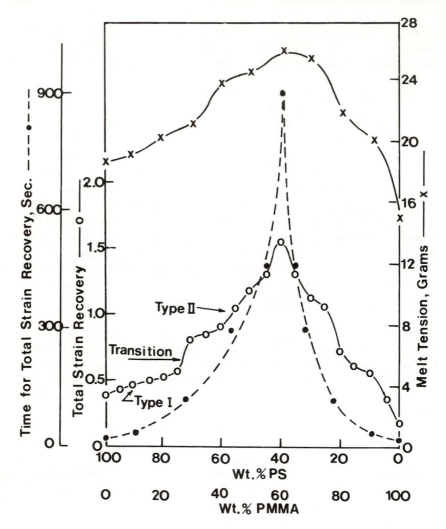

Figure 14. Recoverable strain and time to finish strain recovery for blends of polystyrene and polymethyl methacrylate at 200 °C.

Literature Cited

1. Maxwell, B.; Nguyen, M. *Polym. Eng. Sci.* **1979**, *19*, 1140–50.
2. Graessley, W. W. *Adv. Polym. Sci.* **1974**, *16*.
3. Boyer, R. F. *J. Polym. Sci.* **1966**, *C14*, 267.
4. Nguyen, M.; Maxwell, B. *Polym. Eng. Sci.* **1980**, *20* (*14*), 972–76.
5. Markes, T. E., Ph.D. Thesis, Princeton Univ., Princeton, NJ, 1979.
6. Thornton, B. A.; Villasenor, R. G.; Maxwell, B. *J. Appl. Polym. Sci.* **1980**, *25*, 653–63.

RECEIVED for review November 23, 1981. ACCEPTED June 11, 1982.

A Low Shear, High Temperature Rotational Viscometer

The Viscosity of Ultrahigh Molecular Weight Polyethylene

H. L. WAGNER

National Bureau of Standards, Polymer Science and Standards Division, Washington, DC 20234

J. G. DILLON

Food and Drug Administration, Office of Medical Devices, Silver Spring, MD 20910

To obtain accurate measurements of the limiting viscosity number of solutions of ultrahigh molecular weight polyethylene (UHMWPE), a low shear floating-rotor viscometer of the Zimm—Crothers type was developed to measure viscosities at elevated temperatures (135 °C) and essentially zero shear rate. The zero shear rate measurements for a set of UHMWPE samples were compared with viscosity measurements at moderate and high shear rates (up to 2000 s⁻¹) carried out in a capillary viscometer. The limiting viscosity number of UHMWPE depends, as expected, on shear rate, and the higher shear rate data could not be extrapolated to yield the correct zero-shear rate viscosities.

T HE SIMPLEST AND MOST CONVENIENT TECHNIQUE available for the determination of the relative size of polymer molecules is the measurement of its dilute solution viscosity. The classical techniques of molecular weight measurement such as light scattering or osmometry, as well as size exclusion chromatography (SEC), are either too difficult or impossible with polymers of molecular weight greater than about a million, although they are quite suitable for typical commercially available high-volume polymers with molecular weights in the hundred thousand range. Dilute solution viscosity measurements may be made on ultrahigh molecular weight polymers but by a technique quite different from that used for lower molecular weight polymers. For the latter, the use of a capillary viscometer with shear rates of the order of a few thousand reciprocal seconds is satisfactory because the viscosity is independent of shear rate. This relationship is far from the case with the ultrahigh molecular weight polymers. The non-Newtonian behavior of these polymers is well recognized in the literature as a problem that must be dealt with if accurate measurements of the limiting viscosity number are to be made. As the shear rate increases, the shape and orientation of the molecule change sufficiently so that the shear stress is no longer proportional to the shear rate.

In Figure 1, the data obtained in a multibulb capillary viscometer, in which the flow at several shear rates may be measured, are shown for ultrahigh molecular weight polyethylene (UHMWPE). The dependence on shear rate is apparent and explains, in part, why it is so difficult to obtain good interlaboratory agreement. Not only does the shear rate vary as the inverse cube of the radius of the capillary, but the shear rate is also a function of the flow rate, which will vary with the sample. For T_2 DNA, of molecular weight of about 10^8, viscosities measured at different shear rates may differ by as much as a factor of two (1). Similar behavior has been found for polystyrene (2) and α-methyl polystyrenes (3) of up to 1.4×10^7 in molecular weight.

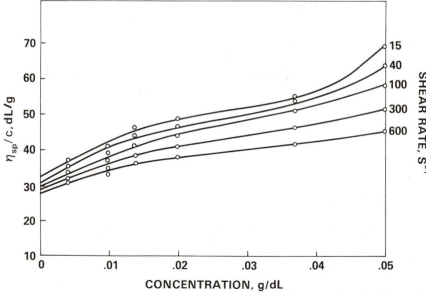

Figure 1. Viscosity number as a function of shear rate for a sample of UHMWPE in decalin at 135 °C in a multibulb capillary viscometer.

Because capillary viscometers are not usually designed to provide a constant or very low shear rate, viscosity data for shear rate-sensitive materials must be obtained in a viscometer where the effect of shear rate is minimal or nonexistent. The Zimm–Crothers floating-rotor low shear viscometer, although used mostly for biological molecules such as DNA (*1*), also has found application for high molecular weight synthetic polymers (*3*). In this viscometer at the low shear rates of a few reciprocal seconds, the viscosity is independent of shear rate, so that the measured viscosity is essentially the zero-shear rate viscosity. Another important advantage is that the degradation due to shear, which can reduce the viscosity substantially, is very unlikely at these low shear rates. This degradation is sometimes observed in a capillary viscometer during a measurement when the flow time decreases as the sample is recycled through the viscometer.

In the past, the viscosity of UHMWPE was measured in capillary viscometers at 135°C in decalin at a single concentration of 0.05% (*4*). It is assumed not only that this datum may be extrapolated to zero concentration using the same viscosity–concentration relationships found for lower molecular weight polymers, but also that such an extrapolation will yield a zero-shear rate limiting viscosity number. However, for molecular weights of the order of 1×10^6 to 10×10^6, which we believe is the range of molecular weights for UHMWPE, and limiting viscosity numbers of about 40, a concentration as high as 0.05% is in the semidilute range (*5*). The viscosity number is no longer expected to be linear with concentration in this range, and consequently, the viscosity–concentration relationships found for lower molecular weight polyethylenes cannot be expected to hold. Figure 2 shows our results for two different UHMWPEs, one at the higher end of the molecular weight range, the other at the lower. The straight line indicates the result obtained making the just mentioned assumptions and gives an idea of the possible error. Although the error is not too large for the lower molecular weight sample, it is quite serious for the higher one.

The limiting viscosity number is used as an indicator of relative but not absolute molecular weights because the presently available parameters for the empirical Mark–Houwink relation for polyethylene were obtained from data on lower molecular weight polymers. Both experiment and theory demonstrate that these parameters do not apply necessarily to polymers in the ultrahigh molecular weight range (*6, 7*).

In the process of characterizing UHMWPE for surgical implants, we assembled a low shear viscometer of the Zimm–Crothers type. This viscometer has some unique features, the principal one being its high temperature capability, a necessary requirement for polyethyl-

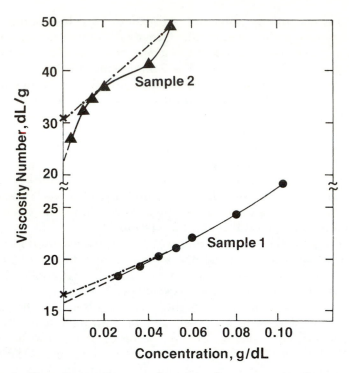

Figure 2. Viscosity number as a function of concentration for samples of UHMWPE in decalin at 135 °C. The same Ubbelohde viscometer was used for all measurements. Shear rate is variable and depends on flow rate. The straight lines are those used for extrapolation to limiting viscosity number, assuming viscosity–concentration relationships observed for lower molecular weight polyethylenes.

ene measurements because this polymer is insoluble at room temperature.

High Temperature, Low Shear Viscometer

The construction of this viscometer is based on the original design of the floating-rotor viscometer first described by Zimm and Crothers (8). The drive system for the rotor is similar to one described elsewhere (9).

The essential features of the viscometer are shown in Figure 3. Solvent or solution is placed between an inner glass cylinder, the rotor, and a fixed outer glass cylinder, the stator. The rotor floats freely in the liquid and is centered within the stator by surface forces; it is subject to a constant torque produced by the interaction of a cylinder of aluminum inserted into the rotor and four surrounding solenoids. The apparatus operates on the principle of an electromagnetic induction motor. There are no mechanical devices attached to the rotor, which is supported by buoyancy. Unlike the Couette viscometer, all the energy loss occurs in the liquid, making possible measurements at low viscosities with good precision. Therefore, the speed of the rotor is inversely proportional to the viscosity of the liquid.

A resistive-capacitance circuit splits the current fed the solenoids into two legs that are equal but 90° out of phase. Each leg energizes a pair of opposite solenoids so that adjacent solenoids are out of phase. Torque on the rotor is produced by eddy currents induced in the aluminum and is held constant by an adjustable constant current supply. The constant current supply is based on a Kepco[1] bipolar operational amplifier capable of supplying 1.5 A at 75 V. The circuit used is the one described by the manufacturer of the unit for operation at constant current. The reference for the amplifier is a precision 60-Hz oscillator stabilized for both frequency and amplitude to better than 0.1%. The root mean square value of the current for the entire circuit is stabilized to better than 0.1%.

To minimize the tendency to wobble at low velocities, the rotor and stator are fabricated from uniform bore glass tubing, about 30 mm in diameter with a 1-mm gap between them for the liquid. The top

[1] Certain commercial equipment, instruments, or materials are identified in this chapter to specify the experimental procedure. In no case does such identification imply recommendation by the National Bureau of Standards, nor does it imply that the material or equipment identified is necessarily the best for the purpose.

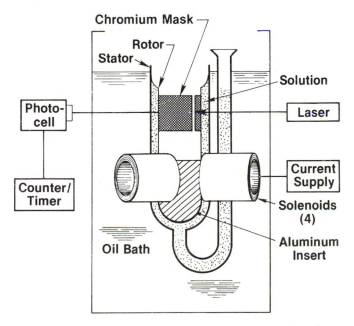

Figure 3. The high-temperature, low shear viscometer, schematic.

edge of the glass rotor is polished and free of nicks, and the bottom is rounded as symmetrically as possible.

The aluminum insert, which must also be symmetrical to reduce wobble, is machined carefully to fit the glass rotor at 135 °C. The total weight of the rotor is adjusted by drilling the proper size hole through the center of the aluminum. Because the rotor is suspended by buoyancy, the size of the hole required depends on the density of the solvent. The length of the insert should be such as to be within the confines of the magnetic field of the solenoids. To avoid shattering the glass, the glass and aluminum are heated separately in a 150 °C oven and then assembled and put into the stator in the bath. This procedure is necessary because of the difference in the rate of expansion of the two materials.

The entire apparatus is bolted rigidly to a frame immersed in the high temperature oil bath. This rigid mounting was necessary to achieve good reproducibility. The bath was fitted with a cover so that measurements could be made under a blanket of argon. Solvent readings are reproducible to about 1% from day to day, even though the bath is cooled overnight and reheated on the morning of a run.

The rotation time, or period, is determined by directing a laser beam onto the rotor which has been coated with a wide band of chromium, except for two vertical clear windows a few millimeters wide and 180° apart. During rotation, the beam passes through these windows and impinges on the photocell, which, in conjunction with a Schmidt trigger assembly, starts a counter-timer to measure the period of the rotor. The relative viscosity is equal to the ratio of the time per cycle for the solution, or the period, to that for the solvent at the same torque or current.

Sample is introduced into the side arm of the stator by means of a funnel. For polyolefins, the funnel must be heated to prevent cooling of the hot solution and precipitation of the polymer. The height of the rotor is then adjusted by adding or withdrawing liquid after it has been floated. The centering effect is achieved by the surface tension of the meniscus. As pointed out by Zimm (1), the shape of the meniscus is crucial for this centering; surface forces act to center the inner rotor if the liquid surface rises up from its rim to the wall of the stator.

The optimum height of the rotor is determined by the height at which the magnetic coupling is a maximum, that is, where the period is shortest for a given current. At this height, the rotor speed is changing very slowly with height, so that small errors in height will lead to minimal error in the period.

Measurement Procedure

Although the data for only one value of the torque, or the current, were finally used, the period was determined for a range of current values from 30 to 150 mA at 10-mA intervals. The periods were first determined for the solvent, then for the solution. The relative viscosity is the ratio of the period for the solution to that of the solvent at a given torque or current setting. The limiting viscosity number then is calculated in the customary way by plotting the viscosity number $(\eta_{rel} - 1)/c$ as a function of concentration and extrapolating to zero concentration. The measurements were restricted to very dilute solutions to make certain the data were in the linear concentration region.

The shear rate may be calculated from the equations given by Zimm and Crothers (6)

$$\dot{\gamma} = \frac{\pi}{P_0} \frac{R_1 + R_2}{R_2 - R_1} \frac{8R_1^2 R_2^2}{(R_1 + R_2)^3(R_2 - R_1)} \ln R_2/R_1 \qquad (1)$$

where R_1 is the outer radius of the rotor, R_2 is the inner radius of the stator, $\dot{\gamma}$ is the shear rate, and P_0 the period. In our case, R_2 was 1.5 cm and R_1 was 1.4 cm so that $\dot{\gamma} = 90.94/P_0$.

One of the major problems encountered during the design of the apparatus was an apparent increase in viscosity number at low velocities. This increase was attributed to wobble of the rotor. To help alleviate this problem, uniform bore tubing was employed and the aluminum insert was machined to fit the rotor closely, as described earlier. This technique eliminated most of the apparent increase in viscosity number at low velocities.

The torque, T, established in the rotor is proportional to the product of the intensity of the inducing magnetic field and the current induced in the rotor. But the latter also depends on the magnetic field—which in turn depends on the current, i, supplied to the coil. Hence,

$$T = k_1 i^2 \qquad (2)$$

But for a Newtonian fluid, the torque acting under constant rotor speed, ω, is

$$T = k_2 \eta \omega,$$

or

$$T = k_2 \eta / P_0 \qquad (3)$$

where η is viscosity and P_0 is the period of rotation. Hence,

$$\eta = k_1/k_2 \, i^2 P_0 \qquad (4)$$

For the solvent, or any Newtonian fluid, this quantity is invariant with shear rate. This relationship was employed to determine how well the apparatus behaved, particularly the extent of the rotor wobble, as shown in Figure 4. Generally, the value of Pi^2 was constant above a

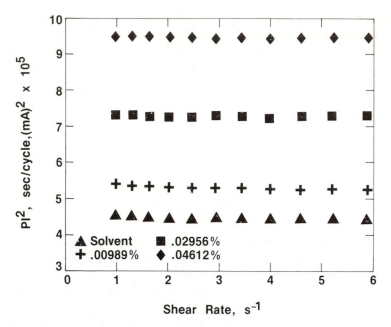

Figure 4. Plot of Pi^2 vs. shear rate for UHMWPE 99716 in decalin at 135 °C.

shear rate of 3 s⁻¹, but at lower shear rates some upturn of the order of 0.5–0.8% at 1 s⁻¹ was noticed, indicating a small amount of wobble. At shear rates greater than 7 or 8 s⁻¹, where the periods are 10 s or less, the possibility of turbulent flow exists. Hence, shear rates in this viscometer were confined to these limits. We note from the figure that the viscosity is also invariant for UHMWPE in this range of low shear rates.

Zimm (1) showed experimentally that if the product of the relaxation time, τ, and shear rate, $\dot{\gamma}$, is less than 0.2, no variation of viscosity with shear rate occurs. The relaxation time, τ, is equal to

$$\alpha\, M[\eta]\, \eta_0/RT_1 \tag{5}$$

where M is the molecular weight of the polymer, $[\eta]$ is its limiting viscosity number, η_0 is the solvent viscosity, and α is a constant that depends on the nature of the molecule. For linear flexible molecules, α is about 40 $[\eta]$ in deciliters per gram. For our highest molecular weight UHMWPE, assuming a molecular weight as high as 10×10^6, we find a value of $\tau\dot{\gamma}$ of less than 0.15, within the region of constant shear rate.

Because no evidence of upturn in the Pi^2 vs. shear rate plots above 110 mA was found, our viscosity calculations were made with data taken at 110 mA, or for a shear rate for the solvent of 3 s⁻¹.

Assessing the Viscometer

To determine whether the instrument was behaving satisfactorily, a series of experiments was run at lower temperature on narrow molecular weight distribution polystyrenes, including two standard reference materials, SRM 705 and SRM 1479. The molecular weights ranged from 18×10^4 to 22×10^6; the absolute molecular weights for the non-SRM materials are estimated and have not been measured here. The values of the limiting viscosity number and the Huggins constant, derived from the slope of the viscosity number concentration curve, are given in Table I. Excellent agreement is found for SRM 705 between our value and the certificate value. For SRM 1479, for which no viscosity is certified, the agreement in the molecular weight derived from the "blob" model (6) using our viscosity result and the weight-average molecular weight determined by light scattering is excellent. The data for SRM 1479 are shown in Figure 5.

Sample Preparation

One of the most difficult problems encountered in characterizing UHMWPE is sample preparation. These materials are difficult to dissolve because of their crystallinity and high molecular weight, and on the other hand, they are extremely sensitive to mechanical forces. A large viscosity decrease occurs if just a small amount of degradation takes place.

Our present technique consists of heating decalin and polymer in a capped bottle. The decalin, 99% pure and composed of *cis*- and *trans*-isomers, is first passed through a silica gel column. An antioxidant, N-phenyl-2-naphthylamine, is added to a concentration of 0.2%, and then nitrogen is bubbled through overnight. The bottle of sample and solvent is flushed with nitrogen and then placed in a 150 °C oven for 1 h and is not disturbed for the first 15 or 20 min. At the first sign of polymer melting which occurs at about this time, the bottle is swirled gently to prevent agglomeration of the particles into a gel-like mass that would be very difficult to dissolve. This gentle swirling is repeated every 15 min for a total heating time of 1 h. All of our UHMWPE samples dissolved under these conditions, but it does not follow that every UHMWPE sample will do so.

At first, tetralin was used as a solvent, but it was almost impossible to achieve satisfactory reproducibility. Tetralin tends to form peroxides that could contribute to degradation, leading to the observed erratic results despite the presence of antioxidant. Consequently, we switched to decalin despite possible complications with isomerism. The results obtained with decalin were more consistent than with tetralin, and the limiting viscosity values were significantly higher, although the values were expected to be about the

Table I. Viscosity of Polystyrene

Polymer	Solvent	T(°C)	$[\eta]$ dL/g	Huggins Constant	Molecular Weight $\times 10^{-6}$	
					Certificate	"Blob" Model
SRM 705	cyclohexane	34.5	0.356 (0.354)[a]	0.62	—	—
SRM 1479	toluene	25	2.42	0.29	1.05	1.05
F 1300	toluene	25	18.5	0.36	—	13.3
IK 1500	toluene	25	23.7	0.42	—	18.8
BK 2500	toluene	25	27.1	0.54	—	21.6

[a] Certificate value.

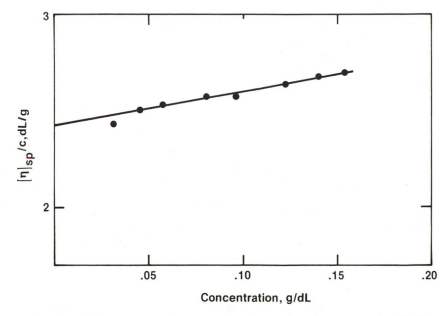

Figure 5. Viscosity number as a function of concentration for SRM 1479, narrow molecular weight distribution polystyrene, molecular weight 10.5 × 10⁶.

same. A value of 24.2 dL/g was obtained for one sample in decalin, but a value of only 18.6 dL/g was found in tetralin. We now suspect that some degradation occurred in tetralin before a measurement could be made.

Results

Figure 6 shows the plots of viscosity number as a function of concentration for three different samples of UHMWPE in decalin at 135 °C. Concentrations were limited to very dilute values because for these high molecular weights, curvature seemed to set in, as already mentioned, at low concentrations. Hence, for sample 99974 the highest concentration was about 0.022% or 0.22 mg/mL. At present, a method of relating limiting viscosity number to molecular weight is not known because the Mark–Houwink parameters in the literature were only determined for polymers of lower molecular weight.

To establish the effect of shear rate on viscosity, measurements were made not only in the low shear viscometer, but also in a multibulb capillary viscometer in which the shear rates ranged from about 50 to about 150 s⁻¹ for the solvent and in a capillary viscometer in which the shear rate is of the order of 2000 s⁻¹. The last viscometer was the ordinary Cannon–Ubbelohde type usually used for control work.

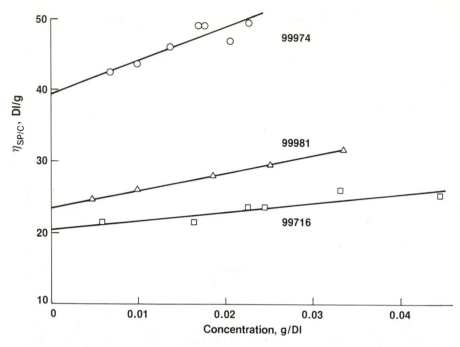

Figure 6. Viscosity number as a function of concentration for three samples of UHMWPE in decalin at 135 °C.

These data are shown in Figure 7. The values of the limiting viscosity number, obtained in the multibulb viscometer for which an attempt was made to extrapolate to zero shear rate, are listed in Table II and are compared with the values obtained in the low shear viscometer. The data obtained with the routine capillary viscometer also are shown. Serious errors in limiting viscosity number can result if measurements are not made at very low shear rate.

Discussion and Conclusion

Zimm (1) discussed the difficulty of trying to predict in detail the non-Newtonian behavior of most large macromolecules but was able to make a useful semiempirical generalization: the extent of distortion of the molecule, which determines the non-Newtonian behavior, is determined by the shear rate relative to the molecular relaxation. This relaxation is the reciprocal of τ given in Equation 5, so that the product of the dimensionless quantity $\dot{\gamma}\tau$ must be below a value, found from experience to be about 0.2. In our case, this value is not exceeded until the shear rate is about 100 s^{-1}, although from Figure 7 it appears to set in somewhat earlier.

We conclude that accurate measurements of the viscosity of

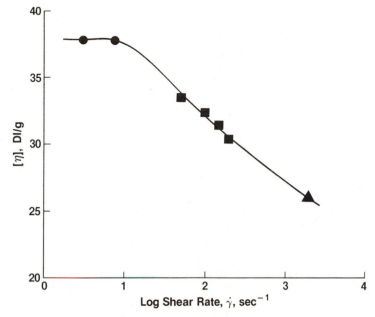

Figure 7. Viscosity number as a function of shear rate using three different viscometers. Key to viscometers: ●, low shear; ■, multibulb capillary; and ▲, regular capillary.

UHMWPE requires viscosity measurements at shear rates no greater than about 50 s^{-1} and preferably closer to 10 s^{-1}. These measurements should be carried out at very low concentrations so that the viscosity number−concentration curve is linear. The freely floating low-shear viscometer described earlier may be used for these measurements at elevated temperatures. We find that the use of this viscometer is quite simple, and measurements may be made rapidly.

Table II. Limiting Viscosity Number Values

	Sample Number		
Method	*99716^a*	*99881^b*	*99974^c*
Zero shear rate (low shear viscometer, initial value)	20.5	23.5	39.5
Zero shear rate, estimated (multibulb viscometer)	18.9	—	34.0
High-shear rate (routine Cannon−Ubbelohde viscometer)	14.6	17.8	26.0

Note: These values were measured in decalin at 135 °C and are given in dL/g.
^{a} Maximum concentration = 0.046%.
^{b} Maximum concentration = 0.03%.
^{c} Maximum concentration = 0.02%.

Acknowledgment

The authors appreciate the assistance of Charles Han and Frederick Mopsik with the design of the electronic components of the apparatus.

Literature Cited

1. Zimm, B. H. In "Procedures in Nucleic Acid Research"; Cantoni, G. L., Ed.; Harper and Row: New York, 1971; Vol. 2, pp. 257–60.
2. Yamaguchi, N.; Suzura, Y.: Okana, I.; Wada, E. *J. Phys. Chem.* **1971,** 75, 1141.
3. Noda, I.; Yamada, Y.; Nagasawa, M. *J. Phys. Chem.* **1968,** 72, 2890.
4. "Determination of Intrinsic Viscosity for 1900 Ultra High Molecular Weight Polyethylene Polymer," Technical Information Report, Hercules, Inc., Wilmington, DE, August 1974.
5. Adam, M.; Delsanti, M. *Polym. Prepr. Am. Chem. Soc. Div. Polym. Chem.* **1981,** 22, 104.
6. Han, C. C. *Polymer* **1979,** 20, 1083.
7. Farnoux, B.; Bowe, F.; Cotton, J. P.; Daoud, M.; Jannink, G.; Nierlick, M.; DeGennes, P. G. *J. Phys.* **1978,** 39, 77.
8. Zimm, B. H.; Crothers, D. M. *Proc. Natl. Acad. Sci. U.S.A.* **1962,** 48, 905.
9. Tsai, B. C.; Meyers, M.; McIntyre, D. *J. Chem. Phys.* **1973,** 58, 4582.

RECEIVED for review October 14, 1981. ACCEPTED January 27, 1982.

10

Morphology and Properties of Styrene and Dimethylsiloxane Triblock and Multiblock Copolymers

SUNIL K. VARSHNEY

University of Florida, Department of Materials Science and Engineering, Gainesville, FL 32611

PUSHPA BAJAJ

Indian Institute of Technology, Polymer and Fiber Science Laboratory, New Delhi, 110016 India

Multiblock copolymers $(BAB)_n$ of styrene and polydimethylsiloxane were synthesized by coupling diphenyldiacetoxysilane with preformed triblock copolymers of siloxane−styrene−siloxane terminated with "living" silanolate ends. Electron microscopy and dynamic measurements were used to examine the phase separation of these copolymers. The dimethylsiloxane content, molecular weight, and polydispersity of these specimens were investigated also. A difference was noted in the tensile behavior of the multiblock copolymers obtained from solution-cast films. The results show that the morphology of the multiblock copolymers studied is defined only when films are cast from their preferential solvents. Characterization of the microdomain structures of these block copolymers revealed polystyrene discrete spheres dispersed in a rubbery polydimethylsiloxane matrix.

SYNTHESIS AND CHARACTERIZATION of AB diblock (*1−4*) and ABA triblock copolymers (*5*), having a hard block of polystyrene (A) and a rubbery polydimethylsiloxane (PDMS) segment (B) have been studied extensively. Recently, we showed (*6*) the dilute solution and morphological behavior of poly(dimethylsiloxane-*b*-styrene-*b*-dimethylsiloxane) elastomers in different solvents and evaluated the solvent−solute interaction parameters. The present investigation extends the synthesis to multiblock copolymers of $(BAB)_n$ type that

0065−2393/83/0203−0179$06.00/0
© 1983 American Chemical Society

should show properties characteristic of thermoplastic elastomers if microphase separation occurs in the bulk state.

Multiblock copolymers $(BAB)_n$ of styrene and PDMS were synthesized by coupling diphenyldiacetoxysilane with preformed triblock copolymers of siloxane–styrene–siloxane terminated with "living" silanolate ends. Films of these block copolymers cast from solution and compression molding had considerable strength, presumably due to the aggregation of polystyrene microdomains, while retaining elasticity owing to their alternating PDMS blocks. The phase separation of these copolymers are examined with electron microscopy and dynamic measurements.

Experimental

Materials.   Styrene was freed from the inhibitor by washing with 5% NaOH solution. It was then washed several times with distilled water to eliminate all traces of alkali and dried over $CaCl_2$. After distillation under vacuum (30 °C/0.05 mm Hg), styrene was finally dried over CaH_2 under vacuum.

Hexamethylcyclotrisiloxane was synthesized by a method described elsewhere (7). It was purified by sublimation under vacuum.

Toluene was dried over sodium wire and distilled over CaH_2. It was distilled directly from the reservoir just before the polymerization. Tetrahydrofuran (THF) was refluxed for several hours over sodium wire and was fractionally distilled into a flask containing sodium and a small amount of naphthalene. Refluxing was continued until the naphthalene complex, which reacts more efficiently with impurities than the alkali metal alone, was formed. THF was flash distilled from the reservoir as needed into a dark-colored flask just before polymerization.

Lithium biphenyl anionic initiator was synthesized according to the method of Scott (8) and was used for the preparation of sodium and lithium naphthalene.

Polymerization.   Block copolymers of poly(dimethylsiloxane-*b*-styrene-*b*-dimethylsiloxane) were synthesized as described previously in the polymerization of hexamethylcyclotrisiloxane-d_3, with "living" α, ω-dilithiopolystyryl species in the presence of bifunctional initiator and THF as a solvent (9). This polymerization yielded BAB triblock copolymer of low polydispersity, terminated at both ends with a lithium silanolate moiety. The living polymer was coupled with diphenyldiacetoxysilane to give multiblock copolymers of styrene–siloxane. After 3 h (10) reaction time, methyl iodide was added to terminate the residual silanolate chain ends, if present. The polymerization scheme is diagramed in Scheme I.

Identification of Block Copolymer.   The characteristic color of living polystyrene anion is dark red, and the living silanolate anion is pure white. When the cyclic trisiloxane was added, only a dilution of the color was noticed immediately. As the polymerization proceeded, the red color of living styrene anion began to fade considerably, and the viscosity of the system increased. After 8 h, the polymer solution was very viscous and slightly pale in color. Therefore, the sequence of polymerization can be described by the color change of the system.

The composition of the block copolymers were determined by proton

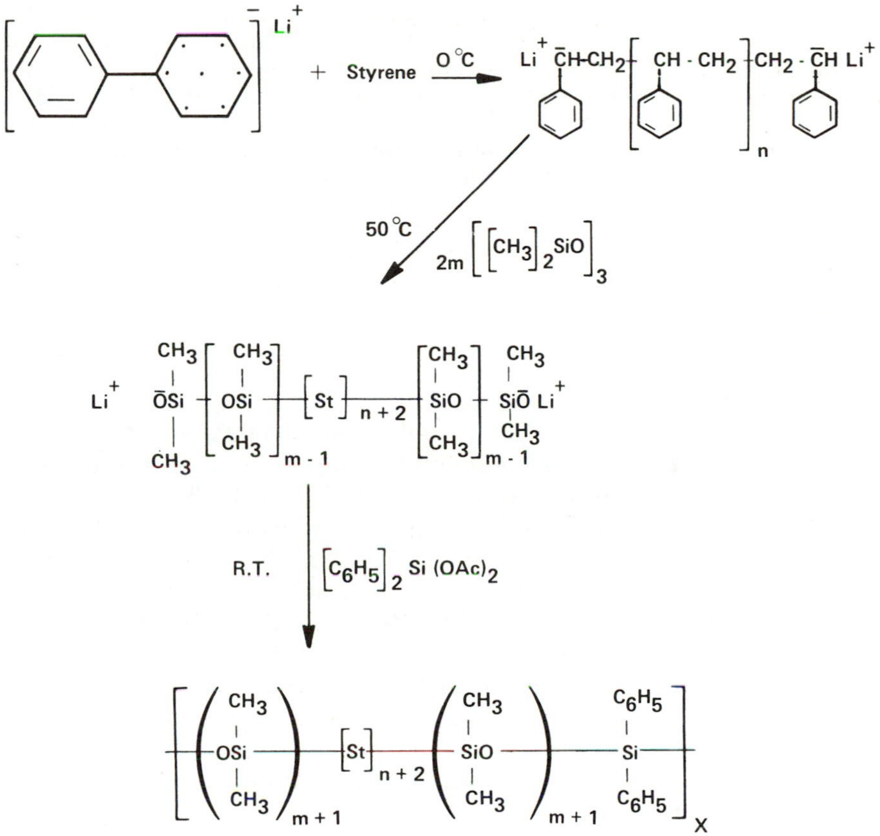

Scheme I. *Polymerization scheme for synthesis of poly(dimethylsilox-ane-b-styrene-b-dimethylsiloxane) multiblock copolymer.*

magnetic resonance (PMR) and IR spectroscopy as described previously (*11*). The molecular characteristics of the copolymers were determined in toluene (which is a solvent for both polymer components) via gel permeation chromatography (GPC).

Film Preparation.   Film specimens about 0.5 mm thick were prepared by casting the copolymers from 10% solutions on glass plate. The solvents chosen were toluene, methyl ethyl ketone, and cyclohexane. In each case, the solvent was evaporated slowly at room temperature (27 °C) over 2−3 days and the cast films were dried subsequently under high vacuum for 3−4 days. Specimen for molding of copolymer was carried out at 140 °C.

Mechanical Measurements.   Stress−strain curves were measured using an Instron tensile tester. Test specimens were 4 cm long and 0.4 cm wide. A cross-head speed of 4 cm/min was used. Dynamic tensile mechanical measurements were made using a Rheovibron dynamic viscoelastometer model DDV−11. Test pieces were about 2 cm long and 0.4 cm wide. The measurements were carried out over the temperature range −120−+110 °C at average intervals of 4 °C. Data were corrected for apparatus compliance using an empirical correction factor measured at near zero length as a function of temperature. Thermomechanical studies were carried out on styrene–siloxane triblock copolymer. Measurements were made using Stanton Redcroft thermomechanical analyzer model TMA691. The probe used was quartz rod, radius 0.25 cm, with a flat end. Samples approximately 0.3 × 0.1 cm were cut and placed under the end of the probe. The probe was adjusted to just touch the surface of the pellet using the probe position controller after cooling the sample in liquid air (−140 °C). The temperature was increased from −140− +250 °C at a heating rate of 10 °C/min.

Electron Microscopy. Multiblock copolymers were dissoved in toluene, methyl ethyl ketone, and cyclohexane and diluted to a concentration of 3 wt %. Films were cast on a carbon-coated copper grid and examined in the instrument at 60 kV.

Results and Discussion

Dimethylsiloxane content, molecular weights, and the polydispersity ($\bar{M}_w/\bar{M}_n$) of block copolymers samples studied in the present work are summarized in Tables I and II. The BAB prepolymers as well as the derived multiblock copolymers were found to be essentially free of homopolymers, as indicated by their solutions' behavior in selective solvents such as cyclohexane or bromobenzane. As shown in Figure 1, the GPC curves of the BAB and (BAB)$_n$ block copolymers indicate unimodal molecular weight distribution. The molecular weight distribution of the triblock copolymers was narrow, i.e., 1.2, although the same measurement showed increases in molecular weight as well as in polydispersity of the coupled multiblock. The broadening of molecular weight distribution in the multiblock copolymers may be attributed to nonuniform coupling of preformed triblock copolymers.

The storage modulus, E', and loss tangent, tan δ, of the multiblock copolymer films cast in toluene (a good solvent for both components of blocks), cyclohexane (better solvent for PDMS), and film prepared by compression molding are plotted against temperature in Figure 2. The dynamic measurements were carried out at a frequency of 110 Hz. At low temperature, −120 °C, the storage modulus, E', is higher as the

Table I. Characterization Data for Polymers Used in This Study

Sample[a]	Mol Fraction $(Me_2SiO)_n$ Incorporated[b]	Stoichiometric Molecular Weight $\bar{M}_n \times 10^{-3} g/mol$	$\bar{M}_w/\bar{M}_n$ GPC
P(DMS−St−DMS)	0.504	11.5−14.0−11.5	1.20
P(DMS−St−DMS)	0.617	18.5−25.0−18.5	1.24
P(DMS−St−DMS)	0.618	22.5−37.0−22.5	1.26

[a] St is styrene; DMS is dimethylsiloxane.
[b] From IR spectral analysis (11).

Table II. Polydimethylsiloxane–Polystyrene $(BAB)_n$ Multiblock Copolymers

(Me_2SiO) Units in Copolymer	Polymer Yield	$\overline{M}_n \times 10^{-3a}$ (g/mol)	Coupling Value[b] n	$\overline{M}_w/\overline{M}_n$
42.3	83	88.3	2.4	1.28
53.7	80	124.0	2.0	1.30
53.0	81	188.6	2.3	1.34

[a] $\overline{M}_n$ measured by membrane osmometry.
[b] Calculated on the basis of initial $\overline{M}_n$ of prepolymer (BAB).

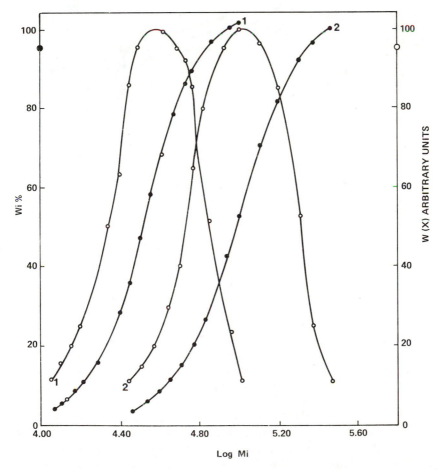

Figure 1. Integral differential MWD curves of BAB and $(BAB)_n$ copolymers. Key: 1, BAB M_w/M_n = 1.2; and 2, $(BAB)_n$ M_w/M_n = 1.28.

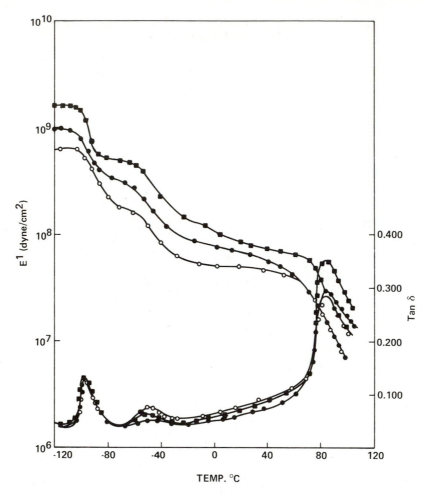

Figure 2. Dynamic mechanical properties of polydimethylsiloxane–polystyrene multiblock copolymer cast from toluene (●), cyclohexane (○), and compression molding (■).

polymer is in the glassy stage. In this region, the thermal energy is insufficient to surmount in the polymer. With increasing temperature, the amplitude of vibrational motions becomes roughly comparable to the potential-energy barriers for segmental motions. At -110 °C, a sharp change in the tan δ was observed related to the glass transition temperature of the PDMS blocks. Furthermore, around -48 °C a smaller peak in tan δ was observed. The change in the tan δ value around -48 °C corresponds to the crystalline melting point of the PDMS block. However, the different sample preparation conditions affect the ease of crystallization of the siloxane moiety (*12–14*). Another sharp change in E' at 80 °C with a maxima in tan δ at 89 °C presumably reflects the response of polystyrene microdomains. This behavior of multiblock copolymer films could be accounted for in terms of a two-phase structure development from microphase separation of the block segments. Furthermore, the value of tan δ in the compression-molded samples was about 1.15 times greater than for the solvent-cast films. This finding shows that more interphase mixing has occurred in the solvent-cast systems than in the molded sample. Other researchers (15) also have found that in poly(dimethylsiloxane−carbonate) block copolymers, the T_g of the polycarbonate phase is shifted to lower temperature and becomes more diffused as the dimethylsiloxane units in the block are increased. A more detailed inspection of the modulus curve of block copolymer films reveals that the value of E' in the rubbery plateau observed in Figure 2 is about 1.5 order of magnitude lower than that measured at low temperatures (when both phases are glassy), indicating that in this range of temperatures the applied load is carried by the rubbery PDMS phase.

Tables III and IV show stress−strain properties measured on multiblock copolymer specimens obtained from solution-cast films and by compression molding. The measurements were performed at 27 °C. Toluene, which is a good solvent for both segments, produces films having higher mechanical properties than those obtained from cyclohexane (characteristic of a plastic rather than a rubbery state). Solution casting from preferential solvents for PDMS is known (1, 2) to improve the rubberlike behavior for ABA and AB polystyrene−PDMS copolymers having a styrene content 50% by weight. Such solvents swell the PDMS phase preferentially and thereby enhance the better dispersion of the polystyrene microdomains that exist, after complete solidification of the material, as a discrete glassy domain in a continuous flexible matrix of PDMS. Compression-molded films show enhanced mechanical properties, as reported in Table IV. Stress−strain behavior of the compression-molded samples exhibits a "knee" in the stress−strain curve. Furthermore, as the amount of polystyrene in the copolymer decreases, the block copolymer becomes increasingly elastomeric, and the stress−strain behavior is rubbery with poor mechanical properties. Furthermore, Morton et

Table III. Mechanical Properties of Multiblock Copolymer Films Cast from Solvent

Solvent	(Me₂SiO) Units in Copolymer	Tensile Strength (psi)	Elongation (%)
Cyclohexane	42.3	1210	175
Cyclohexane	53.0	970	210
Toluene	53.0	1230	210

Table IV. Mechanical Properties of Compression-Molded (BAB)$_n$ Copolymers

$(Me_2SiO)_n$ Units in Copolymer	$\overline{M}_n \times 10^{-3}$ (g/mol)	Tensile Strength (psi)	Elongation (%)	Stress Yield (psi)	Strain Yield (%)
42.3	88.3	1530	180	930	6.5
53.0	188.6	1610	240	810	9.2

al. (16) showed the linear relationship between strength and degree of adhesion in rubber vulcanizates. Poor tensile properties of styrene–siloxane block copolymers are related to the surface free energy of PDMS (24 dynes/cm) compared to that of carbon polymers (32–33 dynes/cm), which leads to poor adhesion.

Figure 3 shows the stress–strain behavior in which the specimen was submitted to two consecutive extension cycles to a strain $\epsilon = 1.2$, which corresponds to approximately half the elongation at break for cyclohexane-cast film. The specimen was relaxed for 10 min between the first and second cycles. As can be seen from Figure 3a, the sample cast from cyclohexane does not exhibit significant stress softening. On the other hand, as shown in Figure 3b, the stress–strain curve measured on compression-molded samples exhibits a well-defined yield at low elongation, indicating plastic deformation caused by the breakdown of continuous polystyrene domains. This phenomenon appears to be similar to the Mullins effect in reinforced rubbers.

Differential scanning calorimetry (DSC) measurements were carried out on block copolymer having 53 wt % dimethylsiloxane units. The glass transition temperatures were chosen as the inflection point of the curves and the melting temperature at the top of the endotherm peak. Before the heating measurement was made, the sample was heated to 100 °C and quenched. The heating curve obtained at 10 °C/min shows a slight shift in baseline in the vicinity of the PDMS glass transition temperature, i.e., −108 °C. Another diffuse peak at around −40 to −45 °C was observed, and it corresponds to the crystalline melting point of PDMS. On the other hand, the cooling curve shows a sharp crystallization endotherm with a maximum at −46 °C. All these features are in good agreement with literature data on pure PDMS (12). Above room temperature, a well-defined shift in baseline was observed around 83 °C (the inflection point of 86°C). The difference between the T_g of the polystyrene block in the copolymer and that of homologous polystyrene homopolymer of the same molecular weight was about 5–6 °C. This behavior of polystyrene–PDMS block copolymer is in sharp contrast with the characteristic behavior of poly-(styrene–butadiene) (17) and poly(styrene–isoprene) (18) diblock

copolymers, where the lowering of the glass transition temperature of the polystyrene block in the copolymer is higher. This behavior of polystyrene–PDMS block copolymers may be considered as
an argument for a complete segregation of the blocks and a two-phase
structure for the copolymers. Such a feature in the polystyrene–
PDMS block copolymers may be explained readily by its high interfacial contact energy owing to the very strong incompatibility of
the two blocks (*19*). The difference observed in the tensile behavior of
styrene–dimethylsiloxane block copolymer film cast from different
solvents made us suspect that differences would exist in their thermomechanical analysis curves (TMA curves are shown in Figure 4 for
block polymer and polystyrene in the form of pellets). An important
difference in behavior is observed between polystyrene and polystyrene–dimethylsiloxane copolymers. Under the applied weight on the
probe, the macroscopic cellular structure is destroyed by an increase
in temperature, and consequently apparent phase-transition results.
In the polystyrene probe, displacement is maximum around 102 °C
with a change in baseline initiated at 60 °C. In block polymers, the
probe displacement shows three phase transitions, without such a
sharp peak maximum at 102 °C as for polystyrene. In the poly(dimeth-
ylsiloxane-*b*-styrene-*b*-dimethylsiloxane) polymer, a well-defined
transition around −110 °C is observed in the vicinity of the second-
order transition temperature of the PDMS segment. Another transition
around −48 °C may be due to the crystallization of the PDMS block.
After this transition, a horizontal plateau is observed followed by a

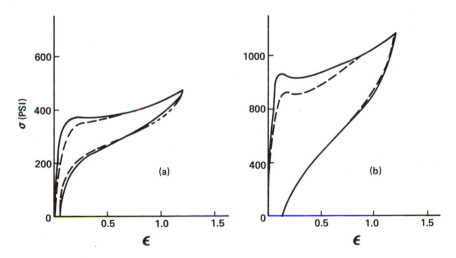

Figure 3. Stress–strain curves for poly(DMS-b-St-b-DMS)$_n$ *copolymers, 53-wt% (Me*$_2$*SiO)*$_n$*. (a) Film cast from cyclohexone; (b) by compression molding.*

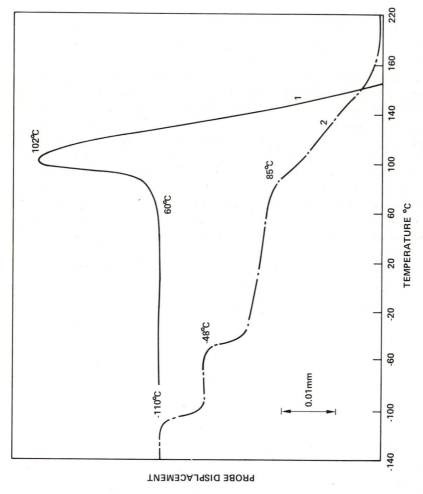

Figure 4. Thermomechanical curves of polymer pellets with a probe load of 5 g. Key: 1, polystyrene; and 2, poly(DMS-b-St-b-DMS) copolymer having 49 wt% $(Me_2SiO)_n$.

change in probe displacement at 85 °C. A shift in the transition to the lower temperature owing to the polystyrene segment in block co-polymers (with respect to polystyrene) may be related to the ease in the deformation of structure due to the siloxane segment. We con-clude from these TMA data that the block polymers may have some interfacial difference regions between polystyrene glassy domains and the PDMS segments that assist in mechanical stress transfer so as to deform the structure at low temperatures as compared to homo-polystyrene.

The transmission electron micrographs of the multiblock copoly-mer are shown in Figure 5a−c. The dark regions are the PDMS, and the light regions are the polystyrene domains. We showed previously (6) that the formation size and shape of domains in poly(dimethylsi-loxane-*b*-styrene-*b*-dimethylsiloxane) copolymers are a function of such molecular properties of the block copolymer as molecular weight distribution of the block and the nature of casting solvents.

When benzene, which is a good solvent for both blocks, was used as solvent, microphase separation with a rodlike structure of PDMS domains in BAB triblock copolymers was seen (6). The use of solvent that selectively solvates either the polystyrene or PDMS block in (BAB)$_n$ copolymers gave films that showed better definition of the microphases. The domains of PDMS cast from methyl ethyl ketone (MEK) solution are cylindrical and dispersed (Figure 5a). This result suggests that MEK is a better solvent for polystyrene than for PDMS, and a clear phase separation exists between these two polymer blocks. Figure 5b shows the micrograph of block copolymer films cast from cyclohexane solution. This block copolymer solution formed a film composed of a dark PDMS matrix with a small, rodlike structure of polystyrene. On the other hand, the phase separation in film cast from toluene solution, Figure 5c, is not as clear as in the film cast from MEK and cyclohexane, which are preferential solvents for polystyrene and polydimethylsiloxane, respectively. Figure 5c shows the PDMS do-mains that are dispersed irregularly in polystyrene domains, indicat-ing considerable mixing of polystyrene and PDMS units in both the domains and continuous matrix. This finding reveals that the mor-phology of these multiblock copolymers is less sensitive to molecular weight, copolymer composition, and casting solvent and is only well defined when films are cast from their preferential solvents.

Conclusions

The study performed on (BAB)$_n$ multiblock copolymers with PDMS end-blocks and with polystyrene hard midblocks indicates that the block copolymer films obtained by different methods influence the stress−strain and dynamic mechanical behavior of the ultimate

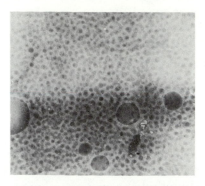

Figure 5a. Electron micrograph of film of multiblock copolymer cast from methyl ethyl ketone.

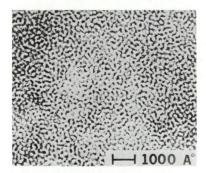

Figure 5b. Electron micrograph of film of multiblock copolymer cast from cyclohexane.

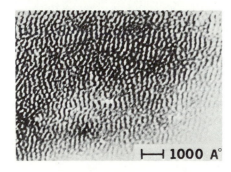

Figure 5c. Electron micrograph of film of multiblock copolymer cast from toluene.

materials. Using a good solvent for polystyrene can result in specimens having mechanical behavior close to that of unfilled vulcanized rubbers. The microdomain structures of these specimens are characterized as polystyrene discrete spheres dispersed in a rubbery PDMS matrix. The specimens obtained by compression molding showed significant stress softening on repeated extensions.

Literature Cited

1. Davies, W. G.; Jones, D. P. *Polym. Prepr. Am. Chem. Soc. Div. Polym. Chem.* **1970,** *11,* 477.
2. Saam, J. C.; Fearon, F. W. G. *Ind. Eng. Chem. Prod. Res. Dev.* **1971,** *10,* 10.
3. Samm, J. C.; Gordon, D. J.; Lindsey, S. *Macromolecules* **1970,** 3, 1.
4. Zilliox, J. G.; Roovers, J. E. L.; Bywater, S. *Macromolecules* **1976,** 8, 573.
5. Greber, G.; Balciunas, A. *Makromol. Chem.* **1964,** 79, 149.
6. Bajaj, P.; Varshney, S. K. *Polymer* **1980,** *21,* 201.
7. Patnode, W.; Wilcock, D. F. *J. Am. Chem. Soc.* **1946,** 68, 358.
8. Scott, N. D. *J. Am. Chem. Soc.* **1936,** 58, 2442.
9. Bajaj, P.; Varshney, S. K.; Misra, A. *J. Polym. Sci., Polym. Chem. Ed.* **1980,** *18,* 295.
10. Dean, J. W. *J. Polym. Sci., Polym. Lett. Ed.* **1970,** 8, 677.
11. Varshney, S. K.; Khanna, D. N. *J. Appl. Polym. Sci.* **1980,** 25, 2501.
12. Lee, C. L.; Johannson, O. K.; Flannigam, O. L.; Hahn, P. *Polym. Prepr. Am. Chem. Soc. Div. Polym. Chem.* **1969,** *10,* 1311.
13. Beatty, C. L.; Karasz, F. E. *J. Polym. Sci., Polym. Phys. Ed.* **1975,** *13,* 971.
14. Sundrarajan, P. R.; Hamer, G. K.; Croucher, M. D. *Macromolecules* **1980,** *13,* 971.
15. Narkis, M.; Tobolsky, A. V. *J. Macromol. Sci., Phys.* **1970,** 4, 877.
16. Morton, M.; Murphy, R. J.; Cheng, T. C. *Polym. Prepr. Am. Chem. Soc. Div. Polym. Chem.* **1974,** *14,* 436.
17. Jean, L.; Leblanc, J. *Appl. Polym. Sci., Chem. Ed.* **1977,** *21,* 2419.
18. Toporowski, P. M.; Roovers, J. E. L. *J. Polym. Sci., Polym. Chem. Ed.* **1976,** *14,* 2233.
19. Samm, J. C.; Gordon Fe ron, F. W. *Polym. Prepr. Am. Chem. Soc. Div. Polym. Chem.* **1970,** *11,* 455.

RECEIVED for review October 14, 1981. ACCEPTED December 30, 1981.

THERMAL METHODS

Differential Scanning Calorimetry of Flexible, Linear Macromolecules

BERNHARD WUNDERLICH and UMESH GAUR

Rensselaer Polytechnic Institute, Department of Chemistry,
Troy, NY 12181

A summary of instrument and application news for differential scanning calorimetry (DSC) and a study of low temperature DSC are presented. From 140 to 300 K, a reasonably critical application range, DSC is shown to be capable of up to 2% precision. This is less than classical calorimetry can provide, but many applications may be served by these measurements. Heat capacities of poly(acrylate) and poly(methacrylate) are presented for the temperature range 220–500 K (specifically, methyl, ethyl, n-butyl, isobutyl, and octadecyl acrylates, and methyl, ethyl, n-butyl, isobutyl, dodecyl, and octadecyl methacrylates). These heat capacities are analyzed in terms of the side-chain heat capacity by comparison with the heat capacities of polyethylene, polypropylene, polybutene-1, polyisobutylene, and polypentene. The latter are taken from our data bank, which contains heat capacities on over 100 different macromolecules and will be the basis of a general addition scheme on heat capacities.

DIFFERENTIAL SCANNING CALORIMETRY (DSC) has become a significant analytical technique over the last 10 years. Because almost any physical or chemical changes occur with a change in enthalpy, all can be followed by calorimetry. Similarly, thermal properties that are expressed through enthalpy, entropy, and Gibbs energy (free enthalpy) can be evaluated by calorimetry using heat capacity and heat of transition measurement from 0 K to the temperature in question. Although calorimetry plays a major role among analytical techniques, it has still not reached its limit. The initial section of this chapter discusses the history of the development of calorimetry and includes a listing of modern DSC apparatus.

0065–2393/83/0203–0195$06.00/0

The main portion of this chapter concentrates on the application of DSC to heat capacity measurements of flexible linear macromolecules, our main research interest. Ultimately, we hope to derive an addition scheme that permits the prediction of heat capacities of macromolecules with the help of a series of tables of group contributions. Differential scanning calorimetry instrumentation is described by using a comparison of different DSC equipment applied to low temperature heat capacity measurements. In addition, heat capacities of poly(acrylate)s and poly(methacrylate)s in the temperature range 220–500 K (methyl, ethyl, n-butyl, isobutyl, and octadecyl acrylates and methyl, ethyl, n-butyl, isobutyl, dodecyl, and octadecyl methacrylates) are presented. These data are used along with literature data on polyacrylates, polymethacrylates, and polyalkenes [polyethylene (PE), polypropylene (PP), polybutene-1 (PBu), polypentene-1 (PPe), polyhexene-1 (PHe), poly(4-methyl-1-pentene) (P4M1P), and polyisobutylene (PIB)] to study the heat capacity contributions due to the side groups. This presents an overview of the utility of DSC using the example of heat capacity of linear macromolecules covering history, instrumentation, data, and data treatment.

History and List of Instruments

Calorimetry has two handicaps that have impeded its application. The first is the lack of perfect insulators. The heat to be measured cannot be contained; it is always in flux, so that loss containment and loss calculations are basic to calorimetry. The second handicap is the lack of a direct heat meter. All calorimetry is done indirectly, either by compensation (e.g., by electrical heating or cooling in case of endotherms or exotherms), or by determination of secondary effects (e.g., the measurement of temperature rise). Differential scanning calorimetry has its roots in twin calorimetry (1, 2) which was developed to minimize the heat loss problem. Next was the development of constant heating rate calorimeters (3) that allowed rapid measurement over a large temperature range without the need of frequent equilibration and loss calibration. The first twin calorimeter operating at constant heating rate was described in 1960 (4). The next step in the development involved invention and commercialization of a modern differential scanning calorimeter for milligram-sized samples based on temperature sensing and electronically regulated heating of the reference and sample (Perkin-Elmer) (5). Currently a variety of additional DSC instruments are available commercially. A number of DSCs involve heating by flux through a controlled leak. The temperature measurement can be done by thermocouple (du Pont) by thermopile (temperature difference) (Mettler), and by resistance thermometer (Heraeus[1]). Additional variations involve the capability to add electrical calibration heat pulses (Netzsch). Finally, there is a DSC based on measurement of heat flux using multiple thermocouple arrangements (Setaram[2]).

All these instruments of DSC are capable of measurement rates of up to 50 K/min and may reach a precision in heat capacity as high as 0.5%. Because of fast heating rates, it is not only possible to measure equilibrium functions of state, but it is also possible to study metastable and unstable states. The latter is of key importance to establish information on thermal, mechanical, and perhaps also electrical history (6).

Instrumentation

We have measured the heat capacity of molten selenium from 500–700 K using the three commercial, widely available DSCs (Mettler TA 2000, du Pont 990, Perkin-Elmer DSC-2) (7). All three DSCs reproduced adiabatic calorimetry data to within 3%. Using a computer coupled DSC, the accuracy of the heat capacity could be improved further to better than 1% (8).

Continuing our comparison of commercial instruments, the subambient accessories for Mettler TA 2000, du Pont 990, and Perkin-Elmer DSC-2 were used for heat capacity measurements of polymers, extending the temperature range of measurements down to liquid nitrogen temperatures.

The du Pont 990 liquid nitrogen accessory is limited in its design. It consists of a small cooler (~150 mL) that is placed over the DSC cell assembly and filled with liquid nitrogen to cool the cell assembly. The cooling of the cell is slow and uncontrolled, precluding precision measurements on cooling. Also, the isotherm at liquid nitrogen temperature is not fully stable.

The Perkin-Elmer liquid nitrogen accessory consists of a tank (~4.5 L) that is filled with liquid nitrogen. This allows for fast, controlled cooling of the sample holders. Cooling rates of as much as 80 K/min are possible. The baseline is quite good. However, isotherms are unstable and show significant drifts, caused by the continuous change of the liquid nitrogen level.

The Mettler TA 2000 is equipped with a sophisticated cooling system for the cell. The furnace is fitted with a heat exchanger for cooling. The liquid nitrogen coolant is stored in a separate tank. An evaporator (heating element) in the liquid nitrogen controls the flow

[1] Heraeus, W. C., GmbH, Postfach 169, 6450 Hanau 1, Federal Republic of Germany.

[2] Setaram, 101-103 rue de Sexe, 69451 Lyon, Cedex 3, France.

of coolant to the heat exchanger. The electric heating and the cooling with liquid nitrogen are separately controlled. The reference and the sample pans are placed on the thin-film sensors. Some difficulties were encountered due to poor contacts between the pans and the sensors. A given scan was found to be reproducible to within $\pm 1\ \mu V$ if the pan was kept in place. However, if the pans were removed and replaced randomly on the sensor, the signal reproducibility was much poorer ($\pm 12\ \mu V$). This problem was avoided by using gold pans that are made up of heavier metal sheet, which resulted in more uniform contact between the pans and the sensor.

The error in heat capacity measurements at liquid nitrogen temperatures using the Perkin-Elmer and the Mettler instruments on Al_2O_3 and PMMA (using benzoic acid as standard are summarized below:

	Perkin-Elmer		Mettler	
Temp (K)	Al_2O_3 (%)	PMMA (%)	Al_2O_3 (%)	PMMA (%)
150	±7	±5	+3	±2
200	±3	±2	±2	±2
250	±1	±2	±2	±2

These data and the general description indicate that low temperature heat capacity measurements are possible with all three instruments with only slightly reduced accuracy, but it is necessary to use considerably more care in avoiding spurious temperature gradients.

Heat capacity measurements were made with a computer coupled Perkin-Elmer DSC-2, fitted with the more reproducible intracooler at approximately 200 K to reach the precision needed for our data bank. This use of mechanical refrigeration limits the low temperature. Details of the instrumentation, calibration, and computations are given in Reference 8.

Results

The acrylic polymers used in this study were secondary standards obtained from Scientific Polymer Products, Inc. The molecular weights provided by the manufacturers are listed in Table I. The manufacturers provided PM-1, PM-2, PM-4i, and PM-18 in granular form. All the other acrylic polymers have T_g values below room temperature. For ease of handling these samples were provided as 40% solution in toluene and were later dried in vacuum at 330–350 K for 24–48 h. Each sample (10–25 mg) was transferred into a hermetically sealed pan for heat capacity measurements. All the measurements were done on heating at 10–20 K/min and Al_2O_3 was used as reference

Table I. Characterization of Polymers

Polymer	Abbreviation	M_w	$\overline{M}_w/\overline{M}_n$
Poly(methyl acrylate)	PA-1	200,000	3.2
Poly(ethyl acrylate)	PA-2	125,000	3.2
Poly(n-butyl acrylate)	PA-4	119,000	3.6
Poly(isobutyl acrylate)	PA-4i	116,000	3.7
Poly(octadecyl acrylate)	PA-18	23,300	1.8
Poly(methyl methacrylate)	PM-1	60,000	1.82
Poly(ethyl methacrylate)	PM-2	340,000	2.7
Poly(n-butyl methacrylate)	PM-4	320,000	4.4
Poly(isobutyl methacrylate)	PM-4i	300,000	2.14
Poly(dodecyl methacrylate)	PM-12	113,000	1.5
Poly(octadecyl methacrylate)	PM-18	671,000	6.9

material. The average data of two to five measurements (within ±1%) are listed in Tables II and III.

Discussion

The newly measured heat capacities of acrylic polymers shown in Tables II and III have been combined with the literature data (mainly at low temperatures) on the same acrylic polymers to derive a set of recommended data for each acrylic polymer (9). These recommended data, which now cover a wider range than the data reported here, have been used to derive the heat capacity contribution of the CH_3 group on the C–C backbone and the contribution of a CH_2 group in the side chain [$(CH_2)CH_3$ to $(CH_2)_{17}CH_3$] below and above the glass transition. Data are given in Tables IV–VII. These contributions have also been derived for polypropylene (10), polybutene (11), polypentene (11), polyhexene (11), and polyisobutylene (11). Also listed in these tables are the corresponding data on polyethylene (12). Heat capacity contributions of the COO– group in polyacrylates and polymethacrylates have also been derived by taking the difference in heat capacity contribution between the acrylic polymer and the corresponding polyalkene. These data below and above the glass transition are listed in Tables VIII and IX.

The discussion of these group contributions is done in stages. First, we look at the theoretical feasibility of an addition scheme for linear macromolecules and then, the possible empirical extensions are analyzed.

A detailed discussion of the heat capacities of linear macromolecules has revealed that, because of the chemical nature of the molecules, the vibrational spectrum can be separated into group and skeletal vibrations (13). Furthermore, the skeletal vibrations are largely intramolecular in nature because of the strong bonding along

Table II. Heat Capacity of Poly(acrylate)s

T(K)	PA-1	PA-2	PA-4	PA-4i	PA-18
220	90.4	114.3	174.7	156.0	—
230	93.3	119.0	217.6	163.0	449.6
240	96.0	123.6	221.5	171.5	476.7
250	98.9	130.2	221.1	197.2	505.0
260	102.1	153.7	223.5	223.9	535.6
270	105.5	173.9	224.9	223.2	570.0
280	110.5	175.2	227.5	226.0	579.7
290	124.2	177.1	230.2	229.0	—
300	153.6	178.9	232.7	232.1	—
310	153.6	180.7	235.5	235.1	—
320	155.1	182.4	238.3	238.1	—
330	156.3	184.5	241.7	241.5	—
340	158.3	187.7	246.0	245.0	694.3
350	160.2	190.2	249.3	248.3	703.5
360	161.3	191.9	252.1	251.6	708.5
370	162.7	193.9	255.6	255.2	718.3
380	163.9	195.8	258.4	258.5	728.4
390	165.1	197.7	261.8	261.2	738.4
400	168.3	200.1	265.4	265.8	747.1
410	170.4	202.3	270.1	273.1	758.9
420	172.6	204.5	273.4	275.1	769.8
430	174.8	206.2	275.7	276.2	771.9
440	175.3	208.6	277.3	277.4	792.8
450	175.6	211.6		284.4	802.4
460	178.5	213.5		278.6	804.5
470	179.8	215.9		287.0	814.7
480	181.1	219.1		293.0	824.4
490	182.9	220.8		295.8	830.9
500	183.8	222.5		300.3	843.2

Note: Heat capacity measurements are given in J mol^{-1}K^{-1}.

the backbone chain of the molecule. Only at temperatures below about 40 K is the influence of the intermolecular skeletal vibrations on the heat capacity dominant.

Thus, a model of linear macromolecules based on this analysis is that of a string of beads. Each bead has the mass of the repeating unit (or single backbone chain atom unit) and is coupled strongly in the chain direction. Each string of beads is, however, only weakly coupled with its neighbor strings. Subtracting the contribution of the group vibrations leaves the heat capacity of an assembly of structureless beads. Because at least all carbon backbone macromolecules have the same bonding between beads and have also similar geometry, their intramolecular heat capacities must be related. This relationship was established (13) using a one-dimensional Debye function D_1 with a θ-temperature proportional to the inverse of the mass of the one carbon backbone bead (M_c). By taking the ratio of M_c to the polyethylene mass (14 g/mol) the universal equation is written by using the polyethylene θ-temperature 540 K

$$C = D_1 [540 (14/M_c)^{1/2}] \tag{1}$$

for the intramolecular skeletal heat capacity.

Table III. Heat Capacity of Poly(methacrylate)s

T(K)	PM-1	PM-2	PM-4	PM-4i	PM-12	PM-18
220					398.1	455.2
230	108.9	132.5	182.7	173.6	467.5	482.9
240	112.8	137.5	192.5	181.5	574.4	514.0
250	116.5	142.3	200.5	188.6	740.0	554.0
260	120.2	147.0	208.8	195.0	525.8	608.2
270	124.5	151.5	218.1	201.6	498.0	
280	128.3	157.7	231.7	207.6	499.3	
290	132.2	162.0	246.3	217.8	504.5	
300	135.9	167.5	259.1	228.5	509.9	
310	140.1	172.9	267.5	240.0	516.2	
320	143.8	179.9	273.6	255.1	523.2	
330	147.7	189.1	277.5	267.4	532.5	720.1
340	151.1	201.8	282.8	278.2	541.7	730.5
350	156.3	215.2	288.1	286.9	549.4	741.1
360	161.4	226.1	294.2	290.4	556.7	749.1
370	167.8	230.0	300.5	293.5	565.2	759.0
380	180.4	233.5	306.1	293.8	572.6	766.8
390	204.3		304.9	295.4	578.1	775.5
400	207.5		311.4	302.5	582.5	789.7
410	209.2		313.7			801.9
420	212.2		318.5			813.3
430	214.8		323.2			825.4
440	217.9		329.1			831.2
450	220.9		337.7			839.5
460	223.6					857.0
470	226.2					873.3
480	228.9					890.1
490	231.4					903.0
500	234.4					913.7

Note: Heat capacity measurements are in J mol^{-1}K^{-1}.

Table IV. Heat Capacity Contribution of CH_3 Group on C–C Backbone Polymers Below the Glass Transition

T(K)	Polyacrylates				Polyalkenes			
	I^a	II^b	III^c	IV^d	V^e	VI^f	VII^g	CH_2
40	2.0				3.8	0.6	2.2	3.2
80	3.2		6.1		5.3	2.8	4.0	7.8
120	5.7	7.1	8.9		9.2	7.1	8.2	10.2
160	8.6	10.6	11.9		12.7	12.2	12.5	13.5
200	13.0	14.6	15.2	9.7	16.5	16.7	16.6	15.6
240	18.2	18.8			19.4			17.9
280					22.0			20.6
320					27.9			23.0
360					32.8			26.5
400					30.6			34.1
440					27.1			42.7

Note: Heat capacity measurements are in J mol^{-1}K^{-1}.
[a] C_p (PM-1) $- C_p$ (PA-1); error ± 0.4–2.6 J mol^{-1}K^{-1}.
[b] C_p (PM-2) $- C_p$ (PA-3); error ± 1.2–2.8 J mol^{-1}K^{-1}.
[c] C_p (PM-4) $- C_p$ (PA-4); error ± 0.6–3.8 J mol^{-1}K^{-1}.
[d] C_p (PM-4i) $- C_p$ (PA-4i); error ± 3.6 J mol^{-1}K^{-1}.
[e] C_p (PP) $- 2C_p$ (PE); error ± 0.2–2.2 J mol^{-1}K^{-1}.
[f] C_p (PIB) $- C_p$ (PE); error ± 0.2–1.2 J mol^{-1}K^{-1}.
[g] [C_p (PIB) $- 2C_p$ (PE)] ÷ 2; error ± 0.2–0.6 J mol^{-1}K^{-1}.

The intermolecular skeletal contribution is not additive, but has to be determined by measurement at low temperature (three-dimensional Debye function). The development of a reliable technique of low temperature heat capacity measurement, preferably to at least 10 K as discussed previously, is thus of key importance. A useful combination of the intramolecular and intermolecular skeletal heat capacities is possible using the Tarasov equation (13).

Based on this analysis it should be possible to develop an addition scheme of heat capacities that covers the intramolecular skeletal vibrations and the group vibrations based on a single atom backbone chain bead. An initial attempt of such an addition scheme showed promising results (14). The temperature range of such simple analysis is estimated to reach from 40 K to the glass transition or the melting transition.

Strict correlation between vibrational frequencies and heat capacities exists only for the heat capacities at constant volume. Heat capacities at constant pressure deviate above 150–200 K increasingly from the heat capacity at constant volume. Within a reasonable temperature range the deviation is, however, proportional to the square of the heat capacity itself with an almost universal constant (15). Therefore, the addition scheme should also apply to heat capacities at constant pressure up to approximately 400–500 K. The early addition

Table V. Heat Capacity Contribution of CH₃ Group on C–C Backbone Polymers Above the Glass Transition

| | Polyacrylates | | | | | Polyalkenes | | | |
T (K)	I^a	II^b	III^c	IV^d	V^e	VI^f	VII^g	VIII^h	CH_2
260						24.0	18.0	21.0	29.2
300			30.1			26.6	21.8	24.2	30.9
340		31.2	36.5	38.8	33.0	29.2	25.5	27.3	32.6
380		37.4	42.8	35.9	40.5	31.7	29.2	30.5	34.4
420	38.2		49.1		47.9	34.3			36.1
460	41.2				55.3	36.9			37.8
500	44.2				62.8	39.5			39.5
540	47.2					42.1			41.3
580						44.7			43.0

Note: Heat capacity measurements are in J mol⁻¹K⁻¹.

[a] C_p (PM-1) – C_p (PA-1); error ± 4.0–4.6 J mol⁻¹K⁻¹.
[b] C_p (PM-2) – C_p (PA-2); error ± 4.4–4.8 J mol⁻¹K⁻¹.
[c] C_p (PM-4) – C_p (PA-4); error ± 5.2–6.4 J mol⁻¹K⁻¹.
[d] C_p (PM-4i) – C_p (PA-4i); error ± 5.6–6.0 J mol⁻¹K⁻¹.
[e] C_p (PM-18) – C_p (PA-18); error ± 14–18.2 J mol⁻¹K⁻¹.
[f] C_p (PP) – 2 C_p (PE); error ± 1.2–2.6 J mol⁻¹K⁻¹.
[g] C_p (PIB) – C_p (PP); error ± 2.0–1.6 J mol⁻¹K⁻¹.
[h] [C_p (PIB) – 2C_p (PE)] ÷ 2; error ± 1.0–1.2 J mol⁻¹K⁻¹.

Table VI. Heat Capacity Contribution of —CH_2 Group in the Side Chain Below the Glass Transition

T (K)	Polyacrylates						Polyalkenes		
	I^a	II^b	III^c	IV^d	V^e	VI^f	VII^g	VIII^h	CH_2
40	5.0					4.7		4.6	3.2
80	10.8	10.6	9.2	11.6	10.2	9.8		9.1	7.8
120	13.0	14.1	12.4	12.0	13.4	13.3		12.9	10.9
160	13.8	18.7	16.7	16.1	17.9	16.6		16.7	13.5
200	16.1	23.3	21.6	20.2	22.3	20.8		21.5	15.6
240	19.4			24.0	26.6	26.4	20.1		17.9
280	23.2			27.8	31.0				20.6
320				31.6					23.0

Note: Heat capacity measurements are in $J\ mol^{-1}K^{-1}$.

[a] C_p (PM-1) − C_p (PM Acid); error ± 0.4–3.2 $J\ mol^{-1}K^{-1}$.

[b] C_p (PA-2) − C_p (PA-1); error ± 1.4–1.8 $J\ mol^{-1}K^{-1}$.

[c] $[C_p$ (PA-4) − C_p (PA-1)] ÷ 3; error ± 1.4–3.0 $J\ mol^{-1}K^{-1}$.

[d] $[C_p$ (PM-2) − C_p (PM-1)]; error ± 1.2–3.6 $J\ mol^{-1}K^{-1}$.

[e] $[C_p$ (PM-4) − C_p (PM-1)] ÷ 3; error ± 1.6–4.4 $J\ mol^{-1}K^{-1}$.

[f] $[C_p$ (PBu) − C_p (PP)]; error ± 0.2–1.6 $J\ mol^{-1}K^{-1}$.

[g] $[C_p$ (PPe) − C_p (PP)] ÷ 2; error ± 1.0 $J\ mol^{-1}K^{-1}$.

[h] $[C_p$ (PHe) − C_p (PP)] ÷ 3; error ± 0.2–0.8 $J\ mol^{-1}K^{-1}$.

Table VII. Heat Capacity Contribution of $-CH_2$ Group in the Side Chain Above the Glass Transition

T (K)	Polyacrylates								Polyalkenes			
	I^a	II^b	III^c	IV^d	V^e	VI^f	VII^g	VIII^h	IX^i	X^j	XI^k	CH_2
260									27.3	25.1	27.2	20.2
300	26.9	27.1							28.7	28.0		30.9
340	29.1	29.3	31.4						30.0	30.9		32.6
380	31.2	31.6	33.2	30.4	33.1	31.8	33.5	33.8	31.3	33.7		34.5
420	33.4	33.8	35.1		36.4	33.7		35.5	32.6	36.6		36.1
460	35.5		36.9					37.6	33.9	39.4		37.8
500	37.7		38.8					39.7	35.3			39.5
540									36.6			41.3
580									37.9			43.0

Note: Heat capacity measurements are in J mol⁻¹K⁻¹.

^{a} C_p (PA-2) − C_p (PA-1); error ± 3.6−4.4 J mol⁻¹K⁻¹.
^{b} [C_p (PA-4) − C_p (PA-1)] ÷ 3; error ± 1.6−1.8 J mol⁻¹K⁻¹.
^{c} [C_p (PA-18) − C_p (PA-1)] ÷ 17; error ± 0.4−1.0 J mol⁻¹K⁻¹.
^{d} C_p (PM-2) − C_p (PM-1); error ± 4.5 J mol⁻¹K⁻¹
^{e} [C_p (PM-4) − C_p (PM-1)] ÷ 3; error ± 2.0−2.2 J mol⁻¹K⁻¹
^{f} [C_p (PM-6) − C_p (PM-1)] ÷ 5; error ± 1.4−1.6 J mol⁻¹K⁻¹
^{g} [C_p (PM-12) − C_p (PM-1)] ÷ 11; error ± 1.0 J mol⁻¹K⁻¹
^{h} [C_p (PM-18) − C_p (PM-1)] ÷ 17; error ± 1.0 J mol⁻¹K⁻¹
^{i} C_p (PBu) − C_p (PP); error ± 2.2−2.4 J mol⁻¹K⁻¹
^{j} [C_p (PPe) − C_p (PP)] ÷ 2; error ± 1.4−2.0 J mol⁻¹K⁻¹
^{k} [C_p (PHe) − C_p (PP)] ÷ 3; error ± 1.0 J mol⁻¹K⁻¹.

Table VIII. Heat Capacity Contribution of COO− in the Side Chain Below the Glass Transition

T (K)	I[a]	II[b]	III[c]	IV[d]	V[e]	VI[f]	VII[g]
40	6.2	10.6					
80	15.6	23.6	23.2		23.4		17.6
120	22.2	28.0	29.5	26.7	28.0		25.1
160	28.3	29.9	33.6	31.1	33.7		31.8
200	33.8	33.2	36.9	34.7	37.1		38.9
240	39.0		40.2	37.2		54.2	
280	44.3						

Note: Heat capacity measurements are in J mol^{-1}K^{-1}.
[a] C_p (PM-Acid) − C_p (PP); error ± 0.2−2.2 J mol^{-1}K^{-1}.
[b] C_p (PM-1) − C_p (P1B); error ± 0.4−2.0 J mol^{-1}K^{-1}.
[c] C_p (PA-1) − C_p (PP); error ± 0.8−2.0 J mol^{-1}K^{-1}.
[d] C_p (PA-2) − C_p (PBu); error ± 1.4−2.4 J mol^{-1}K^{-1}.
[e] C_p (PA-4) − C_p (PHe); error ± 1.4−3.0 J mol^{-1}K^{-1}.
[f] C_p (PA-4i) − C_p (P4M1P); error ± 3.4 J mol^{-1}K^{-1}.
[g] Heat capacity of main chain COO− from Reference 21.

scheme has borne out this analysis (14). As the analysis of the currently developed set of recommended data is completed (9−12), an expanded and improved set of tables will be presented.

In the meantime, two empirical extensions of the addition scheme are attempted. The first deals with liquids (16) instead of solids. In the liquid state, heat capacities are not only caused by vibrations, but have considerable potential energy contributions. Reasonable additivity can be established as long as the increase in heat capacity at the glass transition temperature was normal. The data of Tables V, VII, and IX reaffirm this finding.

The second extension of the addition scheme is tested in Tables IV−IX. Here we try empirically to establish the heat capacity contributions of side-chain groups (disregarding the changes in the nonadditive contributions to the intramolecular skeletal heat capacity). Such

Table IX. Heat Capacity Contribution of COO− in the Side Chain Above the Glass Transition

T (K)	I[a]	II[b]	III[c]	IV[d]
260			60.6	56.2
300			61.9	
340			63.2	
380	73.5	64.6	64.4	
420		65.0	65.7	
460		65.4	67.0	
500		65.9	68.3	

Note: Heat capacity measurements are in J mol^{-1}K^{-1}.
[a] C_p (PM-1) − C_p (PIB); error ± 4.0 J mol^{-1}K^{-1}.
[b] C_p (PA-1) − C_p (PP); error ± 3.2−3.6 J mol^{-1}K^{-1}.
[c] C_p (PA-2) − C_p (PBu); error ± 3.4−4.4 J mol^{-1}K^{-1}.
[d] C_p (PA-4) − C_p (PHe); error ± 4.4 J mol^{-1}K^{-1}.

an approach should be successful for long side chains, which again approach the case of isolated chains, but is less successful for short side chains where the backbone bead change in mass on substitution of a side chain is improperly accounted for.

Before discussing Tables IV—IX it must be remarked that the error limits of the various table entries vary as given in the footnotes to the tables. The error limits are estimated using a 2% error in the heat capacities of the parent data. Thus, the error becomes much larger if the difference in heat capacity needed for the given group is much smaller than the measured heat capacities of the parent polymers.

Taking these error limits into account, one finds that the CH_3 group connected to the carbon backbone substituted for a hydrogen (i.e., inserting a CH_2 between a $C-H$ bond) shows contributions to the heat capacity that are not far from those of CH_2 groups, including skeletal vibrations. The cause of the various deviations is not obvious at present and needs more study in light of the fully developed addition scheme. The liquid CH_3 group data fit a general addition scheme better than the data for the glass. In this case all skeletal vibrations can be assumed to be excited giving the reason for the better agreement. The change in heat capacity, upon introduction of additional CH_2 groups into the side chain, is listed in Tables VI and VII and shows even closer adherence to additivity with the liquid data approaching experimental accuracy. Again, agreement in the liquid state is better than in the glassy state. The heat capacity contributions due to $COO-$ listed in Tables VIII and IX also show that the data are additive within the experimental error limits. Their deviation from the heat capacity contributing ester groups in the main chain also seems small.

Conclusions

Differential scanning calorimetry is one of the basic analysis techniques. With modern instrumentation, accuracies close to classical calorimetry can be reached. Low temperature operation to about 150 K is possible with only slightly reduced precision. The new data on polyacrylate and polymethacrylate heat capacities show that the prior established heat capacity addition scheme of macromolecular solids, which is based on an analysis of the vibrational spectrum, can probably be empirically extended to side group contributions and also to the liquid state. On the basic example of heat capacities, the usefulness and importance of DSC is thus illustrated.

Acknowledgments

This is a publication from our Advanced Thermal Analysis Laboratory. Major financial support for this work was given by the National Science Foundation, Polymers Program DMR 78-15279. The low temperature data were derived with the help of an instrument loan by the Mettler Instrument Corporation (Princeton, NJ).

Literature Cited

1. Joule, J. P. *Mem. Manchester Lit. Phil. Soc.* **1845**, *2*, 559.
2. Pfaundler, L. *Sitzungsber. Akad. Wiss. Wien, Math.–Naturwiss. Kl., Abt. 1* **1896**, *59*, 145.
3. Sykes, C. *Proc. R. Soc. London, Ser. A* **1935**, *148*, 422.
4. Mueller, F. H.; Martin, H. *Kolloid Z.* **1960**, *172*, 97.
5. Watson, E. S.; O'Neill, M. J.; Justin, J.; Brenner, N. *Anal. Chem.* **1964**, *36*, 1233.
6. Wunderlich, B. In "Thermal Analysis in Polymer Characterization"; Turi, E., Ed.; Heyden: New York, 1981.
7. Mehta, A.; Bopp, R. C.; Gaur, U.; Wunderlich, B. *J. Therm. Anal.* **1978**, *13*, 197.
8. Gaur, U.; Mehta, A.; Wunderlich, B. *J. Therm. Anal.* **1978**, *13*, 71.
9. Gaur, U.; Wunderlich, B. B.; Wunderlich, B. *J. Phys. Chem. Ref. Data*, to be published.
10. Gaur, U.; Wunderlich, B. *J. Phys. Chem. Ref. Data*, to be published.
11. Gaur, U.; Wunderlich, B. B.; Wunderlich, B. *J. Phys. Chem. Ref. Data*, to be published.
12. Gaur, U.; Wunderlich, B. *J. Phys. Chem. Ref. Data* **1981**, *10*.
13. Wunderlich, B.; Baur, H. *Adv. Polym. Sci.* **1970**, *7*, 151.
14. Wunderlich, B.; Jones, L. *J. Macromol. Sci., Phys.* **1969**, *3*, 67.
15. Nernst, W.; Lindemann, F. A. *Z. Electrochem.* **1911**, *17*, 817.
16. Gaur, U.; Wunderlich, B. *Polym. Prepr., Am. Chem. Soc., Div. Polym. Chem.* **1979**, *20*, 429.

RECEIVED for review October 14, 1981. ACCEPTED July 28, 1982.

Thermogravimetry Applied to Polymer Degradation Kinetics

BRIAN DICKENS and JOSEPH H. FLYNN[1]

National Bureau of Standards, Polymer Science Division, Washington, DC 20234

The kinetics of polymer degradations (and oxidations) may be represented in the simple general form $d\alpha/dt = f(\alpha)Ae^{-E/RT}$, where α is the extent of reaction, and A and E are Arrhenius parameters. The various attempts to represent $f(\alpha)$ in a simple way are discussed, with the conclusion that none is satisfactory for polymer degradation studies. Therefore, four methods of thermogravimetry have been devised and implemented to avoid any need to model $f(\alpha)$. Three of these methods give values for the activation energy, E, and through it shed some light on the dominant contributors to the kinetic form. The fourth method can be used in favorable cases to examine the importance of competing or successive reactions in the degradation mechanism. The methods are (1) factor-jump thermogravimetry, a series of isothermals requiring only a single sample; (2) isoconversional diagnostic plots, a variable heating rate method applied to a series of samples; (3) analysis of the initial stage of reaction, a variable heating rate method requiring only one sample; and (4) variable heating rate analysis, applied to several samples to examine any change in component reactions in $f(\alpha)$.

T HERMAL METHODS OF ANALYSIS find wide use for three reasons: phase changes can be studied via the heats of transition; temperature affects the rate of reaction of most chemical processes; and complex molecules can be studied from fragments produced by pyrolysis. This chapter is concerned with the use of thermogravimetry, where the sample is heated and weighed continuously, to study polymer degradations. Thermogravimetry is, in principle, a simple technique that

---

[1] To whom correspondence should be sent.

provides kinetic information on the degradation, oxidation, evaporation, or sublimation of samples in any condensed form. The material is presented here in the form of an overview. For more complete treatments, *see* References 1 and 2 and other references cited in this chapter.

Kinetic Analysis of Thermogravimetric Data

The mathematical model that is evoked to describe the kinetics of a system undergoing chemical change is usually expressed in the form

$$d\alpha/dt = f(\alpha) \, k(T) \tag{1}$$

where the rate of change of the conversion or fraction reacted, α, with respect to time, t, is equated to separable functions of α and the absolute temperature, T. Equation 1 represents adequately the kinetics of many reacting systems in the gaseous phase and also applies to some reactions occurring in homogeneous solutions. However, even for homogeneous systems, if the overall kinetic process involves several elementary steps such as opposing, consecutive, parallel, or chain reactions, then Equation 1 is probably not sufficient to describe the rate of reaction. These complexities can be taken into account formally by the addition of a term, $g(\alpha,T)$ (3), to Equation 1 to obtain:

$$d\alpha/dt = f(\alpha) \, k(T) \, g(\alpha,T) \tag{2}$$

where $g(\alpha,T)$ includes all conversion–temperature cross-terms. Although this case may appear to be tractable, a single reaction coordinate, α, is insufficient in many cases such as when the amount of residue or char depends on previous thermal treatment. The usual approach to fitting Equation 2 is to employ specific analytical expressions that usually were obtained from models based on physical and chemical evidence, intuition, and/or experience.

Heterogeneous kinetics pose additional problems and complications over homogeneous kinetics because processes such as diffusion, sorption, evaporation, and chemical reactions among separate phases are involved. If due care is not taken, rates may depend on physical factors such as contact areas and the movement of species through fixed matrices or highly viscous fluids. Therefore, another term must be added to Equation 2 to obtain:

$$d\alpha/dt = f(\alpha) \, k(T) \, g(\alpha,T) \, h(X_i,Y_j,\ldots) \tag{3}$$

where the term, $h(X_i,Y_j,\ldots)$ is an analytical expression of the functional relationships between the chemical rate and all other rate-affecting factors and the cross-terms among the factors themselves as well as with temperature and conversion. The factors, $X, Y, \ldots$, represent the effects of such variables as pressure, gas flow rate, gas composition, physical and geometrical properties of the sample, catalytic impurities, labile groups introduced during preparation, and mechanical stresses.

Many factors that affect the rate of weight loss may change in a noncontrollable manner during the course of an experiment. For example, the temperature in the interior of a large specimen is affected significantly by endothermic or exothermic reactions; the pressure of a gaseous species can be different internally and externally depending on its rate of production or consumption and removal. Such effects set up potential gradients in the systems, and the resultant fluxes affect the overall reaction rate. When the rate is affected or limited by such diffusional processes, it is extremely difficult to model the kinetics adequately. In practice, these complications can be ameliorated considerably by decreasing the driving potential, i.e., by maintaining constant intensities of external factors such as pressure and atmospheric composition, or by slowing down the reaction rate so that internal gradients become less significant.

Some rate-affecting factors are not controlled easily. Standardization of the physical and geometrical properties of the specimens so as to render $h(X_i, Y_j, \ldots)$ in Equation 3 a constant term that may be ignored in the kinetic treatment often is extremely difficult. Control of chemical factors is equally important. Trace amounts of additives and residual catalysts, solvents, and monomers must either be removed or their effects on the kinetics must be taken into account.

The procedure most often followed is to keep the factors just mentioned at controlled levels to incorporate their effects into the expressions for $f(\alpha)$ and $k(T)$. In that case, $g(\alpha, T) h(X_i, Y_j, \ldots) = 1$, and Equation 3 reduces to Equation 1, because only mathematical models for temperature and conversion will be necessary.

The Conversion Function, f(α)

The conversion function, $f(\alpha)$, in general is extremely complicated. In attempting to characterize it, isothermal experiments must be used to separate out the effects of temperature change. A survey of the literature on the isothermal weight-loss kinetics of polymers emphasizes the complexity of the kinetics of these systems. Plots of weight-loss rate vs. conversion fraction do not follow simple reaction orders; also, their overall shape usually changes with temperature. Thus, a particular conversion function is valid only for a limited range of experimental conditions. Because polymer degradations are often chain reactions, $f(\alpha)$ represents the net result of a series of elementary steps. Each elementary step has its own activation energy, which makes each such step respond differently to temperature change. A good example is the existence of a ceiling temperature, above which degradation takes place and below which the material polymerizes.

The traditional radical chain model for the mechanism of vinyl polymer degradation illustrates many of these complexities. This model involves three types of processes: (1) initiation reactions by which radicals are formed, (2) chain propagation reactions that produce scission of the polymer backbone, and (3) termination reactions that remove chain-propagating radicals from the system.

Initiation may result from labile linkages already in the polymer. If such initiation is followed by a depropagation (unzipping) to low molecular weight fragments, then the rate of volatilization will decrease continuously as the reaction proceeds. On the other hand, if the chain breaks come about more or less randomly, either because of random initiation or more usually because of β-scission of radicals formed on the polymer backbone by hydrogen abstraction (transfer), most early fragments will be too large to volatilize, but the rate will increase with conversion as greater numbers of smaller fragments are formed. This process produces an apparent autocatalytic rate of weight loss. If new labile linkages such as unsaturated end groups are formed from transfer and disproportionation reactions, the rate of initiation will increase. Thus, in only a few cases of polymer degradation, such as the thermal degradation of polytetrafluoroethylene (which often is used as an illustrative example), do the weight-loss kinetics conform reasonably closely to simple n^{th} order kinetic models.

Temperature Dependence: The Arrhenius Equation

The Arrhenius equation is the model used almost universally to express the temperature dependence of the rate of reaction, viz.,

$$k(T) = A \exp\left[-E(RT)^{-1}\right] \tag{4}$$

In Equation 4, R is the gas constant and T is the absolute temperature. The energy of activation (E) and the pre-exponential factor (A) are parameters determined by fitting experimental rate data.

The application of the Arrhenius equation to condensed phase kinetics received considerable recent adverse criticism because often values for E change with respect to temperature and conversion and show poor agreement when compared with values obtained using different techniques or under different conditions. Therefore, it is pertinent to give some justification for its use. Much of this criticism results from misconceptions of the theoretical basis for the equation and its role as a mathematical function for fitting and modeling data.

Most models for a kinetic step postulate an energy barrier between the initial state (reactants) and the final state (products). They assume that only a small fraction of the species has sufficient energy to cross this barrier. This assumption is the case whether the model is based on statistical or molecular concepts. The relationship between the energy distribution of the reacting species and temperature is represented by a Boltzmann exponential function, $A \exp(-BT^{-1})$. (Many models predict some sort of mild polynomial functionality with temperature for A, but, in practice, even slight uncertainties in the exponential parameter, E, overwhelm and render futile efforts to measure a temperature dependence for A.) Thus, one finds Arrhenius-type temperature dependence when collision, transition-state, and other models are applied not only to homogeneous chemical processes but also to many physical processes such as the viscous flow of liquids and diffusion in solids.

However, the basic problem in the application of the Arrhenius equation to condensed phase polymer systems is the complexity of the kinetics. A heterogeneous system often is composed of at least several elementary processes, each with its own set of Arrhenius parameters. Whereas in homogeneous systems, these processes can be condensed into a small number of discrete prototype elementary processes, in many heterogeneous systems, such as polymer degradation reactions, the combination of all these processes may approach a continuum. Because of this fact, the more successful approaches to modeling the kinetics of polymer degradation reactions have been statistical in nature, assuming average values for the values of the parameters of the rate constants for an array of similar processes. The radical chain depolymerization models for vinyl polymer degradation fit the description just given. There, whole series of reactions involving homologous species are given an identical rate constant irrespective of molecular size. As a result, those models can usually predict only the general characteristics of isothermal rate curves.

The parameters of the temperature function, $k(T)$, in Equation 1 must, of course, be determined from experiments involving different temperatures. The parameter most amenable to precise determination is E, the activation energy. Methods exist (*see* later discussion) to determine E without knowledge of $f(\alpha)$. The parameter E is useful in that it gives some indication of change in the rate-determining processes during the degradation and, being an energy, can be related to processes, such as diffusion or breaking of a particular type of bond, found in model systems. Although $f(\alpha)$ need not be specified to determine E, a model for $f(\alpha)$ must be used to decompose the activation energy of the overall process into the values of the component steps. The overall value of E is applicable only to the same range of experimental conditions that $f(\alpha)$ applies to, but the individual E values for the component reactions have more universal application.

The other, less temperature-dependent factor in Equation 4, the preexponential factor, A, also may be determined along with the activation energy. For condensed phase kinetics, however, it is impossible to uncouple A and $f(\alpha)$ experimentally, and because $f(\alpha)$ may contain physical and geometrical properties of the reacting species as well as other factors, it is impossible to model it satisfactorily so as to separate it analytically from the preexponential factor.

Kinetic Forms Used in Polymer Thermogravimetry

The simplest and most widely used (and usually incorrect) model for $f(\alpha)$ is taken from homogeneous chemical kinetics, viz.,

$$f(\alpha) = (1 - \alpha)^n \tag{5}$$

where n is the apparent order of reaction. Substituting for $f(\alpha)$ and $k(T)$ in Equation 1 gives

$$d\alpha/dt = (1 - \alpha)^n A e^{-E/RT} \tag{6}$$

In cases where the sample is heated continuously so that the variation, β, of temperature with time is $\beta = dT/dt$, we have

$$d\alpha/dt = \beta \, d\alpha/dT = (1 - \alpha)^n A e^{-E/RT} \tag{7}$$

or, after taking logarithms,

$$\ln (d\alpha/dt) = n \ln (1 - \alpha) + \ln A - E/RT \tag{8}$$

Differential Methods. The simplest method of analysis is to extend the method of van't Hoff (4) to the nonisothermal case, i.e., solve Equation 8 for $\ln A$, n, and E/R from three sets of data points for the rate, fraction reacted, and temperature. Letort (5) extended van't Hoff's method for the isothermal case to any number of data points by plotting $\ln (d\alpha/dt)$ vs. $\ln (1 - \alpha)$ to obtain a slope of n and an intercept of $\ln k$. Freeman and Carroll (6) made a similar extension for thermogravimetric data at constant rate of heating by expressing Equation 8 as a difference equation to remove the quantity $\ln A$, viz.,

$$\Delta \ln (d\alpha/dt) = n \, \Delta \ln (1 - \alpha) - (E/R) \, \Delta(1/T) \tag{9}$$

and dividing by $\Delta(1/T)$ to obtain values for n and E/R from the slope and intercept of a plot of

$$\Delta \ln (d\alpha/dt) \, [\Delta(1/T)]^{-1} \text{ vs. } \Delta \ln (1 - \alpha) \, [\Delta(1/T)]^{-1}$$

This procedure has become the most popular method for the analysis of nonisothermal data. The determination of the order, n, is formally explicit, and gross changes in n and E with changing α may be seen. Unfortunately, Equation 4, on which this method is based, does not apply to the complete range of α in polymer degradations. For example, activation energies determined from Freeman–Carroll plots for the degradation of poly(methyl α-phenylacrylate) compared poorly with those determined from isothermal experiments (7). The isothermal kinetics suggest (7) that a random scission-type model in which the rate goes through a maximum is more appropriate than Equation 5 for this system.

Many differential methods make use of the point of inflection of the thermogravimetric curve. If the derivative of Equation 6 with respect to time is set equal to zero, we can factor out the quotient $E/(nR)$, i.e.,

$$E/(nR) = T_m{}^2 (d\alpha/dt)_m / \beta_m (1 - \alpha)_m \qquad (10)$$

where the subscript, m, refers to the value at the maximum. To obtain E, we must determine or estimate n. We stress that the values of n and E are dependent on the model chosen for the kinetics. Methods in which E and n are determined in this way have limited applicability to polymer degradation kinetics. No test can be made of the validity of the kinetic parameters obtained.

Initial kinetic parameters are significant both in establishing the mechanism and in investigating the slow aging of polymeric materials. Methods at continuous rates of temperature change have an advantage over isothermal methods in the determination of initial rates as they minimize the problems of establishing the true time origin, $t = 0$, during the warmup time. Such methods of analysis, based on a single experiment, can be adapted usefully to the determination of initial Arrhenius parameters. Equation 7 may be differentiated with respect to α (remembering that $T^2 d\alpha/dT = -d\alpha/d(1/T)$ and that T is a function of α because this method is nonisothermal) to give (8)

$$\frac{d}{d\alpha}\left[T^2 \frac{d\alpha}{dT} \right] = \frac{E}{R} + 2T + \frac{T^2}{f(\alpha)} \frac{df(\alpha)}{d\alpha} \frac{d\alpha}{dT}$$

which simplifies to

$$\frac{d}{d\alpha}\left[T^2 \frac{d\alpha}{dT} \right] = \frac{E}{R} + 2T \quad (\alpha << 1, \beta = \text{constant}) \qquad (11)$$

Similar approximations of the integrated rate equation give, at low conversion (9−11),

$$\ln (1 - \alpha)/T^2$$
$$\text{or } (1 - \alpha)/T^2 \cong -E/RT + \text{constant} \quad (\alpha \ll 1, \beta = \text{constant}) \quad (12)$$

Figure 1 shows thermogravimetric curves for $-d\alpha/d(1/T)$ vs. α for linear polyethylene (8). The initial activation energy is calculated from the slope, $-d\alpha/d(1/T)$, to be ~60 kcal/mol for a sample that was preheated in a vacuum at 200 °C for 1 h. A sample with no pretreatment yielded $E \cong 30$ kcal/mol over the first 2% conversion range. This value corresponds to the latent heat of vaporization of hydrocarbons but is also near the activation energy for oxidation of hydrocarbons.

Integral Methods. When $f(\alpha)$ and $k(T)$ are considered to be separable and $k(T)$ is represented by the Arrhenius equation, Equation 4, then Equation 1 may be integrated to obtain

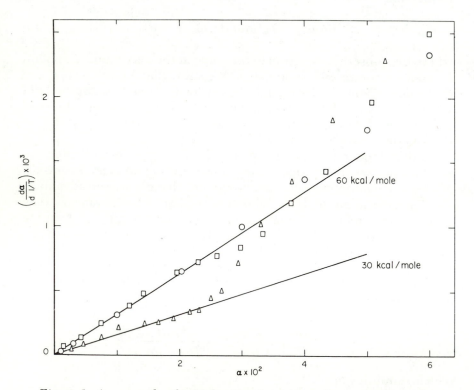

Figure 1. An example of initial activation energy determination: $-d\alpha/d(1/T)$ vs. α (10 mg of polyethylene at 1.8°/min in vacuum). Key: $\triangle$, untreated sample; $\square$, preheated in vacuum at 200 °C for 1 h; and $\bigcirc$, a theoretical case.

$$F(\alpha) = \int_0^\alpha d\alpha/f(\alpha) = (A'\beta') \int_0^T T^a \exp{(-E/RT)}dT \qquad (13)$$

The parameter T^a is included in this equation to represent both the temperature dependence of the preexponential factor, A, (if the preexponential factor is considered to be temperature dependent) and any temperature dependence of the heating rate, β.

If a is zero, the integration of the right side of Equation 13 involves the exponential integral

$$\int_{-x}^\infty (e^x/x)dx, \text{ where } x = -E/RT$$

This integral has been evaluated for various values of x and approximated in many ways (12–18). If a in Equation 13 equals two, then the temperature integral is integrated easily in a closed form. If A is suspected of being temperature dependent, or one wishes to avoid approximating the integral, then a temperature-dependent heating rate may be used to provide a total exponent of 2 for T. In practice, however, this step is not necessary because the term exp $(-E/RT)$ dominates the results, and, because of experimental uncertainties in rate and temperature measurements, any of the many approximations of the temperature integral will usually suffice.

A large number of methods of kinetic analysis were developed to use the integrated form of Equation 4, i.e.,

$$F(\alpha) = \int_0^\infty \frac{d\alpha}{(1-\alpha)^n} = \frac{(1-\alpha)^{1-n} - 1}{1-n} \quad (n \neq 1 \text{ or } 0)$$

$$= \ln{(1-\alpha)} \qquad\qquad (n = 1) \qquad (14)$$

but, as with differential methods, the methods based on this equation are generally inappropriate for the complex kinetic systems found in polymer degradation.

The inability of both differential methods based on Equation 6 and integral methods based on Equation 14 to determine reliable kinetic parameters has been demonstrated by many investigators (12–14, 19, 20). Simple order of reaction type kinetics cannot represent a maximum in reaction rate at other than 0% reaction, and fitting such a process with a combination of order of reaction kinetics and Arrhenius-type temperature dependence requires a large effect from the Arrhenius equation to force the order of reaction kinetic behavior into the right pattern. The essential problem is illustrated in Figure 2 where residual fraction, $1 - \alpha$, is plotted against temperature for three calculated curves: Curve 1 for $n = 1$ in Equation 7 and $E = 80,000$ cal/mol, and Curves 2 and 3 for typical polymer degradation reactions

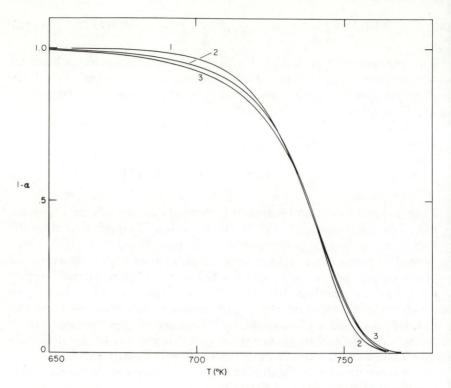

Figure 2. Fraction weight-loss vs. T. *Key: 1,* n = *1,* E = *80,000 cal/mol;*
2, maximum rate at α = *0.5,* E = *40,000 cal/mol; and 3, maximum rate at*
α = *0.5,* E = *35,000 cal/mol.*

in which the rate goes through a maximum at 50% conversion, and E =
40,000 and 35,000 cal/mol, respectively. The closeness of fit of these
three curves shows why investigators who use methods based on the
assumption of an order, n, for analyzing polymeric systems may end
up with absurdly large calculated activation energies.

Efforts have been made to fit more realistic models for $f(\alpha)$ to thermogravimetric data for degrading polymeric systems. Ozawa (21) took the
classical random depolymerization equation of Simha and Wall (22)
and constructed families of master curves for various values of L, the
length of the smallest chain of carbon atoms that does not evaporate. A
weight-loss curve may be compared with such a family of curves to
determine a best value for L. However, few systems, if any, follow
simple random depolymerization kinetics. Also, it is difficult to decide
on the appropriate form of $f(\alpha)$.

Thermogravimetric Methods for Determining Activation Energies

Methods were developed at the National Bureau of Standards for handling two possible ways of taking temperature-dependent data: experiments that consist of a series of isothermals, and other experiments in which the temperature is increased continuously. The general aim was to determine activation energies without the need to define the function $f(\alpha)$. Another method of processing the data gives some insight into the processes that $f(\alpha)$ is supposed to represent. These methods will now be described.

Factor-Jump Thermogravimetry, a Series of Isothermals

In the factor-jump method, a polymer specimen is subjected to a series of temperature isothermals while the temperature and its weight are recorded continuously. The rates of weight loss and the temperatures for adjacent isothermals are extrapolated to halfway between the isothermals in terms of time or in terms of the associated parameter, extent of reaction. The activation energy then is estimated from the Arrhenius equation, Equation 4, as

$$E = \frac{RT_1T_2}{\Delta T} \ln \frac{r_2}{r_1} \qquad (15)$$

where the r and T variables are extrapolated rates and temperatures from two adjacent plateaus and $\Delta T = T_2 - T_1$. Because both rates and both temperatures are estimated at the same extent of reaction, the term containing the extent of reaction and other temperature-independent factors cancel out (*see* Reference 23).

The strong points of the method are that activation energies are determined using only one specimen (whereas in multisample techniques one must assume that thermal histories are unimportant), that an activation energy is provided for roughly every 5% of conversion, that the experiment is conducted over a narrow range of rates of weight loss so the concentrations of reactants within the specimen and the products above the specimen are roughly constant, and that the quantities used to calculate the activation energy are obtained from a small (6–10 °C) temperature range so that the Arrhenius equation is probably valid and the preexponential factor is undoubtedly independent of temperature over this small interval. On the other hand, the initial activation energy cannot be determined because the first determination is made between the first and second isothermals, usually at ~5% weight loss. Also, the method is not computationally robust. It requires that the weight–time trend be fitted to a polynomial that then is differentiated and extrapolated ~15% beyond the range of data. The loss of volatiles during the degradation of many polymers results in the bursting of bubbles in the sample. In a derivative-calculating method such as this one, only slight perturbations in the sample weight are needed to mitigate the successful calculation of the derivative. This point is made because wild values can have enormous effects (proportional to the square of their wildness) on the least-squares curve fitting, which tries to minimize the sum of squared deviations.

The factor-jump method is automated (24–28) and computer controlled in our implementation at NBS (29). Such automation is highly desirable in that conditions must be changed often (every 10 min) and equilibration times allowed (~3 min) before the measurements are continued. The computer program determines the activation energy during the experiment so that visual monitoring of the process as well as computerized feedback is possible. The automation provides a method of measuring the sample weight and also provides control of sample temperature, flow rate of N_2 and O_2 over the sample, and the pressure in the sample chamber.

The general scheme is given in Figures 3 and 4. All modules except the furnace are commercially available. The computer sends

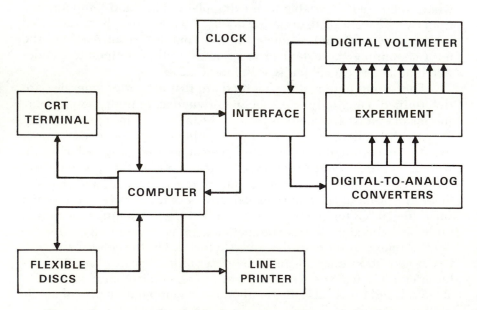

Figure 3. A generalized, block-diagram outline of the factor-jump thermogravimetry apparatus.

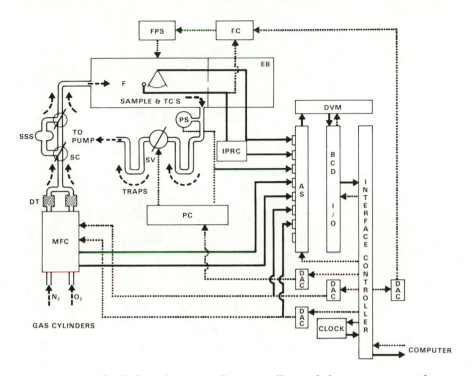

Figure 4. The linkage between the controller and the experiment. Abbreviations used: AS, analog scanner, BCD I/O, digital input/output; DAC, digital to analog converter; DT, drying tubes; DVM, digital voltmeter; EB, electrobalance; F, furnace; FC, furnace controller; FPS, furnace power supply; IPRC, ice point reference cell; MFC, mass flow controller; PC, pressure controller; PS, pressure sensor; SC, stopcock; SSS, saturated salt solution; SV, servo-driven valve; and TC, thermocouple.

commands to the interface that generates voltages used to specify the temperature, pressure, and flow rates around the sample as well as reading voltages generated by the apparatus. The programs and data are stored on a disk and the progress of the experiment is displayed on the cathode ray tube computer terminal.

Furnace

The furnace supplied by the manufacturer of the thermobalance was replaced by a rapid-response furnace (Figure 5) that uses many strands of bare nichrome wire strung between circular forms to heat the environment around the specimen. For applications where a flowing gas stream is used, an additional heater heats the gas stream throughout its entire cross-section, thus avoiding the common practice of heating the stream only at its edge. To include the gas stream heater, we had to reverse

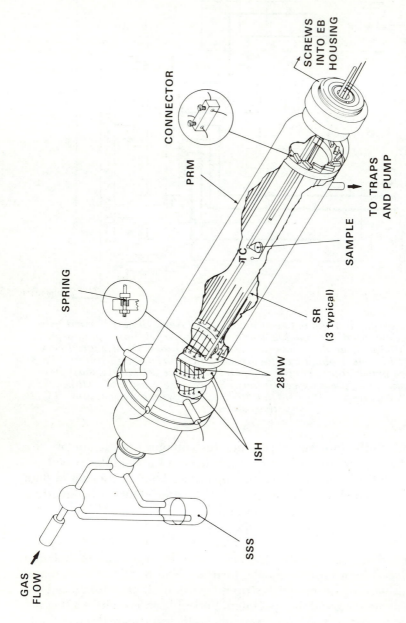

Figure 5. A rapid response furnace featuring in-stream heaters for the gas stream (flow is left to right). Abbreviations used: 28 NW, 28 gauge nichrome wire; EB, electrobalance; ISH, in-stream heater; PRM, Pyrex reaction manifold; SR, support rod; SSS, saturated salt solution.

the direction of gas flow, which now enters on the side of the sample remote from the balance housing and exits before the balance housing through a tube in the borosilicate reaction manifold. The balance is a horizontal beam type.

The thermocouple in the apparatus is type-E, which gives a large ($\sim$80 μV/°C) change of electromotive force (emf) with temperature in the range of interest and has a reasonably linear response. Type-K is not particularly suitable for our application; it has roughly half the temperature coefficient of type-E and suffers from nonlinearity because of an order–disorder transition (*see* Reference 30 and references therein).

It can be shown (26) that the factor-jump method allows an appreciable offset between the real temperature of the sample and the value measured by the thermocouple. The fractional error in the activation energy due to this thermal offset, T_0, is given (26) approximately by $2T_0/T$, which for $T \sim 500$ K and an error of 1% in E allows a temperature offset of 2.5 °C. For a more generous error of 1 kcal in an activation energy of 30 kcal/mol, the tolerable temperature offset at 500 K is 8.3 °C.

This surprisingly large offset may be viewed in two ways: (1) the thermocouple need not be inconveniently close to the sample, and (2) appreciable furnace inhomogeneity can be tolerated. We placed the thermocouple as near to the sample cup as is convenient (<2 mm away), and, at our maximum inhomogeneity of 3° cm^{-1}, expect a maximum temperature offset of <1.5° between the thermocouple and the middle of the sample cup, which would introduce maximum errors of $\sim$0.6% into the calculated values of E.

Sample Considerations

The sample is contained in a quartz spoon on the end of a quartz rod, with the thermocouple centered under the spoon. The spoon bowl is made in the form of a right cylinder, about 4 mm i.d. Typically, 15–30 mg of sample are spread evenly over the floor of the spoon. Samples usually are preconditioned under program control for 10 min at the temperature of the first isothermal so that the temperature stabilizes and the sample attains a modest rate of weight loss via the degradative process of interest.

Computer Program

The philosophy used in designing the computer program to operate the apparatus was to use the computer as a source of active control, so that it can assess the course of the experiment during the experiment and take appropriate action (Figure 6). The computer program is written almost entirely in FORTRAN, is modular, and can be adapted

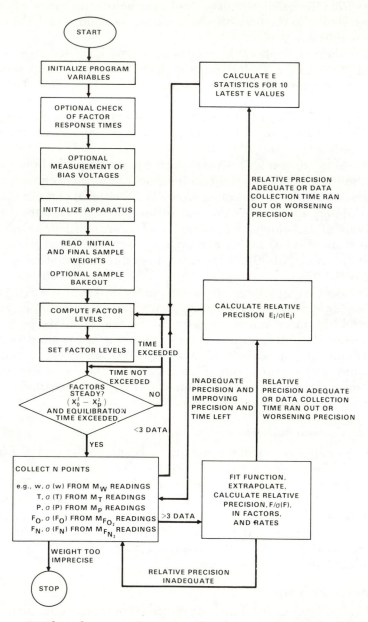

Figure 6. Flow diagram of program for automated control of factor-jump thermogravimetry.

readily to other configurations of apparatus and to other computers. It contains regression and extrapolation routines so that it can process raw data to obtain final derived quantities and can calculate standard deviations in the derived quantities to estimate the attained precision. The computer program also sets up the initial conditions, gives the operator the opportunity to make further changes after displaying all his currently chosen values, and later changes some user chosen options if necessary.

Weight-Loss Polynomial

Extended tests (27) showed that the small portions of the weight loss–time curve used here are best fitted by a second-degree polynomial, and that such a polynomial can be extrapolated successfully beyond the range of its determination to give sufficiently viable estimates of its derivative, the rate of weight loss with respect to time.

Temperature Polynomial

Although the temperature–time behavior is ideally linear and independent of time, in practice there is a complex damped oscillation approach to equilibrium. Therefore, we fit its trend with time with a first-degree polynomial. Generally, higher coefficients than first are not statistically significant and confer instability rather than improve the precision.

Polynomials for Other Factors

The remaining factors, the pressure and flow rates, usually are held at constant levels during the experiment. Thus, it is appropriate to fit them with a zero-order polynomial, that is, to a constant value.

Data Collection

The requirements of data collection are that isothermals should be fairly short so that the weight–time curve can be fitted well with a second-degree polynomial, the behavior of the sample should become steady under a new set of conditions before data collection is started, and that the program should fit polynomials to the weight and factor (temperature, pressure, gas flows) trends with time during the sample equilibration time. A compromise must be struck between many closely spaced data and long computational time-outs and, on the other hand, few widely spaced data with imprecise but rapid polynomial fits.

Applications

The method provides estimates, E_i, of the activation energy for whatever combination of processes is rate limiting. The distribution of

E_i with time and extent of reaction is examined for regions in which the same process is dominant. From applying the propagation of error procedure with no correlations between the errors in T and r, the program also provides estimates of $\sigma_{E_i}^2$, the variance of E_i. These quantities, E_i and σ_{E_i}, are used in the statistical data handling programs to compare the distribution of $(\bar{E} - E_i)/\sigma_{E_i}$ with the normal distribution to seek out aberrant values of E_i; here $\bar{E}$ is the average value of E_i. Satisfactory values of E_i and σ_{E_i} are used to provide reliable estimates of the average activation energy and its standard deviation for regions of conversion, α, where the same rate-limiting process applies. The technique and the methods of statistical evaluation were used to study the thermal degradation and oxidation of polystyrene (30) and in other investigations now in preparation for publication. The statistical techniques in particular are described in detail with examples in Reference 29.

Dynamic Heating Rate Methods

These methods fall into two groups, one where the activation energy of the initial degradation process is estimated from one experiment, and the other where comparison of a series of experiments is necessary before an activation energy can be estimated. The conditions of a particular experiment are kept constant, except for the temperature, which is increased continuously, and the processing of the data is straightforward. Therefore, in contrast to our application of the factor-jump method, there is no need of automation to constantly change experimental conditions and little need of feedback to change the course of the experiment.

A Thermogravimetric Instrument for Isothermal and Constant Heating Rate Experiments

A thermogravimetric instrument for determining weight change of polymers heated at programmed rates was designed and built at the National Bureau of Standards and utilizes over 30 years of experiences in this area. It may be operated under vacuum or mild flow of gas and is equipped with accessories for measurement and control of flow rate, pressure, and vacuum. Cold traps and outlets for the collection of volatiles are available. Weight is determined by a glass-enclosed electrobalance of 10^{-6} g sensitivity so that weight change in a 10-mg polymer specimen may be determined with good precision. Experiments in flowing gas are carried out at slow rates of volatilization, so the rate of gas flow sufficient to remove volatile products can be kept slow enough so as not to disturb the weighing process. Long-term base-line stability of the electrobalance permits experiments at very slow heating rates (9°/day) so that the range of temperature for the investigation of the degradation of a typical polymer is over 200°.

The key to the pyrolysis potential of the apparatus is its furnace. It consists of two concentric 1.6-mm thick steel tubes, the outer one, 38.0 mm in diameter, and the inner, 31.5 mm. The tubes are positioned vertically by two 12.7-mm thick pyrophyllite disks at the top and bottom. Narrow vertical slits in the stainless steel tubes are staggered between the inner and outer tubes so that the heating wires do not transmit direct radiation to the pyrex envelope, which may be cooled with liquid nitrogen. The assembly is designed to permit volatiles to diffuse rapidly from the hot zone; most are entrapped on the sides of the envelope.

The No. 20 nichrome heating wires are wound tightly between the pyrophyllite disks, which are mounted on the steel tubes with six 1.6-mm Inconel springs to maintain the tension on the heating wires as they expand or contract. The furnace attains 300 °C in <2 min, 400 °C in 3 min, 500 °C in 5 min, and 600 °C in 12 min, starting at ambient temperatures. Thus, the system combines the fast response necessary for quick establishment of isothermal temperatures with sufficient thermal inertia for maintenance of isothermal temperatures within 1 °C.

The temperature of the type E thermocouple seated 1−2 mm below the sample bucket was calibrated by comparison with thermocouples embedded in specimens of polyethylene and polytetrafluoroethylene over the temperature range 200−600 °C at a heating rate of 2°/min. A temperature correction of about 3° was established to be necessary for the fixed thermocouple.

Isoconversional Diagnostic Plots. In this method, we assume that the Arrhenius equation is valid and that the same form, $f(\alpha)$, of the dependence of rate on extent of reaction, α, is maintained throughout. Experiments are conducted at a series of heating rates, β. Given the conditions just mentioned, we can formulate $(21, 31)$ for a given extent of reaction, α, the equation

$$\Delta \ln \beta = 1.05(E/R) \, \Delta(1/T) \tag{16}$$

The parameter E then can be estimated from the slope of a plot of $\ln \beta$ vs. $1/T$ at a given extent of reaction from runs at several different heating rates. This method can be extended to give E for several extents of reaction, typically every 10%. As does the factor-jump method, this method obviates the need to know $f(\alpha)$. Also, it covers a wide temperature range and is computationally robust because weight-loss curves are utilized directly. The consistency of the activation energy throughout the range of reaction and throughout a wide temperature range is shown visually by parallel lines on the plot of ln β vs. $1/T$.

The method has some disadvantages in that the effects of errors are cumulative, that is, early errors are passed on to later results. A serious restriction is that one must use more than one specimen and must assume that the weight-loss kinetics are independent of the differing thermal histories and of any difference in the physical character of the specimens. This restriction means that, regardless of the thermal history of the specimen, the temperature, and the effects of degradation, the dependence of rate of weight loss on the extent of reaction is assumed to be the same in all specimens for a given extent of reaction. Also, as is the case with all dynamic heating experiments, the possibility exists that the temperature will not be able to equilibrate at the faster heating rates.

If the slopes of the isoconversional lines of a plot of ln β vs. $1/T$ are equal, then a single global activation energy exists for the ranges of α and T covered by the experiment. However, for polymer weight-loss kinetics such cases are the exception. A more typical example is illustrated in Figure 7 in which $\log \beta$ is plotted against $1/T$ for the degrada-

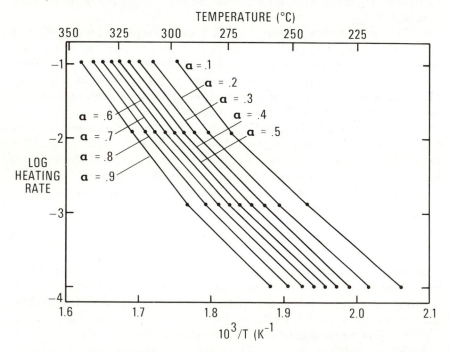

Figure 7. Logarithm of heating rate vs. reciprocal absolute T for polymethyl methacrylate in nitrogen: curves for $\alpha = 0.1, 0.2, 0.3, 0.4, 0.5, 0.6, 0.7, 0.8,$ and 0.9.

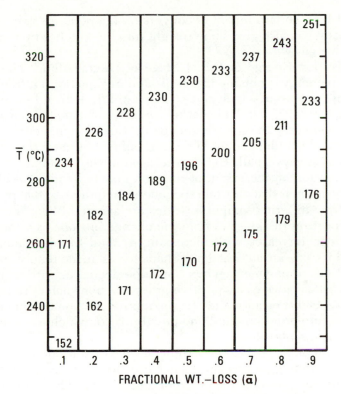

Figure 8. Grid of activation energy kilojoules per mole (kJ/mol) for polymethyl methacrylate in nitrogen as a function of temperature (T) and conversion (α) from data in Figure 7.

tion of polymethyl methacrylate in nitrogen. The range of heating rates from 10^{-4} to 10^{-1} K/s permits an observation of the kinetics over the temperature range from 210 to 340 °C (32). The curvature of the isoconversionals indicates that the activation energy is changing continuously. However, the wide range of temperatures that these reactions cover gives one the precision necessary to determine the activation energies from the slope between each successive pair of points on each isoconversional. From these values a grid is set up as in Figure 8, where the activation energies (in kJ/mol) are given as a function of α and an average temperature, T. Comparison of activation energies parallel to the ordinate shows that E increases with increasing temperature at constant conversion, but parallel to the abscissa, the activation energy appears to be nearly independent of conversion at constant temperature.

This method of treatment of weight-loss data at constant heating rate is not only a good diagnostic to see if the activation energy is constant for a certain region of conversion-temperature space but, in some cases, can also give some insight into the kinetics from the way that E changes with α and T.

Activation Energy of Initial Stage of Degradation. The initial activation energy is obtained, as indicated in Equation 11, from a plot of $T^2 d\alpha/dT$ against α, which gives a slope of $E/R + 2T$ at low extent of reaction ($\alpha \leq 0.05$). This calculation can be done simply from a single thermogravimetric trace of weight against time or temperature (8, 33). This method is independent of the form of $f(\alpha)$ because the extent of reaction is always small. Hence, the form of the dependence of the rate of reaction on the extent of reaction may be ignored safely. Another advantage is that only one specimen is required, so that problems arising from differing sample histories are avoided. We can estimate other aspects of the kinetics from the change in slope as α increases. If the slope increases with increasing α, then the dependence on weight loss has an autocatalytic character as in randomly initiated degradations of polymers. When the slope decreases with increasing α, the kinetics behave as a positive order reaction such as in polymer degradations that occur mainly by unzipping. A slope independent of α implies zero-order kinetics, one possibility for which is evaporation of preformed molecules.

One problem is that the extent of conversion enters into both quantities plotted, so one must be more than usually consistent in picking the beginning of the polymer degradation. The initial rates of weight loss are especially sensitive to volatile contaminants including monomer, solvent, plasticizer, and other small molecules that can be lost without degrading the polymer. Some pretreatment trials will be necessary if the effect of these materials is to be minimized. However, if one wishes to study the kinetics of their loss (usually at some elevated temperature) then this technique may be appropriate. The same reasoning applies to chemical effects from residual catalysts, antioxidants, stabilizers, and labile linkages introduced during synthesis and storage, all of which affect the initial kinetics.

When the kinetics of the main-chain scission process are to be investigated, the effects just noted may mask the early phases of the reaction by the erratic evolution of low molecular weight species as is demonstrated by the curve for the untreated sample in Figure 1. Such complications may be avoided in two ways: (1) the experiments may be conducted at very slow heating rates, when such low E processes will often uncouple as separate low temperature events, and (2) an isothermal pretreatment in vacuum at a temperature ~ 100 °C below that at which the reaction of interest begins to take place at a measurable rate will often remove these complications. The removal of part of the specimen in the faster heating rate experiments in this case requires the calculation of a new fraction reacted, α', based on an initial weight from which the weight lost during the pretreatment has been subtracted. An error in the estimate of this latter quantity results in a curve whose slope will nearly parallel the correct slope but will not extrapolate through the origin.

Varied Heating Rate Analysis.   This method is designed to shed light on some of the elementary reactions making up the composite $f(\alpha)$. It consists of examining the shift of peaks in plots of $d\alpha/dT$ vs. T as the heating rate is varied. Theoretical considerations (*34*) show that peaks corresponding to independent reactions with widely differing activation energies can be resolved at some attainable heating rate. For competitive reactions, one peak or the other will dominate as the heating rate is changed.

The weight-loss process can sometimes be resolved into simple or complex cases, and complicated cases can be resolved into competing and independent reactions. The necessary conditions include a wide range of temperatures and heating rates, so the complete range of the degradation reaction is examined. However, the difference in activation energies must be rather large (~ 20 kcal/mol) to obtain good resolution because the largest range obtainable in heating rates is $\sim 10^4$. Reactions occurring at roughly comparable rates in the same temperature range, and which are realistic alternatives to one another, are often not as different as 20 kcal/mol in activation energy. Experiments at very slow heating rates may provide some indication of which reaction will dominate at service conditions, but the duration of the investigation becomes very long. Nonetheless, the varied heating rate method was applied over a range of heating rates from 10^{-4} K/s ($\sim 9^{\circ}$/day) to 10^{-1} K/s (6°/min) to several polymer degradation reactions (*32*). In some cases, derivative peaks were uncoupled by reduction of heating rate. In others, dispersion effects in peak amplitudes of rate curves were used to diagnose changes in mechanism, and the experimental temperature range was extended significantly to lower temperatures (*1, 2, 32, 35, 36*).

Conclusions

Three of the techniques described in this chapter have a common element—the determination of activation energy for polymer degradation reactions while avoiding the pitfall of applying the more common techniques of kinetic analysis to thermogravimetry—that of having to invoke mathematical models for the elusive $f(\alpha)$, the functionality of the rate with respect to fractional completion. The fourth technique sheds some light on the component reactions that enter into $f(\alpha)$. The techniques were used successfully in several investigations, most of which are cited here.

Literature Cited

1. Flynn, J. H. In "Aspects of Degradation and Stabilization of Polymers"; Jellinek, H. H. G., Ed.; Elsevier: Amsterdam, 1978; Chap. 12, pp. 573–603.
2. Flynn, J. H. In "Thermal Analysis in Polymer Characterization"; Turi, E., Ed.; Heyden: London, 1981; pp. 43–59.
3. Flynn, J. H. In "Thermal Analysis"; Schwenker, R. F., Jr.; Garn, P. D., Eds.; Academic: New York, 1969; Vol. 2, p. 1111.
4. van't Hoff, J. H. In "Studies in Chemical Dynamics" (revised by Cohen, E.; translated by Ewan, T.); Chem. Pub. Co.: Easton, 1896.
5. Letort, M. *J. Chim. Phys. Phys.–Chim. Biol.* **1937**, *34*, 206.
6. Freeman, E. S.; Carroll, B. *J. Phys. Chem.* **1958**, *62*, 394.
7. Cameron, G. G.; Kerr, G. P. *Vac. Microbalance Tech.* **1970**, *7*, 27.
8. Flynn, J. H.; Wall, L. A. *Polym. Lett.* **1967**, *5*, 191.
9. Reich, L. *J. Polym. Sci.* **1965**, *B3*, 231.
10. Coats, A. N.; Redfern, J. P. *J. Polym. Sci.* **1965**, *B3*, 917.
11. Piloyan, G. O.; Ryabchikov, I. D.; Novikova, O. S. *Nature (London)* **1966**, *212*, 1229.
12. Flynn, J. H.; Wall, L. A. *J. Res. Natl. Bur. Stand. (U.S.)* **1966**, *70A*, 487.
13. Cameron, G. G.; Fortune, J. D. *Eur. Polym. J.* **1968**, *4*, 333.
14. MacCallum, J. R.; Tanner, J. *Eur. Polym. J.* **1970**, *6*, 1033.
15. Šesták, J. *Thermochim. Acta* **1971**, *3*, 150.
16. Beigen, J. R.; Czanderna, A. W. *J. Therm. Anal.* **1972**, *4*, 39.
17. Simon, J. *J. Therm. Anal.* **1973**, *5*, 271.
18. Škvará, F.; Šesták, J. *Chem. Listy* **1974**, *68*, 225.
19. Audebert, R.; Aubineau, C. *Eur. Polym. J.* **1970**, *6*, 965.
20. Ozawa, T. *J. Therm. Anal.* **1975**, *7*, 601.
21. Ozawa, T. *Bull. Chem. Soc. Jpn* **1965**, *38*, 1881.
22. Simha, R.; Wall, L. A. *J. Phys. Chem.* **1952**, *56*, 707.
23. Flynn, J. H.; Dickens, B. *Thermochim. Acta* **1976**, *15*, 1.
24. Dickens, B.; Pummer, W. J.; Flynn, J. H. In "Analytical Pyrolysis"; Jones, C. E. R.; Cramers, C. A., Eds.; Elsevier: Amsterdam, 1977; pp. 383–91.
25. Dickens, B.; Flynn, J. H. In "Proceedings of the First European Symposium on Thermal Analysis"; Dollimore, D., Ed.; Heyden: London, 1976; pp. 15–18.
26. Dickens, B. *Thermochim. Acta* **1979**, *29*, 41.
27. Dickens, B. *Thermochim. Acta* **1979**, *29*, 87.
28. Dickens, B. *Thermochim. Acta* **1979**, *29*, 57.
29. Dickens, B. "User's Manual for Factor-Jump Thermogravimetry Apparatus and Associated Computer Programs, Including a General Plotting Program," National Bureau of Standards Internal Report NBSIR 80–2102, 1981.
30. Dickens, B. *Polym. Degrad. Stabil.* **1980**, *2*, 249.
31. Flynn, J. H.; Wall, L. A. *Polym. Lett.* **1966**, *4*, 323.
32. Flynn, J. H. In "Thermal Methods in Polymer Analysis"; Shalaby, S. W., Ed.; Franklin Inst.: Philadelphia, 1978; pp. 163–86.
33. Flynn, J. H.; Pummer, W. J.; Smith, L. E. *Polym. Prepr., Am. Chem. Soc., Div. Polym. Chem.* **1977**, *18* (1), 757.
34. Flynn, J. H. *Thermochim. Acta* **1980**, *37*, 225.
35. Flynn, J. H.; Dickens, B. In "Durability of Macromolecular Materials"; Eby, R. K., Ed.; ACS SYMPOSIUM SERIES No. 95, ACS: Washington, D.C., 1979; pp. 97–115.
36. Flynn, J. H. *Polym. Eng. Sci.* **1980**, *20*, 675.

RECEIVED for review October 14, 1981. ACCEPTED December 28, 1981.

Use of the Single Dynamic Temperature Scan Method in Differential Scanning Calorimetry for Quantitative Reaction Kinetics

THEODORE PROVDER, RICHARD M. HOLSWORTH,
THOMAS H. GRENTZER, and SALLY A. KLINE

Glidden Coatings and Resins, Division of SCM Corporation,
Strongsville, OH 44136

Various reaction kinetics methodologies for differential scanning calorimetry (DSC) are reported and compared experimentally, utilizing the thermal decomposition of calcium oxalate monohydrate and of 2,2'-azobis-(isobutyronitrile) (AIBN) as model systems, to demonstrate the reliability, efficiency, analysis speed, and simplicity of the single dynamic temperature scan method of Borchardt and Daniels. The DSC results for the activation energy of decomposition of AIBN by the Kissinger, Ozawa, and ASTM-E698 methods (all of which utilize numerous thermograms generated at various heating rates) were found to be more than 25% lower than the kinetic results reported by classical techniques (volumetric titration, UV spectroscopy, etc.), yet the single dynamic temperature scan method results were within 2% of the classical results. A comparison of isothermal reaction kinetics with single dynamic temperature scan reaction kinetics for the reaction of phenylglycidyl ether with 2-ethyl-4-methylimidazole indicated that the isothermal method grossly underestimated the total heat of reaction while providing lower values of activation energy and Arrhenius frequency factor than the single dynamic temperature scan method. The single dynamic temperature scan method also was used to provide quantitative reaction kinetics information for some chemical coatings systems (e.g., powder coatings, gel coat resins, and casket varnish).

0065–2393/83/0203–0233$06.25/0

V ARIOUS METHODS HAVE BEEN REPORTED for determining reaction kinetics by differential scanning colorimetry (DSC). Isothermal reaction kinetics require numerous thermograms over a range of reaction temperatures. The Ozawa (1) and Kissinger (2) methods utilize a number of thermograms generated by the use of different heating rates. Methods for obtaining reaction kinetics information from a single differential scanning calorimetry/differential thermal analysis (DSC/DTA) temperature scan greatly decrease the time necessary for analysis and have been reported previously (3–20). In this study, we compared some of these methodologies and applied one of them to the characterization of the reaction kinetics of model systems and some coatings systems.

Methodologies

Multiple Dynamic Temperature Scans. An equation was reported (2) for obtaining reaction kinetics from DTA by scanning a chemical reaction at different fixed heating rates. By plotting $\ln (\phi/T_p^2)$ vs. $(1/T_p)$, where ϕ is the heating rate (K/s) and T_p is the temperature at the peak of the exotherm (K), the activation energy (E) and Arrhenius frequency factor (A) are determined from the slope and intercept, respectively.

Reed et al. (4) compared reaction kinetics results for the decomposition of benzenediazonium chloride by conventional methods reported previously [the DTA (Borchardt–Daniels) method (3) and the Kissinger approach (2)]. The results indicated that the Kissinger approach produced kinetics results that were 42% lower than those obtained by other methodologies. The underlying errors that are associated with the Kissinger method using DTA analysis also were described (4). Prime (8) showed that the application of Kissinger's method to the DSC cure of epoxy resins was in better agreement with other results (isothermal DSC kinetics and dc conductivity) than the DTA application reported by Reed et al.

Another equation to determine reaction kinetics parameters by multiple dynamic DSC scans was reported by Ozawa (1). By scanning the chemical reaction at different fixed heating rates and plotting log ϕ vs. $(1/T_p)$, the energy of activation and the Arrhenius frequency factor can be determined from the slope and intercept, respectively. Corrections for the measured peak temperature and the calculated energy of activation have been applied to this method by the American Society for Testing and Materials (ASTM). The description of this method and its use can be found in ASTM E698–79 (17).

Single Dynamic Temperature Scan Approach [Prime (7, 8)]. Because the fraction reacted $[F(t,T)]$ is a function of time and temperature in a dynamic DSC experiment, the total derivative $dF(t,T)$ must be expressed as a function of time and temperature.

$$dF(t,T) = \left(\frac{\partial F(t,T)}{\partial t}\right)_T dt + \left(\frac{\partial F}{\partial T}\right)_t dt \qquad (1)$$

Then the rate expression $dF(t,T)/dt$ is given by

$$\frac{dF(t,T)}{dt} = \left(\frac{\partial F(t,T)}{\partial t}\right)_T + \left(\frac{\partial F(t,T)}{\partial T^1}\right)_t \phi \qquad (2)$$

where t = time (s); T = absolute temperature (K); and $\phi = dT/dt$ (K/s).

Utilizing Equation 2 and the Arrhenius relationship for the rate constant, Prime (7, 8) showed that the energy of activation, the Arrhenius frequency factor, and the order of the reaction can be calculated from one dynamic DSC temperature scan from Equations 3–5.

$$\ln\left[\frac{dH(t,T)/dt}{(\Delta H_0)Z}\right] = \ln A + \left\{-1/RT + c \ln[1 - F(t,T)]\right\}E_{\text{calc}} \qquad (3)$$

$$Z = 1 + \left(\frac{E_{\text{est}}\,\Delta T}{RT^2}\right) \qquad (4)$$

$$n \cong CE_{\text{calc}} = \left\{\left[\frac{[1 - F(t,T)]_p}{(dH(t,T)/dt)_p}\right]\left(\frac{\Delta H_0\phi}{RT_p{}^2}\right)\right\}E_{\text{calc}} \qquad (5)$$

where the symbols used in the equations are defined here and are illustrated in Figure 1: $dH(t, T)/dt$ = heat flow value (cal/g-s); ΔH_0 = total heat of reaction (cal/g); A = Arrhenius frequency factor (s^{-1}); R =

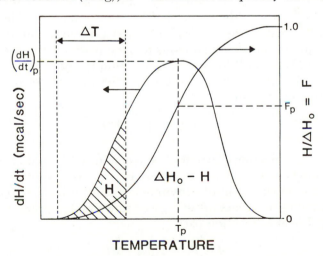

Figure 1. Schematic of an exothermic peak observed by DSC for Prime's calculation.

gas constant (kcal/mol-deg); $F(t,T)$ = fractional extent of reaction $(H/\Delta H_o)$; E_{calc} = calculated energy of activation (kcal/mol); E_{est} = estimated energy of activation (kcal/mol); ΔT = difference between the instantaneous temperature and the initial temperature; $F(t,T)_p$ = fractional extent of reaction at the peak; T_p = temperature at the peak (K); n = reaction order; and $(dH(t,T)/dt)_p$ = heat flow value (cal/g-s) at the peak.

In Prime's method, a value of E_{est} is selected. The left side of Equation 3 is plotted vs. $\{-1/RT + c \ln [1 - F(t,T)]\}$, where E_{calc} is the slope and $\ln A$ is the intercept. The value of E_{calc} then is used in Equations 3 and 4 for E_{est}. This procedure is continued until no difference exists between E_{calc} and E_{est}. Then E_{calc} is used to determine n from Equation 5.

Prime (8) showed that this method provides closer agreement between isothermal and dynamic scan DSC data for epoxy resin/aromatic amine reaction systems.

Single Dynamic Temperature Scan Approach [Borchardt and Daniels (3)]. Kissinger (19), Hill (20), and Simmons and Wendlandt (21) showed that the term $[\partial F(t,T)/\partial T]_t$ in Equation 2 is always zero. Kissinger argues that "Fixing time fixes the position of all the particles in the system." Thus, if time is held constant then F also must be constant. The rate expression then becomes

$$\frac{dF(t,T)}{dt} = \left(\frac{\partial F(t,T)}{\partial t}\right)_T dt \tag{6}$$

The rate expression for the single dynamic temperature scan approach thus is described by the isothermal kinetics equations. The working equations consist of the general n^{th} order rate equation and the Arrhenius equation and are analogous to the equations proposed by Borchardt and Daniels (3) for studying reaction kinetics by DTA. In the present work, autocatalytic reaction kinetics are not considered. The general rate expression is

$$\frac{dF(t,T)}{dt} = k(T) \left[1 - F(t,T)\right]^n \tag{7}$$

where $F(t,T)$ = fractional extent of conversion; $k(T)$ = rate constant for the reaction (s^{-1}); n = reaction order; and T = temperature (K).

The fractional extent of conversion $[F(t,T)]$ is defined as the ratio of the partial heat of reaction, at a given time and temperature, $[H(t,T)]$ to the total heat of reaction (ΔH_o).

$$F(t,T) = H(t,T)/\Delta H_0 \tag{8}$$

The fractional extent of conversion ranges from 0 to 1.0 and is represented in Figure 1.

The rate constant, $k(T)$, can be expressed in terms of observable parameters obtained from the DSC experiment and, subsequently, can be related to the energy of activation (E) and the frequency factor (A) of the Arrhenius equation. Substituting Equation 8 into Equation 7 and solving for the rate constant $k(T)$ in logarithmic form yields the following expression:

$$\ln k(T) = \ln\left\{\left(\frac{dH(t,T)}{dt}\right) \Big/ \frac{[\Delta H_0 - H(t,T)]^n}{\Delta H_0^{n-1}}\right\} \tag{9}$$

where $dH(t,T)/dt$ = heat flow into and out of the sample (cal/g-s); $H(t,T)$ = partial heat of reaction (cal/g); ΔH_0 = total heat of reaction (cal/g); and n = reaction order.

All of the quantities on the right side of Equation 9 are observable parameters except for the reaction order, n. Substituting the Arrhenius expression shown in Equation 10 into Equation 9 yields the working Equation 11.

$$\ln k(T) = \ln A - \frac{E}{RT} \tag{10}$$

where A = Arrhenius frequency factor (s^{-1}); E = activation energy (kcal/mol); and R = gas constant (kcal/mol-deg).

$$\ln A - \frac{E}{RT} = \ln\left\{\left(\frac{dH(t,T)}{dt}\right) \Big/ \frac{[\Delta H_0 - H(t,T)]^n}{\Delta H_0^{n-1}}\right\} \tag{11}$$

The isothermal fractional extent of conversion as a function of time (t) or temperature (T) is obtained by integrating the rate expression to yield the following expressions:

$$F(t,T) = 1 - [(n - 1)k(T)t + 1]^{1/1-n} \tag{12}$$

For the case where $n = 1$

$$F(t,T) = 1 - e^{-k(T)t} \tag{13}$$

In this chapter the basic methodology of Borchardt and Daniels, described earlier, will be used.

Experimental

The du Pont 990 thermal analyzer programmer/recorder and the 910 DSC cell module were used to obtain the experimental results. The sample atmosphere was either nitrogen or compressed air at a flow rate of 50 mL/min. The sample weight was restricted to the 0.5-mg range. Hermetically sealed sample pans were used for reactions exhibiting a weight loss.

Data Analysis Methods

Two data analysis methods were used in this work to obtain n, E, and A. In the first method, the thermograms were digitized manually and the data analysis method of Willard (6) was used. In this method, a series of Arrhenius plots is generated by varying the order of the reaction, n, in Equation 9, from 0.2 through 3.0 in increments of 0.05. The best value of the reaction order is obtained by selecting the reaction order giving the highest correlation coefficient of the linear least-squares fit of the $\ln k(T)$ vs. $1/T$ curve. The value for the reaction order then is substituted into Equation 11 to obtain the activation energy and the Arrhenius frequency factor. During the course of this work, the data acquisition of the DSC was automated by interfacing the instrument to a microcomputer. Data then are transferred over a serial line to a minicomputer system for storage, analysis, report generation, and plotting. Details of the minicomputer–microcomputer system and its organization and operation were reported elsewhere (22, 23). An alternate data analysis procedure was used with the automated DSC system. Rewriting Equation 11 as

$$\ln\left[\frac{1}{\Delta H_0}\left(\frac{dH(t,T)}{dt}\right)\right] = \ln A - \frac{E}{RT} + n\left\{\ln\left[\frac{\Delta H_0 - H(t,T)}{\Delta H_0}\right]\right\} \quad (14)$$

which is of the form

$$Z = a + bx + cy \quad (15)$$

enables the parameters n, E, and A to be obtained simultaneously by a multiple regression technique. Details of the automated thermal analysis system were reported elsewhere (24). Both of the just described data analysis methods produce calculated results with comparable precision allowing for differences in the number of data points used between the manual and the automated data acquisition methods.

Results and Discussion

Model Systems. DECOMPOSITION OF CALCIUM OXALATE MONOHYDRATE. Figure 2 shows the DSC trace for the decomposition of calcium oxalate monohydrate. The endothermic peak, between 125 and 225 °C, represents the dehydration of calcium oxalate monohydrate. The exothermic peak, between 425 and 525 °C, represents the decomposition of calcium oxalate to calcium carbonate. The decomposition of calcium oxalate to calcium carbonate was found to be independent of heating rate and sample mass by Nair and Ninan (25). The reaction kinetics data for the formation of calcium carbonate from the single dynamic temperature scan DSC method along with the kinetics data obtained by thermogravimetric analysis under the same experimental conditions as that of Nair and Ninan (25) are shown in Table I. The results obtained from these two techniques are in good agreement.

DECOMPOSITION OF AIBN. The thermal decomposition kinetics of 2,2′-azobis(isobutyronitrile) (AIBN) in di-n-butyl phthalate were determined by the single dynamic DSC temperature scan method. The

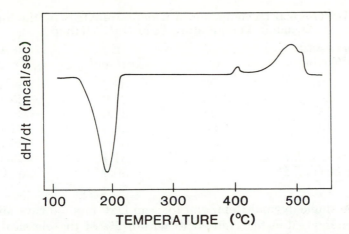

Figure 2. DSC thermogram for the decomposition of calcium oxalate monohydrate. Operating conditions: atmosphere, air; flow rate, 50 mL/ min SCFH; and scan rate, 15 °C/min.

kinetic parameters E, A, and n are reported in Table II at several heating rates along with the literature value determined by Van Hook and Tobolsky (28), who used rate constant data determined by volumetric titration (29–34), UV spectroscopy (35), and other classical techniques to construct a composite Arrhenius plot. Very good agreement between the single dynamic temperature scan DSC results and those obtained by classical methods is shown in Table II. Table III shows thermal decomposition kinetics parameters for AIBN calculated by the multiple dynamic scan approaches of Ozawa, Kissinger, and the modified Ozawa method as described in the ASTM method E−698. For comparison purposes, the data of Van Hook and Tobolsky and the results from the single dynamic temperature scan DSC method also are shown in Table III. The activation energy values obtained by the multiple dynamic scan methods are more than 25% lower than that reported by Van Hook and Tobolsky. On the other

Table I. Thermal Decomposition Kinetics of Calcium Oxalate Monohydrate in a Nitrogen Atmosphere

Parameter	DSC	TGA^a Equation^b	Equation^c
Energy of activation (kcal/mol)	56.6	56.4	57.9
Arrhenius frequency factor	1.76×10^{14}	1.68×10^{14}	5.42×10^{14}
Order of reaction	0.60	0.50	—
Correlation coefficient	0.997	0.996	0.996

^{a} Ref. 25.
^{b} Coats−Redfern equation used to analyze the data, Ref. 26.
^{c} MacCallum−Tanner equation used to analyze the data, Ref. 27.

Table II. Thermal Decomposition Kinetics of AIBN by the Single Dynamic Temperature Scan DSC Method

Scan rate (°C/min)	n	E (kcal/mol)	ln A (s⁻¹)
50	0.95	32.3	36.3
20	1.05	31.9	36.4
15	1.00	32.6	35.0
10	0.90	29.5	33.4
5	0.85	31.1	35.7
Average value	0.95	31.4	35.4
Tobolsky (Ref. 28)	1.00	30.8	35.0

hand, the single dynamic temperature scan DSC method gave an activation energy value that agreed to within 2% of the classical data reported by Van Hook and Tobolsky.

Isothermal DSC Kinetics vs. Single Dynamic Temperature Scan DSC Kinetics.   Isothermal DSC kinetics were studied for the reaction of phenylglycidyl ether and 2-ethyl-4-methylimidazole at a molar ratio of 2:1, respectively. Figures 3 through 5 show the isothermal heat flow response of the empty sample pans and the sample pans containing the reaction mixture at three different isothermal temperature settings. The following procedure must be used for isothermal analysis on the du Pont 990 thermal analyzer. Empty sample pans are placed in the DSC cell; the heat flow response is recorded as a function of time. As shown in Figures 3–5, the sample pans absorb energy and then, after approximately a 1-min time span, equilibrate to a constant heat flow $[dH (t,T)/dT]$ value. The empty sample pan then is removed from the DSC cell and the reactants are placed in the sample pan. The DSC cell is allowed to equilibrate to the isothermal temperature setting.

Table III. Thermal Decomposition Kinetics of AIBN: Comparison of DSC Methods With Classical Data

Source	E (kcal/mol)	ln A (s⁻¹)
Multiple dynamic scan DSC methods[a]		
Ozawa	22.6	28.6
Kissinger	22.2	27.9
ASTM−E698	22.2	21.2
Classical data		
Tobolsky	30.8	35.0
Single dynamic scan DSC method[b]		
	31.4	35.4

[a] The reaction order is assumed to be $n = 1$.
[b] The reaction order was found to be $n = 0.95$.

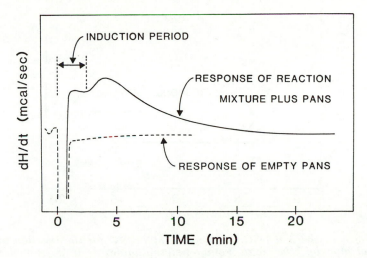

Figure 3. Isothermal DSC thermogram at 90 °C for the reaction of phenyl glycidyl ether with 2-ethyl-4-methylimidazole in a 2:1 molar ratio. Operating conditions: y sensitivity, 0.2 mcal/s-in.; hermetically sealed pan.

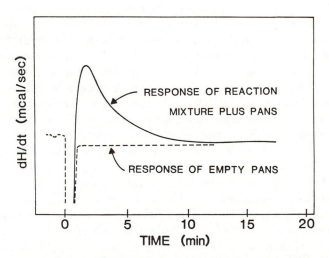

Figure 4. Isothermal DSC thermogram at 100 °C for the reaction of phenyl glycidyl ether with 2-ethyl-4-methylimidazole in a 2:1 molar ratio. Operating conditions: same as in Figure 3.

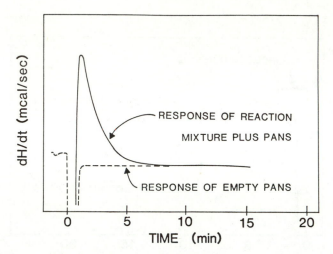

Figure 5. Isothermal DSC thermogram at 110 °C for the reaction of phenyl glycidyl ether with 2-ethyl-4-methylimidazole in a 2:1 molar ratio. Operating conditions: same as in Figure 3.

The sample pans containing the reactants are placed into the DSC cell. The heat flow is monitored as a function of time.

At the 90 °C isothermal temperature setting, as shown in Figure 3, the sample pans absorb energy, equilibrate, the reaction goes through an induction period, and then the reaction begins. For the 100 °C isothermal temperature setting, as shown in Figure 4, again the sample pans absorb energy, equilibrate, but now the reaction begins at the same time that the sample pans are equilibrating. For the 110 °C isothermal temperature setting, as shown in Figure 5, the reaction is taking place before the sample pans have come to thermal equilibrium. These results indicate that part of the reaction exotherm is not being monitored at the higher isothermal temperature setting. This conclusion is confirmed by the calculated heats of reaction at three different isothermal temperature settings, shown in Table IV. As the isothermal temperature setting is increased from 80 to 120 °C, the heat of reaction (ΔH_0) decreases from 105.5 to 69.4 cal/g. The samples that were analyzed isothermally on the DSC then were monitored by DSC at the fixed heating rate of 15 °C/min from 25–200 °C as shown in Figure 6. Next, the samples were cooled quickly to 25 °C and then re-scanned at 15 °C/min to 200 °C as shown in Figure 7. An exothermic peak is observed in Figure 6 in the vicinity of 150 °C. This peak becomes more pronounced as the isothermal temperature setting is lowered; it probably is due to the remaining unreacted material from the isothermal bake schedule. Figure 7 indicates that the reaction is taken to completion by re-scanning the temperature range to 200 °C.

Table IV. Isothermal Heats of Reaction for the Reaction of Phenyl Glycidyl Ether with 2-Ethyl-4-methylimidazole in a 2:1 Molar Ratio

Temperature (°C)	ΔH_0 (cal/g)
80	105.5
100	99.1
120	69.4

These results indicate that if the isothermal reaction temperature is too low, the reaction very possibly will not proceed to completion in the time frame of the experiment because the reacting polymer will vitrify before the reaction is completed at isothermal reaction temperatures below the glass transition temperature of the fully cured polymer. Therefore, the observed heat of reaction will be less than that for the fully cured system. If the isothermal reaction temperature is set too high, the reaction will begin before the sample comes to temperature equilibrium, although the reaction may go to completion. In this case, the observed heat of reaction will be underestimated because the initial part of the reaction exotherm is not being monitored. Widman (36) and Sourour and Kamal (37) provide methods to estimate the amount of heat liberated during the initial heat-up portion of the isothermal DSC experiment. However, these correction methods add complexity, and considerable additional experimental time is required to determine the total heat of reaction. The single dynamic DSC scan

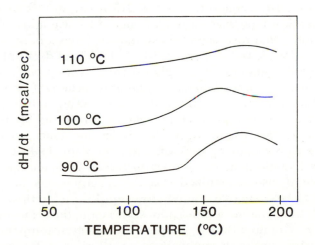

Figure 6. DSC thermograms at 15 °C/min scan rate of reaction mixtures of phenyl glycidyl ether with 2-ethyl-4-methylimidazole after isothermal DSC scans.

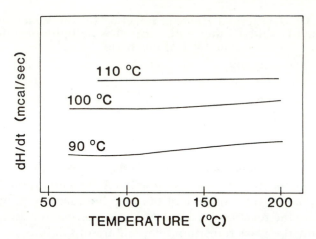

*Figure 7. DSC thermograms of second consecutive dynamic scan at
15 °C/min of reaction mixtures of phenyl glycidyl ether with 2-ethyl-4-
methylimidazole after isothermal DSC scans.*

method provides more accurate values of the total heat of reaction in
much less time than isothermal DSC methods.

Table V contains the results of the single and multiple dynamic
temperature scan DSC methods, and the isothermal method for our
work and is compared to the isothermal and single dynamic temperature scan DSC methods of Barton and Shepherd (*13*). Barton and
Shepherd assumed $n = 1$ throughout their work. In this work for the
isothermal calculations and multiple dynamic temperature scan DSC
methods, we also assumed that $n = 1$. However, in this work for the
single dynamic temperature scan DSC method, n was allowed to vary.

The isothermal results on this system are in excellent agreement
with that of Barton and Shepherd (*13*). Both sets of data indicate that
higher activation energies and higher Arrhenius frequency factors are
obtained from the single dynamic temperature scan DSC method than
obtained from isothermal or multiple dynamic temperature scan DSC
methods. The goodness of fit of the data for Barton and Shepherd and
for our data to the respective models is on the same order of magnitude, with correlation coefficients greater than 0.998 for the isothermal and single dynamic temperature scan DSC methods. Barton
and Shepherd also recognized that the reaction may be occurring
before the sample comes to temperature equilibrium in the isothermal
method. In the dynamic temperature scan mode, the scan is started at
relatively low temperatures, compared to reaction temperatures, so
that the instrument is in temperature equilibrium at the start of the
reaction. Within the time frame of the dynamic temperature scan experiment, the reaction will go to completion. As already discussed,

this behavior is not the case in the isothermal mode at temperatures well below the glass transition temperature of the fully cured polymer because of sample vitrification.

Barton and Shepherd (*13*) attempted to reconcile the discrepancy between the isothermal and dynamic data through Prime's (*7, 8*) method of correction. However, Prime's correction procedure is of doubtful validity based on the arguments of Kissinger (*19*), Hill (*20*), and Simmons and Wendlant (*21*).

The differences in activation energy and Arrhenius frequency factor obtained by Barton and Shepherd and by our group using the single dynamic temperature scan DSC method as shown in Table V at first was quite puzzling. However, DSC reaction kinetics analysis of the reaction mixture prepared and run within 0.5 to 2 h yielded different values of E, A, and n compared to the reaction mixture aged at room temperature for 24 h. Our results for the reaction mixture aged at room temperature for 24 h agreed quite well with those of Barton and Shepherd. We do not know how soon Barton and Shepherd ran their DSC kinetics analysis after the reaction mixture was prepared. However, our DSC results indicate that the chemical reaction was gradually advancing at room temperature as a function of time.

Table V. Comparison of Isothermal DSC Kinetics with Single Dynamic Temperature Scan DSC Kinetics for the Reaction of Phenyl Glycidyl Ether with 2-Ethyl-4-methylimidazole

Scan Rate (°C/min)	E (kcal/mol)	ln A (s^{-1})	n
Isothermal[a]			
This work	20.0	22.3	1.0
Ref. 13	19.6	21.7	1.0
Multiple Dynamic Temperature Scan DSC Method (this work)[a]			
Ozawa	14.5	22.1	1.0
Kissinger	16.0	20.8	1.0
ASTM−E698	16.1	20.9	1.0
Single Dynamic Temperature Scan DSC Method (this work)[b]			
Fresh Reaction Mixture[c] (this work)[b]			
10	27.0	31.7	0.98
15	27.2	31.9	1.01
20	28.7	33.5	0.98
Reaction Mixture Aged 24 h at Room Temperature (this work)[b]			
15	23.7	27.4	1.02
20	23.7	27.3	0.97
Ref. 13[a]			
20	23.8	26.7	1.0

[a] Reaction order n = 1.0 is assumed.
[b] Reaction order n is allowed to vary.
[c] Reaction mixture prepared and run on DSC within 0.5 to 2 h.

This hypothesis was confirmed by high performance gel permeation chromatography (HPGPC) analysis of the individual reactants, of
the reaction mixture within 0.5 h of preparation, of the reaction mixture aged at room temperature for 24 h, and of the reaction mixture
aged at room temperature for 10 days. Details of the HPGPC methodology were described elsewhere (38). Figure 8 confirms that the
reaction mixture slowly reacts at room temperature as a function of
time by the buildup of higher molecular weight components. Even for
the reaction mixture prepared and analyzed within 0.5 h of preparation, the reaction mixture is already advancing, as evidenced by the
small peak at 31.2 mL. After 24 h of aging at room temperature, the
intensity of this peak increased significantly and another peak appeared at 29.5 mL. After 10 days of aging at room temperature, additional higher molecular weight peaks appear. Thus, for the phenylglycidyl ether/2-ethyl-4-methylimidazole reaction, sample prepara-

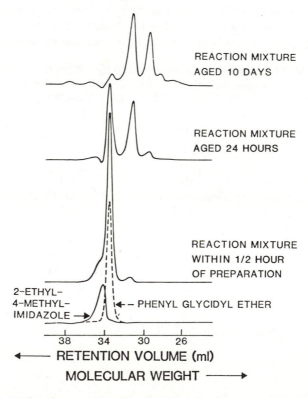

Figure 8. Molecular weight distribution changes occurring at room
temperature as a function of time for the reaction mixture of phenyl
glycidyl ether with 2-ethyl-4-methylimidazole in a 2:1 molar ratio.

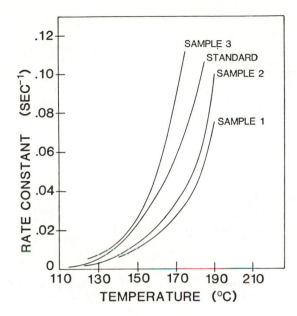

Figure 9. Rate constant vs. temperature profile for powder coating formulations.

tion time must be minimized to avoid inadvertently advancing the re-action mixture. In this work and in that of Barton and Shepherd (*13*), the activation energies and Arrhenius frequency factors obtained from the single dynamic temperature scan DSC method are higher than those obtained from the isothermal method. This result may be due to a change in reaction mechanism in going from an isothermal to a dynamic heating mode.

Coatings Systems.   POWDER COATINGS.   The cure kinetics for epoxy–polyester powder coatings were desired to meet stringent customer bake-time/temperature specifications. The powder coating is applied to metal tubing that is heated inductively. The coating is cured at high temperatures for short time periods (<5 s). Formulation modifications of catalyst type and level were required to match the reactivity of a standard sample. The reactivity of the powder coatings could not be determined by the pellet flow technique (*39*) due to the extremely fast bake schedule. The single dynamic temperature scan reaction kinetics method was used to determine cure kinetics for the powder coating samples. The rate constant vs. temperature profiles are shown in Figure 9. The percent conversion profiles for a number of powder coating formulations within the 5-s cure span at a 220 °C bake temperature are shown in Figure 10. By comparing the conversion curves, cost effective formulations were achieved with regard to the type and

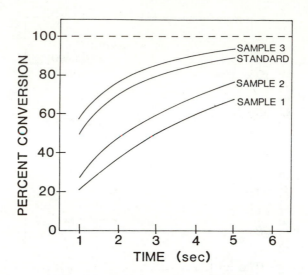

Figure 10. Conversion curves for powder coatings formulations of Figure 8 at a bake temperature of 220 °C.

level of catalyst required to generate the desired cure rate response. In the case shown in Figure 10, Sample 3 came closest to matching the standard.

GEL COAT SYSTEM.   Gel coats are protective/decorative coatings used in sanitary and marine applications. The gel coat is an in-mold integral part of the product. The gel coat system along with catalyst and promoter is sprayed into a mold. The gel coat undergoes free-radical cure at ambient temperature until a tack-free surface is obtained. Then, a fiberglass mat is placed on the gel coat and the system is reinforced by a back-up resin. If the fiberglass mat is applied before the gel coat reaches a tack-free state, the gel coat could pull away from the mold and produce an irregular surface. Therefore, it is desirable to know the cure kinetics of the gel coat system at a variety of application temperatures. The Arrhenius plot obtained from the single dynamic temperature scan method for the curing of the gel coat system is shown in Figure 11. The curing reaction is governed by three different regions. However, because the gel coat system is applied and cured at room temperature, the initial kinetics of the system is described by the low temperature region of the Arrhenius plot. Therefore, the raw DSC thermogram was analyzed only up to 65 °C by the single dynamic temperature scan reaction kinetics method. The conversion curves at four different application temperatures are shown in Figure 12. The time required to produce a tack-free gel coat surface at 22.5 °C is shown in Table VI. The percent cure corresponding to a tack-free gel coat surface was identified by relating physical measurements at 22.5

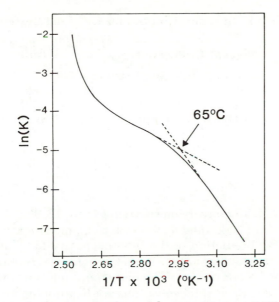

Figure 11. *Arrhenius plot for a gel coat system at 0.1% catalyst level with a reaction order of* n = 0.9.

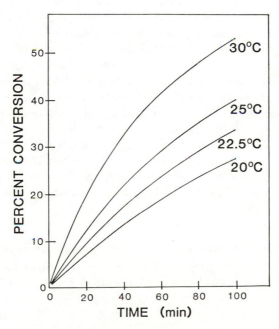

Figure 12. *Conversion curves for a gel coat system at a range of application temperatures.*

Table VI. Reaction Kinetics for Gel Coat System

Temperature (°C)	Percent Conversion	Time for Tack-free Point (min)
From Physical Measurements		
22.5		45
25.0		30
From Single Dynamic Temperature Scan DSC Method		
20.0	19.7	61
22.5	19.7	45
25.0	19.7	33
30.0	19.7	19

°C to the 22.5 °C DSC conversion curve in Figure 12. The 22.5 °C DSC percent conversion is defined as the tack-free point for the curing gel coat system. The prediction of the time necessary to achieve a tack-free gel coat surface at other application temperatures is obtained from Figure 12 at 19.7% conversion. The corresponding times required to reach a tack-free surface at these various application temperatures are shown in Table VI. The calculated tack-free time at 25 °C of 33 min agrees with the experimentally determined tack-free time of 30 min.

CASKET VARNISH EXPLOSION. The single dynamic temperature scan reaction kinetics methodology was used to determine differences in combustibility between production batches of a casket varnish coating. The coating was an alkyd amino-based resin.

An explosion occurred during the application of this coating. Questions arose regarding differences in the relative reactivity toward combustion between production batches of different formulations. The decomposition data for the two production batches are shown in Table VII. A decision as to which production batch is more reactive is obscured by the fact that the sample with the lower activation energy (60 kcal/mol) begins to react at a higher temperature (325 °C). This energy of activation vs. onset temperature conflict is resolved by the calculated isothermal percent conversion profile shown in Figure 13. The sample involved in the explosion reacts faster at the isothermal temperature of 350 °C.

Table VII. Kinetics of Casket Varnish Explosion

Sample	ΔH_0 (cal/g)	E (kcal/mol)	Onset Temperature (°C)
Sample 1	32	92	310
Sample 2 (explosion)	31	60	325

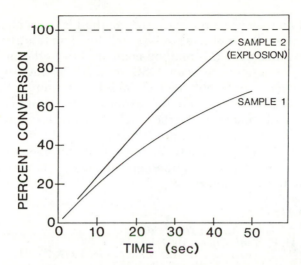

Figure 13. Conversion curves for two production batches of casket varnish coatings at 350 °C bake temperature.

Conclusions

The single dynamic temperature scan DSC method of Borchardt and Daniels was shown to characterize reliably the reaction kinetics for the thermal decomposition of calcium oxalate monohydrate and AIBN. For the thermal decomposition of AIBN, the single dynamic temperature scan DSC method is in much better agreement with the classical data of Tobolsky than are the multiple dynamic temperature scan DSC methods. A comparison of the isothermal DSC kinetics method with the single dynamic temperature scan DSC kinetics method for the reaction of phenylglycidyl ether and 2-ethyl-4-methylimidazole at a 2:1 molar ratio shows that the isothermal method does not observe the total reaction and, thereby, underestimates the total heat of reaction. The isothermal method does indicate the presence of a small induction period (2.5 min) at 90 °C. At temperatures below the glass transition temperature of the fully cured polymer, the isothermal method shows that the reacting polymer is vitrifying. The activation energy and the Arrhenius factor obtained by the single dynamic temperature scan DSC method is higher than that obtained from the isothermal method and may indicate a change in the reaction mechanism in going from an isothermal to a dynamic heating mode.

The single dynamic temperature scan DSC method provides useful quantitative reaction kinetics information regarding the curing of coatings systems such as powder coatings, gel coat systems, and casket varnish and provides insights into the chemical and physical factors affecting the curing process. This method is particularly useful for observing the effects of catalyst type and level on relative kinetics. The major advantages of this method compared to isothermal or multiple dynamic temperature scan DSC methods is the speed with which kinetics data can be obtained, the number of samples that can be analyzed in a given time, and the experimental simplicity of the method. These advantages make the single dynamic temperature scan DSC method a valuable technique for solving production problems, aiding coatings complaints analysis, and establishing useful relationships between end-use application/performance properties and fundamental kinetics parameters.

Literature Cited

1. Ozawa, T. *J. Therm. Anal.* **1970**, *2*, 301.
2. Kissinger, H. E. *Anal. Chem.* **1957**, *29* (*11*), 1702.
3. Borchardt, H. J.; Daniels, F. *J. Am. Chem. Soc.* **1957**, *79*, 41.
4. Reed, R. L.; Weber, L.; Gottfried, B. S. *Ind. Eng. Chem. Fundam.* **1965**, *4* (*1*), 38.
5. Abolafia, O. R. *Soc. Plast. Eng.* [*Tech. Pap.*], *27th* **1969**, *15*, 610.
6. Willard, P. E. *Polym. Eng. Sci.* **1972**, *12* (2), 120.
7. Prime, R. B. *Anal. Calorim.* **1970**, *2*, 201.
8. Prime, R. B. *Polym. Eng. Sci.* **1973**, *13* (5), 365.
9. Willard, P. E. *SPE J.* **1973**, *29*, 38.
10. Slysh, R.; Hettinger, A. C.; Guyler, K. E. *Polym. Eng. Sci.* **1974**, *14* (4), 264.
11. Peyser, P.; Bascom, W. D. *Anal. Calorim.* **1974**, *3*, 537.
12. Kay, R.; Westwood, A. R. *Eur. Polym. J.* **1975**, *11*, 25.
13. Barton, J. M.; Shepherd, P. M. *Makromol. Chem.* **1975**, *176*, 919.
14. Swarin, S. J.; Wims, A. M., presented at the 172nd ACS National Meeting, Anal 061, San Francisco, CA, Aug. 29–Sept. 3, 1976.
15. Peyser, P.; Bascom, W. D. *J. Appl. Polym. Sci.* **1977**, *21*, 2359.
16. Hauser, H. M.; Field, J. E. *Thermochim. Acta* **1978**, *27*, 1.
17. "Standard Test Method for Arrhenius Kinetic Constants for Thermally Unstable Materials," *Annu. Book ASTM Stand.*, Part 41, ASTM E698-79.
18. Draper, A. L. *Proc. Toronto Symp. Therm. Anal.*, *3rd* **1969**, 63.
19. Kissinger, H. E. *J. Res. Natl. Bur. Stand. (U.S.)* **1956**, *57*, 217.
20. Hill, R. A. *Nature (London)* **1970**, *227*, 703.
21. Simmons, E. L.; Wendlant, W. W. *Thermochim. Acta* **1972**, *3*, 498.
22. Niemann, T. F.; Koehler, M. E.; Provder, T. "Personal Computers in Chemistry"; Wiley: New York, 1981; Chap. 7, p. 85ff.
23. Kah, A. F.; Koehler, M. E.; Eley, R. R.; Niemann, T. F.; Provder, T. *Org. Coat. Plast. Chem.* **1981**, *45*, 400.
24. Kah, A. F.; Koehler, M. E.; Grentzer, T. H.; Niemann, T. F.; Provder, T. *Org. Coat. Plast. Chem.* **1981**, *45*, 480.
25. Nair, C. G. R.; Ninan, N. N. *Thermochim. Acta* **1978**, *23*, 161.
26. Coats, A. W.; Redfern, J. P. *Nature (London)* **1964**, *201*, 68.
27. MacCallum, J. R.; Tanner, J. *Eur. Polym. J.* **1970**, *6*, 1033.
28. Van Hook, J. P.; Tobolsky, A. V. *J. Am. Chem. Soc.* **1958**, *80*, 781.
29. Lewis, F. M.; Matheson, M. S. *J. Am. Chem. Soc.* **1949**, *71*, 747.
30. Overberger, C. G.; O'Shaughnessy, M. T.; Shalet, H. *J. Am. Chem. Soc.* **1949**, *71*, 2661.
31. Breitanback, J. W.; Schindler, A. *Monatsh. Chem.* **1952**, *83*, 724.
32. Talat-Erben, M.; Bywater, S. *J. Am. Chem. Soc.* **1955**, *77*, 3712.

33. Russell, R. E.; Tobolsky, A. V. *J. Am. Chem. Soc.* **1954**, *76*, 395.
34. Chapiro, A. *J. Chem. Phys.* **1954**, *51*, 165.
35. Overberger, C. G.; Biletch, H. *J. Am. Chem. Soc.* **1951**, *73*, 4880.
36. Widmann, G. *Thermochim. Acta* **1975**, *11*, 331.
37. Sourour, S.; Kamal, M. R. *Thermochim. Acta* **1976**, *14*, 41.
38. Kuo, C.; Provder, T.; Holsworth, R. M.; Kah, A. F. In *Chromatogr. Sci. 19,* "Liquid Chromatography of Polymers and Related Materials III," Cazes, J., Ed.; Dekker: New York, 1981; p. 169.
39. "Inclined Plate Flow Test," *Annu. Book ASTM Stand.* Part 27, **1980**, 741–742, ASTM Test D3451.17.

RECEIVED for review October 14, 1981. ACCEPTED March 30, 1982.

CHROMATOGRAPHIC METHODS

Field-Flow Fractionation

Promising Approach for the Separation and Characterization of Macromolecules

J. CALVIN GIDDINGS, KATHY A. GRAFF, KARIN D. CALDWELL, and MARCUS N. MYERS

University of Utah, Department of Chemistry, Salt Lake City, UT 84112

Field-flow fractionation (FFF) technology is applicable to the characterization and separation of macromolecules and particulate species over an effective molecular weight range of $10^3 - 10^{18}$. Separations take place in an open flow channel over which a field is applied perpendicular to the flow. Thermal FFF, one of the FFF subtechniques, has been shown applicable to polystyrene, polyisoprene, polytetrahydrofuran, polymethyl methacrylate, and polyethylene polymers in a variety of solvents by utilizing thermal diffusion in a temperature gradient field applied perpendicular to flow. Other FFF subtechniques use electrical, sedimentation, and cross-flow fields to separate macromolecules according to their electrical charges, masses, and Stokes diameters, respectively. Advantages of FFF include flexibility due to the variety of fields and programming techniques, high resolution, minimal adverse effects due to surface interactions and shear degradation, and ability in many cases to characterize macromolecules by calculation from first principles.

FIELD-FLOW FRACTIONATION (FFF) is the general name of a family of separation techniques that use fields or gradients that can interact with solute macromolecules to force them differentially into the slow streamlines of a flowing fluid (1). Separations take place in thin, open channels of rectangular cross-section under conditions of laminar flow. A field is applied perpendicularly to the face of the channel so that solutes form narrow layers against the channel wall based on their field-induced velocity and the counteracting ability to diffuse away from the channel wall (*see* Figure 1). The rate of elution of the layers is

0065–2393/83/0203–0257$06.00/0

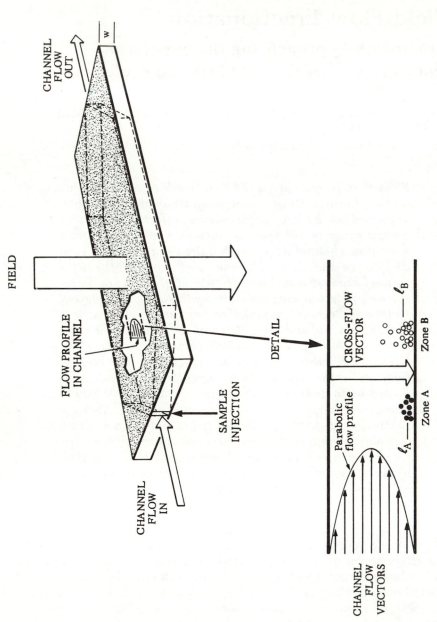

Figure 1. Scheme of the relationship of FFF flow channel to applied field (top) and approximate details of flow profile and macromolecular distribution (bottom).

greater for solutes that, by virtue of weak interaction with the field, can diffuse higher into the parabolic flow profile. In many cases, the degree of interaction of the macromolecules can be calculated precisely by application of known physical laws. In that case, the degree of retention in an FFF channel can be used to characterize a solute based on first principles so that calibration standards are not critical, as in more empirical methods.

Passing the continuous stream of eluate through an appropriate detector gives a fractogram, a plot of detector response vs. elution volume. This fractogram can be converted into a molecular weight distribution (2) for polydisperse materials, or can be used to characterize monodisperse components with respect to molecular weight, diffusivity, and other properties. Altogether, FFF subtechniques cover an effective molecular weight range of $10^3 - 10^{18}$, which includes molecular species, macromolecules, colloids, emulsions, liposomes, and particles up to 100 μm in diameter—larger than typical cells. This chapter will focus on applications and potential applications to macromolecules only, so separations of the larger particles, especially those within the realm of steric FFF (3), will be omitted.

Several advantages are unique to FFF separations. Perhaps foremost is the fact that retention is controlled by external fields that can be varied accurately and rapidly between runs or within runs (programming) to deal best with the materials encountered. Fractograms can give size distributions over a very wide range of molecular weights, especially when programming is used, a strength in the study of polydisperse systems. The open construction of the FFF channel offers the advantages that the dimensions can be used directly in the first-principle calculations of solute properties because the geometry is well defined and the theoretical performance is described rigorously. In addition, the low surface area exposed to the solute (in contrast to packed columns) minimizes surface interactions and the low pressure drop across the channel and lack of extensional shear minimizes shear degradation. Because FFF is a separation technique with high resolution, the apparatus can be used to prepare small quantities of narrow cuts of polydisperse material that may be used for standards or characterized by other methods. A final advantage is that with an FFF apparatus, backflushing to clean the channel is seldom necessary because all sample components can be eluted in a void volume if the external field is removed.

Theory of Separation

The theory of separation common to all FFF subtechniques will be reviewed briefly in this section. More details on both theory and construction of apparatus can be found in a recent publication on FFF technology (4) and references given for the specific subtechniques.

Retention. When a solute is injected into an FFF channel, it starts as a narrow plug in which the concentration of macromolecules is uniform over the channel width, w. As a field is applied, the molecules of the solute drift toward the channel wall to an extent that is determined by their degree of interaction with the field. This effect is counteracted by the tendency to diffuse away from the area of higher concentration. This interaction of two opposing effects results in the formation of an exponential distribution in concentration over the channel width. The process of forming the solute layer is termed relaxation. It usually is accomplished by stopping the flow (the stop-flow procedure) after injecting the sample and applying the field. This solute zone has a mean layer thickness, l, which is characteristic for a given polymer at a given field strength.

The mathematical treatment of retention behavior is simplest if the layer thickness is expressed as a fraction of the channel width w

$$\lambda = l/w \tag{1}$$

where λ is a dimensionless retention parameter. The height of the mean layer thickness and thus the magnitude of λ are determined by two factors: the diffusion coefficient D, which is a function of molecular weight, and the velocity of the solute induced by the field, U. Thus

$$\lambda = D/Uw = R'T/Fw \tag{2}$$

where R' is the gas constant, T is the absolute temperature, and F is the force applied by the field to Avogadro's number of molecules. For λ to be small, the applied field must be strong enough that the energy of interaction, Fw, well exceeds the thermal energy, $R'T$.

As in chromatography, a retention ratio R is defined in FFF as

$$R = V^\circ/V_r \tag{3}$$

where V° is the void volume of the channel, and V_r is the retention volume of a solute. Because an open channel is used in FFF, the void volume can be determined by calculation from the channel dimensions as well as by the more conventional method of injecting a non-retained void marker.

The retention ratio can be calculated by the theoretical expression

$$R = 6\lambda[\coth(1/2\lambda) - 2\lambda] \tag{4}$$

For a solute that is well retained ($\lambda < 0.1$), this expression can be simplified to

$$R = 6\lambda \tag{5}$$

Thus in FFF, retention volume V_r is related directly to field strength, channel dimensions, and physical constants.

Zone Spreading.   As in chromatography, the degree of efficiency is measured by plate height. Because longitudinal diffusion is usually insignificant for macromolecules in the liquids used for carrier solutions in FFF, plate height H is given by

$$H = \chi w^2 \langle v \rangle / D + \Sigma H_i \tag{6}$$

where $\langle v \rangle$ is the mean velocity of the carrier solution, H_i is the plate height due to nonidealities, and χ, the nonequilibrium factor, is a complicated function of λ. For well-retained solutes, χ can be approximated by

$$\chi = 24\lambda^3 \tag{7}$$

The first term of Equation 6 gives the plate height due to nonequilibrium processes. The second term contains contributions to plate height due to nonidealities such as instrument imperfections, injection procedure, incomplete relaxation, and solute polydispersity. For a well-constructed apparatus and good experimental procedures, the second term of Equation 6 can become a direct measure of sample polydispersity for narrow cut polymers (5).

Selectivity and Resolution.   If resolution R_s is defined as

$$R_s = \Delta V_r / 4\bar{\sigma}_v \tag{8}$$

where ΔV_r is the retention volume difference between two monodisperse solute peaks and $\bar{\sigma}_v$ is the average of their standard deviations, separation is considered complete if the resolution $R_s \geqslant 1$. For unit resolution, the peaks must be narrow enough that

$$\bar{\sigma}_v = \tfrac{1}{4} \Delta V_r \tag{9}$$

The number of theoretical plates $\bar{N}$ is given by

$$\bar{N} = (V_r/\bar{\sigma}_v)^2 \tag{10}$$

so that the minimum number of plates necessary for unit resolution is given by combining equations 9 and 10

$$\bar{N} = 16(V_r/\Delta V_r)^2 \tag{11}$$

Because a specific functional relationship between V_r and molecular weight M exists, the following approximation for ΔV_r is valid (6)

$$\Delta V_r = (dV_r/dM)\Delta M \tag{12}$$

Substituting this expression into Equation 11 and rearranging, we find that the relationship for the number of plates required for unit resolution

$$\bar{N} = 16/(d \ln V_r/d \ln M)^2 \, (\Delta M/M)^2 \tag{13}$$

depends on the inverse square of the selectivity, $S = |d \ln V_r/d \ln M|$, and on the fractional difference in molecular weights $\Delta M/M$. For high retention, the maximum selectivity S_{max} for FFF is given by (6)

$$S_{max} = |d \log \lambda/d \log M| \tag{14}$$

The maximum selectivity was calculated to be unity for sedimentation FFF, with flow and thermal FFF having selectivities slightly more than half the value for sedimentation FFF. In contrast, the maximum selectivity of size exclusion chromatography (SEC) was calculated to be less than half that of thermal and flow FFF (6), and it usually is much less than half.

Thermal FFF

Thermal FFF uses a thermal gradient maintained by heating the upper plate of the channel and cooling the channel's lower plate to separate macromolecules by thermal diffusion. Historically, thermal FFF is important because it was the first demonstration of separation in an FFF apparatus: two polystyrene fractions were partially separated in 1967 (7). Since then, resolution and speed were improved considerably so that thermal FFF now is a powerful tool for the separation of polymers. Figure 2 shows the separation of a nine-component polystyrene sample (8) and Figure 3 shows a high-speed separation of three polystyrene fractions (9). The simplicity of the thermal FFF apparatus (which has no moving parts) has made it quite reliable. The channel walls can be formed from copper bars, which may be plated if a different surface is desired. The bars are separated by a spacer that has the outline of the channel cut from it. A typical spacer thickness, w, is of the order of 0.1–0.5 mm. Because of the simple construction, a wide variety of solvents can be used as carrier solutions.

The retention parameter, λ, for thermal FFF derived from Equation 2 is (10, 11)

$$\lambda = D/D_T w(dT/dx) \tag{15}$$

where D_T is the coefficient of thermal diffusion and dT/dx is the temperature gradient. This rather simple expression does not account for distortion of the flow profile caused by the variation of viscosity of the carrier solution over the temperature gradient. This effect perturbs the

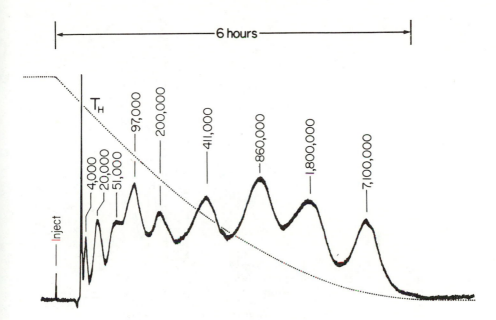

Figure 2. Separation of nine linear polystyrene fractions of the indicated molecular weight by programmed thermal FFF. (Reproduced from Ref. 8. Copyright 1976, American Chemical Society.)

retention expression, Equations 4 and 5, slightly so that corrections for the flow profile distortion or calibration standards must be used for very precise work (*12*).

In addition to the work with polystyrene in a variety of solvents (*13*), retention was demonstrated for other polymers, including polyisopyrene, polytetrahydrofuran, polymethyl methacrylate (*14*), polyethylene, and polypropylene (*15*) using a variety of solvents. Although this preliminary work with polymers other than polystyrene shows considerable promise, it has been hampered by the lack of standards of low polydispersity and data on thermal diffusion for polymers. The situation is encouraging because thermal FFF is a separation technique as well as a characterization tool; an investigator can perhaps prepare small quantities of narrow cuts of a polymer of interest for use as standards.

The lower limit for molecular weights retained on a thermal FFF channel is determined by the magnitude of the temperature difference that can be maintained between the hot and cold walls of the channel. A natural limitation is imposed by the liquid range of the carrier solution. However, this limit can be extended by pressurizing

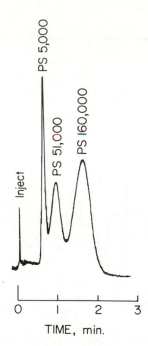

Figure 3. Separation of three linear polystyrenes in a high speed thermal FFF system. Conditions: V°, 0.44 mL (w, 51 μm); V, 0.74 mL/min; ΔT, 60 °C; and T_c, 31 °C. (Reproduced with permission from Ref. 33.)

the channel so that the boiling point is significantly elevated. Compounds of molecular weights less than 1000 were retained using these tactics (*16*). The upper limit in molecular weight, if any, has not been determined. Polymers exceeding molecular weight 10^6 were retained and eluted without difficulty (*see* Figure 2). Although results to date indicate that a wide variety of polymer–solvent systems may be amenable to study by thermal FFF, one of the most important solvents, water, does not appear to lead to thermal FFF retention. Water-soluble macromolecules can be characterized and separated by the remaining three FFF subtechniques, described in the following sections.

Electrical FFF

A second FFF subtechnique, electrical FFF, was first experimentally demonstrated in our laboratory in 1969 (*17*). A semipermeable membrane was used for the channel walls and a voltage was applied to electrodes placed in electrolyte reservoirs on either side of the channel. By applying this electrical field, charged macromolecules could interact with the field and be separated according to the ratio of their electrophoretic mobility and diffusion coefficient.

A mixture of three proteins was separated on this first apparatus (*18*). This subtechnique was an advance on two fronts: the utility of a second field in FFF was demonstrated and separations were extended to aqueous solutions.

The expression for the retention parameter, λ, in electrical FFF is (*19*)

$$\lambda = D/\mu Ew \tag{16}$$

where μ is the electrophoretic mobility and E is the electrical field strength. Although electrical FFF promises to characterize separated components with respect to their μ/D ratios, this subtechnique has not been developed to the extent thermal FFF has. However, it has the potential for achieving both separation and characterization, especially in the area of macromolecules of biological origin.

Sedimentation FFF

A third FFF subtechnique is one using a gravitational or centrifugal field. The realization of sedimentation FFF required the development of a low volume rotor seal so that the channel, fitted into a centrifuge basket, could maintain inflow and outflow connections to the outside.

The equation for the retention parameter in sedimentation FFF is (*20*)

$$\lambda = D/sGw = R'T/GM\,(1 - \bar{v}_s\rho)w \tag{17}$$

where s is the sedimentation coefficient, G is the acceleration, M is the molecular weight of the solute, $\bar{v}_s$ is the partial specific volume of the solute, and ρ is the density of the carrier solution. In the limit of high retention, the relationship between retention volume, V_r, and molecular weight, M, is linear, as can be shown by combining Equations 3, 5, and 17 and rearranging

$$V_r = V^\circ/6\lambda = \quad V^\circ G(1 - \bar{v}_s\rho)Mw/6R'T = \text{const} \cdot M \tag{18}$$

Because of this very straightforward relationship, it was possible to characterize the molecular weight of single particles, as we did with viruses (*21, 22*), and the molecular weight distribution of polydisperse materials.

The sedimentation FFF channel is formed from stainless steel, which may be coated with other material (such as Teflon) for compatibility with the solute or carrier solution. Aqueous solutions were used most frequently as carriers because of the limitations of the materials used in the rotor seal. However, several nonaqueous solvents were used when the seal was modified.

Sedimentation FFF has been applied only to the separation of fairly large particles because of limitations on rotor speed related to a lack of integrity of the rotor seal at high rotation rates. The range of molecular weights retainable on a sedimentation FFF apparatus is given roughly by (23)

$$M \geqslant 10^{10}/g \tag{19}$$

where g is the number of gravities. A limitation of $\sim 10^3$ gravities in our work has until recently restricted the molecular weight of retainable solutes to 10^7 and greater. However, Kirkland et al. recently reported the use of a sedimentation FFF apparatus capable of maintaining a field of 15,000 gravities, reducing the limit on the molecular weight range to $\sim 10^6$ (24). Further refinements in instrumentation should greatly enhance the applicability of sedimentation FFF to macromolecular separation and molecular weight characterization.

Flow FFF

The most recently developed FFF subtechnique applicable to macromolecule characterization and separation is flow FFF (25). An apparatus similar to that used in electrical FFF was constructed with the channel walls formed of semipermeable membranes. A separate pump was used to force carrier solution across the membranes, forming a cross-flow field. This subtechnique is perhaps the most generally applicable because retention is based only on frictional drag related to molecular size or, strictly, Stokes diameter d, which is a universal property.

The equation for the retention parameter in flow FFF is

$$\lambda = DV^\circ/\dot{V}_c w^2 = R'TV^\circ/f\dot{V}_c w^2 \tag{20}$$

where $\dot{V}_c$ is the volumetric cross-flow rate and f is the molar friction coefficient given by Stokes equation

$$f = N \cdot 3\pi\eta d \tag{21}$$

in which N is Avogadro's number; η is the carrier solution viscosity; and d, the Stokes diameter, is the diameter of a sphere having the same f value as the particle under consideration.

Like electrical FFF, flow FFF has been used exclusively for water-soluble species because of the limitations of the membranes employed. However, with appropriate materials, flow FFF probably can be used with other solvent systems. In flow FFF, the lower limit of molecular weights is determined by the permeability of the membrane and the upper limit by the interference of gravitational forces. With materials readily available, separations in the range of $10^4 - 10^9$ molecular weight can be accomplished. Some materials studied with the flow FFF apparatus have included polyacrylic acid, sulfonated polystyrene, viruses, proteins, colloidal silica, and paint pigments (26 – 28).

An interesting characteristic of the flow FFF apparatus arises from the permeability of the channel walls. Besides the inlet and outlet flow streams from the channel, there are an inlet and outlet flow from the cross-flow field. The simplest mode of operation has the inflow and outflow of the channel equal and the inflow and outflow of the cross-flow equal. However, the four flow streams can be balanced in a variety of ways so that the apparatus can be used to concentrate or dilute a solute or, when the channel outflow is very low, the apparatus can function as a pressure dialysis cell (29).

General Experimental Considerations in FFF

A complete FFF apparatus consists of a pump to drive the carrier solution, the FFF channel, the field or gradient generating components, an appropriate detector, chart recorder, flow measuring device, and, if desired, a fraction collector. Samples are injected with a microsyringe or injection valve. Retention ratio, R, in FFF is normally (although there are exceptions) independent of carrier solution flow rate (*see* Equations 2 and 3) so that flow rates may be chosen to give an acceptable peak width and thus resolution in the shortest possible time. By systematically varying the flow rate, the plate height can be plotted vs. the flow velocity to determine diffusion coefficients from the plot's slope. In some cases, the intercept of such a plot can be used to determine solute polydispersity (5).

If a dilute sample is to be analyzed, FFF technology offers several means of on-channel concentration. In sedimentation FFF, river water colloids were concentrated on the head of the channel by pumping the sample through the channel at a low flow rate while applying a high field. The flow was then resumed with a carrier solution at a lower field strength and higher flow rate to elute the colloids and form a fractogram. By using this technique, a sample over ten times the channel's void volume can be loaded onto the head of the channel (30). Recent work suggests that the same method is applicable to flow FFF.

For polydisperse samples with a very large molecular weight range, there are several experimental parameters that can be programmed effectively during an experiment to bring all components out in a reasonable time. The most obvious is field strength; an example is given in Figure 2. Field strength is readily programmable because the field is applied externally. Flow programming is also possible (31). Other types of programming are possible depending on the specific subtechnique. For example, the carrier solution density can be varied in sedimentation FFF to reduce the time of separation (32).

In summary, a number of FFF subtechniques have been developed that have considerable potential for the separation and characterization of many different types of macromolecules. Flexibility is gained by the variety of fields that may be used as well as by programming. FFF appears to have an almost unique ability to handle very polydisperse samples at high resolution.

Acknowledgments

This work was supported by Department of Energy Grant No. De−AC02−79EV10244.

Literature Cited

1. Giddings, J. C. *J. Chem. Ed.* **1973**, *50*, 667.
2. Giddings, J. C.; Myers, M. N.; Yang, F. J. F.; Smith, L. K. In "Colloid and Interface Science"; Kerker, M., Ed.; Academic: New York, 1976; Vol. 4, p. 381.
3. Giddings, J. C.; Myers, M. N. *Sep. Sci. Technol.* **1978**, *13*, 637.
4. Giddings, J. C.; Myers, M. N.; Caldwell, K. D.; Fisher, S. R. In "Methods of Biochemical Analysis"; Glick, D., Ed.; John Wiley & Sons: New York, 1980; Vol. 26, p. 79.
5. Karaiskakis, G.; Myers, M. N.; Caldwell, K. D.; Giddings, J. C. *Anal. Chem.* **1981**, *53*, 13.
6. Giddings, J. C. *Pure Appl. Chem.* **1979**, *51*, 1459.
7. Thompson, G. H.; Myers, M. N.; Giddings, J. C. *Sep. Sci.* **1967**, *2*, 797.
8. Giddings, J. C.; Smith, L. K.; Myers, M. N. *Anal. Chem.* **1976**, *48*, 1587.
9. Giddings, J. C.; Martin, M.; Myers, M. N. *J. Chromatogr.* **1978**, *158*, 419.
10. Thompson, G. H.; Myers, M. N.; Giddings, J. C. *Anal. Chem.* **1969**, *41*, 1219.
11. Hovingh, M. E.; Thompson, G. H.; Giddings, J. C. *Anal. Chem.* **1970**, *42*, 195.
12. Martin, M.; Giddings, J. C. *J. Phys. Chem.* **1981**, *85*, 72.
13. Giddings, J. C.; Caldwell, K. D.; Myers, M. N. *Macromolecules* **1976**, *9*, 106.
14. Giddings, J. C.; Myers, M. N.; Janca, J. *J. Chromatogr.* **1979**, *186*, 37.
15. Brimhall, S. L.; Myers, M. N.; Caldwell, K. D.; Giddings, J. C. *Sep. Sci. Technol.* **1981**, *16*, 671.
16. Giddings, J. C.; Smith, L. K.; Myers, M. N. *Anal. Chem.* **1975**, *47*, 2389.
17. Caldwell, K. D.; Kesner, L. F.; Myers, M. N.; Giddings, J. C. *Science* **1972**, *176*, 296.
18. Kesner, L. F.; Caldwell, K. D.; Myers, M. N.; Giddings, J. C. *Anal. Chem.* **1976**, *48*, 1834.
19. Giddings, J. C.; Lin, G. C.; Myers, M. N. *Sep. Sci.* **1976**, *11*, 553.
20. Giddings, J. C.; Yang, F. J. F.; Myers, M. N. *Anal. Chem.* **1974**, *46*, 1917.
21. Giddings, J. C.; Yang, F. J. F.; Myers, M. N. *Sep. Sci.* **1975**, *10*, 133.
22. Caldwell, K. D.; Nguyen, T. T.; Giddings, J. C.; Mazzone, H. M. *J. Virol. Meth.* **1980**, *1*, 241.

23. Giddings, J. C.; Myers, M. N.; Lin, G. C.; Martin, M. *J. Chromatogr.* **1977**, *142*, 23.
24. Kirkland, J. J.; Yau, W. W.; Doerner, W. A.; Grant, J. W. *Anal. Chem.* **1980**, *52*, 1944.
25. Giddings, J. C.; Yang, F. J.; Myers, M. N. *Anal. Chem.* **1976**, *48*, 1126.
26. Giddings, J. C.; Yang, F. J.; Myers, M. N. *J. Virol.* **1977**, *21*, 131.
27. Giddings, J. C.; Yang, F. J.; Myers, M. N. *Anal. Biochem.* **1977**, *81*, 395.
28. Giddings, J. C.; Lin, G. C.; Myers, M. N. *J. Liq. Chromatogr.* **1978**, *1*, 1.
29. Giddings, J. C.; Yang, F. J.; Myers, M. N. *Sep. Sci.* **1977**, *12*, 499.
30. Giddings, J. C.; Karaiskakis, G.; Caldwell, K. D. *Sep. Sci. Technol.* **1981**, *16*, 725.
31. Giddings, J. C.; Caldwell, K. D.; Moellmer, J. F.; Dickinson, T. H.; Myers, M. N.; Martin, M. *Anal. Chem.* **1979**, *51*, 30.
32. Giddings, J. C.; Karaiskakis, G.; Caldwell, K. D. *Sep. Sci. Technol.* **1981**, *16*, 607.
33. Giddings, J. C.; Fisher, S. R.; Myers, M. N. *Am. Lab. (Fairfield, Conn.)* **1978**, *10*, 15.

RECEIVED for review October 14, 1981. ACCEPTED December 8, 1981.

Foam Fractionation of Polymer Mixtures in a Nonaqueous Solvent System

R. P. CHARTOFF and L. T. CHEN

University of Dayton, The Center for Basic and Applied Polymer Research, Dayton, OH 45469

R. J. ROE

University of Cincinnati, Department of Metallurgy and Materials Science, Cincinnati, OH 45221

The selectivity obtained in the foam fractionation of polymer mixtures in a nonaqueous solvent system was studied and related to a multilayer theory of adsorption of polymers. Two mixtures of polymethyl methacrylate (PMMA) and polystyrene (PS) in xylene were fractionated using a batch foaming technique with and without addition of a surfactant. The two mixtures consisted of (1) two commercial samples with broad molecular weight distributions (MWD) with PS having the higher weight-average molecular weight ($\overline{M}_w$), and (2) two narrow-distribution polymers differing greatly in $\overline{M}_w$ with PS again having the higher value. In case 1, although little separation occurred without surfactant, the trend was toward exclusion of high molecular weight PS from the foamate. This trend was more pronounced and the amount of separation was improved by adding surfactant. In case 2, separation was selective to high molecular weight PS when fractionation was carried out without surfactant. With surfactant added, the foamate was enriched in low molecular weight PMMA. No conclusive results were obtained on the effect of bubble size on separation efficiency, although the data pointed toward an improvement with decreasing bubble size. These observations were consistent with the trends predicted by a multilayer theory of adsorption of polymers from solution.

0065–2393/83/0203–0271$06.00/0

FOAM FRACTIONATION is a technique that uses adsorption of a component or solute at bubble surfaces to separate the component from a solution. Bubbles are produced by supplying gas at the bottom of the liquid pool. The bubbles form a foam, rise to the top of the liquid, and then overflow into a foam breaker. The concentrated liquid collected in the foam breaker is called foamate. By carrying out the foaming process in vertical glass tubes or columns, the dynamics of bubble formation, foam generation, and foamate collection may be observed conveniently.

Because the solute is concentrated at the surface of the bubbles, the size and dynamics of the bubbles are important in foam fractionation. Thus, bubble size, gas flow rate, and column diameter were among the process parameters studied. In addition, solution parameters such as solute concentration and molecular size, choice of solvent, and addition of a surfactant were considered. This chapter reviews the more significant of these factors in terms of the experimental trends observed and their relation to a multilayer theory of adsorption of polymers from solution advanced by Roe (1–3).

Background Information

Foam fractionation of polymers has received relatively little attention from researchers, probably because the process of foam fractionation is complex, and only limited success has been achieved in the studies published previously. A summary of relevant literature is provided in Table I.

Among the various types of separations attempted by foam fractionation are those by molecular size variants including molecular weight, branching, and stereoregularity as well as those by copolymer composition and functionality. The work reported here is unique in that it considers the selectivity in separation of mixtures of a polar and nonpolar polymer in a nonaqueous system. Most of the studies published previously were on aqueous systems and the separation of mixtures of two distinct polymers by foam fractionation has not been reported previously.

Theoretical

One useful property of polymers is that polymer molecules readily become adsorbed to solid surfaces from solution. This tendency for adsorption is both molecular weight dependent and composition dependent so that under different conditions different types of selectivity may be observed. Examples of selective adsorption occur among polymers differing simply in molecular weight, having different chemical compositions, or among polymers of the same type but dif-

Table I. Relevant Literature on Foam Fractionation of Polymers

References	Date	Polymer	Effect Studied
		Aqueous Foam Fractionation	
4	1958	polyvinyl alcohol (PVA)	molecular weight
5,6	1961, 1963	PVA	stereoregularity
7	1962	PVA	molecular weight, stereoregularity, functional groups
8	1972	PVA	stereoregularity, chain branching
9	1970	poly(methacrylic acid–co-methyl methacrylate)	copolymer composition, functional groups
10	1973	poly(methyl methacrylate–co-n,n'-dimethylaminoethyl methacrylate)	copolymer composition, functional groups
11	1974	PVA	chain branching
12	1976	PVA	chain branching, molecular weight
13	1979	alkyd resin	molecular weight
		Nonaqueous Foam Fractionation	
14 15	1968, 1969	polydimethylsiloxane (benzene)	molecular weight
16	1977	polymethyl methacrylate (PMMA) (benzene)	stereoregularity
17	1981	PMMA, polystyrene (xylene)	molecular weight, composition

fering in stereoregularity. Selective adsorption also occurs among copolymers differing in comonomer composition.

In all of these cases, adsorption tends to be highly competitive with one type of molecule being adsorbed almost exclusively at the expense of another even though only minor differences exist in the adsorption affinity of the species. Qualitatively, the selectivity expected for adsorption of polymer molecules from solution onto a solid substrate can be described by theories such as the multilayer theory of adsorption advanced by Roe (1−3). Roe's theory describes adsorption at surfaces in terms of the solvent power, the molecular weight of the polymer solute, the solution composition, the thickness of the adsorbed layer, the shape of the adsorbed polymer chains, etc. The theory confirms the belief that even small differences in segmental adsorption affinities are magnified greatly because of the sheer number of anchoring segments per molecule. This factor leads to the extreme selectivity observed during adsorption.

The same theoretical considerations should apply to adsorption of polymer molecules at the gas−liquid interface formed by the bubbles in a foam. According to the theory, the following trends are predicted:

1. For a single solute with a mixture of molecular weights, higher molecular weight polymer will have a greater affinity for the interface at low-solute concentrations.
2. For multiple solutes, the polymer having stronger interaction with the interface will be adsorbed while the other will be excluded, even if its adsorption affinity is only slightly less.
3. Also for multiple solutes, a polymer having no or little interaction with the interface will be excluded, the higher its molecular weight the greater the exclusion.
4. Poor solvents promote adsorption.

The theory represents the equilibrium limiting case. In practice, although these predictions should indicate the trends to be expected in terms of selectivity, 100% efficiency cannot be achieved in the laboratory. The purpose of this study, however, was to test our original hypothesis by determining if the trends observed in foam fractionation experiments were consistent with the theory, and what effect various process parameters had on the separation efficiencies observed.

Experimental

A schematic drawing of the foam fractionation apparatus used is shown in Figure 1. It is a closed-loop system in which dry nitrogen is saturated with solvent and then circulated. Two glass columns, one with a 25-mm inside diameter (ID) and one with a 20-mm ID are connected to the gas source and a single central foamate collector. Both columns are 65−70 cm long. Foam is

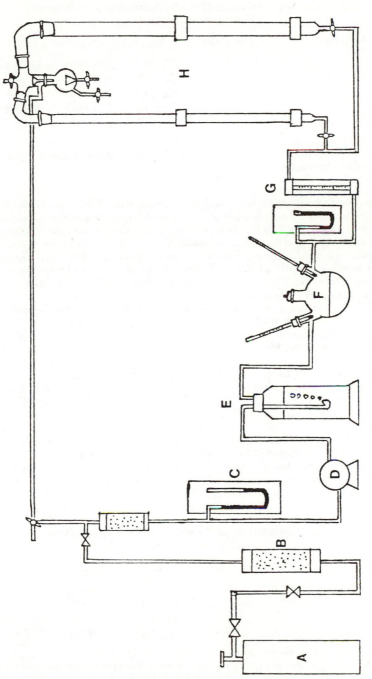

Figure 1. Foam fractionation apparatus. Key: A, N_2 gas; B, desiccant; C, manometer; D, pump; E, saturator; F, saturation monitor; G, rotameter; and H, fractionation column(s).

Table II. Separation of Broad-Distribution Polymers Without Surfactant

Fraction	Time (h)	$\overline{M}_n$	$\overline{M}_w$	$\overline{M}_z$
Original solution[a]	—	42,500	203,000	662,000
1	2	39,900	197,000	654,000
2	4	42,300	199,000	651,000
3	6	32,200	156,000	572,000
Residual solution	—	42,300	197,000	643,000

NOTE: Values given are the average molecular weights of polymer in foamate.
[a] Polymer concentration by weight: 2% PS−1 and 2% PMMA−1; column diameter 20 mm, orifice size 30 μm.

generated by bubbling the saturated N_2 through individual spinarette disks containing precisely formed holes of 15, 30, and 50 μm. Foam fractionation in this type of apparatus is a batch or intermittent process. Liquid is charged, foam is generated, and fractions are collected at selected time intervals. The composition of the liquid pool thus changes with time. The time intervals referred to subsequently in Tables II and III are the times elapsed from the beginning of foaming. The successive fractions listed in Tables IV and V were collected at 10-min intervals.

The polymers used were commercial, broad molecular weight distribution samples of PS and PMMA with $\overline{M}_w$ values of 70,000 (PS−1) and 51,000 (PMMA−1) and narrow distribution samples with $\overline{M}_w$ values of 670,000 (PS−2) and 19,400 (PMMA−2). The solvent selected was xylene. The surfactant used in certain foam fractionations was FC−431, a fluorocarbon compound manufactured by the 3M Company. Analyses of the foamates were performed by size exclusion chromatography (SEC) using a Spectra Physics 8000 high pressure liquid chromatography (HPLC) system equipped with a variable UV detector and a du Pont Zorbax bimodal column set rated at 1000 and 60 Å.

The SEC analyses of various fractions collected during each foaming experiment were carried out by isolating the solute in each sample from the xylene solvent and redissolving it in tetrahydrofuran (THF) for mobile phase compatibility. Molecular weights and concentrations were measured by the appropriate use of elution volumes, peak heights, and peak areas. For the SEC

Table III. Separation of Broad-Distribution Polymers with Surfactant

Fraction	Time (min)[a]	$\overline{M}_n$	$\overline{M}_w$	$\overline{M}_z$
Original solution[b]	—	36,700	192,000	655,000
1	10	16,700	43,000	92,900
2	20	38,100	188,000	636,000
3	30	32,800	167,000	587,000

NOTE: Values given are the average molecular weights of polymer in foamate.
[a] Fresh surfactant was injected after each fraction was taken at 10-min intervals.
[b] Polymer concentration by weight: 2% PS−1 and 2% PMMA−1 with 0.1% FC surfactant; column diameter 20 mm, orifice size 30 μm.

**Table IV. Separation of Narrow-Distribution Polymers—
Effect of Orifice Diameter on Fractionation**

	Orifice Size (μm)		
Fraction	15	30	50
Original solution[a]	29	29	30
1	28	29	29
2	23	25	24
3	29	25	25
4	27	—	—

NOTE: Values given are percent of PS–2 in foamate based on total solute.
[a] Polymer concentration by weight: 0.6% PS–2 and 1.4% PMMA–2 with 0.1% FC surfactant; column diameter 20 mm; fractions were collected at 10-min intervals.

solvent–polymer combination, THF with PS and PMMA, the UV detector was set at 235 nm.

Results and Discussion

Fractionation data were collected for mixtures of PS and PMMA at various concentrations and for different foaming conditions. The results for some of these tests are presented in Tables II–V along with descriptions of the experimental variables. The data on the mixtures of the commercial broad distribution samples in Tables II and III were analyzed by assuming a single MWD peak for the total mixture. The data for the mixtures of narrow MWD samples in Tables IV and V were analyzed for concentrations in terms of individual peak areas and peak intensities.

The data of Tables II and III indicate that fractionation occurs, but efficiency is low without adding a surfactant. Addition of a surfactant (Table III) improves fractionation at shorter foaming intervals. In both cases, the high molecular weight PS tends to be excluded from the foamate. The improvement in fractionation efficiency with surfactant probably arises because of increased foam stability.

Figure 2 contains typical photographs of a foam generated from a

Table V. Effect of Surfactant on Selectivity

Fraction	With Surfactant	Without Surfactant
Original solution[a]	29	28
1	29	30
2	25	32
3	25	38

NOTE: Values given are percent of PS–2 in foamate based on total solute.
[a] Solute concentration 2% with surfactant; and 4% without surfactant; column diameter 20 mm; orifice size 30 μm; fractions were collected at 10-min intervals.

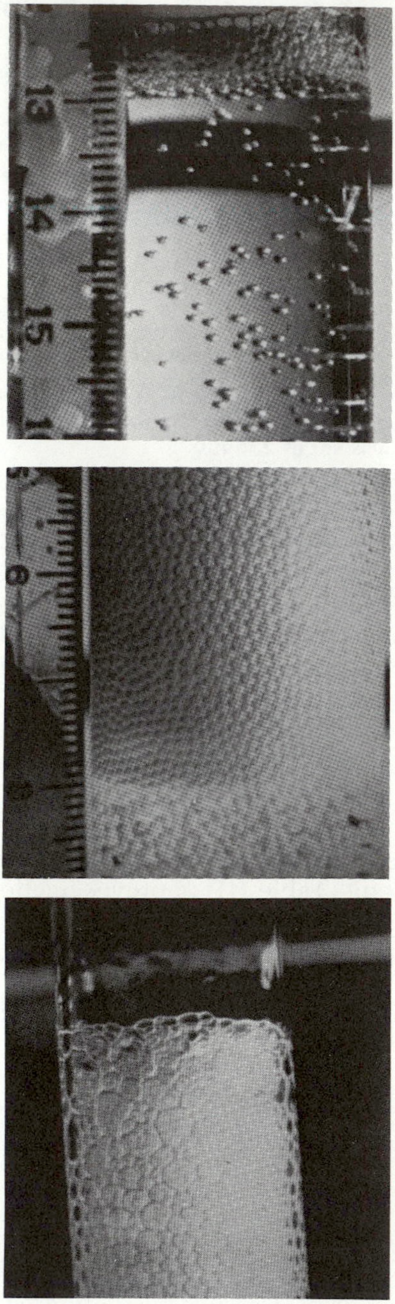

Figure 2. Foam generated during foam
fractionation of a polymer solution
without added surfactant: early stage
of foam buildup (a), steady state foam
structure (b), and upper surface of
foam column showing bubble coal-
escence (c).

polymer solution containing no surfactant: Figure 2a depicts the early stages of foam generation; Figure 2b depicts the cellular structure of a fully developed foam; and Figure 2c is the upper surface of the foam. The upper surface appears to be wet and in practice breaks up readily. When the foams are unstable in this fashion, bubbles coalesce rapidly and collapse so that equilibrium cannot be approached. The low efficiencies achieved in this mode result in little separation as noted earlier. Addition of surfactant promotes foam stability and results in dryer, more uniform foams. The more stable and dry the foams are, the closer we can approach equilibrium. This behavior is the result of both the minimization of foam collapse and the achievement of better enrichment due to improved foam drainage. The latter effect is similar to reflux in a distillation process. Both factors contribute to the improved efficiency of separation observed when surfactant is used.

The data also indicate a great difference in selectivity when a surfactant is used. With the two polymers, PS−2 and PMMA−2, having extremely different molecular weights, the high molecular weight PS−2 normally has the greatest adsorptivity. When surfactant is added, however, selectivity reverses and the PMMA is adsorbed preferentially.

The data of Tables III and IV do not indicate any particular advantage in fractionation by changing bubble diameter (orifice size), although the lowest concentration of PS−2 in any fraction was obtained with the smallest orifice diameter.

In summary, in the mixture that consisted of two commercial samples with broad MWD with PS having the higher weight-average molecular weight, little separation occurred without surfactant and the trend was toward exclusion of high molecular weight PS from the foamate. This trend was more pronounced by adding surfactant, and the amount of separation improved. In the mixture that consisted of two narrow distribution polymers differing greatly in $\bar{M}_w$ with PS having the higher value, separation was selective to high molecular weight PS when fractionation was carried out without surfactant. With surfactant added, the foamate was enriched in low molecular weight PMMA. No conclusive results were obtained on the effect of bubble size on separation efficiency, although one would expect an improvement with decreasing bubble size.

All of the trends observed are consistent with the multilayer adsorption theory discussed earlier. In particular, the reversal of selectivity between PS and PMMA when the perfluorinated surfactant was present is a graphic illustration of the notion that the polymer with the strongest interaction with the interface will be adsorbed.

Acknowledgments

The authors are pleased to have received funding for this research from the National Science Foundation. We thank Paul Karichoff of the Celanese Fibers Company for providing the spinarette disks used to generate foam.

Literature Cited

1. Roe, R. J. *J. Chem. Phys.* **1974**, *60*, 4192.
2. Roe, R. J., presented at the American Physical Society Meeting, Atlanta, GA, March 1976.
3. Roe, R. J. In "Adhesion and Adsorption of Polymers," Part B; Lee, L.-H., Ed.; Plenum: New York, 1980; pp. 629–41.
4. Al-Madfai, S.; Frisch, H. L. *J. Am. Chem. Soc.* **1958**, *80*, 5613.
5. Imai, K.; Matsumoto, M. *J. Polym. Sci.* **1961**, *55*, 355.
6. Imai, K.; Matsumoto, M. *Bull. Chem. Soc. Jpn.* **1963**, *36*, 455.
7. Devin, C.; Minfray, M. *Compt. Rend.* **1962**, *255*, 116.
8. Kikukawa, K.; Nozakura, S.; Murahashi, S. *Polym. J.* **1972**, *3(1)*, 52.
9. Bolewski, K.; Tomaskiewicz, T.; Olbracht, M. *Polimery (Warsaw)* **1970**, *15*, 15.
10. Bolewski, K.; Nalewajko, H.; Maruzewski, I. *Ann. Pharm. (Poznan)* **1973**, *10*, 73.
11. Schildknecht, C. E., Tannebring, J. *Polym. Prepr., Am. Chem. Soc., Div. Polym. Chem.* **1974**, *15(2)*, 439.
12. Morishima, Y.; Irie, Y.; Iimuro, H.; Nozakura, S. *J. Polym. Sci., Polym. Chem.* **1976**, *14*, 1267.
13. Bierwagen, G. P.; Rehfeldt, T. K.; Schewing, D. R., presented at The Water-Borne and Higher Solids Coatings Symposium, New Orleans, LA, February 1979.
14. Gaines, G. L.; LeGrand, D. G. *Polym. Lett.* **1968**, *6*, 625.
15. Gaines, G. L.; LeGrand, D. G. *J. Colloid Interface Sci.* **1969**, *31*, 162.
16. Schröeder, E.; Giesau, K. E. Germ. Patent 125 807, 1977.
17. Chartoff, R. P.; Chen, L. T.; Roe, R. J. *Org. Coat. Plast. Chem.* **1981**, *44*, 607.

RECEIVED for review October 14, 1981. ACCEPTED February 3, 1982.

Orthogonal Chromatography

Polymer Cross-Fractionation by Coupled Gel Permeation Chromatographs

S. T. BALKE and R. D. PATEL
Xerox Research Centre of Canada, Mississauga, Ontario L5L 1J9, Canada

Orthogonal chromatography (OC) is a new combined chromatographic method that can accomplish the high performance liquid chromatography (HPLC) of macromolecules. The theoretical basis for OC is described as the synergistic use of size exclusion with adsorption/partition mechanisms. Fractionation of styrene–n-butyl methacrylate copolymers dominated by composition is obtained readily. However, chromatogram shape and retention time are sensitive to several operating variables, notably the composition of the solvent injected into the second gel permeation chromatograph (GPC). A dynamic method of calibrating for composition utilizing a rapid scanning UV detector is developed and applied to obtain quantitative results. The primary uncertainty in the copolymer composition distributions obtained is the effect of styrene sequence length on the fractionation.

Oathogonal chromatography (OC) is a new combined chromatographic method that can accomplish the high performance liquid chromatography (HPLC) of macromolecules (*1, 2*). It involves the coupling of gel permeation chromatographs (GPCs) and the synergistic utilization of size exclusion and adsorption/partition mechanisms.

OC is directed particularly at polymers that exhibit two or more simultaneous property distributions. Copolymers of styrene–*n*-butyl methacrylate are the specific example in this chapter. Even such simple copolymers can exhibit three simultaneous distributions of important properties: composition, sequence length, and molecular weight. A composition distribution means that some molecules are richer in styrene than others. A sequence length distribution refers to the presence of longer segments of styrene units in a row (before an *n*-butyl methacrylate unit is encountered) in some molecules than in

0065–2393/83/0203–0281$08.50/0

others. A molecular-weight distribution for the free radical polymerized copolymer can mean a range of molecular weights from ~10,000 to over 500,000. Because these property distributions are present simultaneously, interference of the remaining two distributions, while attempting elucidation of one of them, greatly complicates analysis (3–13). For example, in attempts to apply HPLC directly to the problem of composition elucidation, the size exclusion mechanism present in the packings resulted in an interference from the molecular-weight distribution (4). Removal of the size exclusion mechanism by utilizing packing of very small pores (5) greatly restricts the packing surface area accessible to macromolecules for adsorption.

Currently, the most common method of analyzing copolymers is to focus on the copolymer composition distribution and to attach two spectrophotometric detectors to a conventional GPC (7–9). This situation is shown in Figure 1, A and B referring to the different monomer units in the copolymer (styrene and n-butyl methacrylate). Each detector provides one equation in terms of these two concentrations. The solution of the equations often is presented as the ratio of the two concentrations vs. the retention time. The three main problems with this approach are discussed now.

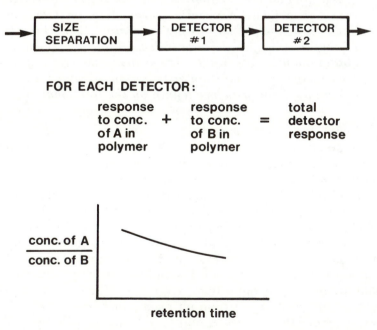

Figure 1. Copolymer analysis by dual detector gel permeation chromatography.

Problem Areas with Dual Detector, Conventional GPC Analysis

Computational. The usual method of presenting the copolymer composition distribution, as shown in Figure 1, has limited utility because retention time is a function of the chromatographic operating conditions. A differential distribution representing copolymer composition can be calculated from polymerization kinetic models and so has a fundamental basis as well as a more general utility (*1, 14*). Details are provided later in this chapter. However, generally the abscissa in this case represents weight fraction of styrene in the copolymer and the ordinate represents the copolymer concentration.

Practical. The coupled problems of axial dispersion within each detector cell, significant time delay between detectors, and very rapid analyses (steeply rising chromatograms) can cause high uncertainty in pairing the equation obtained from one detector with that obtained from the other.

Fundamental. Two fundamental problems associated with dual detector GPC are fractionation and detection. Conventional GPC is directed at obtaining separation with respect to molecular size in solution rather than with respect to composition, sequence length, or molecular weight (*11, 15*). Therefore, at any retention time, each detector can be viewing one molecular size but a wide variety in terms of the three properties of interest. In particular, molecules of many different compositions may be present and must be characterized by one average composition value at each retention time determined by the two detector responses. This has led recently to the suggestion that GPC was not a good method for determining copolymer composition distribution (*11*). A second fundamental difficulty is associated with detector response. In several examples, spectrophotometric detector response has been influenced by both composition and sequence length (*12, 16–19*). Because a variety of sequence lengths as well as compositions is expected within each detector cell, the equation shown in Figure 1 can be invalidated.

The objective of this paper is to show OC development aimed at circumventing the problems described to elucidate copolymer property distributions.

Theory

Fractionation. Figure 2 shows copolymer property distributions illustrated as a three-dimensional situation. The copolymer is represented as a contour map in the ternary diagram where each contour represents a different concentration. The use of a ternary diagram for visualization of copolymer property distributions is unconventional. Usually ternary diagrams are used for such applications as phase

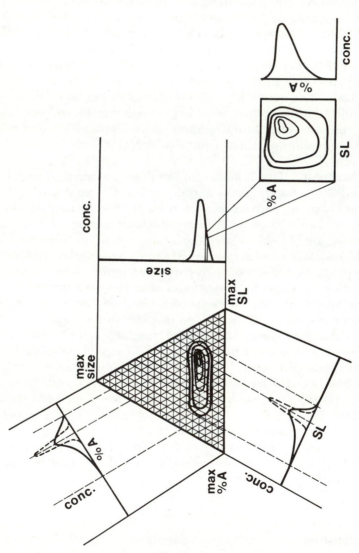

Figure 2. Copolymer property distributions presented as a three-dimensional situation in terms of molecular size in solution, composition, and sequence length.

diagrams, and the values of the coordinates in such cases add to unity. This is not the case here. In fact, for quantitative use, the scale for each of the axes would likely be nonlinear in this case, and/or quite possibly the contours of real copolymers would exceed the boundaries of the triangle. On each axis, composition, sequence length, and molecular size in solution are plotted. Molecular size is plotted rather than molecular weight because it is the basis upon which conventional GPC usually fractionates (15). Each side of this diagram shows the individual property distributions. In dual-detector GPC, fractionation proceeds vertically down the ternary diagram beginning at the vertex marked "Maximum Size." When the copolymer is encountered, the detectors see one molecular size but a variety of compositions. As shown in Figure 2, because this variety in compositions must be represented by a single average obtained from the two detectors at each retention time, a much narrower copolymer composition distribution than is actually present is obtained in this case. In OC, we attempt to remedy the situation by emphasizing desired fractionation. As shown in Figure 2 a segment, or slice of the molecular size distribution is envisioned to be represented by a two-dimensional plot of composition vs. sequence length. The basic idea underlying OC then is to accomplish a molecular size separation and to fractionate each point along the molecular size distribution according to composition to obtain a true copolymer composition distribution at the end. Two complications are that molecular size rather than molecular weight is the basis for the first fractionation and the presence of sequence length may interfere with the composition fractionation and detection.

COMBINING FRACTIONATION MECHANISMS. Figure 3 shows a scheme of an OC system. The effluent from the first GPC flows through the injection valve of the second. At any time, a slice of the chromatogram of the first GPC can be injected into the second. Different solvents are run in each instrument. Copolymer samples alone, or with their monomers in a reaction mixture, and/or with an internal broad molecular weight distribution polystyrene standard are dissolved in tetrahydrofuran (THF) and injected into the first instrument. The objective is to accomplish a molecular size separation with GPC 1 and a composition separation with GPC 2. Various detectors, notably a rapid scanning UV detector, can then be used to interpret the results.

Figure 4 illustrates the basis for the OC cross-fractionation. If, in GPC 1, conventional molecular size separation and adequate size resolution is assumed, then for molecules within the slice (i.e., in the injection loop of GPC 2) many molecular weights, copolymer compositions, or sequence lengths may be present, and all molecules are the same molecular size (hydrodynamic volume). In addition, differences

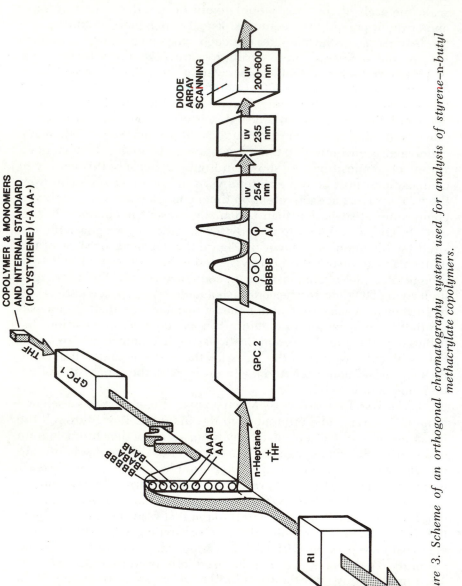

Figure 3. Scheme of an orthogonal chromatography system used for analysis of styrene–n-butyl methacrylate copolymers.

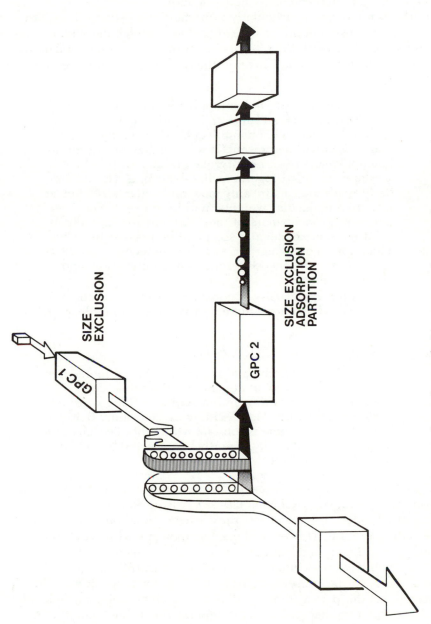

Figure 4. Basis for orthogonal chromatography cross-fractionation.

in molecular weight, if present, are always associated with a difference in composition and/or sequence length.

The last point is a particularly important one because it explains how one dimension of the analytical problem, molecular weight, can be removed although GPC 1 separates on the basis of molecular size and not molecular weight. This point may be explained more clearly in two ways:

1. If we envision a single homopolymer (e.g., polystyrene) dissolved in THF and injected into the first GPC, then within the slice only one molecular size and one molecular weight will be present. If we visualize two different homopolymers [e.g., polystyrene and poly(n-butyl methacrylate)] blended together, dissolved in THF and similarly analyzed, again only one molecular size, but at most two molecular weights, will be present. For three different homopolymers, at most three molecular weights, all of the same molecular size, will be present in a slice. Thus, for a copolymer we say that molecular weight is associated with the difference in composition and/or sequence length.

2. Molecular size usually is considered to be synonomous with hydrodynamic volume and is related to molecular weight by:

$$V = KM^{a+1} \tag{1}$$

where V is hydrodynamic volume, M is molecular weight, and K and a are Mark–Houwink constants. Changes in composition and/or sequence length are expected to affect the Mark–Houwink constants. Because hydrodynamic volume is constant within the slice, changes in these constants will generally be reflected by a change in molecular weight.

GPC columns sometimes exhibit adsorption and partition mechanisms along with size exclusion (20–23). Adsorption and partition are generally undesirable because they cause violations of universal calibration in GPC. However, these mechanisms with GPC columns have been utilized to effect analysis of small molecules (20).

If, in GPC 2, molecular size exclusion, adsorption, and partition are assumed to be present, then, in GPC 2, for a specified column packing, (a) the pore volume available to a given polymer molecule and the surface area that it sees depend on its molecular size in solution; and (b) the size of the molecule in solution, its adsorption coefficient, and its solubility are functions of the molecular weight, composition, and sequence length of the molecule as well as the characteristics of the solvent in GPC 2.

As a result, although molecules are the same size within the slice sampled of GPC 1 (in pure THF), when introduced into the solvent of GPC 2, a size distribution is probable and will influence the separation in GPC 2. Furthermore, because molecular weight is associated with composition and/or sequence length, differences in molecular weight within the slice sampled can actually be desirable because they can be used to enhance the separation in GPC 2.

For example, for the specific case of styrene−n-butyl methacrylate, the solvent mixture in GPC 2 is a much poorer solvent for polystyrene than for poly(n-butyl methacrylate). Thus, styrene-rich molecules are expected to shrink more than n-butyl methacrylate-rich molecules when injected into GPC 2 and, therefore, see more area of the packing. Also, retardation of polystyrene by adsorption partition effects has been demonstrated in this solvent. Thus, the result is that in OC the three mechanisms can be manipulated through solvent choice to work synergistically together to accomplish the HPLC of polymers.

FRACTIONATION COMPLICATIONS. The effect of polymer concentration injected into GPC 1 and the possible influence of the solvent flowing through GPC 1 on the separation obtained in GPC 2 must be considered in applying this type of combined chromatography technique. Too high a concentration of polymer injected into GPC 1 can severely degrade the resolution in that instrument. However, too low a concentration can mean undetectable quantities exiting from GPC 2. This chapter examines the effect of concentration by analyzing a blend of two homopolymers [polystyrene and poly(n-butyl methacrylate)]. In this case, the output from GPC 2 will be two peaks, one for each homopolymer. The resolution of these peaks can be assessed by calculation of a modified resolution index. This index is defined as:

$$R_M = \frac{2\Delta t}{\dfrac{W_1}{h_1} + \dfrac{W_2}{h_2}} \tag{2}$$

where (refer to the insert, Figure 10): Δt is peak separation, W_1 and W_2 are peak widths, and h_1 and h_2 are peak heights.

The difference between this modified index and the conventional one is that the peak widths in each case have been divided by the peak heights. This division was done because the areas under these peaks are generally not the same. The modification is a normalization to unit height for each peak by assuming that the peak height should be proportional to the peak width.

In analysis of small molecules by combined chromatographic methods (particularly GPC with HPLC), compatibility of the mobile phases running in each instrument is a major consideration (24). For example, use of adsorption chromatography in the HPLC attached to a GPC was considered impractical because of the use of THF in the GPC (25). Unlike other methods of combining chromatography, OC utilizes a size exclusion mechanism in GPC 2 along with adsorption and partition. Therefore, it may be possible to separate the GPC 1 solvent from the macromolecules.

Quantitative Detection and Interpretation. Mixed mechanisms in GPC 2 and the coupling of composition with sequence length effects cause calibration referencing retention time to be very difficult. As mentioned previously, the use of sequences of detectors is complicated by practical difficulties. Quantitative use of scanning detectors can potentially eliminate all of these problems (26). A strategy for a dynamic method of calibration is as follows:

SPECTRA ANALYSIS. This involves identifying significant variations in the spectra, attributing a source to these variations, and suitably quantifying the results. Inspection of the homopolymer spectra provides an indication of where the different monomer units will contribute. However, the copolymer spectra can show differences because of sequence length or other microstructure effects (12, 16–19). Also, because very low concentrations are usually exiting from the second GPC, maximizing precision and distinguishing signal variations from noise are important considerations. Use of absorbance values that have been averaged over defined regions of the spectra can significantly improve precision and can allow for minor shifts in peak absorbances. However, use of such averages implies knowledge of all properties affecting absorbance in the region. If, for example, sequence length affects the average absorbance of some samples where only a composition effect has been assumed, a significant error could be introduced. For many comonomers a wavelength region can be found where sequence length has no effect. This region can be determined by examining UV scans obtained either on-line or off-line. In both cases significant spectral shape changes within a region assumed to reflect only the concentration of one of the monomer units would indicate the intrusion of sequence length effects. Such shape changes can be revealed by examining the differences in the normalized absorbances within the region from sample to sample. In practice, the residual absorbance is examined and is defined by:

$$D(\lambda) = A_{N,i}(\lambda) - \bar{A}_N(\lambda) \tag{3}$$

where

$$A_{N,i}(\lambda) = \frac{A_i(\lambda)}{\int_{\lambda_1}^{\lambda_2} A_i(\lambda)d\lambda}$$

$$\bar{A}_N(\lambda) = \sum_{i=1}^{n} \frac{A_{N,i}(\lambda)}{n}$$

and D is residual normalized absorbance, $A_i(\lambda)$ is absorbance at wavelength λ, $A_{N,i}(\lambda)$ is normalized absorbance at λ for spectra i, and $A_N(\lambda)$ is normalized absorbance at λ averaged over n spectra. A plot of the residual normalized absorbance vs. wavelength should show a random pattern if no undesirable shape changes are present.

CALIBRATION FOR EACH PROPERTY.   For the scanning UV detector, the basic calibration equation is Beer's law at each wavelength.

$$A(\lambda) = \left[K_{1,\lambda} W_1 + K_{2,\lambda} (1 - W_1) \right] c \tag{4}$$

where $A(\lambda)$ is the absorbance at λ; $K_{1,\lambda}$ and $K_{2,\lambda}$ are proportionality constants for monomers 1 and 2 in the polymer, respectively; W_1 is the weight fraction of monomer 1 in the copolymer; and c is the concentration of the copolymer in the solution.

Under the following conditions:

1. No sequence length effects on absorbances are used.
2. Absorbances in the wavelength range 254–280 nm reflect only styrene concentration in the copolymer.
3. Absorbances in the range 235–245 nm reflect only styrene and n-butyl methacrylate concentrations in the copolymer.

The following equations are obtained from Equation 4:

$$\frac{A_{235-245}}{A_{254-280}} = b_1 + \frac{b_2}{W_1} \tag{5}$$

$$c = \frac{A_{254-280}}{K_{1,\ 254-280}\ W_1} \tag{6}$$

where $A_{235-245}$ is the absorbance averaged over 235–245 nm inclusive, $A_{254-280}$ is the absorbance averaged over 254–280 nm inclusive, b_1 and b_2 are calibration constants, and $K_{1,\ 254-280}$ is the proportionality constant for styrene averaged over 254–280 nm.

Equation 5 provides an expression for the weight fraction styrene in the copolymer as a function of absorbance ratios. This quantity is the abscissa of the copolymer composition distribution. Equation 6 is an expression for the concentration of the copolymer of each composition and provides the basis for calculation of the ordinate of the copolymer composition distribution. The proportionality constant in Equation 6 does not have to be determined explicitly as it cancels in the following equations.

CALCULATION OF THE PROPERTY DISTRIBUTION ORDINATE FOR THE SLICE. The ordinate for the differential copolymer composition distribution is defined as follows:

$$g_i(W_1)dW_1 = \frac{cdt}{\int_0^\infty cdt} \tag{7}$$

or

$$g_i(W_1) = \frac{\dfrac{cdt}{dW_1}}{\int_0^\infty cdt} \tag{8}$$

where $g_i(W_1)$ is the ordinate of the differential copolymer composition distribution, $g_i(W_1)dW_1$ is the weight fraction of copolymer of styrene weight fraction W_1 to $W_1 + dW_1$, t is the retention time, and cdt is the concentration of copolymer from time t to $t + dt$.

Equation 7 shows the basis for this distribution. That is, the area of a differential increment of the distribution (left side of Equation 7) must equal the differential area of a plot of weight fraction of the polymer vs. time (right side of Equation 7). Calculation of the ordinate from Equation 8 requires estimation of a derivative (dt/dW_1). The derivative was calculated from a linear regression of the absorbance ratio vs. time for each sample, and use of the derivative of Equation 5.

CALCULATION OF THE TOTAL COPOLYMER COMPOSITION DISTRIBUTION. The calculations outlined previously are performed for each slice analyzed by GPC 2. To obtain the total property distribution the results must be summed over all the slices. That is, the ordinate is calculated from

$$g_j(W_1) = \sum_{i=1}^{N} \alpha_i g_i(W_1) \tag{9}$$

where $g_j(W_1)$ is the ordinate for the total differential copolymer composition distribution

$$\alpha_i = \frac{\int_0^\infty cdt}{\sum_{j=1}^{N}\left[\int_0^\infty cdt\right]_j} \tag{10}$$

where N is the number of slices analyzed by GPC 2.

Equation 9 is a weighted sum of the differential copolymer composition distributions of each of the slices. The weighting factor is the concentration of the slice relative to that of the whole polymer. The individual ordinates (g_i) must be at the same weight fraction styrene in the copolymer (W_1) each time Equation 9 is used. This means that the ordinates of the individual copolymer composition distributions must be interpolated at each value of W_1. Lagrangian interpolation was used in this chapter.

Experimental

Both equipment and procedures were similar to that used in previous literature (*1, 2*). In particular, as shown in Figure 15 of Reference 1: GPC 1 (a Waters GPC/ALC 401) was attached to GPC 2 (a Spectra-Physics SP8000) so that the eluent from the first GPC ran through the injection valve of the second. Column switching valves were as shown in that figure and enabled GPC 1 to be operated in a stop-and-go fashion. However, new operating conditions enabled much more rapid analysis than these. Specific conditions were for GPC 1: 10^4, 10^5, 10^3 Å μStyragel, 2 cm³/min THF then stepped at 0.5 cm³/min for each new slice, 0.2–0.5-cm³ injection loop, 0.2–0.4% poly *n*-butyl methacrylate and/or polystyrene–*n*-butyl methacrylate and/or polystyrene; and GPC 2: E1000, E500, E125 μBondagel, 0.5 cm³/min THF/*n*-heptane, 0.1-cm³ injection loop. Three UV detectors were SP8310 (λ = 254 nm), Schoeffel (λ = 235 nm) and HP8450 UV/VIS diode array scanning spectrophotometer (used without stop flow (*10*) operation since it provided scans in ~2 s).

Typically a run would consist of injection of a 50/50 mixture of 0.4% polystyrene–*n*-butyl methacrylate copolymer and NBS-706 polystyrene dissolved in THF. GPC 1 was operated at 2 cm³/min for 9 min. At that time, valves on that instrument are switched to stop the flow. An injection is then made into GPC 2, which is running at 0.5 cm³/min. Just before the sample exits, the scanning spectrophotometer begins to scan for approximately 2 s every 10 s. Analysis times are approximately 15 min in GPC 2. After the GPC 2 run is complete the sample in GPC 1 is advanced by 1 min at a flow rate of 0.5 cm³/min and the operation is repeated. If desired, retention time in figures can therefore be converted to retention volume by multiplying seconds by (2.00/60) for GPC 1 values and seconds by (0.50/60) for GPC 2 values.

Results and Discussion

Fractionation. Figure 5 shows a series of GPC 2 chromatograms obtained at 235 nm. The results from injecting into GPC 2 neighboring slices of the chromatogram of the mixture of three polymers [polystyrene, polystyrene–*n*-butyl methacrylate, and poly(*n*-butyl methacrylate)] injected into GPC 1. The effect of changing the GPC 2 mobile phase concentration of *n*-heptane in THF from 0 to 70%, inclusive, is shown. The following points were evident:

1. Increasing *n*-heptane concentration led to complete resolution showing one whole peak for each type of polymer present.

2. At 0% *n*-heptane, the three polymers exited as only one

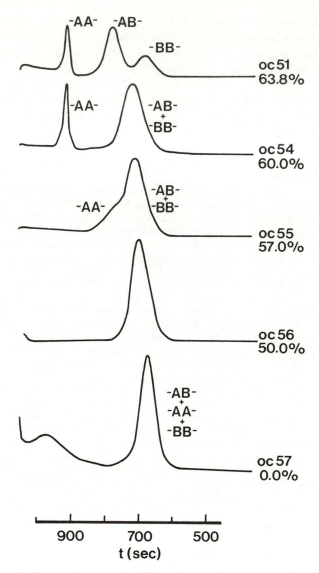

Figure 5a. Series of OC runs showing effect on separation of polystyrene (~AA~), polystyrene–n-butyl methacrylate (~AB~), and poly(n-butyl methacrylate) (~BB~) of n-heptane concentrations ranging from 0 to 63.8% by volume in the THF/n-heptane mobile phase of GPC 2.

peak as expected if a molecular size separation is occurring in both GPCs.

3. Peak shapes and retention times varied not only with the composition of the solvent but also with the location of the slice in the chromatogram of GPC 1.

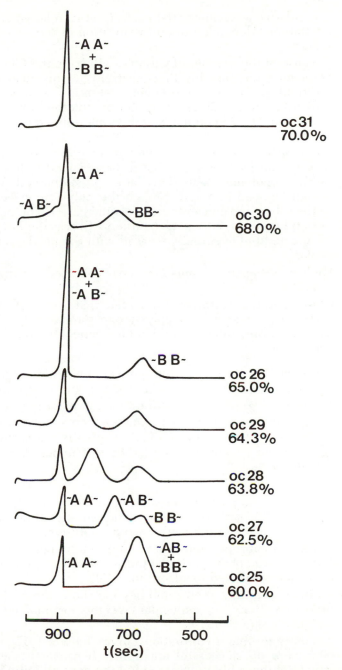

Figure 5b. Series of OC runs showing effect on separation of polysty-rene (~AA~), polystyrene−n-butyl methacrylate (~AB~), and poly(n-butyl methacrylate) (~BB~) of n-heptane concentrations ranging from 60.0 to 70.0% by volume in the THF/n-heptane mobile phase of GPC 2.

4. The relative areas under the peaks (that is, relative concentrations) depended upon the location of the slice in GPC 1.

5. Because of the number of variables affecting the GPC 2 chromatograms (notably the sensitivity to solvent composition) calibration referencing retention volume is difficult (note the different resolution obtained for runs OC-51 and OC-28 nominally at the same solvent composition).

6. As the n-heptane concentration was increased 70%, the polystyrene peak was the first to be resolved followed by the copolymer peak. This latter peak (marked AB) actually moved into and through the polystyrene peak as the n-heptane concentration was increased. The poly-(n-butyl methacrylate) peak (marked BB) was the last to be affected but eventually merged with the polystyrene peak.

7. Higher n-heptane concentrations incurred sharper peaks.

8. Cloud point experiments indicated that some of the higher molecular weight polystyrene standards should have been precipitating at n-heptane concentrations beyond 65%.

To investigate fractionation further, conventional calibration curves were constructed for GPC 2 at different n-heptane compositions of the mobile phase (Figure 6). The calibration curves in Figure 6 were obtained by dissolving monodisperse polystyrene standards in THF and injecting them directly into GPC 2. The calibration curve is a plot of log molecular weight vs. retention time. The points to note regarding this figure are:

1. Increasing n-heptane concentration in the mobile phase from 0 to 70% markedly increases the retention time, particularly for higher molecular weight polystyrenes.

2. There is an abrupt jump in retention time in going from 57 to 60% n-heptane [a similar jump was observed previously and attributed to adsorption (27)].

3. Calibration curves at 60% and beyond show remarkably little sensitivity to molecular weight for molecular weights higher than 10^5.

4. Attempts to obtain calibration curves between 57 and 60% were not successful apparently because the sensitivity of the calibration exceeded the controllability of the mobile phase composition.

5. Although column plugging was not encountered, occa-

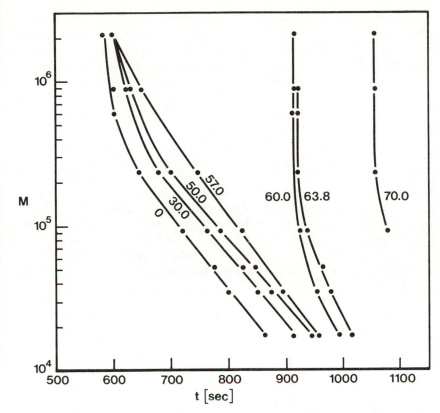

Figure 6. Polystyrene calibration curves for GPC 2: variation with percent n-heptane.

sionally difficulties were encountered because of adsorption effects. For example, very high molecular weight standards beyond 60% *n*-heptane sometimes showed bimodal peaks and were considered invalid data. Also, injection of blank THF samples into the columns would occasionally demonstrate that some polystyrene had adsorbed and could be washed off. Similarly, residual polystyrene could sometimes be seen as part of the next sample's chromatogram. These problems were not encountered with pure poly(*n*-butyl methacrylate) or with any of the copolymer samples. However, they demonstrate that considerable care must be taken in interpreting styrene-rich samples at high *n*-heptane compositions.

6. No precipitation of polymer in the detector cell was evident. Detector response was proportional to polystyrene concentration. All analyses were at 25 °C (detectors and columns were at approximately equal temperatures).

In developing this method, many different column combinations were tried in GPC 2. We recently found an interesting sensitivity to the order of the columns in the second GPC, which possibly accounts for some of the difficulty in column selection for this method. Figure 7 demonstrates the results of analyzing a blend of two homopolymers [polystyrene and poly(n-butyl methacrylate)] injected into GPC 1 and analyzed by GPC 2 at 63.8% n-heptane. When the small pore size column was moved to the front of the column line (near the injection valve), the polystyrene peak moved into the solvent impurity peak. The poly(n-butyl methacrylate) peak was unaffected. Figure 8 shows results that account for both the lack of polymer precipitation in the columns and the sensitivity to column ordering. This figure was obtained by direct injections of polystyrene standards into GPC 2. These standards were dissolved in solvent of composition ranging from 0 to 50% n-heptane in THF. GPC 2 used a mobile phase of 63.8% n-heptane. The figure shows a plot of the polystyrene retention time on the ordinate vs. the composition of the solvent used to dissolve the sample. As the solvent became richer in n-heptane, the polystyrene

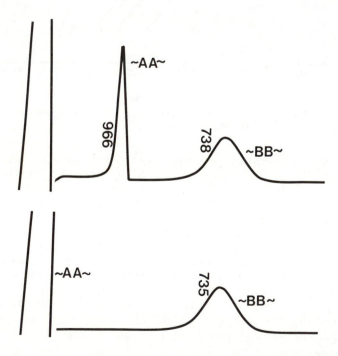

Figure 7. Effect of column reordering on polystyrene (AA) and poly(n-butyl methacrylate) (BB) retention in GPC 2. Key: top, conventional order; and bottom, E125 column first.

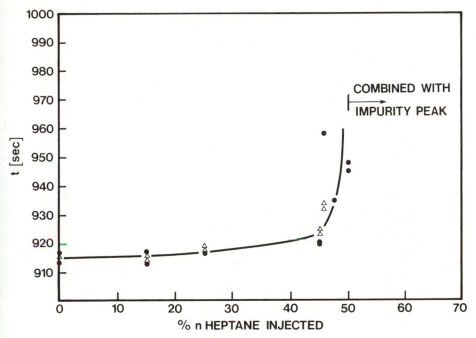

Figure 8. Effect of sample solvent composition on polystyrene reten-tion in GPC 2.

retention time, beginning at about 45% *n*-heptane, rapidly increased until the polystyrene peak combined with the solvent impurity peak. This demonstrated that the solvent in GPC 1 (THF) was participating in the separation obtained in GPC 2. The lack of precipitation of polystyrene at the higher *n*-heptane concentrations then was attrib-uted to the ability of the THF pulse to dissolve the sample off the packing. The sensitivity of the polystyrene retention time to the order of the columns is probably associated with the presence of an exclu-sion mechanism. That is, the polystyrene only manages to separate from the THF when it is able to bypass pores that the THF must enter. On the large pore size columns, it is sufficiently retarded by adsorp-tion to be overtaken by THF. The ability of the polymer to be solvated by THF probably determines the degree to which separation can be obtained.

To investigate the effect of concentration injected into GPC 1 on a resolution obtained in GPC 2, a 50/50 blend of polystyrene–poly-(*n*-butyl methacrylate) was injected at concentrations of 0.1, 0.2, and 0.4% (0.5-cm³ injection loop). Figure 9 shows the peak separation obtained as a function of the location of the slice on the chromatogram of GPC 1. Figure 10 shows the modified resolution indices. These

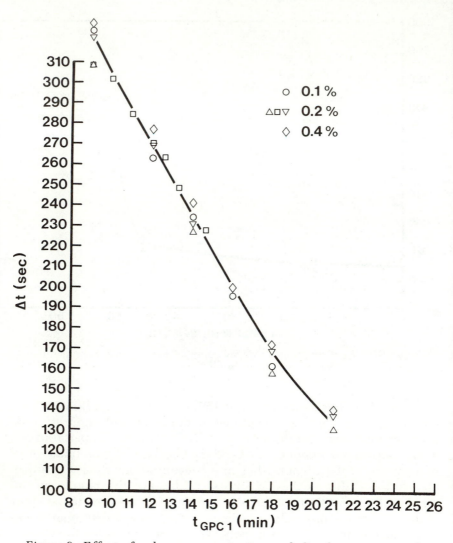

Figure 9. *Effect of polymer concentration and slice location on peak separation in OC analysis of a homopolymer blend.*

results demonstrate that, although peak separation depends on location of the slice on the chromatogram of GPC 1, the values of the
modified resolution index were higher at higher concentrations. This
result was traced to the observation that, while peak widths and peak
separation remain approximately constant, the heights of the peaks
increased greatly as concentration was increased. Thus, in this case,
increased concentrations had only the beneficial effect of improved
detectability limits in GPC 2. Whether this result is due to a concen-

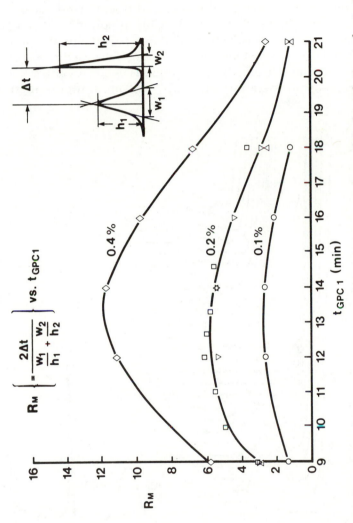

Figure 10. Effect of polymer concentration and slice location on the modified resolution index in OC analysis of a homopolymer blend.

tration tolerance of the columns in GPC 1 or to a tolerance for poor
resolution in GPC 1 by GPC 2, or to the effect of solvent injected with
the sample is yet to be determined.

Quantitative Detection and Interpretation. Figure 11 shows
typical separation of a copolymer from a pure homopolystyrene as it
appears from the scanning UV detector and from a fixed wavelength
254-nm detector. Figure 12 shows UV scans of polystyrene and poly-
(n-butyl methacrylate). Inspection of these scans provides only an
indication of what to expect for the copolymer, because its spectra may
show sequence length effects. Two regions were tentatively identified for composition determination: 254–280 nm as representative of

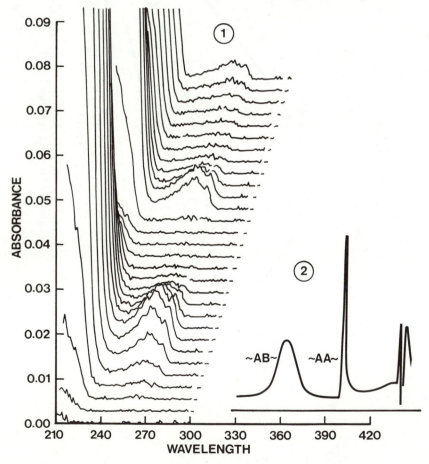

Figure 11. Output of OC analysis utilizing a scanning detector (1) and
a fixed wavelength UV detector (2).

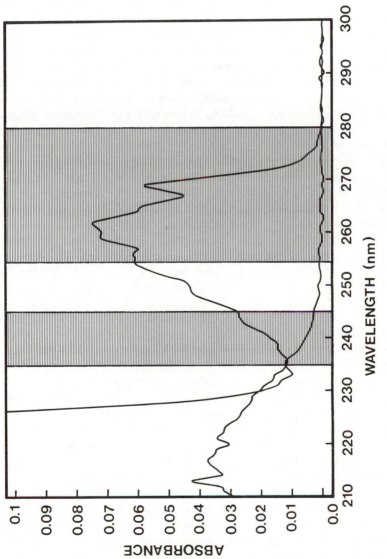

Figure 12. UV scans of polystyrene and poly(n-butyl methacrylate) obtained during OC analysis (samples represent different concentrations).

styrene concentration in copolymer, and 235–245 nm as being the result of both styrene and n-butyl methacrylate concentration in the copolymer. Figure 13 shows estimation of spectral shape changes for the former region by inspection of the residuals of the normalized spectra. From 280 to 300 nm only base line noise is expected. Because the scatter in this region appeared equal to that in the 254–280-nm region, for these samples, changes due to sequence length in this region were not evident. Figure 14 shows a calibration curve obtained for absorbance ratios. Average values of absorbance in the two regions were used to obtain the ratios. As shown in this figure, although a plot vs. the reciprocal weight fraction styrene in a copolymer is linear, the relationship is really a hyperbola with respect to weight fraction alone. Thus, although absorbance ratios can be obtained readily in on-line analysis, their use for polystyrene- or n-butyl methacrylate-rich copolymers can present high experimental error. Also, on-line determination of this calibration curve using copolymer standards of known composition was associated with some significant experimental difficulties. The most desirable way of obtaining the calibration curve was by direct injection of these standards into the second GPC and averaging of the absorbance ratios obtained across the chromatograms. However, upon direct injection, the samples provided broad chromatograms that were usually interfered with by the solvent impurity peak. The calibration curve was finally obtained by averaging results of OC runs.

Figure 15 shows the calculation of a polymer composition distribution for a slice derived from analysis of a blend of a 23.5% styrene–n-butyl methacrylate copolymer and polystyrene. Each pair of points in the chromatogram shown on this figure was obtained from averaging the two respective regions on a UV scan. The absorbance ratios changed only slightly for the copolymer and not at all for the homopolymer. The resulting copolymer composition distributions show a fairly symmetrical peak centering about the average composition for the copolymer and a sharp pulse for the polystyrene. Figure 16 shows the result of summing the contributions of many individual slices to obtain the total copolymer composition distribution for the 23.5% polystyrene–n-butyl methacrylate copolymer. Although the distribution is located near the average composition and the order of the individual components is reasonable (high molecular weight components were richer in n-butyl methacrylate as would be expected because they were produced near the end of the polymerization), the distribution does not agree with the predictions of classical copolymerization kinetics. These kinetics predict a much more skewed distribution beginning at 30% styrene and extending down toward 0% styrene. At this time, uncertainties exist in both the kinetic predic-

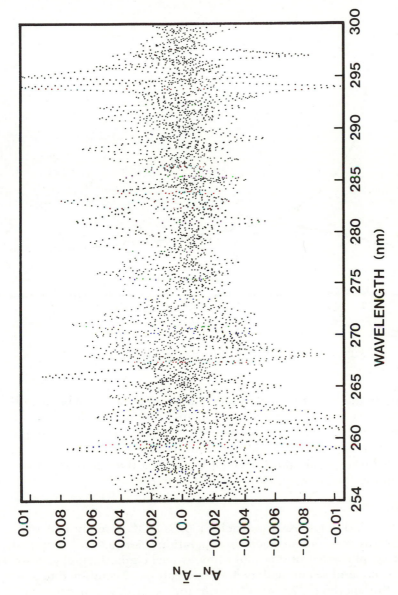

Figure 13. Estimation of spectral shape changes by inspection of residuals of the normalized absorp-tion spectra.

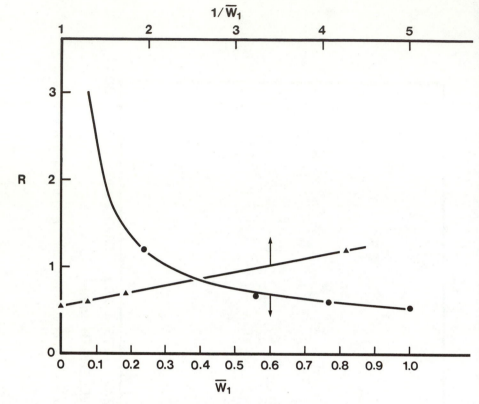

Figure 14. Calibration utilizing absorbance ratios.

tions and in the analytical technique. The kinetics have not been well tested against copolymer composition distributions because of the very limited ability to measure such distributions. Even for average composition values, significant uncertainties have been revealed (*1, 28, 29*). With respect to the analytical technique, the role of sequence length in the fractionation is the main uncertainty.

Conclusions

As detailed in two review articles (*30, 31*), OC is really an association of many previous advances in separation science. Work in mixed chromatographic mechanisms, multidimensional chromatography, cross-fractionation, and scanning detectors have all been combined to synthesize this method. The result is an approach that synergistically unites steric exclusion mechanisms with those of HPLC to obtain the HPLC of macromolecules.

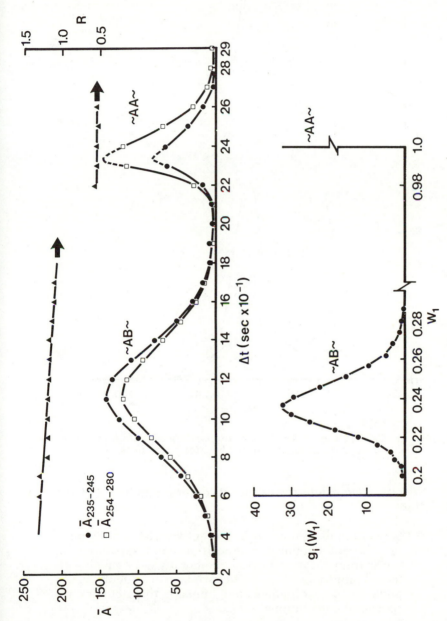

Figure 15. Calculation of copolymer composition distribution for a slice.

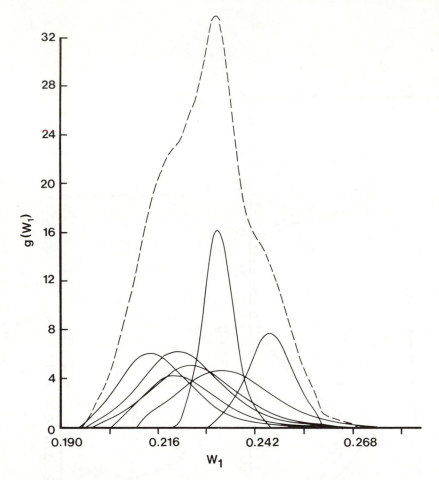

Figure 16. Calculation of copolymer composition distribution for the whole copolymer by summing the slice contributions.

Specific conclusions regarding OC determined in this chapter are as follows:

1. Separation of styrene–*n*-butyl methacrylate copolymers of different chemical compositions and separation of these from their respective homopolymers can be effected rapidly by OC. These various constituents would normally be well-concealed beneath the detector response of conventional GPC.

2. The solvent injected into GPC 2 along with the polymer sample forms a gradient that participates in separation. Evidence for this effect was used to explain the lack of

precipitation of polystyrene at high nonsolvent compositions in GPC 2 and the demonstrated sensitivity of the polystyrene separation to column ordering.

3. An investigation of the effect of concentration on the OC separation of the two homopolymers showed that relatively high concentrations could be used to improve detectability limits without noticeably affecting resolution.

4. A dynamic method of calibrating GPC 2 for copolymer composition that does not depend on established calibrations with retention times is shown based on the use of scanning detectors. An HP8450 UV/VIS scanning detector was used to detect and calculate copolymer composition distribution.

5. No detector response attributable to sequence length was determined with the detector and operating conditions used. Sequence length represents the main uncertainty with respect to the accuracy of the determined copolymer composition distributions.

Acknowledgment

We wish to thank Luis Garcia-Rubio for his helpful and interesting discussions regarding UV analysis of copolymers.

Literature Cited

1. Balke, S. T.; Patel, R. D. In "Size Exclusion Chromatography (GPC)"; ACS SYMPOSIUM SERIES No. 138, ACS: Washington, DC, 1980; p. 149.
2. Balke, S. T.; Patel, R. D. *J. Polym. Sci., Part B* **1980**, *18*, 453.
3. Teramachi, S.; Fukao, T. *Polym. J.* **1974**, *6*, 532.
4. Teramachi, S.; Esaki, H. *Polym. J.* **1975**, *7*, 593.
5. Teramachi, S.; Hasegawa, A.; Shima, Y.; Akatsuka, M.; Nakajima, M. *Macromolecules* **1979**, *12*, 992.
6. Mori, S. *J. Chromatogr.* **1980**, *194*, 163.
7. Runyon, J. R.; Barnes, D. E.; Rudd, J. F.; Tung, L. H. *J. Appl. Polym. Sci.* **1969**, *13*, 2359.
8. Teramachi, S.; Hasegawa, A.; Akatsuka, M.; Yamashita, A.; Takemoto, N. *Macromolecules* **1978**, *11*, 1206.
9. Probst, J.; Cantow, H. J. *Kautsch. Gummi, Kunstst.* **1972**, *25*, 11.
10. Belenkii, B. G.; Gankina, E. S.; Nefedov, P. P.; Lazareva, M. A.; Savitskaya, T. S.; Volchikhina, M. D. *J. Chromatogr.* **1975**, *108*, 61.
11. Ogawa, T. *J. Appl. Polym. Sci.* **1979**, *23*, 3515.
12. Garcia-Rubio, L. H. *J. Appl. Polym. Sci.* **1982**, *27*, 2043.
13. Stojanov, C.; Shirazi, Z. H.; Audu, T. O. K. *Chromatographia* **1978**, *11*, 274.
14. Stejskal, J.; Dratochvil, P. *J. Appl. Polym. Sci.* **1978**, *22*, 2925.
15. Ambler, M. R.; McIntyre, D. *J. Polym. Sci., Part B* **1975**, *13*, 589.
16. Brussau, V. R. J.; Stein, D. J. *Angew. Makromol. Chem.* **1970**, *12*, 59.
17. Gallo, B. M.; Russo, S. *J. Macromol. Sci. Chem.* **1974**, *8*, 521.
18. Stutzel, B.; Miyamoto, T.; Cantow, H. J. *Polym. J.* **1976**, *8*, 247.
19. Gruber, E.; Knell, W. *Makromol. Chem.* **1978**, *179*, 733.

20. Mori, S.; Yamakowa, A. *Anal. Chem.* **1979,** *51,* 382.
21. Bakos, D.; Bleha, T.; Ozima, A.; Berek, D. *J. Appl. Polym. Sci.* **1979,** *23,* 2233.
22. Campos, A.; Soria, V.; Figueruelo, J. E. *Makromol. Chem.* **1979,** *180,* 1961.
23. Dawkins, J. V.; Hemming, M. *Makromol. Chem.* **1975,** *176,* 1975.
24. Majors, R. E. *J. Chromatogr. Sci.* **1980,** *18,* 571.
25. Johnsen, E. L.; Gloor, R.; Majors, R. E. *J. Chromatogr.* **1978,** *149,* 571.
26. Mirabella, F. M., Jr.; Johnson, J. F.; Barrall, E. M., II *Am. Lab.* **1975,** Oct., 65.
27. Tennikov, M. B.; Nefedov, P. P.; Lazareva, M. A.; Frenkell, S. Y. *Vysokomol. Soedin., Ser. A* **1977,** *19,* 657.
28. Johnson, M.; Karmo, T. S.; Smith, R. R. *Eur. Polym. J.* **1978,** *149,* 572.
29. Dionisio, J. M.; O'Driscoll, K. F. *J. Polym. Sci., Part B* **1979,** *17,* 701.
30. Balke, S. T. *Sep. Purif. Methods,* in press.
31. Balke, S. T. *Polym. News,* in press.

RECEIVED for review October 14, 1981. ACCEPTED April 22, 1982.

Size Exclusion Chromatography of Copolymers

LUIS H. GARCIA RUBIO

Xerox Research Centre of Canada, Materials Processing Laboratory, 2480 Dunwin Drive, Mississauga, Ontario L5L 1J9 Canada

J. F. MacGREGOR and A. E. HAMIELEC

McMaster University, Department of Chemical Engineering, Hamilton, Ontario L8S 4L7 Canada

Size exclusion chromatography (SEC) of copolymers has two basic requirements: the capability to detect selectively the concentration of the copolymer components in the eluting solution, and to obtain suitable calibration and interpretation for both size and composition fractionation. To calibrate and interpret copolymer fractionation results by using the existing theory, it is necessary to determine the experimental conditions that guarantee either size or composition fractionation. Sufficient knowledge of the hydrodynamic properties of the copolymer molecules in solution is also required to relate copolymer size to molecular weight. This chapter reviews the SEC fractionation of copolymers and outlines techniques used to ensure size separation for styrene—acrylonitrile copolymers. To test the separation mechanism, a number of solvents and column configurations were investigated. Conditions that give mainly size or mainly composition fractionation are discussed, as well as guidelines for the selection of solvent and column configuration. Extensive characterization of styrene—acrylonitrile copolymers, based on size fractionation and careful selection of the detection system, is reported.

SIZE EXCLUSION CHROMATOGRAPHY (SEC) of copolymers has two basic requirements: the capability of selectively detecting the concentration of the copolymer components from the eluting solution, and the ability to obtain suitable calibration and interpretation for both size and composition fractionation. The problems of concentration detection, detector selection, and the use of UV spectrophotometers for sequence length determination are discussed elsewhere (1). To calibrate and interpret copolymer fractionation results by using existing theory, it is necessary to determine the experimental conditions that ensure either size or composition fractionation. Sufficient knowledge of the hydrodynamic properties of the molecules in solution is also required to relate molecular size to molecular weight. Because changes in microstructure are known to influence solubility strongly, and presumably adsorption, as well as the size of the molecules in solution, sequence length effects may add to these complications. Extensive research (2–5) on polymer–solvent–packing interactions, with pure and mixed solvents, has shown that significant adsorption effects may be present in column fractionation. Furthermore, some remarkable copolymer fractionations in terms of composition have been reported for liquid and thin layer chromatography (6, 46). However, no methodology has been proposed to test for adsorption and interaction effects and to ensure size separation for copolymer molecular weight analysis. This chapter reviews the SEC fractionation mechanism and the limitations of existing calibration and interpretation techniques, and outlines the technique used to ensure size fractionation for styrene–acrylonitrile (SAN) copolymers.

SEC Separation Mechanism

Under ideal conditions in SEC (uniform solvent–polymer–packing interactions) polymer molecules are fractionated according to their size in solution. Some complications may arise, however, due to the interactions between the packing material, the polymer, and the solvent. These interactions will tend to increase the dispersion and decrease the resolution. In copolymers, where there is a significant difference in the chemical nature of the comonomers (e.g., SAN copolymers), interactions are expected to be important, and a fractionation strictly on size or composition cannot be assumed. Anomalous molecular weight–retention volume behavior has been observed over the years for polymers of varying polarity in a number of mobile phase–substrate systems. Systematic studies on the effect of such interactions on the molecular weight interpretation have been reported. The elution behavior of polymers with wide ranges of polarities was studied in dimethylformamide/lithium bromide and a variety of substrates. Styragel and silanized glass were reported as having affinity for polar polymers. Consequently, application of universal calibration curves (based on polystyrene) to SEC analysis of more polar polymers with these substrates led to underestimation of the molecular weights. Ideal SEC behavior was found for a number of polymers on untreated glass surfaces; however, polymers with strong hydrogen bonding functionality appear to be susceptible to marked adsorption in dimethylformamide/glass systems (5). Polystyrene standards in a variety of solvents of different thermodynamic quality were studied (2, 3, 7) with Porasil as stationary phase. A number of interesting observations that have direct bearing on copolymer analysis were drawn from those results:

1. Adsorption is a function of the thermodynamic quality and the selectivity of the solvent.
2. Partition of the polymer solute between the mobile and stationary phases is a function of the molecular weight.
3. The universal calibration procedure based on the hydrodynamic volume of the polymer molecules is directly applicable only for systems for which adsorption/partition dependence is approximately the same as molecular size dependence.

The polymer retention mechanism in gel permeation chromatography (GPC) on active gels was studied by using pure and mixed eluents and narrow PS standards (4). The effect of different eluents on the exclusion volume, total permeation volume, and shifts of the universal calibration curve ($M[\eta]$ vs. RV) were analyzed as a function of the thermodynamic quality of the eluent and the solvent strength. The effective radius of the gel pores (or conversely the hydrodynamic volume) was found to be a joint function of the solvent strength and its thermodynamic quality. This finding was explained in terms of the formation of a layer of a quasi-stationary phase of eluent that interacts with the substrate. The solvent with greater affinity for the substrate will be more concentrated in the neighborhood of the gel. Depending on the thermodynamic quality of the solvent and the compatibility of the solute with the packing material, the partition coefficient will increase to the point of complete adsorption, or decrease to total exclusion. Fractionation of styrene-*n*-butyl methacrylate copolymers was accomplished for composition from mixtures of tetrahydrofuran (THF)/*n*-heptane (8). These remarkable results, although not yet quantitative, indicate not only an alternative to composition analysis, but also, the magnitude of the effects that could be present due to polymer−solvent−gel interactions. Obviously, if reliable copolymer molecular weight results are to be obtained by SEC, the conditions that guarantee exclusive size separation must be met.

Review of Size Exclusion Chromatography Applications to Copolymer Analysis

Under the assumption of size separation the molecular weight distribution (MWD) for any polymer can be calculated from a molecular weight calibration curve that is based on universal calibration, the chromatogram, and the spreading correction.

In the case of copolymers, the hydrodynamic volume depends not only on the molecular weight, but also on the comonomer content. Thus, as illustrated in Figure 1, a mixture of polymer species with a variety of molecular sizes $(V^+ - V^\circ)$ and comonomer contents $(P_1^\circ - P_1^+)$ are eluted at any given time. In this case, the calibration curve cannot be fixed because it must be drawn in accordance with each

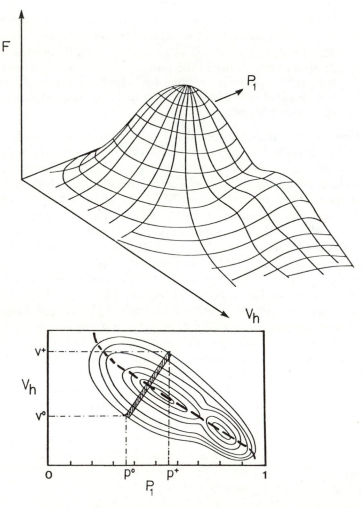

Figure 1. Schematic of the SEC fractionation of copolymers. Key: top, actual bivariate molecular size-composition distribution; and bottom, a possible trajectory of the fractionation (--) illustrating the mixture of sizes and compositions eluted at any one time.

polymer species. Theoretically, by measuring two properties of the copolymer at a given retention time, the problem of molecular weight and composition could be solved.

A variety of detectors and calibration techniques for copolymer analysis have been reported. Some of the difficulties of obtaining good quantitative data for copolymer systems have been discussed, and it was suggested that multiple runs with different solvents, or the use of multiple detectors, might be helpful (9). An IR detector was used on-line with the GPC to detect copolymer compositions (10). This procedure recommended the use of internal standards to avoid the errors arising in the injection loops and preparation of solutions. A UV detector and a refractometer were used in the analysis of styrene–butadiene block copolymers (11). A linear response for both detectors was assumed. The ratio of the responses of the homopolymers in each detector was used to correlate the composition results. The molecular weights were estimated under the assumption that the size of the copolymer molecule is equal to the size of two individual portions (as though each one was a homopolymer) and then interpolated on a log scale. Adams (12) used the same configuration as Runyon to analyze styrene–butadiene copolymers. A linear response was assumed again for the detectors. The use of two detectors was proposed (13) [UV and differential refractometer (DR)] in conjunction with an on-line viscometer. The UV and the refractometer were used to determine the composition and concentration of the fractions while the viscometer was used to obtain the molecular weight of the fractions from the viscosity and the universal calibration curve. A linear correction factor was used on the refractometer response to account for deviations from the overall copolymer composition. These results were found to be in agreement with molecular weight averages determined by osmometry and light scattering. However, the heterogeneity parameters determined from chromatography were unrealistic.

Mirabella and Barral (14, 15, 47) used IR detection and a "stop and go" technique in the analysis of styrene–maleic anhydride copolymers. A full IR spectrum was obtained every time the flow was stopped; therefore, the maximum information for every fraction across the chromatogram was obtained. The MWD was reported as "polystyrene equivalent". A dual calibration (i.e., a calibration curve for M_n and another for $\overline{M}_w$) was proposed (16) for the analysis of styrene–maleic anhydride copolymers of homogeneous composition [5–50 mol % maleic anhydride (MA)]. The universal calibration and the dual calibration yielded the same results. No effect of the copolymers composition on the separation mechanism or the molecular weight was reported.

Vinyl chloride–vinyl acetate copolymers have been studied extensively (17, 18, 48, 49). The validity of the universal calibration was established for this system in THF for the composition range 10–50% of vinyl acetate. Two packing materials were used (silica and Styragel) to test for polymer–solvent–packing interactions. The effect of the copolymer composition on the size exclusion mechanism for these packing materials was found negligible. The universal calibration was also found to be valid for styrene–butadiene block copolymers (6, 19). Tung also compared the universal calibration with the interpolation technique suggested by Runyon (11). The agreement on the molecular weights is within 6%.

The effect of the composition distribution on the elution behavior of ethylene–propylene copolymers was investigated (20, 21). Computer simulations were carried out assuming a bivariate normal copolymer composition distribution (CCD) and log-normal MWD. Results show that even for this system where the two homopolymers behave similarly in solution, and for which the copolymers give the same change in refractive index, molecules of different molecular weights and compositions elute at the same time. This simultaneous elution causes errors in the estimation of molecular weight averages, particularly $\bar{M}_w$, of up to 20%. These errors are a function of the breadth of the CCD.

The molecular weight distribution and the correlation between chemical composition and molecular weight of styrene–methyl acrylate copolymers was studied (22). The composition varies with molecular weight by approximately 20% in high conversion copolymers. However, for the composition range investigated (45–80 mol% MA), no effect of the composition on the separation mechanism was observed.

The elution behavior of styrene–acrylonitrile copolymers in dimethylformamide (DMF) and $CHCl_3$ with Styragel columns has been investigated (23). The universal calibration was found to be valid for $CHCl_3$ solutions but not for DMF and Styragel columns. The long retention times observed for polystyrenes in DMF were attributed to solvent–polymer interactions.

Calibration and Interpretation Techniques

Column calibration (i.e., molecular weight calibration and estimation of column dispersion) is an important step in the SEC analysis of polymeric materials. Traditionally, standards of known molecular weights have been used to calibrate the columns. The calibration curve as a function of the retention volume can be represented as:

$$M(v) = Q(v) \tag{1}$$

The chromatogram [$F(v)$], in turn, is represented as the convolution of the true distribution [$W(y)$] and the spreading function $G(v,y)$ (Tung's equation)

$$F(v) = \int_0^\infty W(y) \, G(v,y) \, dy \tag{2}$$

The solution of Equation 2 in terms of $W(y)$, jointly with Equation 1, allows for the recovery of the true MWD.

The development of new low angle laser light scattering photometers (LALLS) as SEC detectors simplifies the calibration of the columns and the interpretation of the results because the actual weight average molecular weight of each fraction in the eluting stream can be obtained readily. The calibration and interpretation of the SEC experiments for copolymer applications have different initial requirements on the fractionation mechanism, depending on whether LALLS detectors or the traditional calibration and interpretation techniques are used. These requirements depend on the sensitivity of the technique to the molecular size, relative to the sensitivity of the technique to the copolymer composition.

Calibration Methods. Given the lack of well-characterized copolymer standards of varying molecular weight, composition, and microstructure, three different approaches have been suggested for the calibration of SEC columns:

1. Polystyrene equivalent calibration and interpretation.
2. Dual calibration.
3. Universal calibration.

Generally, the first approach, although useful for comparative purposes, does not yield the true molecular weight of the copolymer. Dual calibrations (*11*) require standards for both homopolymers and some interpolation. Interpolation depends on the composition. The copolymer molecular weight is defined at the ith retention volume as:

$$\log M_{c_i} = P_1 \log M_{1_i} + (1 - P_1) \log M_{2_i} \tag{3}$$

where P_1 is the fraction of monomer 1 in the copolymer.

The universal calibrations require only standards for one homopolymer and the calibration is based on the assumption that the separation mechanism is strictly by molecular size. A number of size parameters have been proposed as being appropriate for universal calibration

$[\eta] \cdot M$	hydrodynamic volume (*24*)
V_h	effective hydrodynamic volume (*25*)
$(<L_o^2>)^{1/2}$	end-to-end root mean squared distance (*24*)

of which the hydrodynamic volumes are the most widely used. By replacing the Mark–Houwink expression for the intrinsic viscosity as a function of the molecular weight (Equation 4) in the definition of the hydrodynamic volume (Equation 5), it can be shown that

$$[\eta] = KM^\alpha \tag{4}$$
$$V_h = [\eta]M \tag{5}$$

the dual calibration and the universal calibration are equivalent, and should yield similar results provided there are no interactions between the packing material and the functional groups present in the copolymer molecule. This hypothesis has been verified experimentally for a variety of copolymers (19, 26, 27).

Several theories can be used to calculate the effective hydrodynamic volumes as a function of the polymer molecular weight, solvent quality, and concentration. Among these, two have been used for copolymer measurements: the model of Ptitsyn and Eizener first applied to SEC by Coll and Prosinowski (28), and the model of Rudin and Wagner (25). In the first case, the effective hydrodynamic volume is given by the equation

$$V_h = \frac{[\eta]\cdot M}{\phi(\epsilon)} \tag{6}$$

where $\phi(\epsilon) = 1 - 2.63\ \epsilon + 2.86\ \epsilon^2$, and

$$\epsilon = \frac{2}{3}\left(\alpha - \frac{1}{2}\right).$$

$\phi(\epsilon)$ is a function that reflects the excluded volume effects in terms of the Mark–Houwink parameter α (Equation 4). This approach has been applied to styrene–acrylonitrile copolymers in $CHCl_3$ and DMF. It requires no additional parameters, but it does not account for concentration effects on the hydrodynamic volume. The model developed by Rudin and Wagner (25) accounts for both concentration and polymer solvent interactions. The hydrodynamic volume is given by:

$$V_h = \frac{4\pi\ [\eta]M}{9.3 \times 10^{24} + 4\pi\ N_0\cdot c([\eta] - [\eta]_\theta)} \tag{7}$$

where N_0 is Avogadro's number, c is the concentration of polymer, and $[\eta]_\theta$ is the intrinsic viscosity at θ conditions. This model has been applied to styrene–α-methylstyrene copolymers (29). It has the disadvantage, however, of requiring knowledge of $[\eta]_\theta$ as a function of the copolymer composition.

The universal calibrations given by Equations 5 and 6 are effective for a variety of polymers provided there are no significant interactions with the packing material. Thus far, the limitation for the analysis of homogeneous copolymers has been the lack of knowledge of Mark–Houwink parameters as functions of the copolymer composition (30), that would allow for both effective testing of the dispersion parameters, and an evaluation of the interaction effects between molecules of different composition.

Low Angle Laser Light Scattering Detectors (LALLS).   Light scattering provides one of the most important methods for the determination of the molecular weights and dimensions of polymer molecules. Unfortunately, its application to copolymers is not straightforward (31–33). The basic equation for the estimation of molecular weight at low angles is given by:

$$\frac{K \cdot c}{\overline{R}_\theta} = \frac{1}{\overline{M}_{wapp}} + 2A_2(P_1)c \tag{8}$$

where

$$K = \frac{4\pi^2 n_0^2}{N_0\, \lambda^4}\, \nu_0^2 \left(\frac{1 + \cos\theta}{2} \right) \tag{9}$$

$$\overline{M}_{wapp} = \overline{M}_w + 2P\left(\frac{\nu_A - \nu_B}{\nu_0} \right) + Q\left(\frac{\nu_A - \nu_B}{\nu_0} \right)^2 \tag{10}$$

$$\nu_A = \left(\frac{dn}{dc} \right)_A$$

$$\nu_0 = \left(\frac{dn}{dc} \right)_{cop}$$

$$\nu_B = \left(\frac{dn}{dc} \right)_B$$

Where the apparent molecular weight ($\overline{M}_{wapp}$) is a function of the average composition, the spread of the composition distribution through the refractive index increments (ν_0, ν_A, ν_B), and the heterogeneity parameters P and Q that are characteristic of each particular polymer sample. The second virial coefficient A_2 is also a function of the molecular weight and the composition; however, this dependence is sometimes small and it is sufficient to approximate it with the value for the total polymer. The heterogeneity parameters P and Q in Equation 10 are estimated in the standard light scattering experiments, by varying the solvent and, therefore, the refractive index increments. Given the difficulties in quantifying the effect of different solvents on the fractionating mechanism, varying the refractive index increments is not a practical approach for the estimation of the molecular weight and the heterogeneity parameters in a coupled SEC/LALLS experiment. It is more convenient to fractionate the copolymer in terms of composition (i.e., obtain fractions with uniform composition for which $P=Q=0$) prior to the light scattering measurements. The application of light scattering to copolymers of homogeneous composition is straight-forward, and should provide sufficient information for the estimation of the Mark–Houwink parameters and the evaluation of column dispersion as a function of the copolymer composition.

Column Dispersion and Chromatogram Interpretation. To solve Tung's dispersion equation, it is necessary to estimate the unknown spreading function $[G(v,y)]$ and its corresponding parameters. The parameter estimates are conditional upon the definition of the calibration curve when molecular weight information is used in the estimation. Alternatively the $G(v,y)$ can be estimated by experimentally eliminating the broadening caused by the polydispersity of the sample (e.g., by using reverse flow techniques when skewing is negligible). The estimation of the $G(v,y)$ parameters as a function of the retention volume is dependent on the adequacy and the structure of the model used. A wide variety of models for the spreading function, as well as solutions to Equation 2 have been proposed and are reviewed elsewhere (34, 35).

Proper evaluation of the spreading function and spreading parameters is important in SEC of copolymers because molecules with the same size, but different composition or microstructure, may experience different interaction effects, and therefore, different spreading through the columns (Figure 1). These spreading fluctuations may lead to significant errors in the estimation of the molecular weights or heterogeneity of the copolymer sample.

Among the chromatogram recovery techniques and solutions to Tung's integral equation reported in the literature, the technique proposed by Yau, Stocklosa, and Bly (GPCV methods) (24) is particularly useful because it allows for the direct calculation of the average molecular weights in the detector's cell. This approach is based on the assumption that the contribution from neighboring species to the molecular weight falls off rapidly when integrating the contents of the detector cell. Under this assumption, the local values of the calibration curve and dispersion parameters can be used with negligible error (37). The equations apply to the general situation where the molecular weight calibration curve is nonlinear and the broadening parameters change with molecular weight or retention volume.

$$\frac{\bar{M}_K(v,uc)}{M(v)} = \frac{F\left[v - (K-1)D_2(v)\sigma(v)^2\right]}{F\left[v - (K-2)D_2\sigma(v)^2\right]} \exp\left\{(2K-3)\left[D_2(v)\sigma(v)\right]^2/2\right\}$$

$$(11)$$

$$\frac{\overline{[\eta]}\,(v,uc)}{\overline{[\eta]}\,(v)} = \frac{F\big[v - \alpha D_2(v)\sigma(v)^2\big]}{F(v)}\,\exp\{[\alpha D_2(v)\sigma(v)]^2/2\} \qquad (12)$$

where $\overline{M}_K(v,uc)$ is the kth average molecular weight of the contents of the detector cell at retention volume v and $\overline{\eta}(v,uc)$ is the intrinsic viscosity. $M(v)$ and $\eta(v)$ are given by:

$$M(v) = D_1(v)\,\exp\big[-D_2(v)\cdot v\big] \qquad (13)$$

$$[\eta(v)] = K\,M(v)^\alpha \qquad (14)$$

where $D_1(v)$ and $D_2(v)$ are the local slope and intercept of the calibration curve, respectively, and $\sigma(v)^2$ is the variance of the Gaussian spreading function. Under the assumption of the validity of the universal calibration curve, Equations 11 and 12, together with on-line intrinsic viscosity measurements, can be used to determine the Mark–Houwink parameters for copolymers of homogeneous composition. This of course, requires that the variance of the spreading function be independent of the chemical nature of the copolymer. Experimental results by other researchers (35, 38) on a variety of polymers and copolymers indicate that for those cases where the universal calibration is valid, the spreading parameter is independent of the chemical nature of the polymers.

Universal Calibration and Evaluation of the Fractionation.   Given the experimental observation that the Mark–Houwink equation (Equation 4) is valid for copolymers of homogeneous composition, as discussed previously, the intrinsic viscosity–molecular weight relationship can be used as an indication of the presence of polymer–solvent–packing interactions. A plot of $[\eta]$ vs. M for the homopolymers defines a region about which the copolymers are expected to fall. Strong deviations from this region, for molecular weights calculated based on the universal calibration, would be an indication of polymer–solvent–packing interactions. Figure 2 shows the application of this criterion to styrene–isoprene copolymers analyzed in independent laboratories (13, 19). Both sets of data are consistent for unfractionated anionic copolymers of varying compositions (11–93% styrene). The unfractionated copolymers fall within the lines for polystyrene and polyisoprene, whereas the SEC fractionation data (13) deviate significantly as the content of isoprene in the copolymer increases, thereby indicating the presence of interactions. This deviation may explain the differences reported between the heterogeneity parameters P and Q obtained from light scattering (Equation 5), and those obtained from SEC analysis. This consistency test is particularly useful when the copolymers and the homopolymers are soluble in

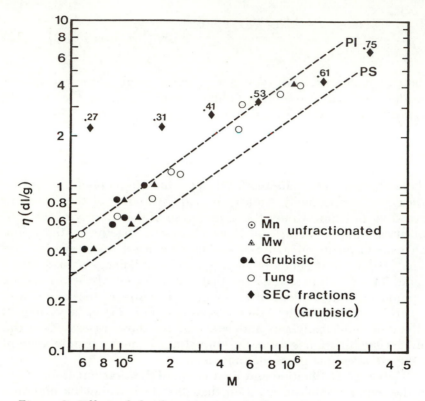

Figure 2. Effect of the fractionation on the intrinsic viscosity—mo-lecular weight relationship for styrene—isoprene copolymers in the composition of the fraction in brackets (13, 19). Key: PI, polyisoprene; and PS, polystyrene.

the same solvent system. When one of the homopolymers or some of the copolymers are insoluble in the desired solvent two approaches can be used to evaluate the fractionation mechanism:

1. Plots of $[\eta]-M$ can be used as a consistency test if it is further assumed that the K and α parameters for the insoluble polymers are those corresponding to the unperturbed dimensions of the polymer coils (i.e., $\alpha = 0.5$, $K = K_\theta$).

2. The solubility and polar characteristics of the groups present in the polymer molecule can be used to enhance interactions with the packing materials, such that composition fractionations are obtained. Cross-size fractionation should then indicate significant compositional effects.

The difficulties in applying the first approach are exemplified in Figures 3—5 where the complex nature of the Mark—Houwink param-

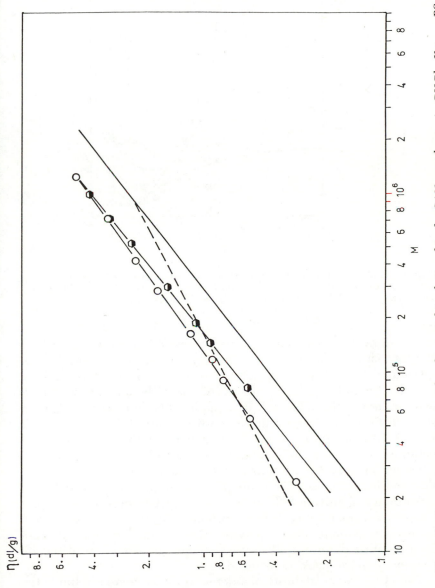

Figure 3. Intrinsic viscosity–molecular weight relationship for SAN copolymer in CHCl₃. Key: –, PS; -- PAN (θ); ○, SAN (25% acrylonitrile); and ◓, SAN (12% acrylonitrile). (Data are from Reference 23.)

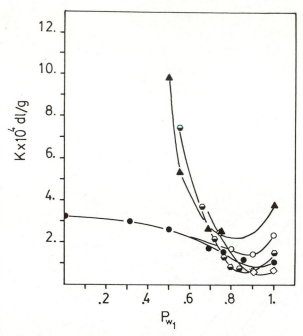

Figure 4. The Mark—Houwink parameter K for SAN copolymers as a function of the solvent and the copolymer composition. Key to conditions: ○, *DMF 20 °C;* ●, *DMF 30 °C;* ▲, *MEK 20 °C;* ◇, *CHCl₃ 20 °C; and* ⬮, *THF 25 °C.*

eters is shown for SAN copolymers in several solvents. Clearly, θ conditions can be reached at significantly high styrene compositions. Superimposed on this are the effects of the microstructure and the molecular weight on the solubility of the polymer. These effects greatly influence the value of K in Equation 4 (Figure 4). Consistency tests with Equation 4 under these conditions are difficult unless the behavior of the intrinsic viscosity—molecular weight relationship as a function of the composition is known for the solvent of interest. The application of the cross-fractionation approach on the other hand requires careful selection of the solvent and the packing materials.

Column and Solvent Selection. The selection of solvent systems for copolymer analysis is dictated primarily by two factors: solvent transparency in the regions of polymer absorption, and solubility considerations that determine the coil size and the direction of the interactions with the packing material.

The choice or availability of a detection system greatly limits the number of solvents that can be used. In the case of styrene copolymers, solvents transparent in the region from 230 to 270 nm are highly desirable (strong absorption of the styryl group). Unfortunately, for

SAN copolymers most transparent solvents are nonsolvents for polyacrylonitrile (PAN) or acrylonitrile rich copolymers. Solubility parameters and considerations of intermolecular forces can be used to estimate the solubility characteristics of the copolymers and to indicate the direction of the interactions, i.e., polymer–solvent or polymer–packing. For example, comparison of the three-dimensional solubility parameters for the copolymers and for THF and DMF (Table I) indicates that the polar forces (δ_p) play a significant role in the dissolution of acrylonitrile rich copolymers. Therefore, small additions of a transparent polar solvent, like methanol, are expected to enhance solubility for acrylonitrile rich copolymers and decrease the interactions between acrylonitrile and the polar packings, as shown in Figure 6 for unfractionated SAN copolymers and silica columns. Excessive addition of methanol increases the hydrogen bonding (δ_H) significantly which causes the mixture to become a nonsolvent for both homopolymers and copolymers.

The quasi-stationary phase concepts discussed previously (2–4) can also be used to guide the solvent selection. If polymer–packing interactions are to be diminished (size separation) a nonsolvent that

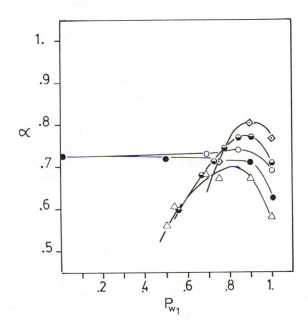

Figure 5. The Mark–Houwink exponent α for SAN copolymer as a function of the solvent and the copolymer composition P_{w_1} weight fraction of styrene in the copolymer. Key to conditions: ○, DMF 20 °C; ●, DMF 30 °C; △, MEK 30 °C; ◇, ChCl₃ 20 °C; and ◓, THF 25 °C.

Table I. Solubility Parameters (δ) for SAN Copolymers in Various Solvents

Mole Fraction of Styrene	Solvent	δ_o	δ_d	δ_p	δ_H	Reference
1.	—	—	9.1	0.3	0.10	39
0.74	—	—	9.1	1.4	1.0	39
0.62	—	—	9.1	2.1	1.2	39
0.48	—	—	9.1	3.0	1.5	39
0.0	—	—	9.0	8.4	3.1	39
—	THF	9.52	8.22	2.8	3.9	40
—	DMF	12.14	8.52	6.7	5.5	40
—	MeOH	14.28	7.42	6.0	10.9	40

Note: Subscripts are as follows: o, overall; d, dispersion; p, polar; and H, hydrogen bonding. The units for δ are $(\text{cal/cm}^3)^{1/2}$.

preferentially wets the packing is to be selected. If interactions are to be enhanced (composition fractionation) a nonsolvent highly miscible in the solvent is to be preferred. These guidelines are of course qualitative, and it is important to determine the solubility limits and intrinsic viscosities for each solvent combination to define both the operating range and the effect of the nonsolvent on the coil size.

A wide variety of porous packing materials is available for SEC of polymers (40), most of which have been tested for solvent compatibility from the viewpoint of the stability of the packing material. Very few attempts have been made to select a packing material on the basis of its interactions with the polymer molecules (6, 8). As discussed previously, polymer packing interactions can be enhanced (for polystyrene and possibly for styrene copolymers) on silica and Styragel packings by proper choice of the solvent system. Silica substrates however, show the strongest interactions and therefore are the most likely candidates for composition fractionations.

Experimental Methods

Styrene–acrylonitrile copolymers were synthesized by bulk free-radical polymerization at 60 and 40 °C over a wide range of initial monomer compositions (1). Copolymer compositions were determined by gas chromatography and verified by ¹H-NMR spectroscopy. The copolymers were purified prior to the analysis by precipitation from absolute methanol (Table II).

The SEC is a modified room temperature Waters GPC equipped with a differential refractometer and a Waters 440 dual UV detector. The injection valve, detectors, and volume counter were interfaced with a Data General Nova minicomputer that was used primarily for data logging, base line corrections, and data synchronization. The first detector downstream from the columns was used as reference. The refractometer and UV spectrophotometers were calibrated to absolute units using benzene–carbon tetrachloride (CCl₄) solutions. Four types of columns were investigated: Waters microBondagel (E-Lincar), du Pont S.E. silica columns (60, 100, 500, 1000, and 4000 A°), Porasil (69 A°), and Styragel (500, 10^3, 10^4, 10^5, and 10^6 A°).

Intrinsic viscosities were measured with Cannon Ubbelohde viscometers at 20 and 25 °C.

Results and Discussion

Column Calibration. The columns were calibrated using 18 narrow polystyrene standards (Pressure Chemicals) as reference materials. The variance of the instrumental spreading function $\sigma^2 (v)$ in Equation 11 was represented as a third-order polynomial in v and the parameters of the polynomial were estimated by minimizing the de-

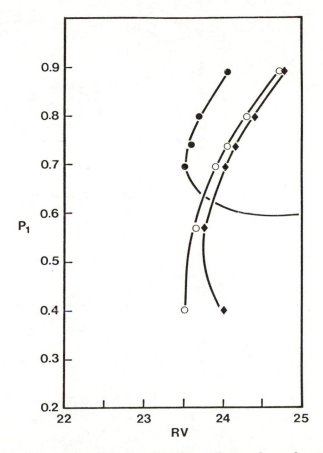

Figure 6. Adsorption effects on SE silica columns for unfractionated SAN copolymers. Key: RV, retention volume (cm^3); P_1, mole fraction of styrene in the copolymer; ●, M_0; ○, M_1; and ◆, M_2. Conditions: M_0, THF; M_1, THF/MeOH (90:10) (v/v); and M_2, THF/MeOH (80:20) (v/v).

Table II. Molecular Properties of SAN Copolymers Used in this Study

| Sample | P_{Iw} | $[\eta]$ (dL/g) | | $\bar{M}_w$ | $\bar{N}_s$ | SEC Results | | |
		THF at 25 °C	DMF at 20 °C	(Viscosity)	(NMR)	$\bar{M}_w$	$[\eta]$	$\bar{N}_s$
R0201A	0.68	1.048	1.072	141323				
R0202	0.873	0.406	0.439	80031				
R0204	0.837	0.533	0.553	92955				
R0206	0.80	0.588	0.729	120151				
R0208	0.74	0.790	0.826	124305				
R0210	0.69	0.963	0.838	107264				
R0401	0.55	1.582	2.798	351924	1.10	348000	1.51	1.13
R0403	0.67	1.520	1.427	210037	1.16	220080	1.53	1.22
R0405	0.72	1.574	1.442	236993	1.37	245400	1.52	1.41
R0406	0.75	1.346	1.383	255916	1.56	263123	1.90	1.58
R0407	0.79	0.935	1.210	221689	1.90	218650	1.15	2.10
R0408	0.84	0.904	0.904	184943	2.60	190000	0.95	2.67
R0409	0.90	0.697	0.723	161550	4.50	154200	0.73	4.36

R0501	0.56	1.479	2.518	314563
R0502	0.63	1.390	1.461	190689
R0503	0.66	1.387	1.294	188789
R0504	0.68	1.369	1.277	180101
R0505	0.72	1.279	1.227	183518
R0506	0.74	0.964	1.001	162734
R0507	0.79	0.755	0.922	167105
R0508	0.83	0.762	0.771	148082
R0601	0.56	1.555	2.739	341816
R0602	0.63	2.069	2.274	348332
R0603	0.56	2.137	2.046	346517
R0605	0.72	1.959	1.968	363871
R0606	0.74	1.403	1.438	270048
R0607	0.79	1.179	1.407	301770
R0608	0.84	1.156	1.138	254985
R0609	0.89	1.153	1.150	310504

Note: The numbers recorded for the properties, in particular $\bar{M}_w$, are a direct output of the estimation programs. The accuracy of the characterization methods does not warrant all the significant figures.

viations between the calculated averages ($\hat{\bar{M}}_n$, $\hat{\bar{M}}_w$) and the molecular weights given by the manufacturer. That is, by:

$$\min \sum_{i=1}^{N_s} \left(\frac{\bar{M}_{n_i} - \hat{\bar{M}}_{n_i}}{\bar{M}_{n_i}} \right)^2 + \left(\frac{M_{w_i} - \hat{\bar{M}}_{w_i}}{\bar{M}_{w_i}} \right)^2 \qquad (15)$$

where N_s is the number of standards used.

Figure 7 shows the calibration curve and the chromatograms of some of the PS standards used. Figures 8 and 9 show the corresponding variance of the chromatograms and spreading function as a function of the retention volume. Because of the interpolations required by Equation 11, some oscillations can be expected when calculating the molecular weights at the tails of the chromatograms where the signal-to-noise ratio is low. This problem can be removed easily by smoothing the data. Smoothing is accomplished by fitting the chro-

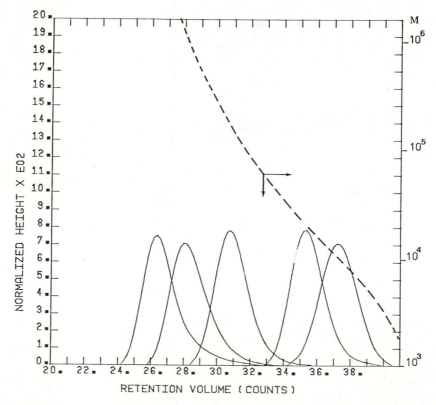

Figure 7. Fractionation of PS standards in THF Styragel columns.

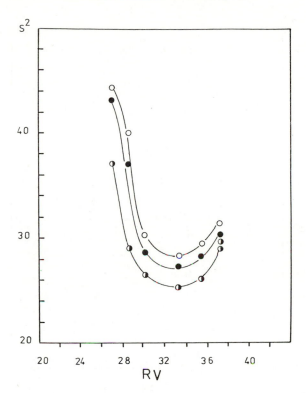

Figure 8. Variance of the raw chromatograms for PS standards as a function of the polymer concentration (THF). Key to concentration (g/mL): ○, 0.00100; ●, 0.00050; and ◐, 0.00025.

matograms to a sum of Gaussian functions (*41*). Normally one to four functions were required. Figures 10 and 11 show the number and weight average molecular weights calculated as a function of the retention volume for two narrow standards.

Size and Composition Fractionations. The initial column selection was done on the basis of adsorption effects observed on silica columns for SAN copolymers having acrylonitrile contents greater than 40 mol %. Figure 6 shows the peak retention volume for unfractionated SAN copolymers of homogeneous composition in THF–methanol mixtures. Clearly, addition of methanol reduces the polymer packing interaction for rich acrylonitrile copolymers, as well as the size of the polystyrene molecules. Superimposed on the compositional effects shown in Figure 6 are the hydrodynamic size effects. To study these, three fractions were collected (Figure 12) and reinjected into a system of five SEC columns, plus two 69-Å Porasil columns. The composition calibration curves obtained are shown in Figure 13. Injection of these fractions into the Styragel columns indicates

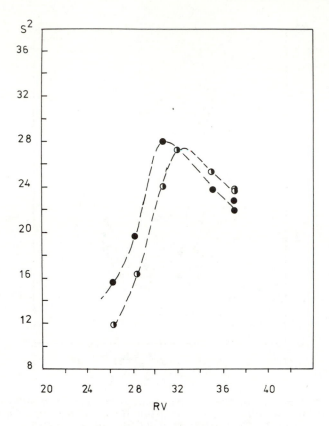

Figure 9. Estimated spreading parameter for PS standards in THF as a function of the polymer concentration. Key to concentration (g/mL): ●, 0.00050; and ◑, 0.00025.

a reduction of the hydrodynamic volume only with increasing methanol content (Figure 14). No significant composition effects can be observed on the Styragel fractionation. The calibration curves shown in Figure 13 indicate that magnitude of the composition effects should not be underestimated in copolymer fractionations. Although the resolution of the composition fractionation is not enough for evaluation of the copolymer composition distribution, it is sufficient to demonstrate that the main separation mechanism for SAN copolymers in Styragel columns, using THF or THF—methanol mixtures, is mainly size exclusion. Therefore, the use of the universal calibration curve appears to be justified for the estimation of molecular weights.

Estimation of the Mark—Houwink Parameters. The weight average molecular weights for SAN copolymers of homogeneous compositions were estimated from intrinsic viscosity measurements in DMF at 20 °C and the correlations obtained by Bauman and Lange (*42*). The molecular weights obtained by this method (Table II), and the corresponding intrinsic viscosities in THF at 25 °C were used to estimate K and α as a function of the copolymer composition for SEC analysis (Table III). The estimation of the weight average molecular weights from viscosity data is supported by an investigation on the effect of the polydispersity on the molecular weight–intrinsic viscosity relationship (*43*). This study indicated that the weight average molecular weights calculated from $[\eta]/\bar{M}_w$ correlations provide a good approximation to the actual molecular weights, or alternatively, that good estimates of K and α can be obtained if the polydispersity of the polymer is moderately low. Table III and Figures 4 and 5 summarize the behavior of K and α as a function of the solvent and the copolymer composition. As the acrylonitrile content increases, the exponent α rapidly approaches a value of 0.5 for $CHCl_3$, THF, and methyl ethyl

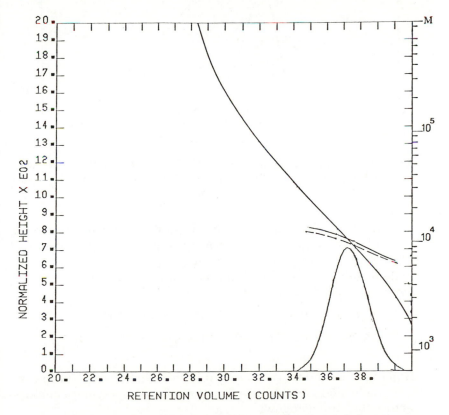

Figure 10. Average molecular weights as a function of the retention volume standard PS4 in THF ($\bar{M}_w$ = 10,300). Key: –, $\bar{M}_w$ (v); and --, $\bar{M}_n$ (v).

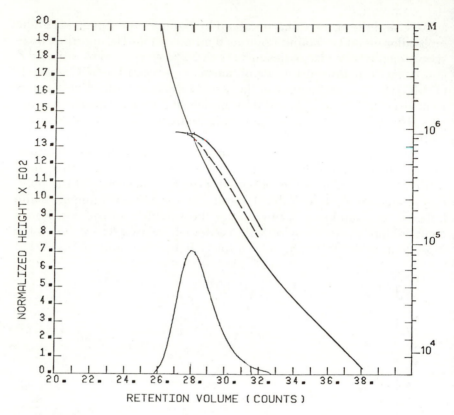

Figure 11. Average molecular weights as function of the retention volume standard PS12 in THF ($\bar{M}_w$ = 867,000). Key: −, $\bar{M}_w$ (v); and --, $\bar{M}_n$ (v).

ketone (MEK), which are nonsolvents for polyacrylonitrile. The value of *K*, on the other hand, rises sharply approaching the value of K_θ = 25.0 × 10⁻⁴ reported for polyacrylonitrile *(30)*. The values of *K* and α are adequately correlated with the weight fraction of styrene in the copolymer by Equations 16 and 17.

$$\alpha(P_{1_w}) = -0.6109 + 3.2274\ (P_{1_w}) - 1.9037\ (P_{1_w})^2 \qquad (16)$$

$$K(P_{1_w}) = 5.082 \times 10^{-3} - 1.1488 \times 10^{-2}(P_{1_w}) + 6.5612 \times 10^{-3}(P_{1_w})^2 \qquad (17)$$

Equations 16 and 17 can be used with Equations 11−14 to estimate the molecular weights as a function of the copolymer composition across the chromatogram.

Concentration, Composition, and Sequence Length Detection. Traditionally, copolymer concentrations and compositions have been the only variables that were measured in SEC experiments in addition to the molecular weight determinations. That is, two mass detectors plus the molecular weight measurements have been considered sufficient to characterize the copolymer molecules. However, two aspects of the mass detection system require special consideration (*1*).

1. In the case of size fractionation, the contents of the detector cell will be formed by molecules of the same size but not necessarily the same composition or microstructure.
2. Spectrophotometers and refractometers that are used traditionally as mass detectors are sensitive to the microstructure.

Therefore, depending on the selectivity of the detection system to one or more copolymer properties, additional measurements may be required to characterize the material properly. In the case of SAN copolymers (*1*), the refractive indices in THF follow the ideal behavior (i.e., linear and additive) with respect to the styrene content in the copolymer. Therefore,

$$\Delta n = v \cdot c \tag{18}$$

$$v = v_1 - N\,(1 - P_{1_w}) \tag{19}$$

where v_1 is 0.1928, and N is 0.1235.

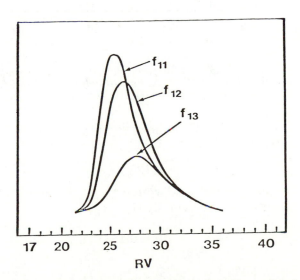

Figure 12. Fractions collected to analyze the effect of molecular size on the composition fractionation. Conditions: sample, National Bureau of Standards polystyrene 706; solvent, 90:10 THF/MeOH; and RV, in milliliters.

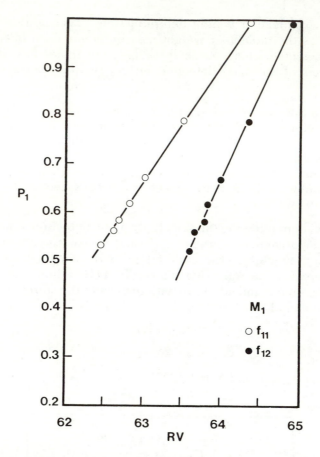

Figure 13. Calibration curves for fractionated SAN copolymers on sil-ica columns (4SE plus 2 (4-ft) 60-Å₄ Porasil columns). Conditions: sol-vent, 90:10 THF/MeOH, RV in milliliters.

The UV absorption in THF, however, is sensitive to the average length of the styrene sequences, i.e.,

$$A_\lambda = \epsilon_\lambda \, l \, P_{1_w} \, c \tag{20}$$

where l is the path length, and where the extinction coefficient at several wavelengths (ϵ_λ) is given by (1)

$$\left(\frac{\epsilon}{\epsilon_{p_s}} \right)_{254} = 1.0 - 0.1831 \, (1/\bar{N}_s) \tag{21}$$

$$\left(\frac{\epsilon}{\epsilon_{p_s}} \right)_{261} = 1.0 - 0.3081 \, (1/\bar{N}_s) \tag{22}$$

$$\left(\frac{\epsilon}{\epsilon_{p_s}}\right)_{269} = 1.0 - 0.4075\ (1/\bar{N}_s) \tag{23}$$

Combining Equations 18–23, explicit expressions for the average sequence length, the composition, and the copolymer concentration in the solution can be obtained as function of the refractive index and two absorption measurements.

$$\bar{N}_s = \frac{0.4075\ (A_{254}/A_{269}) - 0.1831\ (\epsilon_{p_s\ 254}/\epsilon_{p_s\ 269})}{(A_{254}/A_{269}) - (\epsilon_{p_s\ 254}/\epsilon_{p_s\ 269})} \tag{24}$$

$$P_{1_w} = \frac{(\nu_1 - N)\ A_{254}}{\Delta n\ \epsilon_{254}\ l\ - N\ A_{254}} \tag{25}$$

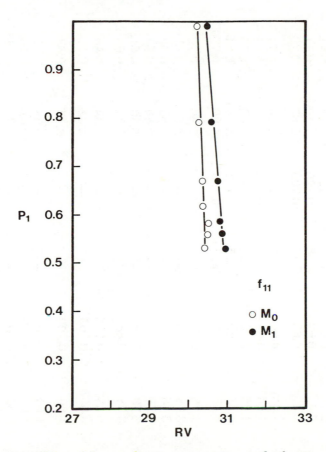

Figure 14. *Effect of the copolymer composition on the fractionation of SAN copolymers on Styragel columns. Conditions: M_0, THF; and M_1, 90:10 (v/v) THF/MeOH.*

Table III. Mark–Houwink Parameters for SAN Copolymers

P_{lw}	PS		PAN		SAN		Solvent (Temperature)	Reference
	$K \times 10^4$	α	$K \times 10^4$	α	$K \times 10^4$	α		
0.90	2.4	0.63	3.25	0.725	1.45	0.71	DMF(20 °C)	44
0.75	2.4	0.63	3.25	0.725	1.80	0.71	DMF(20 °C)	42
0.50	2.4	0.63	3.25	0.725	2.65	0.72	DMF(20 °C)	44
0.84	1.05	0.69	3.25	0.725	1.20	0.74	DMF(30 °C)	45
0.76	1.05	0.69	3.25	0.725	1.62	0.73	DMF(30 °C)	45
0.69	1.05	0.69	3.25	0.725	1.72	0.73	DMF(30 °C)	45
0.76	1.05	0.69	3.25	0.725	2.944	0.646	DMF(30 °C)	45
0.54	1.05	0.69	3.25	0.725	1.20	0.770	DMF(30 °C)	46
0.90	3.78	0.58	—	—	1.50	0.70	MEK(30 °C)	44
0.75	3.78	0.58	—	—	2.50	0.67	MEK(30 °C)	42
0.50	3.78	0.58	—	—	9.80	0.56	MEK(30 °C)	44
0.76	3.78	0.58	—	—	3.60	0.62	MEK(30 °C)	46
0.54	3.78	0.58	—	—	5.30	0.61	MEK(30 °C)	46
0.69	3.78	0.58	—	—	2.54	0.68	MEK(30 °C)	45
0.88	0.657	0.765	—	—	0.681	0.80	CHCl$_3$(20 °C)	23
0.75	0.657	0.765	—	—	2.327	0.711	CHCl$_3$(20 °C)	23
0.76	1.60	0.706	—	—	2.15	0.68	THF(25 °C)	30
0.90	1.60	0.706	—	—	0.68	0.77	THF(25 °C)	a
0.84	1.60	0.706	—	—	0.815	0.768	THF(25 °C)	a
0.79	1.60	0.706	—	—	0.860	0.755	THF(25 °C)	a
0.75	1.60	0.706	—	—	1.342	0.74	THF(25 °C)	a
0.72	1.60	0.706	—	—	2.207	0.71	THF(25 °C)	a
0.67	1.60	0.706	—	—	3.65	0.68	THF(25 °C)	a
0.55	1.60	0.706	—	—	7.44	0.60	THF(25 °C)	a

a These values are a result of the work presented in this chapter.

$$c = \frac{A_{269}}{\epsilon_{269}\, l\, P_{1_w}} \tag{26}$$

If sequence lengths are not important the traditional analysis can be done by using Equations 18–20. The complete algorithm for the SEC characterization of SAN copolymers is shown in Figure 15.

SAN molecular weight characterization

↓

Read in chromatograms for RI, UV$_{254}$, UV$_{269}$

↓

Synchronize chromatograms
(Important for high resolution columns)

↓

Calculate copolymer average sequence
length, composition and concentration
from Equations 24 to 26

↓

Normalize concentration vector
to obtain weight fractions

↓

Calculate Mark-Houwink parameters
at each retention volume
from the composition and Equations 16 and 17

↓

Correct intrinsic viscosity values
(Equation 12)

↓

Calculate molecular weight averages at
each retention volume (Equation 11)

↓

Integrate properties calculated at each
retention volume to calculate total properties

Figure 15. Algorithm for the SEC characterization of SAN copolymers.

SEC of Low Conversion Copolymers. To verify the fractionation mechanism, the copolymer composition and the styrene average sequence length measurements, low conversion SAN copolymers were fractionated. Fractions were collected every dump (5 ml). The intrinsic viscosities and absorbances at 269 nm were measured off-line (the Waters 440 spectrophotometer has no filters for wavelengths between 254 and 280 nm). Some results are shown in Figures 16 and 17. Generally, the agreement with the overall molecular weight, viscosity, and composition is quite good. The average sequence lengths measured across the chromatogram do not deviate significantly from the mean value for the total polymer. The deviations observed at the tails can be attributed to the propagation of errors in Equations 24−26.

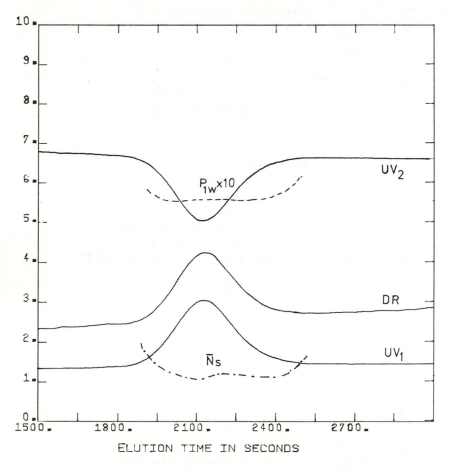

Figure 16. SEC characterization of copolymer sample R04-01.

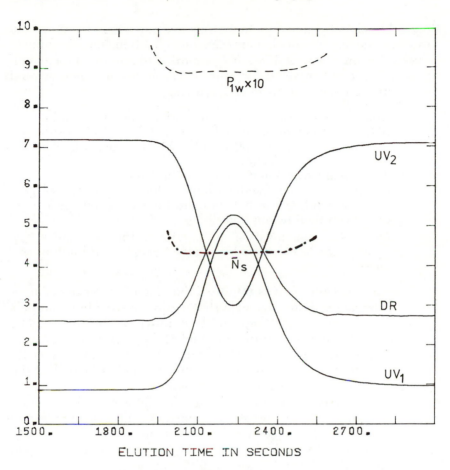

Figure 17. SEC characterization of copolymer sample R04-09.

Summary and Conclusions

The difference in the chemical nature of the comonomers can cause significant polymer−solvent−packing interactions and therefore, can lead to significant errors in the estimation of molecular weights by SEC. It is necessary to determine the experimental conditions that would yield mainly size or mainly composition fractionations. The selection of such conditions is largely determined by the detection system used for the measurement of molecular weights, composition, concentration, and sequence lengths. Two methods have been proposed for the verification of size fractionation, and the selection of the solvent system to achieve size or composition fractionation. Low conversion SAN copolymers have been fractionated according to composition on silica columns, and according to size on Styragel columns using THF−methanol mixtures. These results open several possibilities for the analysis of copolymers:

1. Coupled size exclusion−composition chromatography in the form suggested by Balke and Patel (8) can be used with a single solvent (90:10 THF:MeOH) for both size and composition fractionation.
2. The use of LALLS molecular weight detectors is simplified if composition fractionation is achieved prior to size fractionation because the standard theory for light scattering can then be applied.
3. Once size fractionation is obtained, the molecular weights can be calculated using the universal calibration curve.

The first two alternatives require good resolution on the composition fractionation. Extensive experimentation on packing materials and solvent systems is still required to improve the resolution. The third possibility however, can be applied readily, once size fractionation has been demonstrated. The Mark−Houwink parameters for SAN copolymers in THF have been estimated as a function of the composition. These correlations, along with other equations developed in this chapter, have been used for the estimation of molecular weights, average sequence length, composition, and concentration of the copolymers in the detector cell (1). Although SEC results agree well with the overall copolymer properties, the procedure developed must be applied with caution for the following reasons:

1. The propagation of experimental errors through the calculations has yet to be investigated; therefore, very little can be said regarding the sensitivity of the method to experimental errors stemming from the spectrophotometers and the universal calibration.
2. The parameters contained in the equations have been estimated with a limited set of standards, especially with regard to molecular weight. Clearly improved accuracy can be achieved as more data become available.

However, the results obtained are very encouraging. Rather complete copolymer characterizations can be obtained from SEC experiments by carefully selecting the fractionation and detection system in accordance with the characteristics of the copolymer investigated.

Literature Cited

1. Garcia Rubio, L. H., Ph.D. Thesis, McMaster Univ., Hamilton, Ontario, Canada, 1981.
2. Bakos, D.; Berek, D.; Bleha, T. *Eur. Polym. J.* **1976**, *12*, 801.
3. Bakos, D.; Bleha, T.; Ozima, A.; Berek, D. *J. Appl. Polym. Sci.* **1979**, *23*, 2233.
4. Campos, A.; Soria, V.; Figueruelo, J. *Makromol. Chem.* **1979**, *180*, 1961.
5. Dubin, P.; Koontz, S.; Wright, K. *J. Polym. Sci.* **1977**, *15*, 2047.
6. White, J. et al. *J. Appl. Polym. Sci.* **1979**, *24*, 953.
7. Berek, D.; Bakos, D.; Soles, L. *Makromol. Chem.* **1975**, *176*, 391.
8. Balke, S. T.; Patel, R. D. In "Size Exclusion Chromatography (GPC)" Provder, Theodore, Ed.; ACS SYMPOSIUM SERIES No. 138; ACS: Washington, D.C., 1980; pp. 149–82.
9. Owens, F. G.; Cobler, J., presented at the 4th Intl. Seminar on GPC, 1967.
10. Terry, S. L.; Rodriguez, F. *J. Polym. Sci., Part C* **1968**, *21*, 103.
11. Runyon, J. R. et al. *J. Appl. Polym. Sci.* **1969**, *13*, 2359.
12. Adams, H. E. In "Gel Permeation Chromatography"; Altgelt, K.; Segal, L., Eds.; Dekker: New York, 1971.
13. Grubisic-Gallot, Z. et al. *J. Appl. Polym. Sci.* **1972**, *16*, 2933.
14. Mirabella, F.; Barral, E.; Johnson, J. *J. Appl. Polym. Sci.* **1975**, *19*.
15. Ibid., **1976**, *20*.
16. Chow, C. D. *J. Appl. Polym. Sci.* **1976**, *20*, 1619.
17. Chen, H. R.; Blanchard, L. P. *J. Appl. Polym. Sci.* **1972**, *16*, 603.
18. Janca, J.; Kolinsky, M. *J. Appl. Polym. Sci.* **1972**, *16*, 83.
19. Tung, L. H. *J. Appl. Polym. Sci.* **1979**, *24*, 953.
20. Ogawa, T.; Inaba, T. *J. Appl. Polym. Sci.* **1977**, *21*, 2979.
21. Ogawa, T. *J. Appl. Polym. Sci.* **1979**, *23*, 3515.
22. Teramachi, S. et al. *Macromolecules* **1978**, *11* (b), 1206.
23. Kranz, D.; Pohl, H. U.; Baumann, H. *Angew. Makromol. Chem.* **1972**, *26* (101), 67.
24. Yau, H.; Stocklosa, J.; Bly, D. D. *J. Appl. Polym. Sci.* **1977**, *21*, 1911.
25. Rudin, A.; Wagner, R. A. *J. Polym. Sci.* **1976**, *20*, 1483.
26. Chang, F. S. C. *J. Chromatogr.* **1971**, *55*, 67–71.
27. Morris, M. C. *J. Chromatogr.* **1971**, *55*, 203–10.
28. Coll, A.; Prusinowski, L. R. *J. Polym. Sci., Part B* **1967**, *5*, 1153.
29. Samanta, M. C., Ph.D. Thesis, Univ. of Waterloo, Waterloo, Ontario, Canada, 1978.
30. "Polymer Handbook"; Brandrup, J.; Immergut, E. H., Eds.; Wiley: New York, 1975.
31. Benoit, H.; Froelich, D. In "Light Scattering from Polymer Solutions"; Huglin, M. B., Ed.; Academic: New York, 1972; Chap. 11.
32. Bushuk, W.; Benoit, H. *Can. J. Chem.* **1958**, *36*, 1616.
33. Spatorico, A. L. *J. Appl. Polym. Sci.* **1974**, *18*, 1793.
34. Friis, N.; Hamielec, A. E. *Adv. Chromatogr.* **1975**, *13*.
35. Tung, L. H. "Fractionation of Synthetic Polymers: Principles and Practices"; Dekker: New York, 1977.
36. White, J. et al. *J. Appl. Polym. Sci.* **1972**, *16*, 2811.
37. Hamielec, A. E. *J. Liq. Chromatogr.* **1980**, 3(3), 381.
38. Probst, J. J.; Unger, K.; Cantow, H. J. *Angew. Makromol. Chem.* **1974**, *35*, 177.
39. Blanks, R. F.; Shah, B. N. *J. Polym. Sci.* **1976**, *14*, 2589.
40. Yau, W.; Kirkland, J.; Bly, D. "Modern Size-Exclusion Liquid Chromatography: Practice of Gel Permeation and Gel Filtration Chromatography" Wiley: New York, 1979.
41. Sorenson, W. H.; Alspach, D. L. *Automatika* **1971**, *7*, 465.

42. Baumann, H.; Lange, H. *Angew. Makromol. Chem.* **1969,** 9, 16.
43. Quackenbos, H. M. *J. Appl. Polym. Sci.* **1980,** 25, 1435.
44. Lange, H.; Baumann, H. *Angew. Makromol. Chem.* **1970,** 14, 25.
45. Reddy, C. R.; Kalpagam, V. *J. Polym. Sci.* **1976,** 14, 749.
46. Shimura, Y.; Mita, I.; Kambe, H. *J. Polym. Sci., Polym. Lett. Ed.* **1964,** 2, 403.

RECEIVED for review October 14, 1981. ACCEPTED August 17, 1982.

Aqueous Size Exclusion Chromatography

J. E. ROLLINGS—Worcester Polytechnic Institute, Department of Chemical Engineering, Worcester, MA 01609

A. BOSE—Battelle Columbus Laboratories, Columbus, OH 43201

J. M. CARUTHERS and G. T. TSAO—Purdue University, School of Chemical Engineering, West Lafayette, IN 47907

M. R. OKOS—Purdue University, Department of Agricultural Engineering, West Lafayette, IN 47907

The important aspects of aqueous size exclusion chromatography are reviewed. The molecular size of polyelectrolytes in dilute solutions depends on the molecular weight and the ionic strength of the solution. Size exclusion chromatography calibration procedures based on the dilute solution conformation statistics of polymers have been proposed previously. The applicability of these calibration procedures to aqueous size exclusion chromatography is examined critically. In aqueous size exclusion chromatography secondary separation mechanisms such as adsorption, ion exclusion, and ion inclusion can also be important. The criteria to be considered in the design of an efficient size exclusion chromatographic system are discussed.

SIZE EXCLUSION CHROMATOGRAPHY (SEC) is an important technique for determining the molecular weight and the molecular weight distribution of polymers in dilute solution (1, 2). Of particular interest is the development of aqueous SEC to be used in the characterization of water-soluble natural polymers and synthetic polyelectrolytes. Many polymers of biological interest, such as proteins, polysaccharides, and nucleic acids, are water soluble (3, 4); thus, the development of aqueous SEC as an analytical tool is of considerable importance for both biochemical and medical research. This chapter reviews the important aspects of aqueous SEC.

For polymers soluble in organic solvents, SEC is a common analytical technique (1); however, SEC methods for water-soluble polymers have not yet reached the same level of development as SEC techniques for polymers soluble in organic solvents. Some of the reasons for the less advanced state of aqueous SEC are (1) a lack of readily available, monodisperse, water-soluble polymer standards; (2) difficulties in obtaining chromatographic supports for aqueous systems that possess the necessary separation characteristics; (3) inherent difficulties in the description of the dilute solution conformation statistics of polyelectrolytes; and (4) the presence of additional separation mechanisms, such as ion inclusion and ion exclusion, that are usually unimportant in organic SEC. High performance column supports that are suitable for aqueous SEC, such as Aquapore and TSK-GEL (5, 6) have recently become available, as well as monodisperse polyelectrolytes (7) and dextrans (8) with relatively narrow molecular weight distributions. As a result of these two key developments, aqueous SEC should now be able to realize its full potential as an analytical technique.

In the next section we will discuss the fundamental separation mechanisms in aqueous SEC. First we will consider separation by molecular size, examining the effects of molecular weight and ionic strength on the size of the macromolecule followed by discussions of various polymer–solvent–support interactions that can occur in addition to the separation by molecular size. Subsequently, techniques for calibration of aqueous SEC using secondary standards will be presented; and finally, the experimental methods required to implement aqueous SEC effectively will be reviewed.

Separation in Aqueous SEC

Separation by Molecular Size. The principal separation mechanism in SEC is differential migration of molecules between the flowing solvent and the solvent within the porous matrix of an SEC column packing (1, 2). Separation occurs because the total accessible volume of the column (i.e., inside and outside the pore matrix) varies with the size of the polymer molecules in solution. Smaller macromolecules will, on the average, see more pore volume and spend a longer period of time in the relatively stationary solvent inside the porous matrix than larger macromolecules. The larger macromolecules have a smaller pore volume available to them and thus elute from the column earlier than the smaller macromolecules. Because the principal separation mechanism is governed by the size of the macromolecules—size being correlated with molecular conformation—dilute solution conformation statistics of polyelectrolytes is of paramount importance.

The principal objective of SEC is to determine the molecular weight and/or molecular weight distribution of macromolecules. However, the separation process in SEC depends primarily on the size of the polymer. To interpret SEC elution data effectively, the relationship between the size of the macromolecule in solution and the molecular weight must be available. Flory has proposed that the molecular weight M is related to the molecular volume $(r^2)^{3/2}$ as follows (9)

$$[\eta] = \Phi_0 \, (r^2)^{3/2}/M = \Phi_0 \, \alpha^3 (r_0^2)^{3/2}/M \qquad (1)$$

where $[\eta]$ is the intrinsic viscosity, (r^2) is the mean-square end-to-end distance, (r_0^2) is the unperturbed mean square end-to-end distance, α is the expansion factor, and Φ_0 is a constant equal to 3.6×10^{21} dL/cm^3. Because the intrinsic viscosity can be measured independently for a given polymer sample, Equation 1 is the desired relationship between molecular weight and the size of the macromolecule in solution.

In the derivation of Equation 1, excluded volume effects were not considered in the calculation of the front factor, Φ_0. Other treatments of the intrinsic viscosity that include the contributions of excluded volume have been summarized by Yamakawa (10). Representative of these theories is the one developed by Ptitsyn and Eizner (11)

$$[\eta] \cdot M = \Phi \, (r^2)^{3/2} = \Phi_0 \, f(\epsilon) \, (r^2)^{3/2} \qquad (2)$$

where $f(\epsilon) = 1 - 2.63 \, \epsilon + 2.86 \, \epsilon^2$, $\epsilon = (2a - 1)/3$, where a is the Mark–Houwink exponent. The Mark–Houwink exponent qualitatively indicates the thermodynamic interaction between the polymer and the solvent (12). The exponent a increases as the quality of the solvent increases, reflecting the increase in hydrodynamic volume, and α as the polymer–solvent interaction increases (12). The proposed theories that account for excluded volume in computing the intrinsic viscosity front factor predict that Φ is a monotonically decreasing function of the expansion factor α.

The size of an uncharged, isolated macromolecule in solution as specified by the mean-square end-to-end distance depends on the molecular weight, the interaction between the polymer and the solvent, and intramolecular polymer–polymer interactions. The conformation statistics of neutral polymers in nonpolar solvents are well understood (12); however, this is not the case for polyions in polar solvents (13). For polyelectrolytes the molecular conformation depends on the amount and type of charged species, the ionic strength of the solvent, and the molecular weight. Thus, the separation of polyelectrolytes by SEC depends on the ionic strength of the solvent.

For polyelectrolytes in solution, electrostatic repulsion between the charged moieties on the mainchain backbone will cause an expansion of the macromolecule and increase the local chain stiffness (13). This chain expansion can be considered to be like an increase in the excluded volume. Numerous theoretical developments have been proposed to describe the conformations of polyions in solution and have been summarized elsewhere (13). These theoretical treatments are of two major types: one group employs a random-coil chain on which discrete charges are located, while the other group models the macromolecule as a sphere in which the charges are uniformly distributed. Most of the theoretical descriptions of the conformation of polyelectrolytes can be cast into a form that relates the expansion factor to the excluded volume. A typical relationship of this kind is that of Flory (14)

$$\alpha^5 - \alpha^3 = 2.60 \, Z_{el} \tag{3}$$

where Z_{el} contains the effects of intramolecular electrostatic interactions.

To determine the molecular weight of polyelectrolytes by SEC, we again need to relate the molecular weight to the molecular size of the macromolecule. Electrostatic interactions were not considered explicitly in the derivations of Equations 1 and 2. If, however, the expansion of polyelectrolytes can be described by the excluded volume, and the effect of excluded volume on the intrinsic viscosity is the same for neutral polymers and polyelectrolytes, the relationships between molecular weight and molecular size given in Equations 1 and 2 should also be valid for polyelectrolytes. The proposal of Ptitsyn and Eizner (11) as given in Equation 2 assumes that the factor Φ changes with expansion of the macromolecule. The predictions of Ptitsyn and Eizner along with experimental data for both neutral polymers and polyelectrolytes are shown in Figure 1 (15). For neutral polymers the value of Φ/Φ_0 initially decreases as the macromolecules begin to expand, then Φ/Φ_0 gradually increases again with additional chain expansion. However, the changes in Φ/Φ_0 are small over the limited range of chain expansions accessible to organic polymers. Because Φ/Φ_0 is nearly constant for organic polymers, the relationship between molecular weight and the molecular size given in Equation 1 is probably sufficient to describe the SEC of polymers soluble in organic solvents. For polyelectrolytes the ratio Φ/Φ_0 changes significantly as the macromolecules expand. Experimental data for sodium polystyrene sulfonate (NaPSS) is described reasonably well by the Ptitsyn−Eizner theory (11); however, the ratio Φ/Φ_0 for other polyelectrolytes is not always described accurately by the Ptitsyn−Eizner theory. For

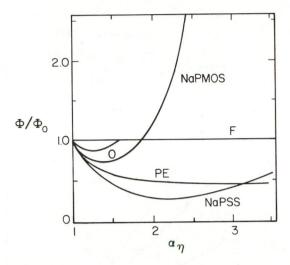

Figure 1. Intrinsic viscosity constant, Φ/Φ_0 vs. α_η. Theoretical predictions of Flory (F) and Ptitsyn−Eizner (PE). Experimental data for sodium polystyrene sulfonate (NaPSS), sodium poly(3-methacrylogloxypropane-1-sulfonate) (NaPMOS), and nonionic polymer−solvent systems (O). (Reproduced with permission from Ref. 15. Copyright 1975, John Wiley & Sons, Inc.)

NaPSS it is anticipated that the SEC data would be more effectively described by Equation 2 than by Equation 1.

In addition to molecular size, separation by size exclusion is also strongly dependent on the molecular shape. The more extended the conformation of a macromolecule, the more it will be excluded from the pores of SEC packings (Figure 2) (16). For a given molecular weight, a rod shaped molecule elutes earlier than a random coil polymer of the same molecular weight. The more compact hard sphere elutes even later than a random coil polymer. Polyelectrolytes in solutions of low ionic strength have been predicted to attain an asymmetrical shape (17, 18), and these shape effects must be considered in describing the SEC response of polyions (19).

Solute–Support Interactions. In addition to the effect of molecular size on the separation mechanism, other effects can influence the separation process. Many of these are important for polyelectrolytes, where ionic interactions between the polymer, solvent, and support can be significant. This section addresses these secondary separation mechanisms and discusses their relative significance.

Adsorption of a polymer on the chromatographic resin may affect the separation. Nonpolar polymers, such as polystyrene, can adsorb on a polystyrene support (20, 21) as well as on inorganic supports (20, 22,

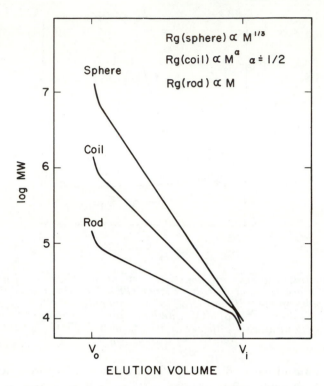

Figure 2. Effect of molecular geometry on SEC calibration curves. (Reproduced from Ref. 16. Copyright 1980, American Chemical Society.)

23). Specific adsorption effects of this type depend on the solvent used and/or the addition of cosolutes (21−25). In systems where the polymer and the support are oppositely charged, the polymer elutes later as the electrostatic attractions retard movement of the polyion through the porous matrix (26). In addition to affecting the average elution volume from the column, sorption phenomena may also cause the elution profile to exhibit multiple peaks for polymers with a monomodal molecular weight distribution (25, 27). In extreme cases, the polymer may be irreversibly adsorbed to the surface of the support (25).

For polymers and chromatographic supports that do not interact electrostatically, the penetration of the polymer into the pores is restricted only by the molecular size. However, if charged groups are present on the surface of the resin, electrostatic repulsion prevents the diffusion of polyelectrolytes of like charge into the pore (28−30). This would diminish the effective pore volume, and hence, the polymer would elute from the column earlier than a neutral polymer of the same size. This phenomenon is called ion exclusion, and has been discussed in connection with ion-exchange chromatography (*31*). Ion exclusion is attributed to the early elution from control pore glass packing of NaPSS in low ionic strength aqueous phosphate solutions (*32*). Addition of small amounts of cosolutes ($\sim 10^{-2}$ M) to the eluent will suppress ion exclusion (*28*).

When a solution contains two or more ionic solutes and one of the ionic species is excluded from some region of the gel or membrane that can be penetrated by the other ionic species, a Donnon equilibrium is established (*3, 29, 33–35*). In the chromatography of the polyelectrolytes, the size of the pore opening can prohibit free passage of the polyion; however, the pore is completely permeable to simple electrolytes. Thus, the larger ions exterior to the pore will cause the smaller ions of like charge to migrate into the pore to minimize electrostatic repulsion. This phenomenon is called ion inclusion (*29*). Using a conductometric detector coupled with a refractive index detector, Domard et al. observed a salt peak for polyelectrolytes in DMF and DMF with added electrolyte. They attributed the observed results to ion inclusion and determined that the phenomena could be suppressed if the eluent contained 5×10^{-2} M of added electrolyte (*36*).

In any particular aqueous SEC system, these nonsize related separation mechanisms may exist. Adsorption and association effects are a consequence of the specific choice of polymer, solvent, and support employed. It may not be possible to suppress these effects; therefore, one must critically examine the data to account for adsorption and association. Ion inclusion and ion exclusion can in general be eliminated by appropriate addition of simple electrolytes to the eluting media.

Calibration Procedures

SEC can be used for the routine characterization of polymers, provided the relationship between polymer molecular weight and retention volume can be established. Direct calibration can be employed in the analysis of those polymers for which well-characterized standards are available. Such calibration schemes have been employed for the analysis of globular proteins (*37, 38*) and low molecular weight polysaccharides (*39*). Unfortunately, for most water soluble polymers of interest, well-characterized standards are not available. Hence calibration curves usually must be constructed with commercially available standards that are different than the polymer of interest. The molecular weight and molecular weight distribution can be calculated from the secondary calibration curve if the relationship between molecular weight and molecular size is known for both the polymers.

Various secondary calibration schemes have been proposed for relating SEC elution data to molecular characteristics of the polymer. The universal calibration procedure proposed by Grubisic et al. (40) has found the widest applicability. This technique is based on the predictions of Equation 1. The product $M[\eta]$ is proportional to the hydrodynamic volume; therefore, a plot of log $M[\eta]$ vs. the SEC elution volume should yield a common curve for a given chromatographic column irrespective of the chemical structure of the polymer. The calibration curve can be constructed from SEC and intrinsic viscosity data of polymer standards with known molecular weights, typically monodisperse polystyrene. Because the intrinsic viscosity can be measured independently for any polymer, the molecular weight of an unknown polymer can be determined from the elution volume and the SEC calibration curve. The validity of this calibration method has been demonstrated extensively for neutral polymers in nonpolar solvents (41, 42). Little work has been done to extend Grubisic's technique to SEC in polar solvents. Spatorico and Beyer (43) have shown that $[\eta]M$ could describe the elution data for NaPSS and dextrans in aqueous solutions of 0.2 M and 0.8 M Na$_2$SO$_4$ as shown in Figure 3. Further support for this concept has been provided by Rochas et al. (33) for dextrans, poly(sodium glutamate), and NaPSS in 0.1 M NaNO$_3$. However, as shown in Figure 4 this method proves invalid for NaPSS and dextrans in 0.005–0.877 N NaOH solutions and 0.005–1.0 M NaCl solutions (19).

An alternate calibration scheme for SEC has been proposed by Coll and Prusinowski (44). It is based on Ptitsyn–Eizner's theoretical development (Equation 2), which accounts for excluded volume effects. In this method, log $\{M[\eta]/f(\epsilon)\}$ is plotted against elution volume. This procedure is valid for neutral polymers in nonpolar solvents (45, 46) and has also been shown to be applicable for NaPSS in 0.097–0.877 N NaOH solutions (Figure 5).

Several researchers have analyzed SEC data by both the procedure proposed by Grubisic and the Coll–Prusinowski method (19, 45, 46). No difference was found between the two methods for neutral polymers in organic solvents. If the Mark–Houwink exponents are similar for the various polymer–solvent systems examined by SEC, Equations 1 and 2 are identical up to a multiplicative constant and the two proposed calibration techniques will be indistinguishable. Molecular conformation of polyelectrolytes in polar solvents is strongly dependent on the salt concentration and, thus, the Mark–Houwink exponents for polyelectrolytes exhibit significant changes with the ionic strength of the solvent. Comparing the calibration curves shown

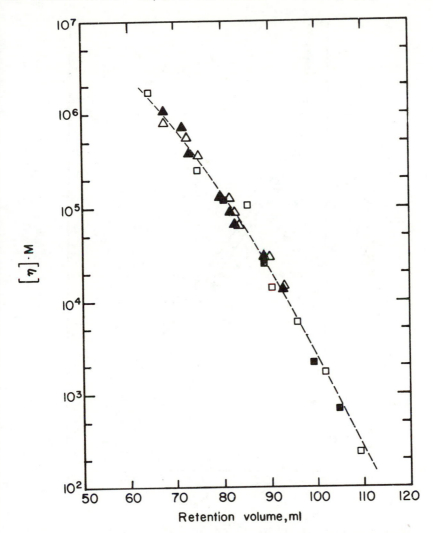

Figure 3. Plot of log [η]·M vs. elution volume for NaPSS and dextrans. Key: □, NaPSS in 0.2 M sodium sulfate; △, dextrans in 0.2 M sodium sulfate; ■, NaPSS in 0.8 M sodium sulfate; and ▲, dextrans in 0.8 M sodium sulfate. (Reproduced with permission from Ref. 43. Copyright 1975, John Wiley & Sons, Inc.)

in Figures 4 and 5, we observe that for NaPSS and dextrans in NaOH solutions of varying ionic strength the Coll–Prusinowski procedure describes the SEC data better than the universal calibration technique proposed by Grubisic et al. The Coll–Prusinowski procedure is unable to describe the SEC data of NaPSS in very low ionic strength solutions of NaOH, and this difficulty is probably related to changes in

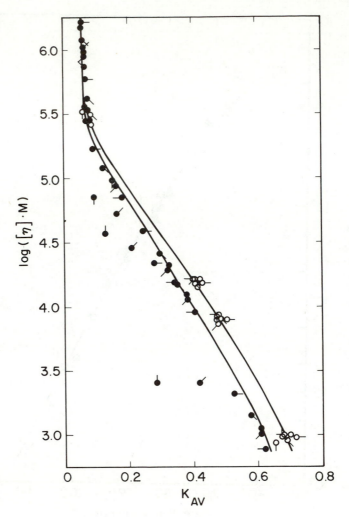

Figure 4. Plot of log $[\eta] \cdot M$ *vs.* K_{AV} *for NaPSS and dextrans. NaOH concentrations of 0.005 N, 0.0185 N, 0.075 N, 0.287 N, 0.501 N, 0.671 N, and 0.877 N are indicated by pips starting upward and moving clockwise at 45° angles, respectively (19).*

the shape of the macromolecule in low ionic strength solutions or solute−support interactions. The Coll−Prusinowski technique was not completely successful in describing the SEC data for NaPSS and dextrans in NaCl solutions, although the technique was better than the universal calibration procedure (*19*).

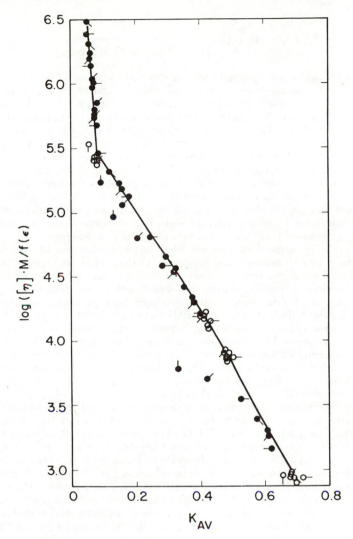

Figure 5. Coll–Prusinowski calibration curve for NaPSS and dextrans. NaOH concentrations are the same as indicated in Figure 4 (19).

Experimental

The efficiency of separation in a SEC system depends on the properties of the packing material and the flow rate. The column efficiency is related to plate height with lower plate heights corresponding to a more efficient column. The plate height H depends on the properties of the packing materials as follows

$$H = C_{SL}\frac{vd_p}{D_{SL}} + \frac{1}{1/ad_p + D_M/C_M vd_p^2} \qquad (4)$$

where C_M and a are constants for a given SEC system, d_p is particle packing diameter, C_{SL} is a constant associated with the stationary liquid phase mass transfer, D_{SL} is solute diffusion coefficient in stationary liquid phase, D_M is solute diffusion coefficient in mobile phase, and v is eluent linear velocity (1). Equation 4 shows that the most efficient column operation is realized when the eluent velocity is low and the packing is composed of small particles. A narrow distribution of particle sizes is also required for efficient column operation (47). The operation of any real SEC system is limited by numerous practical considerations. The use of small particles for the column packing causes a high pressure drop across the column, which the packing material may not be able to tolerate or the pump may be unable to attain. Extremely slow flow rates lengthen analysis time; hence, some loss of separation efficiency is generally sacrificed for experimental convenience.

The development of aqueous SEC has been limited by a lack of high performance chromatographic supports. As discussed previously, it is desirable that the column have sufficient mechanical strength to withstand a large pressure drop across the column because the time required to obtain a chromatogram can be decreased significantly by high pressure operation. Traditional aqueous SEC packing materials have only been able to separate polymers over a relatively narrow molecular weight range, and these columns have not possessed the necessary mechanical integrity (1). The composition and properties of these traditional aqueous SEC packing materials have already been discussed in detail (48, 49). Recently, high performance, aqueous chromatographic supports have become available (5, 6): TSK-GEL (Biorad) and Aquapore (Chromatix). Both column packings are silica based, have excellent mechanical strength, and are available in small particle sizes. Separation efficiency is reported to be four to five times better than with conventional supports (5). The TSK-GEL has been used successfully for fractionation of proteins, polysaccharides, and water-soluble synthetic polymers (26, 50). However, polymers such as polyacrylamide, sodium polystyrene sulfonate, and polyethyleneimine showed delayed responses, which were presumably due to selective adsorption on the support (26). These packing materials have a limited range of pH stability and are available only in a few pore sizes. Although these new columns have not been characterized completely, they represent a significant improvement over the traditional aqueous SEC packing materials.

Column selection for SEC depends on the range of molecular sizes to be separated, subject to the restrictions imposed by chemical properties of the solvent. The traditional approach for extending the fractionation range of a SEC system has been to combine several columns in series, each of which separates a small range of molecular weights. We have combined two Sepharose columns with different molecular weight ranges of separation (47). Using the composite column, SEC data for NaPSS and dextrans in various ionic strength NaOH solutions were obtained and are plotted according to the Coll–Prusinowski procedure in Figure 6. The range of molecular weights that can be resolved has increased considerably with this composite column. As an added advantage, the lengths of the individual columns can be adjusted so that the calibration curve is a linear function of the elution volume.

The traditional procedure of combining in series many small columns of

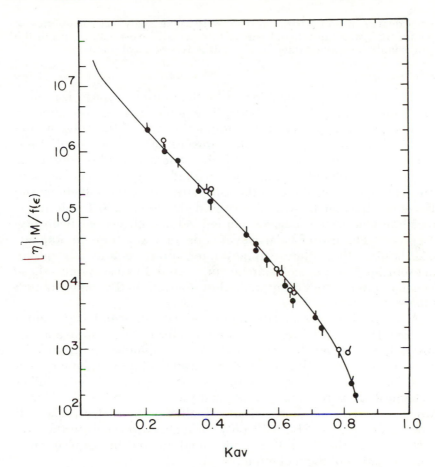

Figure 6. Coll–Prusinowski calibration curve for NaPSS and dextrans on Sepharose CL-6B and Sepharose CL-2B series column. NaOH concentrations are 0.185 N (pip directed upward) and 0.501 N (pip directed downward) (52).

various pore sizes into a single SEC system is not the best means for analyzing a polymer sample with a broad molecular weight distribution (51). These composite columns often result in longer analysis times and lower separation efficiency. The most efficient separation will result if narrow pore size distribution support particles are used and their molecular weight separation ranges do not overlap (51). The effectiveness of the bimodal pore-size distribution concept has been demonstrated in the fractionation of polystyrenes on a pair of porous silica microsphere columns (51).

We have employed aqueous SEC in the study of polysaccharide hydrolysis (52). Analysis of the hydrolysis products by aqueous SEC is more accurate, can be used to resolve a wider range of molecular weights, and is implemented more easily than traditional bulk chemical assays. Bulk assays are insensitive above a degree of polymerization of 20–25, whereas aqueous SEC can easily ascertain quantitative differences between the hydrolysis products with a degree of polymerization of 5×10^4. The analytical capabilities of aqueous SEC were an integral part of this overall investigation, and in the future should be applied effectively to other biochemical problems.

Conclusions

Aqueous SEC is a valuable tool for the routine characterization of biopolymers and polyelectrolytes. The molecular conformation of polyelectrolytes depends on the ionic strength of the solvent. Because the principal separation mechanism in SEC is differential migration, governed by the size of the macromolecules, the ionic strength of the solvent can influence markedly the elution behavior of polyelectrolytes. The effect of the ionic strength of the solvent on the SEC of polyions can be treated as excluded volume. A SEC calibration procedure that acknowledges excluded volume effects has been proposed (44). The use of polar solvents or ionic solutions in aqueous SEC requires that solute–support interactions such as adsorption, ion exclusion, and ion inclusion be considered. In most systems these secondary effects can be suppressed by controlling the ionic strength of the solvent.

Advances have been made in our understanding and use of aqueous SEC. However, there is still the need for further improvements, especially in the development of chromatographic supports and the synthesis of new, well-characterized, water-soluble, polymer standards. Desirable characteristics in support materials include higher mechanical strength, compatibility with a wide variety of solvents, narrow pore-size distributions, and smaller particle sizes. The full potential of aqueous SEC will only be realized through further advances in both the theoretical understanding, and the experimental implementation of the technique.

Acknowledgment

Part of this research was sponsored under DOE Grant No. DE-ACO2-78CS 40071.

Literature Cited

1. Yau, W. W.; Kirkland, J. J.; Bly, D. D. In "Modern Size Exclusion Liquid Chromatography"; Wiley: New York, 1979; p. 1.
2. Tung, L. H.; Moore, J. C. In "Fractionation of Synthetic Polymers— Principle and Practice"; Tung, L. H., Ed.; Dekker: New York, 1977; p. 545.
3. Tanford, C. In "Physical Chemistry of Macromolecules"; Wiley: New York, 1962; p. 1.
4. Eisenberg, H. In "Biological Macromolecules and Polyelectrolytes in Solution"; Oxford Univ.: Oxford, England, 1976; p. 1.

5. Chromatix Technical Bulletin, Sunnyvale, CA.
6. Biorad Technical Bulletin, Richmond, CA.
7. Pressure Chemical Co., Pittsburgh, PA.
8. Pharmacia Fine Chemicals, Piscataway, NJ.
9. Fox, T. G.; Flory, P. J. *J. Am. Chem. Soc.* **1951**, *73*, 1904.
10. Yamakawa, H. In "Modern Theory of Polymer Solutions"; Harper & Row: New York, 1971; p. 299.
11. Ptitsyn, O. B.; Eizner, Y. E. *Sov. Phys.—Tech. Phys.* (*Engl. Transl.*) **1960**, *4*, 1020.
12. Flory, P. J. In "Principles of Polymer Chemistry"; Cornell Univ.: Ithaca, NY, 1953; p. 1.
13. Eisenberg, A.; King, M. In "Ion-Containing Polymers—Physical Properties and Structure"; Academic: New York, 1977; p. 228.
14. Flory, P. J. *J. Chem. Phys.* **1953**, *21*, 162.
15. Tan, J. S.; Gasper, S. P. *J. Polym. Sci., Polym. Phys. Ed.* **1975**, *13*, 1705.
16. Yau, W. W.; Bly, D. D. In "Size Exclusion Chromatography (GPC)"; Provder, Theodore, Ed.; ACS SYMPOSIUM SERIES No. 138, ACS: Washington, DC, 1980; pp. 197–206.
17. Kurata, M.; Yamakawa, H. *J. Chem. Phys.* **1958**, *29*, 311.
18. Kurata, M. *J. Polym. Sci., Polym. Lett. Ed.* **1966**, *15*, 347.
19. Bose, A.; Rollings, J. E.; Caruthers, J. M.; Okos, M. R.; Tsao, G. T. *J. Appl. Polym. Sci.* **1982**, *27*, 795.
20. Darkins, J. V. *Polym. Prepr., Am. Chem. Soc., Div. Polym. Chem.* **1977**, *18* (2), 198.
21. Booth, C.; Forget, J. L.; Georgii, I.; Li, W. S.; Price C. *Eur. Polym. J.* **1980**, *16*, 255.
22. Bakos, D.; Bleha, T.; Ozima, A.; Berek, D. *J. Appl. rolym. Sci.* **1979**, *23*, 2233.
23. Campos, A.; Soria, V.; Figueruelo, J. E. *Makromol. Chem.* **1979**, *180*, 1961.
24. Coppola, G.; Fabbri, P.; Pallesi, B. *J. Appl. Polym. Sci.* **1972**, *16*, 2829.
25. Siebourg, W.; Lundberg, R. D.; Lenz, R. W. *Macromolecules* **1980**, *13*, 1013.
26. Fukano, K.; Komiya, K.; Sasaki, H.; Hashimoto, T. *J. Chromatogr.* **1978**, *166*, 47.
27. Cha, C. Y. *J. Polym. Sci., Polym. Lett. Ed.* **1969**, *7*, 343.
28. Neddermeyer, P. A.; Rogers, L. B. *Anal. Chem.* **1968**, *40*, 755.
29. Stenlund, B. *Adv. Chromatogr.* **1976**, *14*, 37.
30. Rinaudo, M.; Desbrieres, J. *Eur. Polym. J.* **1980**, *16*, 849.
31. Wheaton, R. M.; Bauman, W. C. *Ind. Eng. Chem.* **1953**, *45*, 228.
32. Cooper, A. R.; Matzinger, D. P. *J. Appl. Polym. Sci.* **1979**, *23*, 419.
33. Rochas, C.; Domard, A.; Rinaudo, M. *Eur. Polym. J.* **1980**, *16*, 135.
34. Buytenhaus, E. A.; Van der Maeden, F. P. B. *J. Chromatogr.* **1978**, *149*, 489.
35. Lindstrom, T.; de Ruvo, A.; Soremark, C. *J. Polym. Sci., Polym. Chem. Ed.* **1977**, *15*, 2029.
36. Domard, A.; Rinaudo, M.; Rochas, C. *J. Polym. Sci., Polym. Phys. Ed.* **1979**, *17*, 673.
37. Blagrove, R. J.; Fenkel, M. J. *J. Chromatogr.* **1977**, *132*, 399.
38. Chang, S. H.; Gooding, K. M.; Regnier, F. *J. Chromatogr.* **1976**, *125*, 103.
39. John, M.; Dellweg, H. *Sep. Purif. Methods* **1973**, *2*, 231.
40. Grubisic, Z.; Rempp, R.; Benoit, H. *J. Polym. Sci., Polym. Lett. Ed.* **1967**, *5*, 753.
41. Dawkins, J. V.; Hemming, M. *Polymer* **1975**, *16*, 554.
42. Ambler, M. R.; McIntyre, D. *J. Polym. Sci., Polym. Lett. Ed.* **1975**, *13*, 589.
43. Spatorico, A. L.; Beyer, G. L. *J. Appl. Polym. Sci.* **1975**, *19*, 2933.

44. Coll, H.; Prusinowski, P. *J. Polym. Sci., Polym. Lett. Ed.* **1967**, *5*, 1153.
45. Samay, G.; Kubin, M.; Podesva, J. *Angew. Makromol. Chem.* **1978**, *72*, 185.
46. Otocka, E. P.; Hellman, M. Y. *J. Polym. Sci., Polym. Lett. Ed.* **1974**, *12*, 331.
47. Rollings, J. E.; Bose, A.; Okos, M. R.; Tsao, G. T. *J. Appl. Polym. Sci.* **1982**, *27*, 2281.
48. Yau, W. W.; Kirkland, J. J.; Bly, D. D. In "Modern Size Exclusion Liquid Chromatography"; Wiley: New York, 1979; p. 419.
49. Cooper, A. R.; Van Derveer, D. S. *J. Liq. Chromatogr.* **1978**, *1*, 693.
50. Imamura, T.; Konishi, K.; Yokoyama, M.; Konishi, K. *J. Liq. Chromatogr.* **1981**, *4*, 613.
51. Yau, W. W.; Ginnard, C. R.; Kirkland, J. J. *J. Chromatogr.* **1978**, *149*, 465.
52. Rollings, J. E., Ph.D. Thesis, Purdue Univ., Lafayette, IN, 1981.

RECEIVED for review October 14, 1981. ACCEPTED July 6, 1982.

ELECTRON MICROSCOPY

Organic Coatings Analysis by Scanning Electron Microscopy and Energy Dispersive x-Ray Analysis

RICHARD M. HOLSWORTH

Glidden Coatings and Resins, Strongsville, OH 44136

The use of the scanning electron microscope and the energy dispersive x-ray analyzer in the study of coatings/substrate defects and problems is demonstrated. Examples discussed include the morphology and elemental analysis of metal pretreatments and substrates, both unexposed and exposed to various environments and chemical tests. Miscellaneous coatings/substrate problems such as mildew growth on an exterior exposure panel, interior can corrosion, and defects in clear coatings on metal substrates are shown.

Scanning electron microscopy (SEM) was first used by us to study the surface of paint films and substrates in 1962. This work was performed at the Pulp and Paper Research Institute of Canada, at that time, one of the few locations where SEM equipment was available. Pigment volume concentration ladders and pigmentation were studied. Later, untreated and bonderized steel panels and electrocoated panels were studied. One of the first publications concerning SEM investigation of coatings appeared in 1967 (*1*). This excellent work demonstrated the usefulness of SEM in coatings analysis and characterization. A symposium on "Scanning Electron Microscopy of Polymers and Coatings" (*2*) was held in Toronto, Ontario, in 1970 under sponsorship of the Chemical Institute of Canada and the American Chemical Society. Since that time, many articles have been published concerning either in whole or in part SEM analysis of coatings (*3, 4*). The coatings section of *Analytical Chemistry's* "Application Reviews 1979" (*5*) listed 43 references to the use of microscopy, 27 references to the use of surface analysis, and 18 references to the use of x-rays in the analysis and characterization of coatings. There have, of course, been many additional publications in this area in recent years.

0065–2393/83/0203–0363$06.00/0
© 1983 American Chemical Society

The use of SEM has increased dramatically in the past several years, not only in the coatings industry, but other surface-oriented disciplines as well. Probably the most significant reason for this is the relatively low price of SEM today, compared to the cost several years ago. Specialty SEMs that give good resolution at low and intermediate magnification ranges have been commercialized and, thereby, put SEM within the reach of many industrial labs. Such a specialty SEM is the MINI-SEM-II (6); it has made the SEM not only a research tool at the Dwight P. Joyce Research Center, but a tool for routine analysis and problem solving.

Experimental

The SEM used in our lab is a MINI-SEM Model II (International Scientific Instruments, Inc.) with a 16-step magnification range of 30×−40,000× and resolution of 250 Å. The accelerating voltage is fixed at 15 kV. The samples are coated in an Akashi vacuum evaporator with palladium−gold (40−60) or carbon. A KEVEX 5100 x-ray energy spectrometer is used for the analysis of heavy elements found on surfaces, substrates, and in organic coatings.

Results and Discussion

Pigment Volume Concentration (PVC) Ladders. One of the early studies utilizing SEM was to observe the morphological changes that occur in a paint film as the pigment volume concentration (PVC) is increased through the critical pigment volume concentration (CPVC). Many of the physical and mechanical properties of paint films such as tensile strength, elongation, and moisture permeability change dramatically at the CPVC. The CPVC was defined (7) as the PVC at which the binder or resin just fills the voids between the packed pigment particles. When the PVC lies below the CPVC, the film is composed of pigment dispersed in a continuous resin phase. Above the CPVC there is insufficient binder to fill the spaces between the packed pigment particles. The result is a void-filled film. It was anticipated that this phenomenon could be studied by SEM. Some results of this study are shown in Figures 1−4. Figure 1 at 770× magnification shows the surface of a 30 PVC latex paint. The surface appears to be binder rich in this case. Figure 2 at 750× magnification shows the surface of a 60 PVC latex paint. The surface appears to be very porous and binder deficient in this case. The latex paint is most likely above the CPVC or the point of optimum pigment loading to maintain an optimum balance of properties. Figure 3 shows the surface of a 50 PVC latex paint at 500× magnification. The surface looks very similar to that of Figure 2. Figure 4 shows this paint surface at 25,000× magnification. The smooth binder matrix can be seen along with several voids. The pigment particles that are seen readily are most likely single and agglomerated particles of TiO_2.

Figure 1. 30 PVC Latex paint (770×).

Figure 2. 60 PVC Latex paint (750×).

Figure 3. 50 PVC Latex paint (500×).

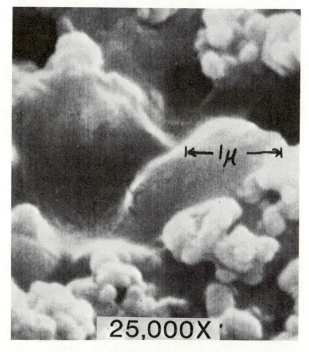

Figure 4. 50 PVC Latex paint (25,000×).

Mildew Defacement.   During a routine evaluation of exposure panels, a panel from the exposure fence appeared to have a very heavy dirt pickup. There were tiny black spots on the surface that apparently did not wash off. The black spots did and did not behave like dirt, and upon visual observation, they did not look like mildew. The SEM photomicrograph at 400× magnification, Figure 5, shows the surface of the film to be broken, and mildew growing out onto the surface of the coating. Figure 6, at 700× magnification, shows this even better. The coating is being pushed off or broken by the eruption and growth of the mildew from beneath the top coat. Figure 7 is a 1300× magnification of the same panel. The coating can be seen to be fractured by the eruption of the mildew growth. Figure 8, at 700× magnification, is of a similar spot that has been washed free of the mildew. Figure 9 at 3000× magnification, shows residual mildew within the defect in the paint film. Whether the spore got into the primer by way of small foam or air voids through the top coat to the substrate, or was on the plywood substrate before it was painted is not known. The mildew does not seem to be starting its growth on the surface but rather from underneath the top coat of latex paint.

Aluminum Siding Complaint.   Figure 10 is a micrograph at a magnification of 100× of aluminum siding that was investigated as the result of a complaint. The siding had become very dirty. Heavy dirt

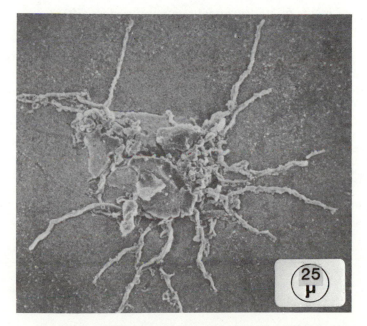

Figure 5. Mildew growth on panel surface (400×).

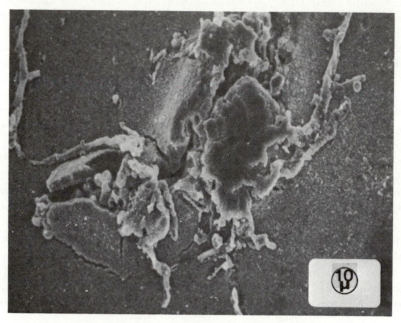

Figure 6. Mildew growth on panel surface (700×).

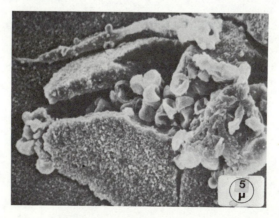

Figure 7. Mildew growth on panel surface (1300×).

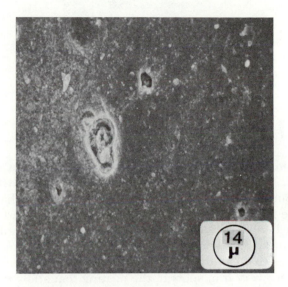

Figure 8. Scrubbed mildew covered panel (700×).

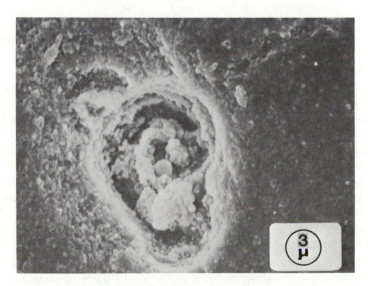

Figure 9. Scrubbed mildew covered panel (3000×).

Figure 10. Aluminum siding complaint (100×).

pickup and small pimples or bumps had developed across the surface of the panel. This micrograph shows several of these little bumps. Around the bumps can be seen apparent boundary lines that are really defects in the film resulting from solvent evaporation and the formation of Benard cells. This leaves a fairly regular pattern of resin-rich areas on the surface that degrade at a rapid rate. It is assumed that the boundary lines that can be seen are, in some areas, open through to the metal or are very thin spots in the coating that allow easier penetration of moisture or other contaminants that can lead to corrosion of the aluminum substrate. Figure 11 is a close-up of a cross-section of the bumps on the panel at 350× magnification. It shows that underneath the raised portion of the coating, a crystal or salt growth formation has developed between the aluminum substrate and the coating. This area was studied by energy dispersive x-ray analysis (EDXRA), and the crystals proved to be aluminum salts including sulfur and chlorine.

Clear Coatings on Reflective Metal Substrates. The inspection of clear coatings on highly reflective metal substrates is an annoying problem in optical microscopy. The reflection of the incident light from the metal obscures the fine detail of the surface morphology that needs to be seen. A styrene–acrylic basecoat for a quality as received plate was observed under the optical light microscope from 10× to 200× magnification. The only defects that could be seen were occasional gas bubbles that had been trapped in the film. There was no visible sign of the 2% wax that had been added to the coating as a lubricant under the optical microscope. Although the coating passed visual inspection, when the material was placed in a copper sulfate bath, metallic copper was formed on the iron substrate by a displacement reaction, indicating that the defects in the coating are open to the iron substrate. Flower-like copper deposits could be seen over a wide area of the panel. Figure 12 shows one of the copper deposits on the panel surface at 50× magnification. Many other defects on the surface of the

Figure 11. Cross-section of aluminum siding defect (350×).

panel can be seen in the micrograph. Area A of Figure 12 is shown at 600× magnification in Figure 13. Three types of defects can be seen in Figure 13; open bubbles from entrapped gas bubbles (B), craters probably resulting from the wax lubricant on the surface (C), and craters from insufficient flow of a broken bubble (F). The large crater in the center of Figure 13 is shown at 3000× magnification in Figure 14. The defects are labeled the same as in Figure 13. The material in the bottom of the large crater is probably residual wax or other residue from the bath. Analysis of a similar type of structure by EDXRA showed the material to be iron salts. A very small copper crystal formation is seen in Figure 15 at 3000× magnification.

To determine which type of defect was open to the substrate and allowed the copper crystals to grow, the copper deposit was dissolved with nitric acid and the resulting clear area studied in the SEM. Figure 16 at 560× magnification shows the dark irregular area that was

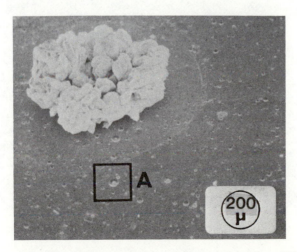

Figure 12. Metallic copper deposit on panel surface (50×).

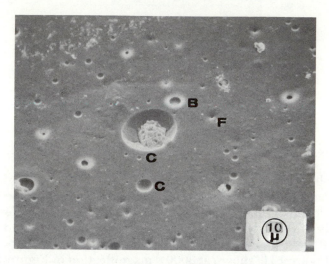

Figure 13. Surface defects in clear coating (600×).

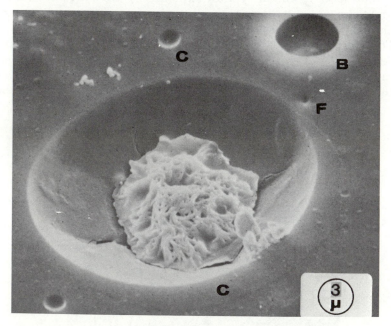

Figure 14. Surface defects in clear coating (3000×).

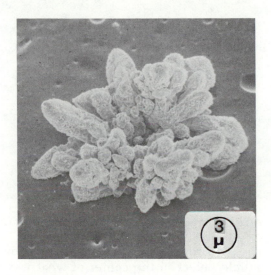

Figure 15. Copper deposit on panel surface (3000×).

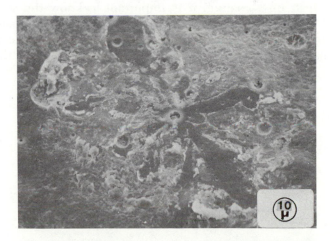

Figure 16. Area of copper deposit after acid treatment (560×).

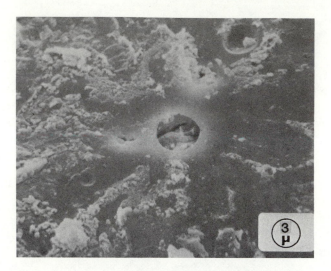

Figure 17. Area of copper deposit after acid treatment (2400×).

covered by the copper, residue from the acid dissolution of the copper, and a gas bubble defect in the center of what was the copper. Figure 17 at 2400× magnification shows the gas bubble defect and what looks like a crack or split in the bottom of the crater. This could be the opening to the substrate.

Metal Substrate Pretreatment. When running accelerated or natural in-use tests on a coating, it is important to know the substrate being used. The following are several examples where confusion developed somewhere between the initial pretreatment of the panels and the analysis of the surface. Figures 18 and 19 show the two

Figure 18. Zinc phosphate with calcium rinse pretreatment (700×).

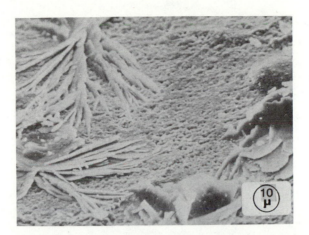

Figure 19.   Zinc phosphate without calcium rinse pretreatment (700 ×).

phosphate-treated panels in question. Physically they appear similar.
The zinc phosphate appears as small almost microcrystals and as large
plate-like crystals, mostly with a vertical orientation. The sample in
Figure 18 was to have received a calcium rinse, while the sample in
Figure 19 did not. The EDXRA spectra in Figures 20 and 21 show the
panels to be similar in surface treatment. In fact, both samples show
the presence of calcium. In Figures 22 and 23 are the familiar crystals
of zinc phosphate pretreatment. The sample in Figure 22 was to have
received a chrome rinse, while the sample in Figure 23 received no
rinse. The spectra of these two samples, shown in Figures 24 and 25,
exhibit no chromium. Only zinc, phosphorus, iron, and nickel from a
nickel fluoride rinse are present.

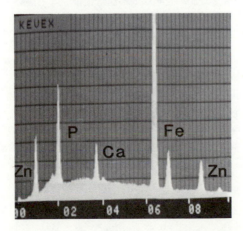

Figure 20. EDXRA spectrum of Figure 18.

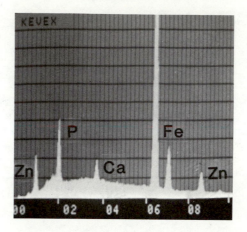

Figure 21. EDXRA spectrum of Figure 19.

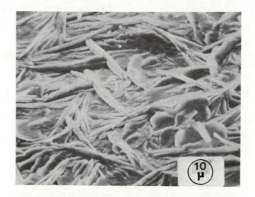

Figure 22. Zinc phosphate pretreatment with chrome rinse (700×).

Figure 23. Zinc phosphate pretreatment without chrome rinse (700×).

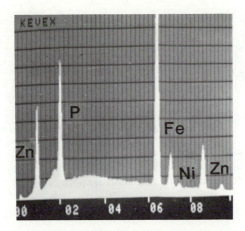

Figure 24. EDXRA spectrum of Figure 22.

Figures 26 and 27 show the surface of two samples with a calcium zinc phosphate surface treatment. The EDXRA spectra in Figures 28 and 29 show the overall surfaces to be chemically similar. However, when the surfaces of the two samples are compared, the test panel in Figure 26 shows small crystals tightly packed on the surface, while the plant metal substrate in Figure 27 shows crystals larger than in Figure 26, plus some plate-like crystals with mostly vertical orientation. Of more importance is the obvious fact that the crystals are so loosely packed on the surface that the metal substrate is exposed. This is confirmed by EDXRA which shows only iron to be present in the exposed areas (Figures 28 and 29).

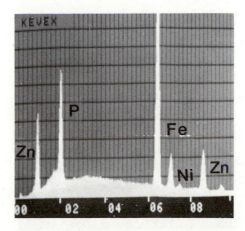

Figure 25. EDXRA spectrum of Figure 23.

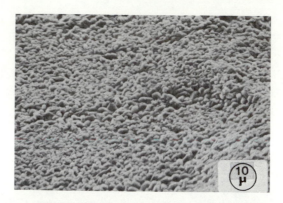

Figure 26. Calcium zinc phosphate pretreatment (700×).

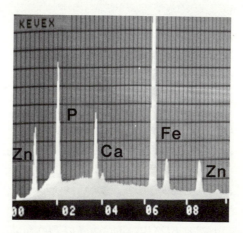

Figure 27. Calcium zinc phosphate pretreatment (700×).

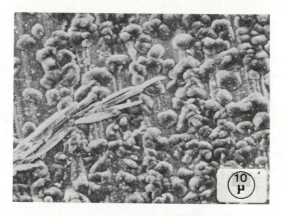

Figure 28. EDXRA spectrum of Figure 26.

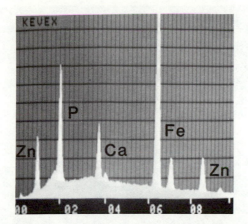

Figure 29. EDXRA spectrum of Figure 27.

Can Coating Substrate Corrosion.   The use of waterborne coatings for metal substrates, such as can coatings, has increased the occurrence of a problem, not unique to waterborne systems. This problem occurs as a result of entrapped foam or poor flow out of foam or bubble induced defects and is characterized by blistering and corrosion. A film defect such as seen previously in Figure 17 can, upon accelerated lab testing, produce an area such as that seen in Figure 30. This SEM photomicrograph at 200× magnification shows a small crater-like defect in the center of what appears to be a blister. Under the optical microscope the blister appeared to be filled, in part, with a

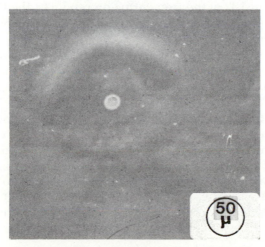

Figure 30. Coating defect and blister over tin-plated steel (200 ×).

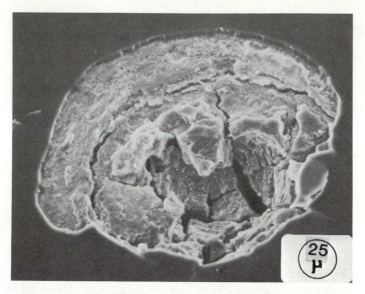

Figure 31. Corroded area under blister in Figure 30 (400×).

rust brown liquid over the still shiny tin plate surface. After being exposed to room conditions for several days, the blister was taped off, i.e., removed by adhesive tape. The results are shown in Figures 31 and 32 at 400× magnification. Figure 31 is the area immediately underlying the blister. The metal substrate is corroded through almost completely. EDXRA indicates iron throughout the corroded area and high levels of phosphorus and chlorine. Figure 32 is the taped off material that consists of the iron corrosion products, the tin plate, and the organic top coat. A mechanism for corrosion of this type has been described elsewhere (8).

Figure 33 shows a line of entrapped blisters or bubbles in a clear coating on tin plated steel substrate. The closed and partially open blisters showed no corrosion within the blister, although moisture had permeated the film and was trapped inside the bubbles. The exposed area in the center was filled with a rusty liquid. The top was carefully removed, including corrosion products from the center, exposing the pit and a covering of the clear coating over most of the surface. Figure 34 shows a similar bubble with the top removed, exposing the corrosion plug in the center of the bubble, plus corrosion products on the interior wall. Apparently, a defect like that in Figure 30 was present inside the bubble and corrosion started after permeation of liquid through the bubble wall.

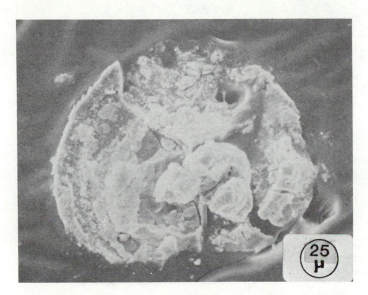

Figure 32. Backside of taped off blister in Figure 30 (400×).

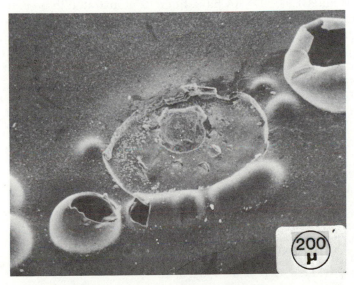

Figure 33. Corrosion area within blister, corrosion removed (50×).

Figure 34. Corrosion area within blister (200×).

Conclusions

SEM and SEM/EDXRA have been demonstrated to be extremely useful tools for characterizing coatings, substrates, and interfaces and for organic coatings failure analysis. Sample preparation is minimal compared with the wealth of information that can be obtained by these methods. The SEM is a logical extension of the low power stereomicroscope for coatings analysis, and the great depth of field available is invaluable for coatings problem analysis. The SEM/EDXRA system is a fast, easy method for rapid identification of small inclusions or defects, and for the quick comparison of coatings formulations.

Literature Cited

1. Brooks, L. E.; Sennett, P.; Morris, H. H. *J. Paint Technol.* **1967**, *39*, 472.
2. Princen, L. H. *Appl. Polym. Symp.* **1971**, *16*.
3. Princen, L. H.; Baker, F. L. *Paint Varn. Prod.* **1971**, *16*, 21.
4. Princen, L. H. *Appl. Polym. Symp.* **1974**, *23*.
5. Anderson, D. G.; Vandeberg, T. *Anal. Chem.* **1979**, *31*, 80R.
6. Evins, D. J.; Engle, R. J. *Am. Lab. (Fairfield, Conn.)* **1974**, *6*, 51.
7. Asbeck, W. K.; Van Loo, M. *Ind. Eng. Chem.* **1949**, *41*, 1470.
8. Funke, W. *Prog. Org. Coat.* **1981**, *9*, 29.

RECEIVED for review October 14, 1981. ACCEPTED March 12, 1982.

20

Molecular Optical Laser Examiner (MOLE)
Application to Problems Encountered
by Electron Microscopists
in the Analysis of Polymers

MARK E. ANDERSEN
Walter C. McCrone Associates, Inc., Chicago, IL 60616

Individual polymer fibers, layers in laminates, inclusions, and contaminants have been studied with the Raman microprobe, MOLE. The layers of a 14-μm thick five-layer laminated film were identified and molecular maps were obtained. The central polyester layer was determined to be of low crystallinity. Polarized Raman spectra were recorded from a 16-μm nylon 6/6 fiber. The calcium stearate reaction product of a pharmaceutical solution with a rubber stopper coating was identified using the energy dispersive x-ray analysis system of a scanning electron microscope (SEM) and the Raman spectrum. A 40-μm inclusion of polystyrene in a high-impact polystyrene copolymer was identified. The cause of a plating defect in a circuit board connector was traced to polystyrene packing material.

T HE MOLECULAR OPTICAL LASER EXAMINER (MOLE) is a Raman microprobe that was developed in France (*1, 2*) and is now available commercially. Contemporaneous with this development, a similar instrument was designed and built at the U.S. National Bureau of Standards (*3*). Both of the instruments yield molecular information, nondestructively, from both organic and inorganic samples or regions of samples as small as 1 μm in diameter. The MOLE is designed around a research quality optical microscope allowing the analyst to utilize microscopical techniques to orient and characterize the particular sample being analyzed (*4*). The MOLE has the additional capability to form images of a sample and define the locations of various molecular species. Both the high spatial resolution and the molecular mapping capability (analogous to x-ray elemental maps) are of interest to electron microscopists.

Macro-Raman spectroscopy has proved to be a useful technique

0065–2393/83/0203–0383$06.00/0

in relating the physical properties of polymers to their molecular structures (5–12). The mechanical properties of polymers, often studied by electron microscopy, are also related to their molecular properties. With the Raman microprobe, much smaller sample sizes can now be studied, and the interaction of electron microscopy with Raman spectroscopy can only be expected to grow.

Industrial production of polymer products usually involves mixing solvents, fillers, pigments, antioxidants, and other compounds with the polymer. Product failures may arise from inadequate mixing or dissolution of these additives, improper temperature control during product formation, or introduction or improper removal of contaminants and by-products. Tracking down the source of a failure is often a difficult analytical problem, frequently requiring the application of a variety of analytical tools including Raman microprobe and electron microscopy.

The purpose of this chapter is to show the unique capabilities of the MOLE as applied to polymer analysis, especially in conjunction with electron microscopy.

Experimental

The 514.5-nm line of a Spectra Physics Model 164 argon ion laser served as the illumination source. Plasma lines were removed with a 10-Å HBW interference filter. A mica halfwave plate was incorporated in the light path to control the polarization orientation of the laser illumination source. Both 50× (0.85 N.A.) and 100× (0.90 N.A.) Leitz objectives have been used to focus the laser beam and to collect the scattered radiation.

A scheme of a portion of the light path is shown in Figure 1. An axis is designated to define the polarization directions. A field diaphragm or pinhole in a sample image plane serves as a spatial filter, while an aperture diaphragm in an image of the back focal plane of the objective can be used to limit the solid collection angle, or numerical aperture of the objective. For polarization studies an analyzer and polarization scrambler are inserted in the light path.

The light passing through the monochromator is detected with a Type R 374 Hamamatsu photomultiplier cooled to −30 °C with a Products for Research thermoelectric housing (Model TE-175RF). Measurements were made with a Model 1140A quantum photometer from Princeton Applied Research.

Nonconductive samples for electron microscopy were mounted on aluminum stubs with double-sided tape. These samples were shadowed with carbon and gold. The metallic sample was adhered to a stub using silver paint. Micrographs were recorded with a Cambridge Mark IIA scanning electron microscope (SEM).

Laminated Polymer Study

A polymeric film was cross-sectioned with a razor blade and examined by SEM (Figure 2). The 14-μm thick film was composed of five laminated layers. The other half of the cross-section was ex-

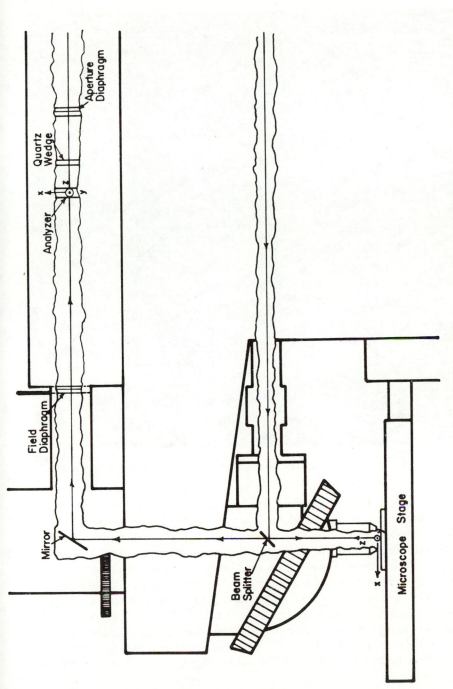

Figure 1. Schematic of MOLE microscope showing optical components incorporated in the light path.

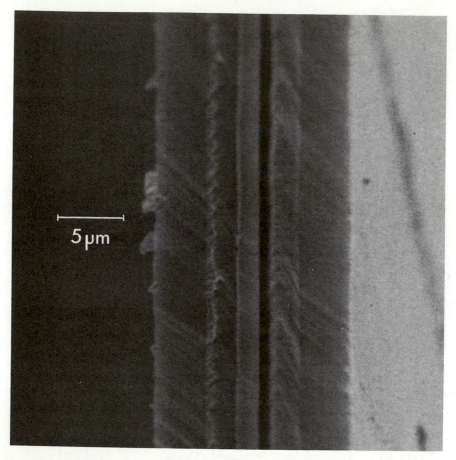

Figure 2. SEM micrograph of 14-μm thick, five-layer polymer laminate.

amined by MOLE. A pinhole limited the spectral collection area of
the sample to approximately a 5-μm diameter. The most intensely
illuminated region of the sample (approximately 1-μm diameter) gives
rise to the majority of the collected Raman scatter radiation. The outer
layers were about 3 μm thick and were identical in composition
(Figure 3). This material was identified as predominantly polypropylene. The central layer was less than 2 μm thick and was identified
as a polyester similar to polyethylene terephthalate (Figure 4). The
last two layers were identical and were identified as polyethylene
or a similar material (Figure 5). Some spectral leakage from the surrounding layers occurred and an unambiguous result could not be
obtained.

A molecular map of the sample was produced with the radiation

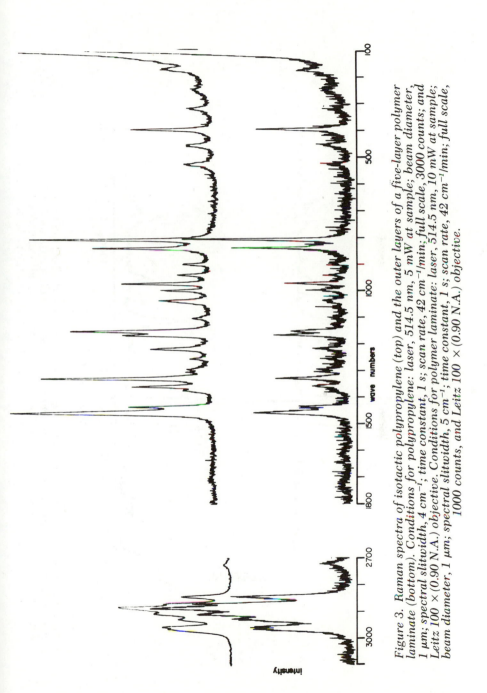

Figure 3. Raman spectra of isotactic polypropylene (top) and the outer layers of a five-layer polymer laminate (bottom). Conditions for polypropylene: laser, 514.5 nm, 5 mW at sample; beam diameter, 1 μm; spectral slitwidth, 4 cm⁻¹; time constant, 1 s; scan rate, 42 cm⁻¹/min; full scale, 3000 counts; and Leitz 100 × (0.90 N.A.) objective. Conditions for polymer laminate: laser, 514.5 nm, 10 mW at sample; beam diameter, 1 μm; spectral slitwidth, 5 cm⁻¹; time constant, 1 s; scan rate, 42 cm⁻¹/min; full scale, 1000 counts, and Leitz 100 × (0.90 N.A.) objective.

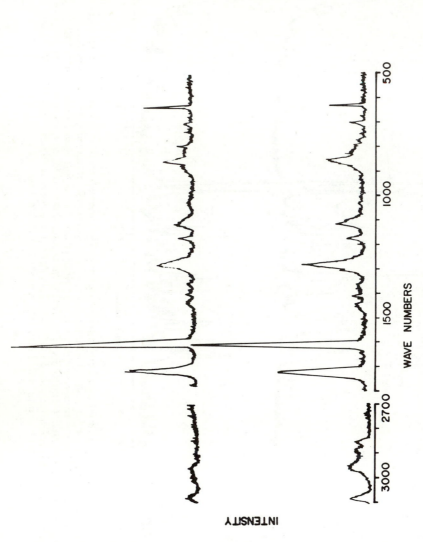

Figure 4. Raman spectra of polyethylene terephthalate (Scientific Polymer Products, Inc.) (top) and the middle layer of a 14-μm thick polymer laminate (bottom). Conditions: laser, 514.5 nm, 5 mW at sample; beam diameter, 1 μm; time constant, 1 s; scan rate, 42 cm⁻¹/min; full scale, 3000 counts; and Leitz 100 × (0.90 N.A.) objective. Spectral slitwidth: top, 4 cm⁻¹; and bottom, 5 cm⁻¹.

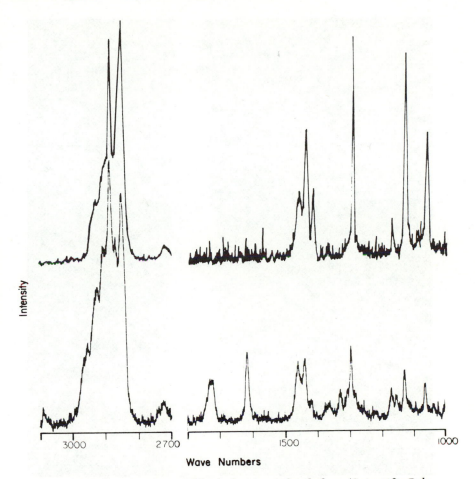

Figure 5. Raman spectra of high density polyethylene (Scientific Poly-
mer Products, Inc.) (top) and the intermediate layer of a 14-μm thick
five-layer polymer laminate. Conditions for polyethylene: laser, 514.5
nm, 5 mW at sample; beam diameter, 1 μm; spectral slitwidth, 4 cm⁻¹;
time constant, 1 s; scan rate, 42 cm⁻¹/min; full scale, 1000 counts; and
Leitz 100 × (0.90 N.A.) objective. Conditions for polymer laminate: laser,
514.5 nm, 15 mW at sample; beam diameter, 1 μm; spectral slitwidth, 6
cm⁻¹; time constant, 3 s; scan rate, 17 cm⁻¹/min; full scale, 1000 counts;
and Leitz 100 × (0.90 N.A.) objective.

emitted by the sample at 1615 cm^{-1} (Figure 6). The polyester layer
is clearly delineated. At 2884 cm^{-1} the outer layers emit Raman scattered radiation more intensely than does the central polyester layer,
which is dark compared to the outer layers (Figure 7).

The crystallinity of the polyester layer is determined by measuring the band width of the 1730-cm^{-1} Raman band (6, 8). This polyester
layer has a very low degree of crystallinity, indicated by the broadness
of this band (half band width = 27 cm^{-1}).

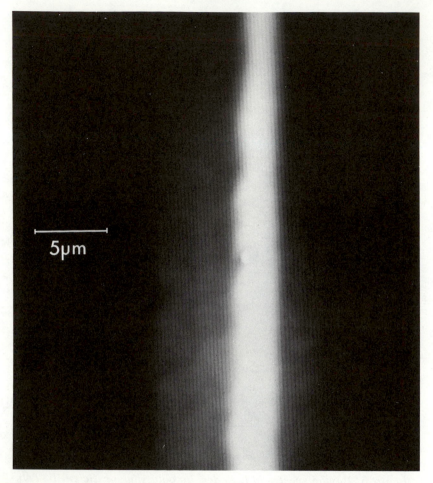

Figure 6. MOLE molecular map of five-layer polymer laminate at 1615 cm^{-1} showing 2-μm polyester middle layer.

Polarization Study

The properties of a polymer depend in part upon the orientation of its molecules, which changes as a polymer is stretched or extruded as fibers or sheets. Information about the molecular orientation is determined by measuring the polarization properties of the various Raman bands. Although such studies may be performed on relatively large samples with standard Raman instruments (9–11), the Raman microprobe has only recently been applied (4). The dielectric beam splitter in the MOLE (Figure 1) inevitably alters the polarization state of light impinging on its surface. If such effects can be measured and the derived data can be corrected for the effects, useful information can be obtained about the molecular orientation of much smaller samples than have been studied previously.

Carbon tetrachloride is used widely as a reference material for studying the polarization capabilities of a Raman system (*13, 14*). Carbon tetrachloride was placed in a glass capillary and examined using the 50× objective. The depolarization ratios of the asymmetric vibrational modes at 218 and 314 cm^{-1} are expected to be 0.75, while the totally symmetric vibrational mode at 459 cm^{-1} is expected to have a depolarization ratio near zero (*14*). The measured spectra for the two positions of the analyzer designated by the Porto convention (*15*) are shown in Figure 8. Both the raw data and the values corrected for the polarizing components in the MOLE are listed in Table I, along with a recent determination. Errors quoted arise from signal noise only. In this instance the correction factor measured happened to be near unity over the 200–500-cm^{-1} spectral region.

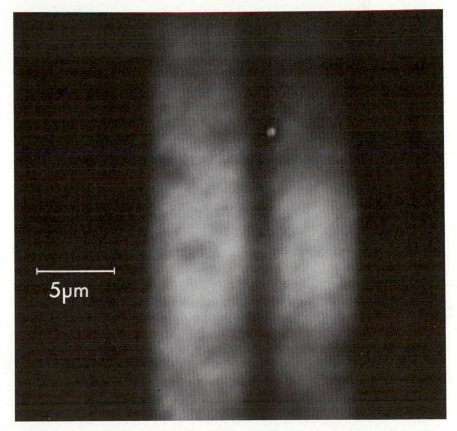

Figure 7. MOLE molecular map of five-layer polymer laminate at 2884 cm^{-1}.

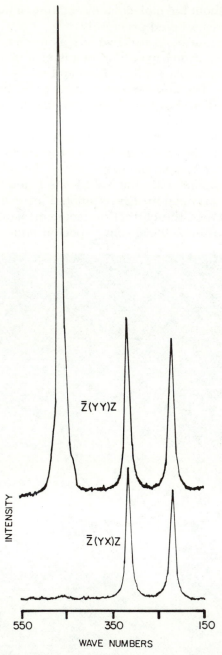

Figure 8. Polarized Raman spectra of CCl₄ in a glass capillary. Conditions: laser, 514.5 nm, 10 mW at sample; beam diameter, 1 μm; spectral slitwidth, 8 cm⁻¹; time constant, 1 s; scan rate, 42 cm⁻¹/min; full scale, 10,000 counts; and Leitz 50× (0.85 N.A.) objective.

Table I. Depolarization Ratios of CCl$_4$
Using a 50× (0.85 N.A.) Objective

Raman Band (cm^{-1})	$\rho(\pm2\sigma)$ meas.	$\rho(\pm2\sigma)$ corrected	ρ^a
218	0.72 (±0.03)	0.74 (±0.03)	0.75
314	0.76 (±0.03)	0.77 (±0.03)	0.75
459	0.006 (±0.002)	0.006 (±0.002)	0.003

^{a} Reference 13.

A pleochroic, birefringent, 16-μm fiber of nylon 6/6 was oriented parallel to the y-direction as defined in Figure 1. Polarized Raman spectra of this fiber are shown in Figure 9 and display distinct intensity differences in the various Raman bands. Peak heights ratios, $\overline{Z}(YY)Z/\overline{Z}(YX)Z$, are subject to a correction factor varying smoothly between 1.4 at 900 cm^{-1} and 2.8 at 1800 cm^{-1}, and between 1.4 at 2700 cm^{-1} and 1.0 at 3000 cm^{-1}. Errors in ratios arising from sample birefringence are reduced when using a high numerical aperture objective, because the sampled depth is minimized to a few micrometers (2, 16). A more complete study of this fiber would allow one to determine the symmetry of the Raman bands and the orientation of the molecules in the fiber. Potentially this technique may allow a more complete understanding of the skins of some types of polymer fibers.

Analytical Problem Solving

A translucent 200-μm thick sheet of high-impact polystyrene/polybutadiene copolymer was marred by a bulge or fisheye. When the sheet was sectioned through the defect region with a razor blade, a transparent, 40-μm inclusion was revealed (Figure 10). The Raman spectra of the two materials were nearly identical except in the region near 1650 cm^{-1}. The matrix material had two polystyrene bands near 1600 cm^{-1} (Figure 11A) and a small peak near 1650 cm^{-1}, which is indicative of *cis*-polybutadiene (7). The inclusion lacks this peak (Figure 11B) and is nearly pure polystyrene. Inclusions such as undissolved antioxidant or extraneous polymers as small as 2 μm in size and buried as much as 10 μm beneath the surface of transparent polymer sections have been identified in situ with MOLE.

Solutions contaminated with particles are found occasionally in pharmaceutical samples. Often these particles are created by a reaction between the solution and a bottle stopper that is usually composed of rubber containing zinc and sulfur compounds used as curing accelerants and filled with carbon black, anatase, calcite, quartz, and

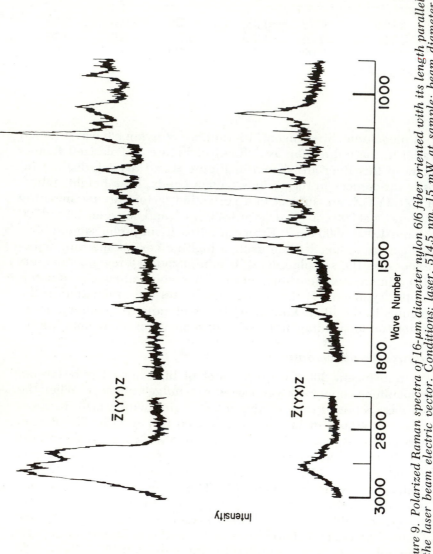

Figure 9. Polarized Raman spectra of 16-μm diameter nylon 6/6 fiber oriented with its length parallel to the laser beam electric vector. Conditions: laser, 514.5 nm, 15 mW at sample; beam diameter, 1 μm; spectral slitwidth, 8 cm⁻¹; time constant, 3 s; scan rate, 17 cm⁻¹/min; full scale, 1000 counts; and Leitz 50× (0.85 N.A.) objective.

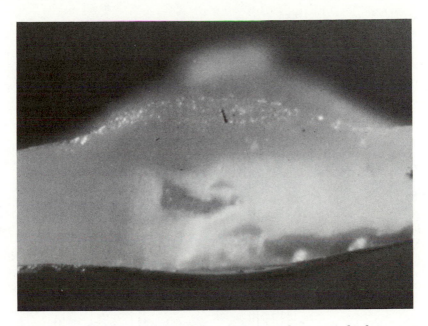

Figure 10. Photomicrograph of a 40-μm inclusion in high-impact polystyrene.

other minerals. The stoppers are formed in molds often coated with a waxy material that serves as a mold release agent.

One common example of such a contamination problem is shown in Figure 12, which is a micrograph of a stopper surface covered with crystals that have apparently grown from solution. Energy dispersive x-ray analysis of these crystals and of particles suspended in the solution indicated that the crystals contained a small amount of calcium but were mostly organic. MOLE analysis of these particles identified them as calcium stearate (Figure 13). The crystals probably formed by a reaction between calcium ions in the solution and residual stearic acid mold release agent on the stopper surface.

A tooth from a defective circuit board connector is shown in Figure 14 where gold plating has incompletely covered the tooth (Figure 15). This defect region had a thin transparent layer of material adhering to it, identified by MOLE as polystyrene. The connectors, which had been shipped in boxes with polystyrene "noodles" used as a packing material, were degreased before gold plating. Apparently residual particles of polystyrene remained on the connectors during the degreasing operation and partially dissolved forming a thin layer on the metal surface which hindered the gold plating.

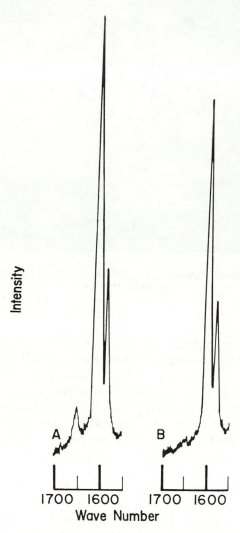

Figure 11. Raman spectrum of high-impact polystyrene (A) and in-clusion in high-impact polystyrene (B). Conditions: laser, 514.5 nm, 10 mW at sample; beam diameter, 1 µm; spectral slitwidth, 4 cm⁻¹; time constant, 3 s; scan rate, 17 cm⁻¹/min; full scale, 3000 counts; and Leitz 100× (0.90 N.A.) objective.

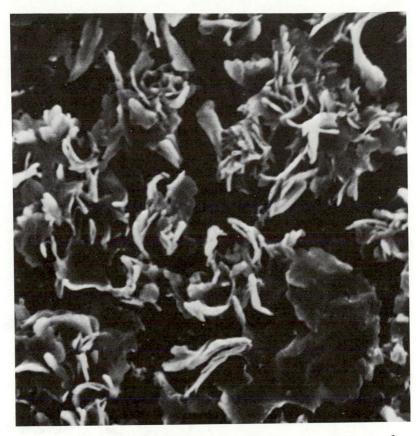

Figure 12. SEM micrograph of crystals (2–20 μm in size) on surface of rubber stopper.

Conclusion

With its high spatial resolution and molecular mapping capabilities, MOLE is a unique tool for analyzing polymers and problems created in their production and use. As with all analytical tools, it can best be utilized in conjunction with other instruments such as the electron microscope.

Acknowledgments

The author thanks Paul Dhamelincourt for his assistance in designing the polarization experiments; and Fran Adar, Leroy Pike, Walter Kramer, and Robert Muggli for providing helpful discussions. Special thanks are extended to Fran Einbinder and Jim Gerakaris for their patience and help in preparing this manuscript.

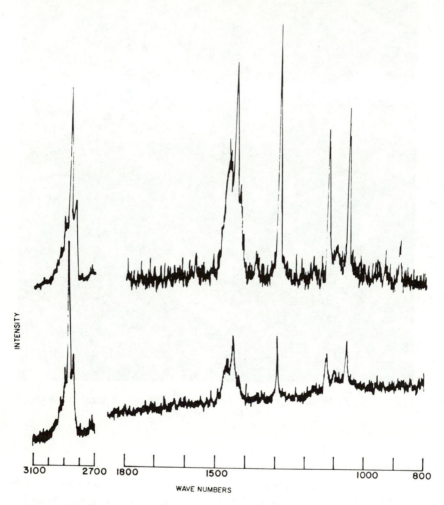

Figure 13. Raman spectra of calcium stearate (top) and a crystal removed from the stopper (bottom). Conditions for calcium stearate: laser, 514.5 nm, 20 mW at sample; beam diameter, 1 μm; spectral slitwidth, 3 cm⁻¹; time constant, 1 s; scan rate, 42 cm⁻¹/min; full scale, 1000 counts; and Leitz 100 × (0.90 N.A.) objective. Conditions for crystal from stopper: laser, 514.5 nm, 10 mW at sample; beam diameter, 1 μm; spectral slitwidth, 5 cm⁻¹; time constant, 1 s; scan rate, 42 cm⁻¹/min; full scale, 1000 counts; and Leitz 100 × (0.90 N.A.) objective.

Figure 14. Photomicrograph of a poorly plated tooth of a circuit board connector.

Figure 15. Elemental map showing gold plating distribution on connector tooth.

Literature Cited

1. Delhaye, M.; Dhamelincourt, P. *J. Raman Spectrosc.*, **1975**, *3*, 33.
2. Dhamelincourt, P., Ph.D. Thesis, Lille Univ., Lille, France, 1979.
3. Rosasco, G. J.; Etz, E. S.; Cassatt, W. A. *Appl. Spectrosc.*, **1975**, *29*, 396.
4. Andersen, M. E.; Muggli, R. Z. *Anal. Chem.*, **1981**, *53*, 1771.
5. Tadokoro, H. "Structure of Crystalline Polymers"; Wiley: New York, 1979.
6. Melveger, A. J. *J. Polym. Sci.*, **1972**, *10*, 317.
7. Sloane, H. J.; Bramston-Cook, R. *Appl. Spectrosc.*, **1973**, *27*, 217.
8. Ogilvie, G. D.; Addyman, L. *Actual. Chim.* **1980**, *1980*, 51.
9. Bailey, R. T.; Hyde, A. J.; Kim, J. J.; McLeish, J. *Spectrochim. Acta*, **1977**, *33A*, 1053.
10. Jasse, B.; Koenig, J. L. *J. Polym. Sci.*, **1980**, *18*, 731.
11. Kobayashi, M.; Tadokoro, H. *J. Chem. Phys.*, **1980**, *73*, 3635.
12. Koenig, J. L.; Boerio, F. J. *J. Chem. Phys.*, **1970**, *52*, 4170.
13. Kiefer, W.; Topp, J. A. *Appl. Spectrosc.*, **1974**, *28*, 26.
14. Porto, S. P. S. *J. Opt. Soc. Am.*, **1966**, *56*, 1585.
15. Damen, T. C.; Porto, S. P. S.; Tell, B. *Phys. Rev.*, **1966**, *142*, 570.
16. Rosasco, G. J. In "Advances in Infrared and Raman Spectroscopy"; Clark, R. J. H.; Hester, R. E., Eds.; Heyden: London, 1980, Chap. 4.

RECEIVED for review October 14, 1981. ACCEPTED January 18, 1982.

Micromechanical Measurements of Polymers by Transmission Electron Microscopy

EDWARD J. KRAMER

Cornell University, Department of Materials Science and Engineering and the Materials Science Center, Ithaca, NY 14853

A method for measuring the micromechanical properties of crazes and plane stress plastic zones in thin polymer films has been developed. Cast films about 1-μm thick are bonded to annealed copper grids and are crazed by plastically deforming the grid in tension, either in air or in a gaseous or liquid environment. A series of transmission electron micrographs is exposed along the length of the craze or zone. The local extension ratio λ is then determined from the optical densities (on the plate) of the craze or zone, the solid film, and a hole through the film. Because the local craze thickness (in the film plane) can be measured accurately from the plate, the total craze opening displacement $2w = \tau(1 - 1/\lambda)$ can also be determined to within ±20 nm at intervals as small as 100 nm along the craze. These surface displacements are converted into craze surface stresses using the Fourier transform method and these surface stresses are multiplied by λ to determine the true stresses in the craze fibrils or the web of the plastic zone.

GLASSY POLYMERS exhibit a number of localized mechanisms of plastic deformation. Particularly important for subsequent fracture behavior is craze formation. Crazes are planar, crack-like features that consist of two surfaces of undeformed polymer with small (<10 nm diameter) polymer fibrils connecting them at normal incidence as shown in the transmission electron micrograph (TEM) in Figure 1. The fibrils cause the craze to be load bearing, unlike a true crack. However, when cracks initiate, they invariably start within existing crazes by breakdown of the fibril structure and grow slowly until they reach critical size for catastrophic propagation. Crazes also play an

0065−2393/83/0203−0401$06.00/0

Figure 1. The microstructure of an air craze in a thin film of polystyrene.

important role in the environmental stress cracking of polymers because they can grow and break down to form cracks under greatly reduced stresses in the presence of many environments. To achieve an understanding of crazing it is necessary to be able to measure the stresses along the craze surface $S(x)$ as a function of position x along its length, the volume fraction of craze fibrils $v_f(x)$ or, because the fibril formation process is one of plastic deformation at constant volume, the fibril extension ratio $\lambda(x)$ where $\lambda(x) = 1/v_f(x)$. The value of $S(x)$ can be computed by the Fourier transform method of Sneddon (1) if the craze surface displacement profile $w(x)$ is known, i.e.,

$$S(x) = \Delta S(x) + \sigma_\infty \tag{1}$$

where $\Delta S(x)$ is the self stress of the craze and σ_∞ is the applied tensile stress normal to the craze. The self stress is given by

$$\Delta S(x) = -\frac{2}{\pi} \int_0^\infty \bar{p}(\xi) \cos (x\xi) d\xi \tag{2}$$

where

$$\bar{p}(\xi) = (\xi E^*/2) \int_0^a w(x) \cos (x\xi) dx$$

E^* is an effective Young's modulus of the polymer, and a is the craze half length. Because $w(x)$ is given by $[1 - v_f(x)]\tau(x)/2$, measurements of the craze thickness $\tau(x)$ and fibril volume fraction profiles will suffice to determine all other profiles. Once $S(x)$ and $v_f(x)$ are known, the true tensile σ_t stress in the fibrils can be determined by $\sigma_t = S(x)/v_f(x)$. All these measurements can be made on crazes in polymer thin films by using transmission electron microscopy.

Experimental

Films of the glassy polymer, between 0.5 and 1 μm thick, are bonded to ductile copper grids. The film is then subjected to a tensile stress by straining the copper grid. If large strains are used, air crazes nucleate and grow in each grid square. At strains below the critical strain for crazing in air, crazing occurs only on exposure to a gaseous or liquid environment. The copper grid deforms plastically and holds the polymer film under tension even when the applied force on the grid is removed. A typical grid square can be selected using the optical microscope and cut from the grid using a razor blade without disturbing the applied strain in the film over that grid square. This grid square can be examined subsequently in the transmission electron microscope. Crazes may also be grown from cracks introduced by either indenting the film with a diamond pyramid hardness indentor while the film is still on the glass slide or by burning a slot in the film using the electron beam of an electron microprobe. A series of TEM micrographs is taken such that a composite micrograph, spanning the entire craze, can be pieced together and $\tau(x)$ can be measured from this composite. Local values of v_f are determined from measurements of the optical density ϕ of the electron image plates using the following equation

$$v_f = 1 - \frac{\ln (\phi_{\text{craze}}/\phi_{\text{film}})}{\ln (\phi_{\text{hole}}/\phi_{\text{film}})} \tag{3}$$

The derivation of this equation, as well as considerations of the effects of multiple scattering, is given elsewhere (2).

Results

The craze thickness profile for a craze that nucleated at a dust particle in a polystyrene (PS) film and grew to a total length of 189 μm is shown in Figure 2. Representative micrographs from selected regions along the craze are also shown. The dense interconnected network of fibrils typical of air crazes is seen, as is the so-called midrib, as a region of lower fibril density running along the central plane of the craze.

Figure 3 shows the fibril extension ratio profile of the craze as derived from the calculated values of v_f. These values of λ also can be determined by restricting the densitometer probe to a small portion of the craze image. Values of λ in the midrib, measured in this way, are shown in Figure 3. The craze opening displacements are shown in Figure 4, which shows the excellent resolution possible with this technique. Near the base (center) of the craze the resolution in $w(x)$ is

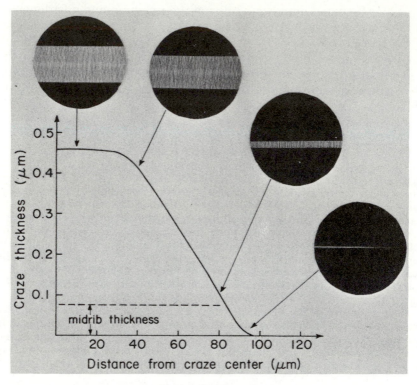

Figure 2. The thickness profile for an isolated air craze in polystyrene. (Reproduced with permission from Ref. 2. Copyright 1979, Taylor & Francis, Ltd.)

±10 nm. Closer to the tip the displacement resolution is even better and the spatial frequency of the data is sufficient to allow the determination of the surface stresses on a scale of ~1 μm.

The surface stress profile is shown in Figure 5. In the base region the surface stress lies below the applied stress level but then rises to a small stress concentration at the craze tip. The small amount of stress relief in the base, and the correspondingly small stress concentration at the tip, support the observation that air crazes are quite strong, exhibiting high fracture stresses relative to environmental crazes (3). Indeed the surface stress and extension ratio data at various points along the craze may be combined to produce the fibril true stress-extension ratio curve. Evidently fibrils in all regions of the craze have experienced severe strain hardening with the greatest extension ratio and greatest strain hardening occurring in the region of highest surface stress just behind the craze tip.

The technique has been applied to crazes at crack tips (4, 5), environmental crazes (6), and intersecting crazes (7). The technique

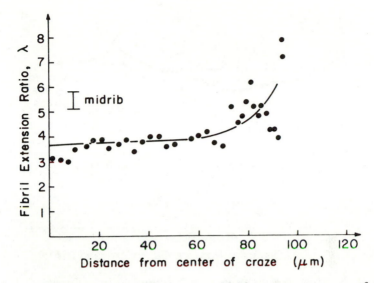

Figure 3. *Measured values of the craze fibril extension ratio as a function of the distance along the craze shown in Figure 2. (Reproduced with permission from Ref. 2. Copyright 1979, Taylor & Francis, Ltd.)*

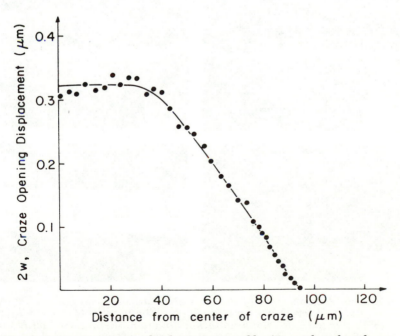

Figure 4. *Craze opening displacement profile. (Reproduced with permission from Ref. 2. Copyright 1979, Taylor & Francis, Ltd.)*

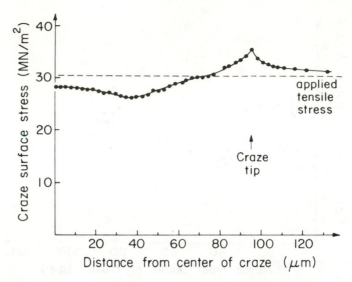

Figure 5. Craze surface stress profile computed from Equations 1 and 2. (Reproduced with permission from Ref. 2. Copyright 1979, Taylor & Francis, Ltd.)

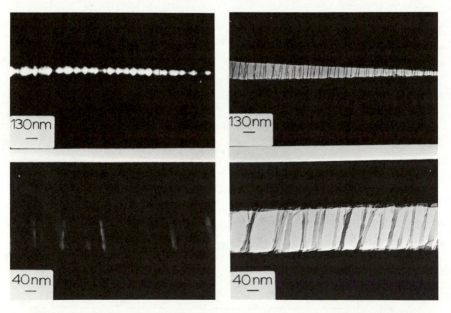

Figure 6. A comparison of the microstructure of air crazes in oriented polystyrene films grown by stressing the film parallel to (left) and perpendicular to (right) the orientation direction. (Reproduced with permission from Ref. 10. Copyright 1981, Butterworth's & Co.)

can be combined with optical microscopy to measure stress distributions in more ductile polymers, such as polycarbonate, in which plane stress plastic zones grow from crack tips (8, 9). A particularly interesting application is to oriented polystyrene (10). Here crazes grown under a tensile stress directed parallel to the orientation direction have a higher v_f ($v_f = 0.45$, $\lambda = 2.2$) than those grown under a tensile stress perpendicular to the orientation direction ($v_f = 0.05$, $\lambda = 20$). The craze microstructure is shown for each case in Figure 6 (*10*). As expected the surface stress profile for a perpendicular craze appears crack-like with a high stress concentration at the tip and a large stress relief at the craze base. These crazes also break down readily to produce cracks. On the other hand, the surface stress profile for a parallel craze shows it to be strongly load bearing, with only a small stress concentration. We believe that this marked anisotropy of craze microstructure and mechanical properties underlies the corresponding strong anisotropy of the fracture properties in oriented glassy polymers.

Acknowledgments

The financial support of the U.S. Army Research Office–Durham and the use of the facilities of the Cornell Materials Science Center, which is funded by the National Science Foundation are appreciated.

Literature Cited

1. Sneddon, I. N. "Fourier Transforms"; McGraw-Hill: New York, 1951; p. 395.
2. Lauterwasser, B. D.; Kramer, E. J. *Phil. Mag., Part A* **1979**, *39*, 469.
3. Krenz, H. G.; Ast, D. G.; Kramer, E. J. *J. Mater. Sci.* **1976**, *11*, 2198.
4. Donald, A. M.; Chan, T.; Kramer, E. J. *J. Mater. Sci.* **1981**, *16*, 669.
5. Chan, T.; Donald, A. M.; Kramer, E. J. *J. Mater. Sci.* **1981**, *16*, 676.
6. Yaffe, M. B.; Kramer, E. J. *J. Mater. Sci.* **1981**, *16*, 2136.
7. King, P. S.; Kramer, E. J. *J. Mater. Sci.* **1981**, *16*, 1843.
8. Donald, A. M.; Kramer, E. J. *J. Mater. Sci.* **1981**, *16*, 2967.
9. Donald, A. M.; Kramer, E. J. *J. Mater. Sci.* **1981**, *16*, 2977.
10. Farrar, N. R.; Kramer, E. J. *Polymer* **1981**, *22*, 691.

RECEIVED for review October 14, 1981. ACCEPTED August 13, 1982.

Electron Crystal Structure Analysis of Linear Polymers—An Appraisal

DOUGLAS L. DORSET and BARBARA MOSS

Medical Foundation of Buffalo, Inc., Electron Diffraction Department, Buffalo, NY 14203

Many microcrystalline linear polymers have been found to give good single crystal electron diffraction patterns; yet such intensity data are used only rarely for crystal structure analysis. Although the earliest assumptions made for such data are not rigorously true, the kinematical diffraction approach used in x-ray crystallography often will yield correct structural results. Thus, standard phasing methodologies—for example, use of Patterson maps, direct phasing,—can be quite satisfactory. Two important perturbations to diffracted intensities, resulting from n-beam dynamical scattering and elastic crystal bending, must be recognized and minimized to guarantee the derivation of real crystal structures. The first-named perturbation is minimized by restricting crystal thickness and/or selecting an appropriate electron wavelength. Because elastic bends of a few degrees are always present in thin molecular crystals, the latter perturbation is most noted when the unit cell length in the direction of the incident beam is large. Fortunately, a short unit cell repeat in this (fiber axis) direction is often found for polymer crystals. The influence of n-beam dynamical scattering and elastic crystal bending is demonstrated by model calculations on cytosine and representative polymer structures.

ELECTRON DIFFRACTION is used frequently in polymer physics to determine the unit cell dimensions and symmetry for molecular packing in single microcrystals. As frequently pointed out (1), the enhanced scattering cross-section of matter for electrons (compared to x-rays or neutrons) allows the acquisition of information unobscured by the overlap of reflections with near-reciprocal spacing found in powder diffraction patterns.

0065–2393/83/0203–0409$06.00/0

Despite its popularity, electron diffraction is underutilized by polymer scientists. Quantitative use of intensity data for crystal structure analysis is attempted only rarely and then by only a very small number of research groups in the world. This situation is difficult to comprehend given the potential for a crystal structure analysis. Polymer crystals, after all, are not often grown to sizes large enough for single-crystal x-ray experiments.

Reluctance to use such intensity data reflects a basic misunderstanding of their domain of validity. Early use of such data for organic crystal structure analysis (2), which demonstrated the promise of the technique, adopted too many assumptions used for x-ray crystallography. Resultant analytical procedures seemed to vary from compound to compound. Despite this situation, the agreement between observed and calculated structure factor moduli often remained poor. Furthermore, most crystal structures reported had been determined previously using x-ray data. These factors all contributed to widespread mistrust of the technique.

Over the past few years, we have sought a realistic understanding of the perturbations that limit the use of such data for ab initio crystal structure determination. Although this work continues, a self-consistent overview emerges that shows why some of the conceptual models used in x-ray crystallography are not appropriate for electron crystallography. On the other hand, minimization of data perturbations allows the use of electron diffraction intensities for crystal structure analysis, as is also demonstrated by the increasing number of polymer crystal structures derived from such data (3-11).

n-*Beam Dynamical Scattering*

High-energy electron beams have small wavelengths compared to x-rays (at 100 kV, 0.037 Å vs. 1.54 Å for CuKα x-rays), resulting in a Ewald sampling sphere that is approximately a plane (i.e., many diffracted beams are excited simultaneously instead of one). Given the larger scattering cross-section of matter for electrons (2) (f_{el}/f_{x-ray} = 10^3), a dynamical description of many beam interactions should be more appropriate than the kinematical one, which assumes the independence of all diffracted beams (12). Unfortunately, adherence to the kinematical approximation allows the most direct determination of a crystal structure from diffraction data. Experimental conditions must be established to approach this ideal to ensure the success of the analysis.

Originally, researchers thought that thin microcrystals used for electron diffraction would conform to the mosaic model used in x-ray crystallography. Each microblock in a mosaic was imagined to be so oriented that only one strong reflection was excited from it (13) and was used as a justification for the kinematical scattering approximation. If mosaic block sizes were large enough to cause interaction between the individual diffracted beams and the incident beam, then a two-beam dynamical correction was made (i.e., a primary extinction correction). However, experimental application of this correction was inconsistent and showed a wide variability in the relation between corrected structure factor magnitudes and the measured diffraction intensities. Sometimes mosaic block shapes were incorporated as an additional variable (*14*). Certainly, organic microcrystals contain defects that would account for a mosaic of domains within them, but their concentration does not appear large enough to justify this model (*15*). The crystals used for electron diffraction experiments are largely perfect; the mosaic block size is generally larger than the coherence width of the incident beam (*12*).

Consistent with the notion of crystal perfection, n-beam interactions may be demonstrated in thin molecular crystals, for example, in experiments with monomolecular layers of orthorhombic n-paraffins, which may be regarded as an oligomeric form of polyethylene. Use of such crystals has established the presence of n-beam effects in four ways (to be discussed individually in the following sections).

Fit of Observed Data with Calculation. n-Beam dynamical calculations of the multislice type (*12*) or using the phase grating approximation (*12*) give the best explanation of *specific* diffraction intensities that differ from their kinematical values. This has been shown for paraffins as well as short- and long-chain polyethylenes (*16, 17*).

Intensity Change with Increasing Electron Accelerating Voltage. Electron diffraction intensities from paraffin taken with beam accelerating voltages up to 1000 kV are accordant with the n-beam dynamical theory (*18*). There is no actual realization of kinematical diffraction at high voltage as would be predicted by the two-beam diffraction theory.

Behavior of Continuously Excited Diffracted Beams with Change of Crystal Orientation.   If a row of diffraction beams coincides with a tilt axis for the crystal, then the continuously excited intensities will change during tilt only if n-beam dynamical interactions are present. This behavior was shown for thin crystals of a wax (*17*).

The demonstration of n-beam interactions has been important for the conception of a realistic crystal model. However, their influence on ab initio crystal structure analyses also is important. Rigorous n-beam dynamical calculations were carried out for a representative molecular crystal structure, cytosine, considering a projection down the shortest unit cell axis at two accelerating voltages. Such data were sampled at increasing crystal thicknesses and used naively as input diffraction data for an automated direct phasing computer program. At

100 kV, these diffraction data yielded false crystal structures for thicknesses above 75 Å; at 1000 kV the maximum acceptable crystal thickness is around 300 Å (19).

The importance of n-beam dynamical scattering also can be assessed for polymer structures solved from electron diffraction data. Four such structures are summarized in Table I. Multislice n-beam calculations generally do not change the agreement between observed and calculated structure factors appreciably. The polymers listed have far fewer repeats within the reported crystal thicknesses than polyethylene (40 repeats for a 100-Å thickness). Therefore, the diffraction data are not changed appreciably by dynamical scattering, even though this description of the scattering process is the most rigorous. An exception is found for poly(ε-caprolactone) for which best agreement with experiment was obtained at a thickness of 156 Å or nine unit cell repeats. However, for the $hk0$ zone, there is effectively a subcell of length 2.47 Å in the beam direction, so that the structure factors at 156 Å thickness are indicative of 63 repeat units. This finding explains the great difference between the dynamical ($R = 0.16$) and kinematic ($R = 0.27$) results.

Elastic Crystal Bending

Although thin crystals used for electron diffraction experiments are often nearly perfect in terms of defect concentration, they are also commonly deformed by the substrate surface. Elastic bending of such crystals is observed easily in low-magnification diffraction contrast electron micrographs taken at low-beam doses. Because such bends are only over a few degrees, this property can cause a severe change in the diffracted intensities, particularly if the unit cell repeat along the incident beam direction is large (20).

Using an analytical procedure proposed by Cowley (20), which considers a Patterson function altered by a Gaussian term dependent on the crystal curvature and the unit cell length along the incident beam, bending was shown to explain apparent diffraction coherence restrictions from various paraffinic crystals for which the true unit cell is much longer than the chain zig-zag repeat (17, 21). These calculations show that solution growth of molecular crystals gives the least favorable projection for an electron diffraction experiment because the longest unit cell axis is commonly normal to the best-developed crystal face. Epitaxial crystal growth, which forces a shorter axis in this direction, allows the collection of diffraction data that are more representative of the total unit cell (17). Maximum Patterson vector components in the beam direction, and thus the attenuation of them caused by the crystal bending, are smaller. Consequently, bends have less impact on the diffracted intensity for these zones.

Table I. Attempted Intensity Data Corrections for Published Polymer Structures

Structure	Multislice n-Beam Calculation	Correction for Crystal Bending
α-Poly[3,3-bis(chloromethyl) oxacyclobutane] $a = 17.85$ Å, $b = 8.15$ Å, c(fiber) $= 4.78$ Å	In agreement with original finding, only a slight improvement; however, only 14 monomer repeats for cited crystal thickness	Also, no real improvement; crystals described as being particularly rigid
Anhydrous nigeran $a = 17.76$ Å, $b = 6.0$ Å c(fiber) $= 14.62$ Å	No significant improvement; however, only six monomer repeats for cited crystal thickness; best $R = 0.25$	Best agreement ($R = 0.21$) at bend ±2° and temperature factor $B = 6$ Å²; originally cited $R \simeq 0.25$
Polytrimethylene terephthalate $a = 4.64$ Å, $b = 6.27$ Å, c(fiber) $= 18.64$ Å $\alpha = 98.4°$, $\beta = 93.0°$, $\gamma = 111.1°$	Best agreement $R = 0.36$ at $t = 56$ Å (estimated thickness of 80 Å only four monomer repeats); $R = 0.34$ comparing data to kinematical model	Best agreement $B = 12$ Å² at bend ±8°, $R = 0.25$; significant improvement
Poly(ε-caprolactone) $a = 7.496$ Å, $b = 4.974$ Å, c(fiber) $= 17.297$ Å	A considerable improvement over the kinematic model; $R = 0.16$ compared to $R = 0.27$ at $t = 156$ Å; nine unit cell repeats	Best agreement $B = 5$ Å² at bend ±3°, $R = 0.23$

Fortunately, the fiber repeat of many polymer structures is small enough to allow collection of useful diffraction-intensity data. However, for the three structures in Table I with longest fiber repeats, a correction for elastic bending produced a significantly improved fit of observed and calculated data.

Crystal Structure Analysis

If the caveats mentioned earlier are observed, the electron diffraction intensity data from thin organic crystals are useful for structure analysis. In fact, the tendency of polymers to form extremely thin microcrystals of the order of 100 Å makes them particularly suitable. The experiment also implies a minimization of radiation damage by use of fast photographic films and low electron beam doses, as is discussed often.

Preliminary "Lorentz factor" corrections were not found useful (16). Intensities are taken directly from integration of densitometer scans across a diffraction film with due attention paid to the possible errors caused by nonlinearities, as described by Wooster (22).

Phasing of diffraction data can be approached in as many ways as used in x-ray crystallography. Both Patterson techniques and trial and error have been successful. If there are an adequate number of large normalized structure factors in the experimental diffraction data for each atom in the asymmetric unit, direct phasing methods can be used (19, 23).

Zonal diffraction data from linear polymer crystals, however, often represent a small number of reflections per atom. To create a realistic number of variable parameters for such limited diffraction data, known skeletal structures of monomers or oligomers are assumed to be unchanged in the polymer, but their mutual conformations are varied by rotations about linkage bonds. In such a phasing procedure, a match of packing energy minimum with R-value minimum is used to specify the best structural model (9).

Refinement of the structures must incorporate corrections for either n-beam scattering or crystal bending, depending on which perturbation will have the most effect on the experimental intensities. Such corrections allow construction of structural models with physically realistic thermal parameters (see Table I), a feature disallowed in earlier kinematical treatment of such data (7−11, 16), and stress the importance of their inclusion in a crystallographic analysis. For this refinement, however, additional information should be known about the crystals, i.e., there must be some estimate of crystal thickness and also diffraction contract micrographs to estimate crystal curvature. Both corrections imply that the structure is known in three dimensions. Currently, this third dimension is inferred from a structural model but should be directly obtainable from three-dimensional diffraction data resulting from a tilt experiment.

Future Investigations

Several directions for future work are clearly indicated. One of these areas is a combined treatment of bending and dynamical diffraction to assess how these perturbations to intensity data may interact. Calculation of the n-beam dynamical interactions from a curved paraffin crystal and representative linear polymers, now in progress, may give insight into these interactions.

Another very important consideration is the optimized collection of three-dimensional electron diffraction intensities. This gathering would, of course, allow a much more accurate determination of a crystal structure and, with an increased number of data, may allow the general use of direct phasing methods for organic crystals, including linear polymers. A short unit cell axis in the beam direction, posited by epitaxial growth for molecular organic crystals, or provided already by the fiber repeat of many polymers, gives more confidence in the intensity data from slightly curved crystals. Tilt of a crystal up to the commonly found 60° goniometer limit could double the crystal repeat along the beam. If this value remains tolerably small, then a three-dimensional data set could be used. The utility of such three-dimensional data has been demonstrated already for thin protein crystals (*24*).

Acknowledgments

Research was supported by Public Health Service Grant No. GM–21047 from the National Institute of General Medical Sciences and by National Science Foundation Grant Nos. PCM78–16041 and CHE79–16916.

Literature Cited

1. Brisse, F.; Marchessault, R. H. In "Fiber Diffraction Methods"; French, A. D.; Gardner, K. H., Eds.; ACS SYMPOSIUM SERIES No. 141, ACS: Washington, DC, 1980; p. 267.
2. Vainshtein, B. K. "Structure Analysis by Electron Diffraction"; Pergamon: Oxford, 1964.
3. Tatarinova, L. I.; Vainshtein, B. K. *Vysokomole. Soedin.*, **1962**, *4*, 261.
4. Vainshtein, B. K.; Tatarinova, L. I. *Sov. Phys. Crystallogr.* **1967**, *11*, 494.
5. Claffey, W.; Gardner, K.; Blackwell, J.; Lando, J.; Geil, P. H. *Philos. Mag.* **1974**, *30*, 1223.
6. Claffey, W.; Blackwell, J. *Biopolymers* **1976**, *15*, 1903.
7. Roche, E.; Chanzy, H.; Boudeulle, M.; Marchessault, R. H.; Sundararajan, P. *Macromolecules* **1978**, *11*, 86.
8. Poulin-Dandurand, S.; Perez, S.; Revol, J. F.; Brisse, F. *Polymer* **1979**, *20*, 419.
9. Perez, S.; Roux, M.; Revol, J. F.; Marchessault, R. H. *J. Mol. Biol.* **1979**, *129*, 113.

10. Day, D.; Lando, J. B. *Macromolecules* **1980,** *13,* 1483.
11. Noe, P., Mc.Sc. Thesis, Univ. of Montréal, Montréal, Canada, 1979.
12. Cowley, J. M. "Diffraction Physics"; North-Holland: Amsterdam, 1975.
13. Vainshtein, B. K. *Sov. Phys. Crystallogr.* **1956,** *1,* 15.
14. Lobachev, A. N.; Vainshtein, B. K. *Sov. Phys. Crystallogr.* **1961,** 6, 313.
15. Dorset, D. L. *J. Polym. Sci., Polym. Phys. Ed.* **1979,** *17,* 1797.
16. Dorset, D. L. *Acta Crystallogr.* **1976,** *A32,* 207.
17. *Ibid.,* **1980,** *A36,* 592.
18. Dorset, D. L. *J. Appl. Phys.* **1976,** *47,* 780.
19. Dorset, D. L.; Jap, B. K.; Ho, M.-S.; Glaeser, R. M. *Acta Crystallogr.* **1979,** *A35,* 1001.
20. Cowley, J. M. *Acta Crystallogr.* **1961,** *14,* 920.
21. Dorset, D. L. Z. *Naturforsch.* **1978,** 33A, 964.
22. Wooster, W. A. *Acta Crystallogr.* **1964,** *17,* 878
23. Dorset, D. L.; Hauptman, H. A. *Ultramicroscopy* **1976,** *1,* 195.
24. Unwin, P. N. T.; Henderson, R. *J. Mol. Biol.* **1975,** *94,* 425.

RECEIVED for review October 14, 1981. ACCEPTED December 16, 1981.

NUCLEAR MAGNETIC RESONANCE

NMR Spectroscopy in the Characterization of Polymers
Introductory Material

E. G. BRAME, JR.

E. I. du Pont de Nemours & Co., Inc.,
Experimental Station Laboratory, Wilmington, DE 19898

AMONG THE EARLIEST APPLICATIONS OF NMR spectroscopy was the highly successful structure determination of polymers, including both homopolymers and copolymers. Many examples can be cited from the early literature including those attributed to Bovey and his colleagues. Bovey found the use of ^{1}H NMR to be a fertile field to characterize the microstructure of hydrocarbon polymers. He used ^{19}F NMR for characterization of fluoropolymers. In addition, Bovey pioneered the use of statistics to explain the numbers and relative abundances of lines in spectra. From those early days to the present, NMR continues to be one of the most important instrumental methods for characterizing the microstructures of polymers. Currently, ^{13}C NMR, because of its larger dispersiveness, has mostly taken over this task from ^{1}H NMR.

In addition to the usual examination of polymers in solution and in the molten state, we are seeing a new surge of interest in the application of ^{13}C NMR to the examination of polymers in the solid state by magic angle spinning (MAS). This technique is providing new and sometimes different structural information about polymers because it is examining them in their natural states. Thus, with the use of MAS, along with the more common methods of examination, NMR has taken on an even more important role in its application to the determination of polymer structure.

The chapters in this section cover the latest applications of NMR to polymer analyses. In addition, they show and even stress the importance of ^{13}C NMR in polymer studies. Each chapter covers one or more aspects of ^{13}C NMR in its application to polymers. Bovey shows how ^{13}C NMR can be used to determine the structure of a commonly known material, polyethylene, as well as a less well-known material, Hytrel polyester elastomer. Polyethylene is examined in solution and Hytrel is examined in the solid state by MAS. Tonelli reports on the applica-

0065−2393/83/0203−0419$06.00/0

tion of ^{13}C NMR in the analysis of fluoropolymers. He discusses the use of simultaneous decoupling of ^{1}H and ^{19}F from ^{13}C to obtain very useful structural information about fluoropolymers. Fleming shows that a variable temperature system can be useful in examining polymers in the solid state by MAS. Mandelkern describes the use of ^{13}C NMR relaxation parameters to determine nonordered regions in semicrystalline polymers. Finally, Ford states that ^{13}C NMR spin lattice relaxation times along with linewidths, nuclear Overhauser effects, and relative signal areas can be used to monitor cross-linking in copolymers of styrene.

ACCEPTED October 14, 1981

Structural and Dynamic Characterization of Polymers by ^{13}C and ^{19}F NMR

F. A. BOVEY, R. E. CAIS, L. W. JELINSKI, F. C. SCHILLING, W. H. STARNES, JR., and A. E. TONELLI

Bell Laboratories, Murray Hill, NJ 07974

^1H and ^{19}F NMR spectroscopy can provide significant structural and motional information for synthetic polymers. More recently, Fourier transform (FT) instrumentation has permitted natural abundance ^{13}C NMR spectroscopy, giving great sensitivity to tacticity, comonomer sequence, regiospecificity, branching, and other structural features. ^{19}F NMR also shows high sensitivity, particularly to tacticity and regiospecificity. High resolution ^{13}C NMR spectroscopy in the solid state, achieved using proton dipolar decoupling, magic angle spinning, and $^1H-^{13}C$ cross-polarization, is also of great interest with respect to both chain structure and dynamics. Examples of tacticity studies by NMR include ^{13}C NMR spectroscopy of polypropylene (and the prediction of spectral fine structure using the "γ-effect" model), polyvinyl chloride, polyvinyl bromide, and (by ^{19}F NMR) polyfluoromethylene $(-CFH-)_n$. The detailed branch structures of polyethylene and polyvinyl chloride are discussed, together with their interpretation in terms of polymerization mechanisms. The solid state ^{13}C NMR spectroscopy of Hytrel polyester thermoplastic elastomers is described and discussed with regard to domain structure and chain motion. The chain mobilities in these segmented copolymers appear to range over several orders of magnitude.

$\mathbf{F}$OR OVER 20 YEARS ^1H (*1–3*) and ^{19}F (*4*) NMR spectroscopy has supplied important information concerning polymer structure, particularly stereochemical configuration (*2–4*) and copolymer sequences (*2*), as well as dynamic information concerning polymers in solution

0065–2393/83/0203–0421$06.00/0

(2). More recently, ^{13}C NMR spectroscopy has assumed increasing importance as the development of Fourier transform (FT) instrumentation with spectrum accumulation has permitted the practical observation of carbon-13 in natural abundance. Carbon spectroscopy has now developed into a method of fairly high sensitivity that can be used to detect and measure very minor but significant features of polymer structure such as end-groups, branches, and head-to-head monomer sequences. It has now largely displaced ^{1}H spectroscopy because of the much greater range of carbon chemical shifts and the consequent greater sensitivity to details of structure.

High resolution ^{13}C spectroscopy in the solid state, achieved using high power proton decoupling, $^{1}H-^{13}C$ cross-polarization, and magic angle spinning (MAS) also has generated much interest and excitement. (Reference 5 is a useful review.)

^{19}F NMR is also highly sensitive to details of microstructure, including head-to-head:head-to-tail isomerism and tacticity, and is very valuable where applicable.

We shall illustrate briefly all these applications of NMR by illustrations drawn from our work.

The Dependence of ^{13}C and ^{19}F Shifts on Stereochemical Configuration and Its Conformational Dependence

In the observation of configurational sequences, pentads usually can be resolved by ^{13}C NMR, particularly at superconducting frequencies, and the discrimination of heptad sequences is not uncommon. An example is shown in Figure 1, which exhibits the methyl spectrum of an atactic polypropylene observed in three solvents at varying temperatures (6). The spectrum shows little solvent dependence but a fairly marked temperature dependence.

The configurational fine structure in this spectrum, and the spectra of the α- and β-carbons of the main chain, can be predicted very accurately by a simple "γ-effect" model (7−10), which postulates that when two carbon atoms separated by three bonds are in a *gauche* conformation, they shield each other by −5.2 ppm compared to their resonance frequencies in the *trans* conformation. The CH_3 chemical shift then is determined by the *gauche* content of the main chain bonds flanking the α-carbon to which the CH_3 group is attached. This effect is illustrated in Scheme I. When the main chain bonds are *trans*, the methyl carbons experience *gauche* interactions, but when the main chain bonds are *gauche +*, they experience no *gauche* interactions (*gauche−* is populated very sparsely). The chirality of neighboring α-carbons, even when removed by as many as seven intervening bonds, influences the CH_3 resonance position by affecting this local *gauche* content, which can be accurately calculated using

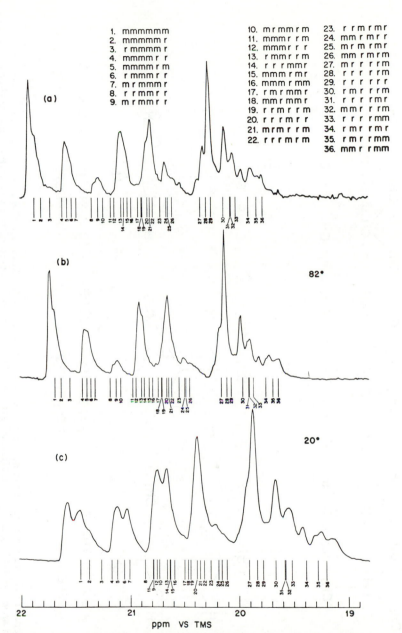

Figure 1. The 90-MHz ¹³C NMR spectra of atactic polypropylene: a, observed in heptane at 67 °C; b, in 1,2,4-trichlorobenzene−dioxane-d₈ at 82 °C; and c, in 1,2,4-trichlorobenzene−dioxane-d₈ at 20 °C. Calculated "stick" spectra (of arbitrarily equal line intensity) are shown below each spectrum.

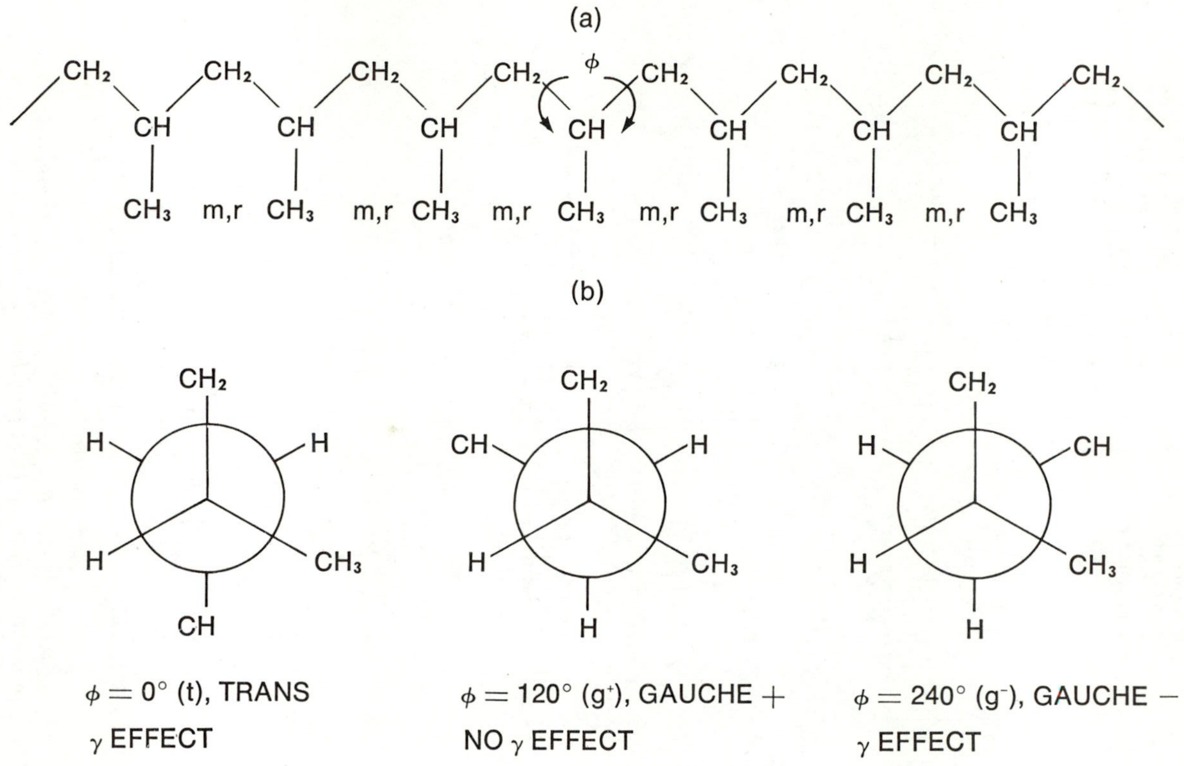

Scheme I. Heptad configurational sequences in a polypropylene chain (1). The central CH_3 carbon experiences trans (II), gauche + (III), and gauche-(IV) interactions (below) with neighboring α-carbons about the bonds indicated by the arrows.

the rotational isomeric state model for polypropylene developed by Suter and Flory (*11*). Thus, there is no need to invoke any mysterious long-range influences in the polymer molecule.

¹³C NMR is effective in determining the stereochemistry of polyvinyl chloride (*8, 12−14*) and polyvinyl bromide (*14*) and their dependence on polymerization temperature and the presence of aldehydes, two much disputed questions. Figure 2 (*14*) shows the 22.6-MHz ¹³C spectra of polyvinyl chloride (a) and polyvinyl bromide (b). In spectrum a, the α-carbon (left) and β-carbon (right) resonances are well separated and the spectrum can be interpreted in a straightforward manner; spectrum b is more complex and requires observation of the *meso* and racemic 2,4-dibromopentane models for a satisfactory interpretation.

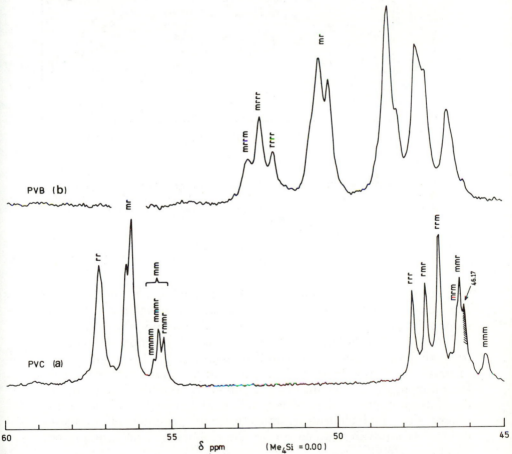

Figure 2. The 22.6-MHz ¹³C NMR spectra of polyvinyl chloride (a) and polyvinyl bromide (b) in 1,2,4-trichlorobenzene at 100 °C.

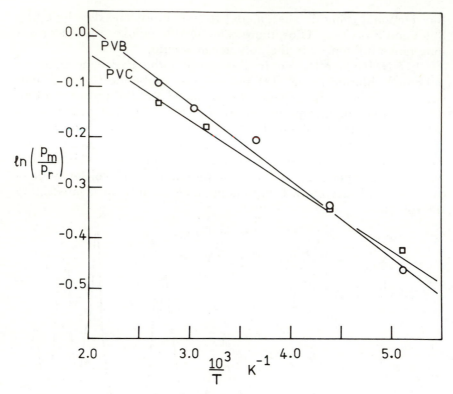

Figure 3. Arrhenius plot of P_m/P_r *[* $=P_m/(1 - P_m)$ *] for polyvinyl chloride (PVC) (□) and polyvinyl bromide (PVB) (○). Points corresponding to polymers prepared in* n-*butyraldehyde are shown as CA1(PVC) and BA1(PVB).*

The temperature dependence of the stereochemical configuration can be expressed as $\Delta H^{\ddagger}_{meso} - \Delta H^{\ddagger}_{racemic}$, which has the value 260 cal/mol for polyvinyl chloride and 300 cal/mol for polyvinyl bromide. Polymerization in *n*-butyraldehyde increases the probability of syndiotactic propagation to a perceptible degree (*15*). This point is demonstrated in Figure 3 (*14*), which shows an Arrhenius plot of the configuration of polyvinyl chloride and polyvinyl bromide. The points for the aldehyde-modified polymers fall well below the plots for the others. For vinyl chloride propagation, the presence of *n*-butyraldehyde is equivalent to lowering the polymerization temperature by about 120 °C.

^{13}C NMR is not always the best means for measuring stereochemical configuration. For example, the ^{1}H and ^{13}C spectra are quite ineffective in resolving steric sequences in polyfluoromethylene, $(-CFH-)_n$, but the ¹⁹F spectrum (Figure 4) exhibits quite rich detail to the pentad level spread over the unusually large range of nearly 20 ppm (*16*). The computer simulation in Figure 4 is based on Bernoullian propagation with a P_m of 0.42. In this system, asymmetric centers are being added two at a time to the growing chain, so a Bernoullian model does not have to be obeyed even though this mode is nearly universally found for free-radical propagation. However, this finding seems to hold within experimental error.

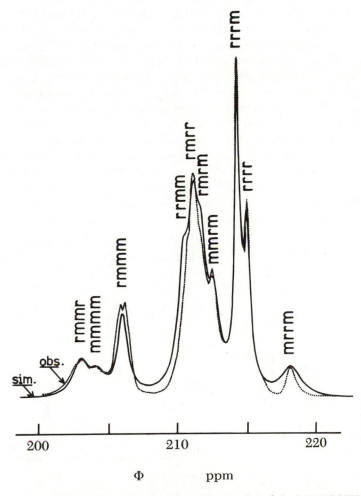

Figure 4. Observed (....) and computer-simulated (—) 84.67-MHz ¹⁹F spectra of polyfluoromethylene. The experimental spectrum was observed in acetone solution. In the simulated spectrum, a P_m of 0.42 was assumed.

The *cis*- and *trans*-1,2-difluoroethylene monomers yield identical polymers. The ¹⁹F chemical shifts, like those of ¹³C, can be rationalized on the basis of a *gauche* γ-effect model (*17*).

Branching in Polyethylene and Polyvinyl Chloride

Branching can be observed in detail in polyethylenes by ¹³C NMR and, under proper conditions, can be measured quantitatively (*18*). Figure 5 shows the 50-MHz ¹³C spectrum of a typical high-pressure polyethylene (*19*). The observed composition is shown in Table I. The dominant butyl and amyl branches presumably are formed by intramolecular chain transfer or backbiting. The origin of ethyl branches is uncertain. Spectral complications not observed in ethylene–butene-1 copolymers clearly occur (*20, 21*). The resonance at 39.5 ppm for the branch carbons of isolated ethyl branches is much smaller than the corresponding methyl intensity near 11 ppm. A dozen or so small, unassigned resonances in the 37–39-ppm branch carbon region are seen, and the ethyl branch methyl carbon resonance is composed of numerous overlapping peaks. The Roedel (*22*) mechanism for short-branch formation was extended by Willbourn (*23*) to account for the formation of ethyl branches and could at least partially account for the complexity of their NMR spectrum, although not so intended. These reactions are summarized in Scheme II. The presence of isolated ethyl branches, the content of which is evidently variable but not zero, is not accounted for by these mechanisms but may simply

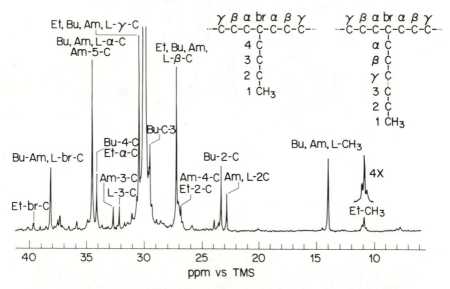

Figure 5. A 50-MHz ¹³C NMR spectrum of high-pressure polyethylene, observed in 1,2,4-trichlorobenzene at 110 °C (F. C. Schilling).

Table I. Branch Composition of High-Pressure Polyethylene

Branch Type	Branches per 1000 CH_2 Groups
Ethyl	2.5
n-Propyl	0.0
n-Butyl	10.9
n-Amyl	3.7
Long ($\geq C_6$)	4.5
Total	21.6

NOTE: $\bar{M}_n$ = 18,400; and $\bar{M}_w$ = 129,000, where $\bar{M}_n$ and $\bar{M}_w$ refer to the number-average molecular weight and the weight-average molecular weight, respectively.

result from backbiting via a four-membered cyclic transition state. The absence of propyl branches, however, is entirely puzzling.

Branching in polyvinyl chloride can be determined by ^{13}C NMR, but so far not on the polymer itself, the spectrum of which is very much complicated by configurational sequences, as already shown. It is advantageous to abolish these complexities by reductively dehalogenating the polymer with lithium aluminum hydride (24–26) or preferably with the more convenient and effective reagent tri-n-butyltin hydride (27). The resulting hydrocarbon then is examined in the same manner as polyethylene. Figure 6a shows the 25-MHz spectrum of reduced polyvinyl chloride prepared at 100 °C; for the reduced polymer, $\bar{M}_n$ = 5600 and $\bar{M}_w$ = 11,900. The principal short branch (2–3 per 1000 main chain carbons) is CH_3, with a smaller frequency of ethyl, butyl, and long branches. By reduction with (n-butyl)$_3$SnD, each chlorine atom is replaced by deuterium instead of hydrogen. The positions of the chlorine atoms in the original structure can be established because each deuterium splits the resonance of the directly bonded carbon into a 1:1:1 triplet (J_{CD} = 19.0 Hz). A shielding effect of about 0.5 ppm is noted also. (Very small resonances essentially disappear, another useful clue.) These features can be observed in the spectrum shown in Figure 6b, which represents the same polyvinyl chloride as in 6a. The bottom (low gain) spectrum shows the singlet resonance of the main chain CH_2 carbons and the triplet of the CHD carbons. The upper spectrum at higher gain reveals that the methyl branches are CH_2Cl groups in the original polymer. Observations on a polymer from vinyl chloride-α-D established (28) that the methylene carbon of the chloromethyl branch arises from the β-carbon of the monomer and that the mechanism suggested by Rigo et al. (29) is correct (see later discussion).

Careful analysis of the spectra in Figure 6 has led to the identification of the branches and chain ends shown in Scheme III. Tertiary chlorine atoms are present at several points. Because tertiary chlorine is known to be highly labile (30), this finding strongly suggests that

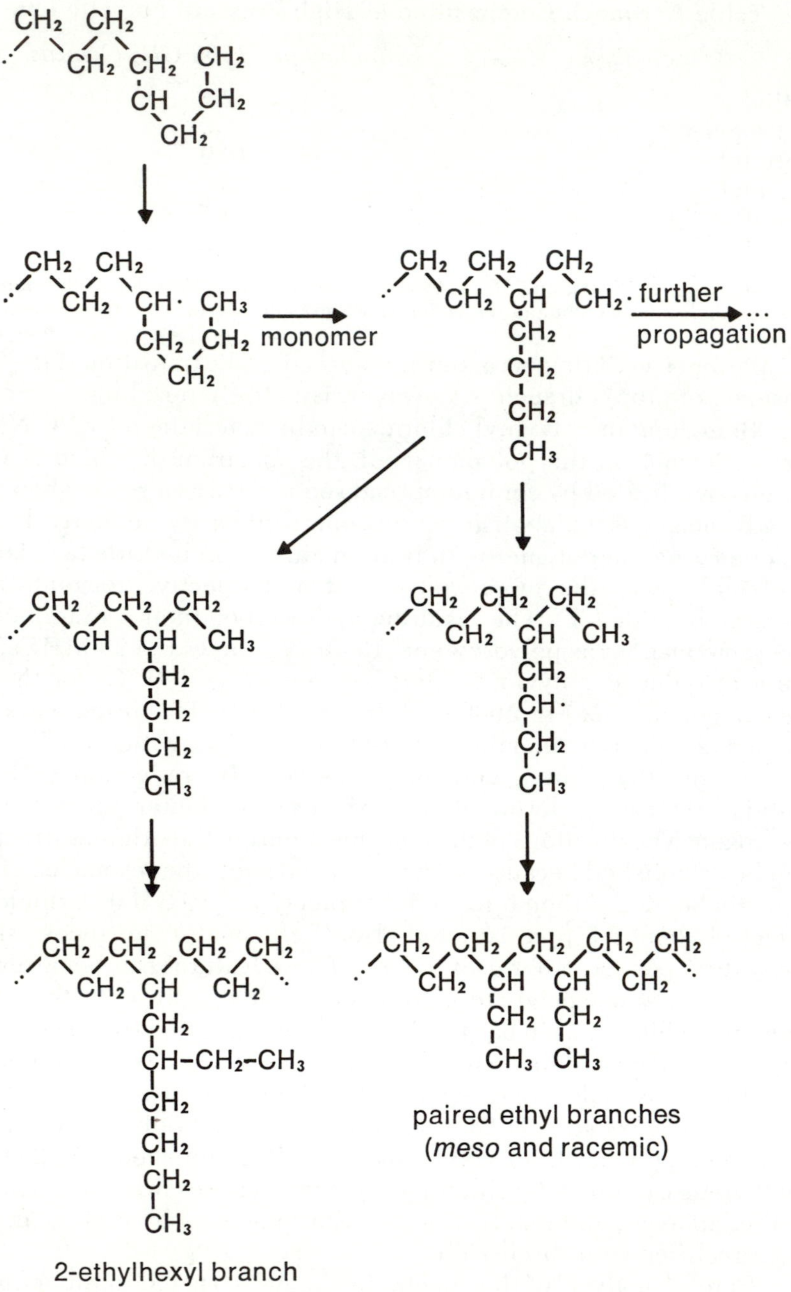

Scheme II. Branch formation in high-pressure polyethylene.

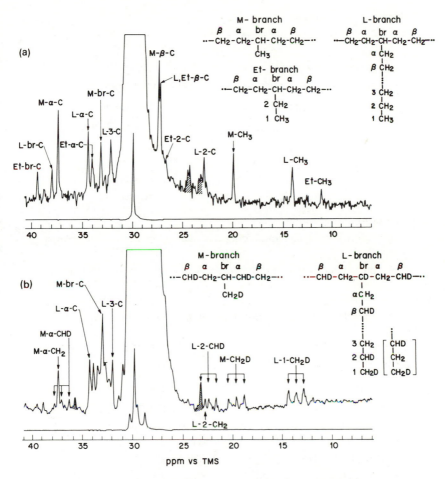

Figure 6. The 25-MHz ¹³C NMR spectra of polyvinyl chloride reduced with Bu₃SnH (a) and Bu₃SnD (b); observed in 1,2,4-trichlorobenzene solution at 110 °C (F. C. Schilling).

these structures may be a major contributor to the thermal instability of polyvinyl chloride. One of the principal purposes of this work is to identify such weak points.

The formation of chloromethyl branches and the occurrence of 1,2,4-trichloro and 1,3-dichloro chain ends provide important information concerning the polymerization process. The reactions that were deduced are shown in Scheme IV. The occasional head-to-head addition (b) results in a primary radical that rearranges (c) by a 1,2-chlorine migration, the Rigo mechanism (29). This radical mainly propagates further, leading to chloromethyl branch formation (d−i), but also may undergo β-scission (d−ii) to give a terminal chloroallyl

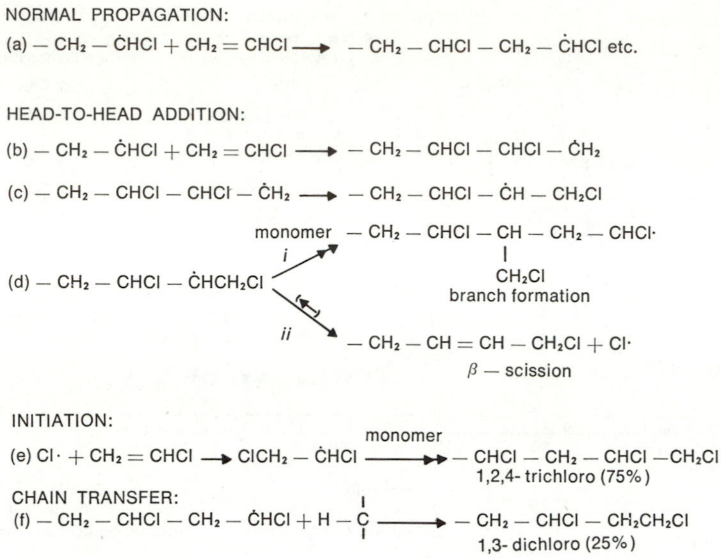

NORMAL PROPAGATION:

(a) $- CH_2 - \dot{C}HCl + CH_2 = CHCl \longrightarrow - CH_2 - CHCl - CH_2 - \dot{C}HCl$ etc.

HEAD-TO-HEAD ADDITION:

(b) $- CH_2 - \dot{C}HCl + CH_2 = CHCl \longrightarrow - CH_2 - CHCl - CHCl - \dot{C}H_2$

(c) $- CH_2 - CHCl - CHCl - \dot{C}H_2 \longrightarrow - CH_2 - CHCl - \dot{C}H - CH_2Cl$

(d) $- CH_2 - CHCl - \dot{C}HCH_2Cl$

$\xrightarrow{\text{monomer}}$ (i) $- CH_2 - CHCl - CH - CH_2 - CHCl\cdot$
                                                    |
                                                  CH_2Cl
                                            branch formation

(ii) $- CH_2 - CH = CH - CH_2Cl + Cl\cdot$
                        β — scission

INITIATION:

$\quad\quad\quad\quad\quad\quad\quad\quad\quad\quad$ monomer
(e) $Cl\cdot + CH_2 = CHCl \longrightarrow ClCH_2 - \dot{C}HCl \xrightarrow{\quad\quad} - CHCl - CH_2 - CHCl - CH_2Cl$
$\quad\quad\quad\quad\quad\quad\quad\quad\quad\quad\quad\quad\quad\quad\quad\quad\quad\quad\quad\quad\quad$ 1,2,4- trichloro (75%)

CHAIN TRANSFER:

(f) $- CH_2 - CHCl - CH_2 - \dot{C}HCl + H - \overset{|}{\underset{|}{C}} \longrightarrow - CH_2 - CHCl - CH_2CH_2Cl$
$\quad\quad\quad\quad\quad\quad\quad\quad\quad\quad\quad\quad\quad\quad\quad\quad\quad\quad\quad$ 1,3- dichloro (25%)

Scheme III. Branches and chain ends in polyvinyl chloride.

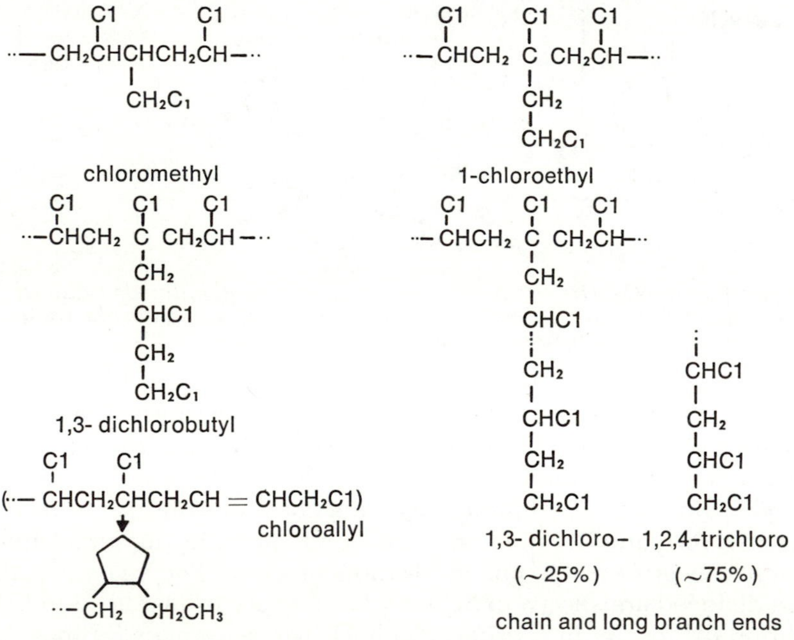

$\quad\quad$ Cl $\quad\quad$ Cl
$\quad\quad$ | $\quad\quad\quad$ |
$\cdots- CH_2\overset{|}{C}HCHCH_2\overset{|}{C}H -\cdots$
$\quad\quad\quad\quad$ |
$\quad\quad\quad\quad CH_2Cl$

chloromethyl

$\quad\quad$ Cl $\quad\quad$ Cl $\quad$ Cl
$\quad\quad$ | $\quad\quad$ | $\quad$ |
$\cdots-\overset{|}{C}HCH_2\;\overset{|}{C}\;CH_2\overset{|}{C}H -\cdots$
$\quad\quad\quad\quad$ |
$\quad\quad\quad\quad CH_2$
$\quad\quad\quad\quad$ |
$\quad\quad\quad\quad CH_2Cl$

1-chloroethyl

$\quad\quad$ Cl $\quad\quad$ Cl $\quad$ Cl
$\quad\quad$ | $\quad\quad$ | $\quad$ |
$\cdots-\overset{|}{C}HCH_2\;\overset{|}{C}\;CH_2\overset{|}{C}H -\cdots$
$\quad\quad\quad\quad CH_2$
$\quad\quad\quad\quad$ |
$\quad\quad\quad\quad CHCl$
$\quad\quad\quad\quad$ |
$\quad\quad\quad\quad CH_2$
$\quad\quad\quad\quad$ |
$\quad\quad\quad\quad CH_2Cl$

1,3- dichlorobutyl

$\quad\quad$ Cl $\quad$ Cl $\quad$ Cl
$\quad\quad$ | $\quad$ | $\quad$ |
$\cdots-\overset{|}{C}HCH_2\;\overset{|}{C}\;CH_2\overset{|}{C}H-\cdots$
$\quad\quad\quad\quad CH_2$
$\quad\quad\quad\quad$ |
$\quad\quad\quad\quad CHCl \quad\quad\quad CHCl$
$\quad\quad\quad\quad$ | $\quad\quad\quad\quad$ |
$\quad\quad\quad\quad CH_2 \quad\quad\quad CH_2$
$\quad\quad\quad\quad$ | $\quad\quad\quad\quad$ |
$\quad\quad\quad\quad CHCl \quad\quad\quad CHCl$
$\quad\quad\quad\quad$ | $\quad\quad\quad\quad$ |
$\quad\quad\quad\quad CH_2Cl \quad\quad CH_2Cl$

1,3- dichloro- 1,2,4-trichloro
$\quad$ (~25%) $\quad\quad$ (~75%)

$\quad\quad$ Cl $\quad$ Cl
$\quad\quad$ | $\quad$ |
$(\cdots- CHCH_2CHCH_2CH = CHCH_2Cl)$
$\quad\quad\quad\quad \downarrow$
$\quad\quad\quad\quad$ [cyclopentane ring structure]
$\cdots- CH_2\quad CH_2CH_3$

chloroallyl

chain and long branch ends

Scheme IV. Initiation, propagation, chloromethyl branch formation, and chain transfer in the polymerization of vinyl chloride.

group and a free chlorine atom. The chlorine atom then initiates another chain (e). Some chains are terminated by transfer involving the removal of hydrogen atoms (f); if this removal occurs from another polymer chain, the new radical site will lead to the formation of a long branch.

The validity of the mechanism requires the occurrence of terminal chloroallyl groups and these have not been observed. Starnes et al. (31) proposed that such groups in the presence of the tributyltin hydride will be reduced and cyclized to terminal ethylcyclopentyl groups, for which evidence is found in the ^{13}C spectrum.

1-Chloroethyl and 1,3-dichlorobutyl branches may arise by hydrogen shifts from chlorine-bearing carbons, that is, backbiting reactions parallel to those occurring in ethylene polymerization. When polymerization occurs in a solvent, the formation of 1,3-dichlorobutyl branches, normally scarcely observable, is enhanced markedly, but ethyl branch formation is not affected similarly (32).

Solid State ^{13}C NMR Studies of Segmented Polymer

Hytrel thermoplastic elastomers (du Pont) are segmented copolymers available in a range of compositions of m-tetramethylene terephthalate hard units and n-polytetramethyleneoxy terephthalate soft units that are assumed to be random in distribution (33). It has been shown that these polymers have a morphology consisting of continuous and interpenetrating crystalline and amorphous phases (34), rather than the discrete and ordered domain structure seen for other segmented copolymers (35).

$$\left[O-\overset{O}{\overset{\|}{C}}-\underset{\bigcirc}{\bigcirc}-\overset{O}{\overset{\|}{C}}-O-CH_2CH_2CH_2CH_2-O \right]_m \cdots \left[O-\overset{O}{\overset{\|}{C}}-\underset{\bigcirc}{\bigcirc}-\overset{O}{\overset{\|}{C}}-O\left\{(CH_2)_4-O\right\}_{12} \right]_n$$

Solid state ^{13}C NMR can provide information concerning the nature of phase separation in these segmented copolymers, and concerning molecular motion in both the mobile (36) and rigid domains (37).

Mobile Domains

The carbons comprising the mobile domains in Hytrel samples containing 0.80−0.96 mol fraction hard segments can be observed selectively using low-power proton decoupling (scalar decoupling; $\gamma H_2/2\pi = 4$ kHz) and a standard $90° − \tau$ pulse sequence. Representative spectra are shown in Figure 7 (a−d). The peak centered at 29 ppm is due to the $−CH_2−$ carbons flanked by methylene units on either side; the peak at 73 ppm arises from the $−OCH_2−$ carbons.

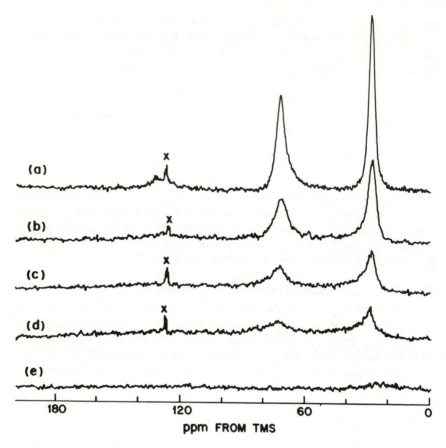

*Figure 7. Comparison of scalar decoupled ($\gamma H_2/2\pi$ = 4 kHz) 50.3-MHz
^{13}C NMR spectra of Hytrel containing 0.80 (a); 0.87 (b); 0.94 (c); and
0.96 (d) mol fraction hard segments, with the spectrum of polybutylene terephthalate (e). All spectra were obtained at 34 °C under
identical conditions.*

Spin-counting indicates that all of the soft-segment carbons and about
10% of the hard-segment carbons contribute intensity to these scalar
decoupled spectra. The additional signal intensity occurs primarily in
the protonated aromatic carbon region, and apparently arises from
hard-segment units that are too short to crystallize. In contrast, the
polybutylene terephthalate homopolymer has no discernible peaks
under the same spectral accumulation conditions [Figure 7 (e)]. These
experiments indicate that all soft-segment carbons in Hytrel undergo
motions that are rapid compared to the dipolar interaction (ca. 10^5 s^{-1}).
However, the following experiment suggests that these motions are
not isotropic. Magic angle sample spinning, in conjunction with scalar

decoupling and a $90° - \tau$ pulse sequence, causes an approximately fourfold reduction in the line widths for the soft-segment aliphatic carbons, indicating that the primary sources of line broadening in these spectra are residual dipole–dipole interactions and chemical shift anisotropy. With the sources of line broadening thus established, the line widths can be interpreted in terms of phase separation and angular range of reorientation. The line widths for both types of soft-segment aliphatic carbons are a linear function of the average hard-block length of the polymer, indicating that increasing hard-segment content restricts the angular range over which soft-segment reorientation can take place. Furthermore, the linear relationship between the average hard-block length and the line width suggests that the distribution of hard and soft segments is random. The soft-segment carbon line widths are nearly independent of temperature over an approximately 80° range.

The relaxation parameters [spin–lattice relaxation time (T_1) and nuclear Overhauser enhancement (NOE)] of the soft-segment carbons can be interpreted in terms of rate of motion. The T_1 values for both types of aliphatic soft-segment carbons are about 0.2 s at 50.3 MHz and 34 °C. These T_1 values are independent of the hard-segment content of the polymer but increase with increasing temperature and increasing magnetic field strength, indicating that the motions for these carbons are on the short-correlation-time side of the T_1 curve. The NOE values for the two types of aliphatic carbons also are independent of the hard-segment content of the polymer.

These results again support the conclusion that the angular range of reorientation of the soft-segment carbons is inversely related to the hard-segment content of the polymer, but that the rate of these motions is independent of the amount of hard segment present. These findings are consistent with a model for Hytrel structure in which phase separation occurs with negligible mixing of the two phases at the domain boundaries.

Rigid Domains

Various combinations of dipolar decoupling ($\gamma H_2/2\pi \approx 45$ kHz), cross-polarization, and MAS can be used to study the rigid domains in Hytrel. The proton-enhanced static powder spectrum of a Hytrel sample containing 0.96 mol fraction hard segments is shown in Figure 8a and the corresponding proton-enhanced MAS spectrum in Figure 8b. In Figure 8c the sidebands were removed. The chemical shifts in the solid state spectrum are readily assignable by comparison with the solution spectrum (Figure 8d). In order of increasing shielding, the resonances are assigned to the carbonyl, nonprotonated aromatic, protonated aromatic, soft segment $-OCH_2-$, hard segment $-OCH_2-$,

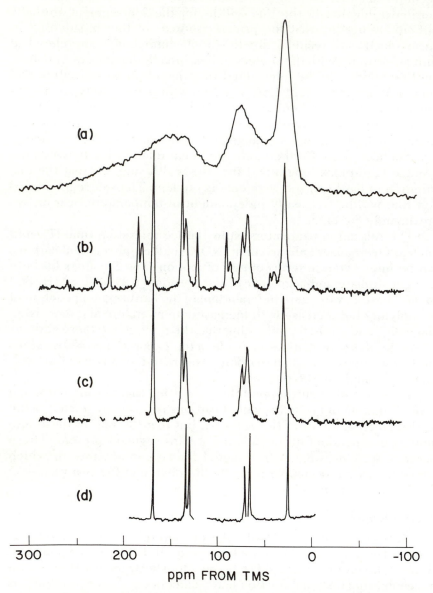

Figure 8. The 50.3-MHz ^{13}C NMR spectra of Hytrel containing 0.96 mol fraction hard segments. Key: a, proton-enhanced, dipolar decoupled static spectrum; b, same as a but with MAS at 2.3 kHz; c, spectrum b with sidebands deleted; and d, solution spectrum in m-cresol; resonances artificially broadened and solvent peaks deleted.

and overlapping hard and soft segment central $-CH_2-$ carbons. The spectrum in Figure 8b shows that resonances unique to the hard and soft segments can be resolved using dipolar decoupling and MAS.

The range of motions present in these polymers is illustrated by the series of dipolar decoupled, MAS spectra shown in Figure 9. These spectra were obtained as a function of the pulse repetition rate,

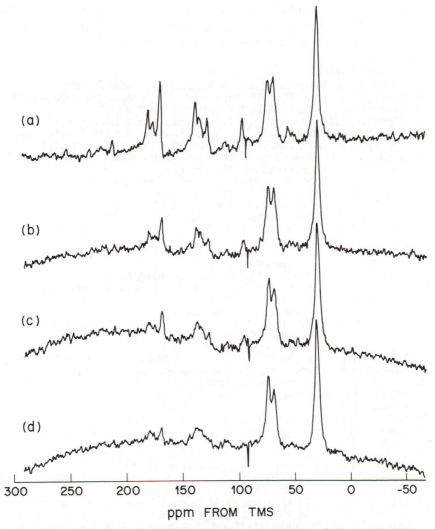

Figure 9. Dipolar decoupled, Overhauser-suppressed 50.3-MHz ¹³C NMR spectra of Hytrel containing 0.96 mol fraction hard segments. These spectra were obtained with MAS at 2.1 kHz using a standard 90°−τ pulse sequence, with pulse repetition rates of 30 s (a); 6 s (b); 4 s (c); and 2 s (d).

using a standard 90° − τ pulse sequence without Overhauser enhancement. At a 2-s repetition rate (Figure 9d), the intensities for all of the aliphatic carbons already have attained nearly their full intensity, whereas the carbonyl and aromatic resonances require much longer pulse repetition times to reach full intensity. These results indicate that the T_1 values for the protonated aromatic and the aliphatic carbons differ by at least an order of magnitude. In addition, these results indicate that the aliphatic carbons of the hard-segment units have short T_1 values, suggesting that the aliphatic carbons that reside in the rigid domains have substantial motional freedom.

Literature Cited

1. Saunders, M.; Wishnia, A. *Ann. N.Y. Acad. Sci.* **1958,** *70,* 870.
2. Bovey, F. A.; Tiers, G. V. D.; Filipovich, G. *J. Polym. Sci.* **1959,** *38,* 73.
3. Bovey, F. A.; Tiers, G. V. D. *J. Polym. Sci.* **1960,** *44,* 173.
4. Tiers, G. V. D.; Bovey, F. A. *J. Polym. Sci., Part A* **1963,** *1,* 833.
5. Schaefer, J.; Stejskal, E. O. "Topics in Carbon-13 NMR Spectroscopy"; Levy, G. C., Ed.; Wiley: New York, 1979; Vol. 3; pp. 284–324.
6. Schilling, F. C.; Tonelli, A. E. *Macromolecules* **1980,** *13,* 270.
7. Carman, C. J. *Macromolecules* **1973,** *6,* 725.
8. Provasoli, A.; Ferro, D. R. *Macromolecules* **1977,** *10,* 874.
9. Tonelli, A. E. *Macromolecules* **1978,** *11,* 565.
10. Tonelli, A. E.; Schilling, F. C. *Acc. Chem. Res.,* in press.
11. Suter, U. W.; Flory, P. J. *Macromolecules* **1975,** *8,* 765.
12. Carman, C. J. W.; Tarpley, A. R., Jr.; Goldstein, J. H. *Macromolecules* **1971,** *4,* 445.
13. Tonelli, A. E.; Schilling, F. C.; Starnes, W. H., Jr.; Shepherd, L.; Plitz, I. M. *Macromolecules* **1979,** *12,* 78.
14. Cais, R. E.; Brown, W. L. *Macromolecules* **1980,** *13,* 80.
15. Rosen, I.; Burleigh, P. H.; Gillespie, J. F. *J. Polym. Sci.* **1961,** *54,* 31.
16. Cais, R. E. *Macromolecules* **1980,** *13,* 806.
17. Tonelli, A. E. *Macromolecules* **1980,** *13,* 734.
18. Bovey, F. A.; Schilling, F. C.; Starnes, W. H., Jr. *Polym. Prepr., Am. Chem. Soc., Div. Polym. Chem.* **1979,** *20*(2), 160.
19. Schilling, F. C., unpublished data.
20. Dorman, D. E.; Otocka, E. P.; Bovey, F. A. *Macromolecules* **1972,** *5,* 574.
21. Randall, J. C. *J. Polym. Sci., Polym. Phys. Ed.* **1973,** *11,* 275.
22. Roedel, M. J. *J. Am. Chem. Soc.* **1953,** *75,* 6110.
23. Willbourn, A. H. *J. Polym. Sci.* **1959,** *34,* 569.
24. Cotman, J. D., Jr. *Ann. N.Y. Acad. Sci.* **1953,** *57,* 417.
25. Cotman, J. D., Jr. *J. Am. Chem. Soc.* **1955,** *77,* 2790.
26. George, M. H.; Grisenthwaite, R. J.; Hunter, R. F. *Chem. Ind. (London)* **1958,** 1114.
27. Starnes, W. H., Jr.; Schilling, F. C.; Abbås, K. B.; Plitz, I. M.; Hartless, R. L.; Bovey, F. A. *Macromolecules* **1979,** *12,* 13.
28. Starnes, W. H., Jr.; Schilling, F. C.; Abbås, K. B.; Cais, R. E.; Bovey, F. A. *Macromolecules* **1979,** *12,* 556.
29. Rigo, A.; Palma, G.; Talamini, G. *Makromol. Chem.* **1972,** *153,* 219.
30. Berens, A. R. *Polym. Eng. Sci.* **1974,** *14,* 318.
31. Starnes, W. H., Jr.; Villacorta, G. M.; Schilling, F. C. *Polym. Prepr., Am. Chem. Soc., Div. Polym. Chem.* **1981,** *22*(2), 307.
32. Starnes, W. H., Jr.; Schilling, F. C.; Plitz, I. M.; Cais, R. E.; Bovey, F. A. *Polym. Bull.* **1981,** 555.

33. Wolfe, J. R. *Polym. Prepr., Am. Chem. Soc., Div. Polym. Chem.* **1978,** *19(1),* 5.
34. Cella, R. J. In "Encyclopedia of Polymer Science and Technology"; Wiley: New York, 1977; Suppl. Vol. 2, p. 485.
35. Morèse-Séguéla, B.; St.-Jacques, M.; Renaud, J. M.; Prud'homme, J., Jr. *Macromolecules* **1980,** *13,* 100.
36. Jelinski, L. W.; Schilling, F. C.; Bovey, F. A. *Macromolecules* **1980,** *14,* 581.
37. Jelinski, L. W. *Macromolecules* **1981,** *14,* 1341.

RECEIVED for review October 14, 1981. ACCEPTED December 7, 1981.

^{13}C- and ^{19}F-NMR Chemical Shifts and the Microstructures of Fluoropolymers

A. E. TONELLI, F. C. SCHILLING, and R. E. CAIS

Bell Laboratories, Murray Hill, NJ 07974

Simultaneous ^{1}H- and ^{19}F-decoupled ^{13}C-NMR and ^{1}H-decoupled ^{19}F-NMR spectra are recorded for poly(vinylidene fluoride) (PVF$_2$), poly(fluoromethylene) (PFM), poly(vinyl fluoride) (PVF), and poly(trifluorethylene) (PF$_3$E). Observed ^{13}C- and ^{19}F-NMR chemical shifts are compared to those calculated as a function of stereoregularity and/or defect structure, i.e., head-to-head:tail-to-tail (H−H:T−T) addition of monomers. Calculated chemical shifts are obtained through enumeration of the numbers and kinds of γ effects (three-bond gauche arrangements) that involve each carbon or fluorine atom as dictated by the conformational characteristics of each polymer. Agreement between measured and predicted ^{13}C- and ^{19}F-NMR chemical shifts is observed for each fluoropolymer when the following γ effects are assumed: $\gamma_{C,C\ or\ F} = -2$ to -5 ppm; $\gamma_{F,F} = 10$ to 25 ppm; and $\gamma_{F,C} = 20$ to 35 ppm, where $\gamma_{a,b}$ is the upfield shift at carbon or fluorine a produced by carbon or fluorine b when in a gauche arrangement with a. This agreement permits detailed assignments of ^{13}C- and ^{19}F-NMR resonances in the spectra of all four fluoropolymers including identification of those resonances belonging to the carbon and fluorine atoms in the H−H:T−T defects present in PVF$_2$, PVF, and PF$_3$E, whose abundance increases in this order.

$\mathbf{A}$NALYSIS OF THE NUMBERS AND KINDS of γ interactions involving each carbon atom type in a polymer chain can allow for the prediction of ^{13}C-NMR spectra of vinyl homopolymers and copolymers (1). Each nonhydrogen γ substituent in a three-bond *gauche* arrangement (*see* Figure 1) with a given carbon atom produces an upfield chemical shift

0065−2393/83/0203−0441$06.00/0

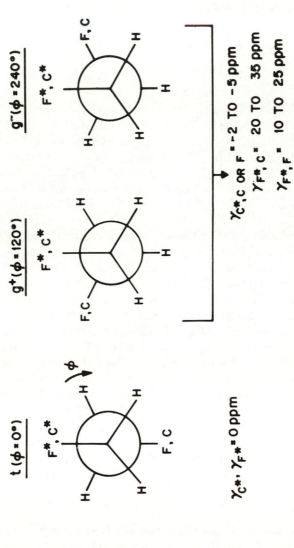

Figure 1. Scheme of γ-gauche interactions between observed (F*;C*) and γ substituent (F,C) fluorine and carbon nuclei in fluoropolymers (17).

of that carbon resonance relative to their *trans* arrangement. The frequency with which these γ-*gauche* interactions occur can be evaluated from the conformational characteristics of the polymer chain as manifested by calculated bond rotation probabilities.

The sensitivity of vinyl homopolymer and copolymer ¹³C-NMR chemical shifts to stereomonomer and comonomer sequence results from the dependence of bond rotation probabilities on the same microstructural features, and leads, via the γ-*gauche* effect, to a dispersion of ¹³C-NMR chemical shifts reflecting the different possible microenvironments found along the vinyl polymer chain. All that is needed to calculate the effects of polymer microstructure on ¹³C-NMR chemical shifts are the bond rotation probabilities obtained from the polymer's conformational characteristics and the magnitudes of the upfield γ effects produced by the *gauche* arrangements of a given carbon atom with its γ substituents.

This approach, which fully utilizes all the microstructural information provided by ¹³C-NMR spectroscopy, has been applied successfully to the calculation of the ¹³C-NMR chemical shifts in a variety of vinyl homopolymers and copolymers (*1*). Polypropylene oligomers (*2, 3*) and homopolymer (*4*) and its copolymers with ethylene (*5, 6*) and vinyl chloride (*7*), poly(vinyl chloride) oligomers (*8*) and homopolymer and its copolymers with ethylene (*9*) and propylene (*7*), and polystyrene (*10*) and its oligomers have all been treated via the γ effect method of predicting ¹³C-NMR chemical shifts in polymers.

This chapter extends the application of the γ effect method to the calculation of the ¹³C- and, for the first time, ¹⁹F-NMR chemical shifts expected for the carbon and fluorine atoms in fluoropolymers. Poly-(vinylidene fluoride) (PVF₂), poly(vinyl fluoride) (PVF), poly(fluoromethylene) (PFM), and poly(trifluorethylene) (PF₃E) are treated.

This study was prompted for three principal reasons. First, the polymers studied possess fluorine atoms whose ¹⁹F nuclei exhibit magnetic resonance and whose chemical shifts are at least as sensitive to polymer microstructure as their ¹³C nuclei. We sought to determine if the sensitivity of ¹⁹F-NMR chemical shifts to microstructure has its origin, as do ¹³C-NMR chemical shifts, in γ-*gauche* interactions. Second, the magnitude of the upfield ¹³C-NMR chemical shift produced at a carbon atom by a γ fluorine substituent when in a *gauche* arrangement was unknown. Third, with the exception of PVF₂ (*11*), the conformational characteristics of these fluoropolymers had never been studied experimentally. We hope to test the recently predicted conformational characteristics (*12*) of PVF, PF₃E, and PFM by comparing their observed and predicted ¹³C- and ¹⁹F-NMR chemical shifts, the latter of which are derived from bond rotation probabilities as obtained (*13*) from their conformational models.

Materials

The synthesis of the PFM employed in this study has been reported previously (14). PF₃E was provided through Pennwalt Corporation. PVF was purchased from Aldrich Company, and the PVF$_2$ employed is Kynar 301 (Pennwalt Corporation).

Experimental

Generally, spin−spin coupling between directly bonded ^{13}C and ^{19}F nuclei obliterates the detailed structure of the dispersion of ^{13}C-NMR chemical shifts produced by different fluoropolymer microstructures. Only in the methylene carbon regions of the usual ^{1}H-decoupled ^{13}C-NMR spectra of PVF$_2$ and PVF can we begin to separate the effects of ^{19}F coupling and microstructure on the ^{13}C-NMR chemical shift dispersion. Consequently, we have performed ^{13}C-NMR measurements on these fluoropolymers with simultaneous decoupling of both the ^{1}H and ^{19}F nuclei. The details of this triple-resonance experiment are published separately (15).

The NMR spectra were recorded on Bruker WH-90 and Varian XL-200 spectrometers at ^{13}C and ^{19}F frequencies of 22.62, 84.6, and 188 MHz. Complete experimental details for the ^{13}C-NMR spectra (16) and ^{19}F-NMR spectra (17) are published elsewhere (15).

Results and Discussion

Comparisons of calculated ^{13}C-NMR chemical shifts with the ^{13}C-NMR spectra observed for PVF$_2$, PFM, PVF, and PF₃E are presented in Figures 2−5. The agreement is generally good, with the calculated ^{13}C-NMR chemical shifts reproducing faithfully the effects of stereosequence and monomer addition defects (18, 19) [head-to-head:tail-to-tail (H−H:T−T)] observed in the ^{13}C-NMR spectra. The γ effects in the range −2−−5 ppm upfield, reflecting the shielding produced at the observed carbon by γ-carbon and/or fluorine substituents in a *gauche* arrangement (*see* Figure 1), permit this agreement.

Prior to our triple-resonance (15) ^{13}C-NMR study of PVF$_2$, the H−H:T−T CH$_2$ defect resonance 4 observed (16) at −0.8 ppm upfield from H−T CH$_2$ carbons (*see* Figure 2) was obscured by long-range ^{19}F−^{13}C-spin−spin coupling in the usual ^{1}H-decoupled ^{13}C-NMR spectrum of PVF$_2$ (20). In addition, the CF$_2$ region of the ^{13}C-NMR spectrum is a continuum of resonances without ^{19}F-decoupling resulting in the complete obliteration of H−H:T−T defect resonances A, B, and C, so clearly apparent in Figure 2.

On the other hand, even the triple-resonance ^{13}C-NMR spectrum of PFM reveals a near continuum of chemical shifts resulting from the long-range sensitivity to stereosequence (*see* Figure 3). ^{13}C-NMR chemical shifts calculated for the pentads in PFM are sensitive to nonad stereosequences leading to a ^{13}C-NMR chemical shift disper-

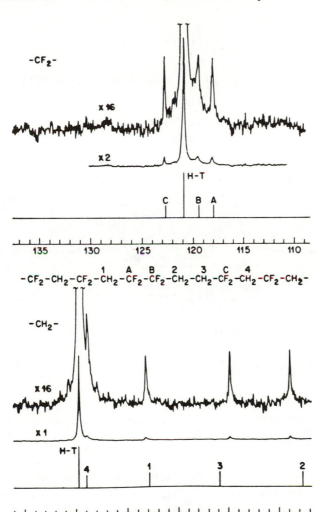

Figure 2. ¹³C-NMR spectrum (22.62 MHz) and calculated chemical shifts for PVF₂ (16).

sion of 0.2–0.5 ppm calculated for each pentad. A comparison of observed and calculated ¹³C-NMR chemical shifts shows that the sample of PFM can be characterized (*14*) as predominantly atactic (discussed later).

Without recording the ¹³C-NMR spectrum of PVF in the triple resonance mode (*see* Figure 4) the sensitivity of the H–T CHF and CH₂ resonances to stereosequence would not be revealed. Sensitivity

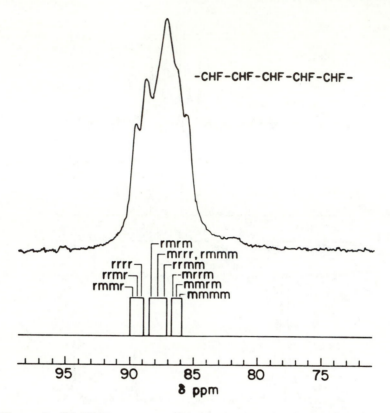

Figure 3. ^{13}C-NMR spectrum (22.62 MHz) and calculated chemical shifts for PFM (16).

to triad stereosequences in the CHF region and tetrad sensitivity in the CH$_2$ region was observed. With increasing field the three resonances observed in the H–T CHF region are assigned to the rr, mr, mm triads by comparison to our calculated shifts. Thus, our PVF sample is predominantly atactic with P_m = 0.43 (17).

The ^{13}C-NMR spectrum of PF$_3$E in Figure 5 does not permit a separation of the effects of stereosequence and defect structure on the observed ^{13}C-NMR chemical shifts. A comparison of observed and calculated ^{13}C-NMR chemical shifts enables us to conclude that H–H:T–T addition in our PF$_3$E sample is more prevalent than in our PVF$_2$ and PVF samples.

The agreement achieved between the observed and calculated ^{13}C-NMR chemical shifts implies that the bond rotation probabilities used to evaluate the frequencies of the various γ effects occurring for the carbon atoms in each of the possible microstructural environments

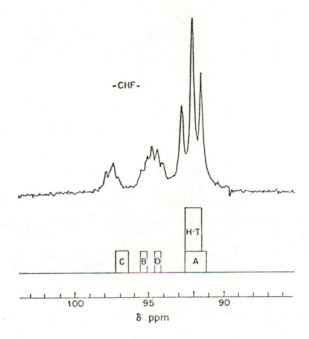

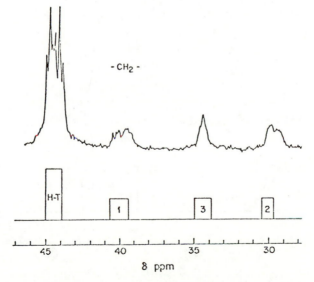

Figure 4. *¹³C-NMR spectrum (22.62 MHz) and calculated chemical shifts for PVF (16).*

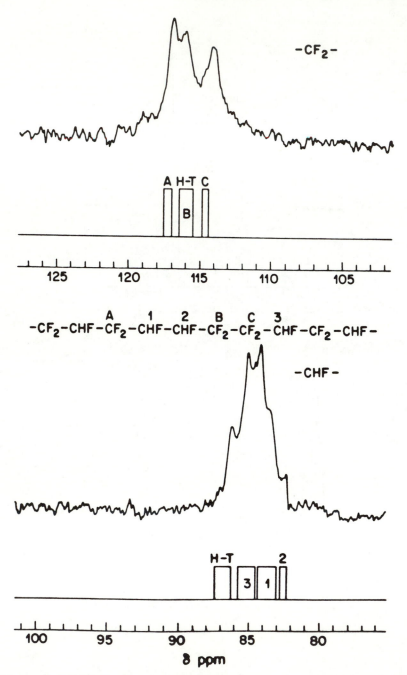

Figure 5. ^{13}C-NMR spectrum (22.62 MHz) and calculated chemical shifts for PF$_3$E (16).

in these polymers are realistic. This provides, for the first time, direct microstructural support for the conformational models (*11, 12*) derived previously for these chains (*17, 21*).

Traditional measures (*13*) of the conformational characteristics of polymer chains, such as end-to-end distance and dipole moment, were predicted (*11, 12*) to be relatively insensitive to microstructure (both stereosequence and defect content) for the fluoropolymers considered (*11, 12*). Thus, the successful testing of their conformational models as provided by the correct prediction of their ¹³C-NMR chemical shifts takes on added importance.

With the confidence gained in the conformational models of these fluoropolymers through the successful prediction of ¹³C-NMR chemical shifts, ¹⁹F-NMR chemical shifts were also calculated, and the results are presented in Figures 6−9. The agreement between observed and calculated ¹⁹F-NMR chemical shifts is good and leads to γ-*gauche* ¹⁹F-shielding interactions of 10 to 25 ppm for fluorine and 20 to 35 ppm for carbon γ substituents. The γ effects on fluorine nuclei are nearly an order of magnitude larger than those experienced by carbon nuclei (*see* Figure 1).

The larger magnitude of the γ effects operating on ¹⁹F nuclei, as compared to ¹³C nuclei, results in a greater sensitivity of the ¹⁹F-NMR spectra of fluoropolymers to their microstructure. In Figure 6 we see ¹⁹F resonances (188-MHz spectrum) due to defect structures other than H−H:T−T addition in PVF$_2$ (resonances 5, 6, and 7). Instead of a continuum of chemical shifts as observed in the ¹³C-NMR spectrum of PFM (Figure 3) the ¹⁹F-NMR spectrum of PFM (Figure 7) manifests detailed information regarding the stereosequences present in our sample. Adoption of Bernoullian polymerization statistics and comparison of observed and calculated chemical shifts permits a faithful simulation (*14*) of the observed spectrum and indicates clearly the predominantly atactic nature (P_m = 0.42) of our PFM.

Comparison of Figures 5 and 9 reveals the greater sensitivity of ¹⁹F-NMR, relative to ¹³C-NMR spectroscopy, to the microstructure of PF$_3$E. This greater sensitivity enables us to estimate the content of monomer units added in a H−H:T−T manner.

The detailed assignment of resonances in the ¹³C- and ¹⁹F-NMR spectra of PVF$_2$, PVF , and PF$_3$E achieved by comparison to predicted ¹³C- and ¹⁹F-NMR chemical shifts permits a quantitative estimate of the H−H:T−T defect content in each fluoropolymer. Integration of defect (H−H:T−T) and normal (H−T) resonances to obtain peak areas leads to estimates of the defect content in each fluoropolymer. We observed 3, 12, and 20 mol% H−H:T−T monomer addition in the samples of PVF$_2$, PVF, and PF$_3$E, respectively.

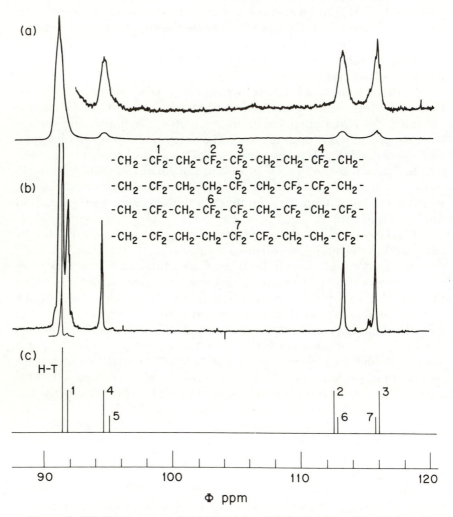

Figure 6. ¹⁹F-NMR spectra (84.6 MHz) (a), (188 MHz) (b), and calculated chemical shifts for PVF₂ (see Ref. 22) (17).

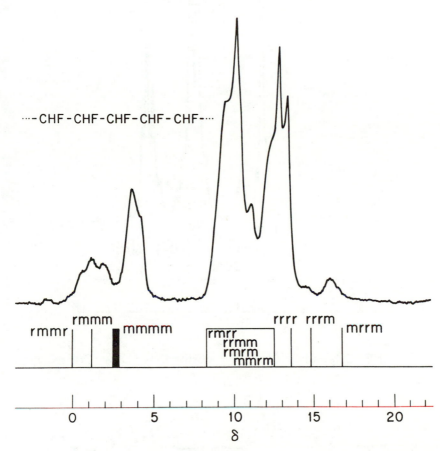

Figure 7. ¹⁹F-NMR spectrum (84.6 MHz) and calculated chemical shifts for PFM (17).

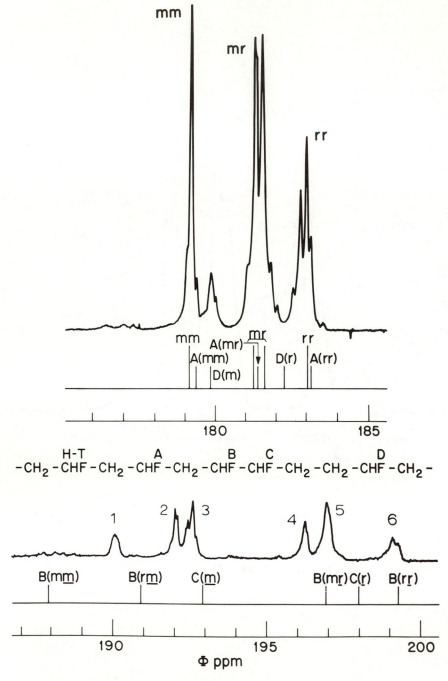

Figure 8. *¹⁹F-NMR spectrum (188 MHz) and calculated chemical shifts for PVF (17).*

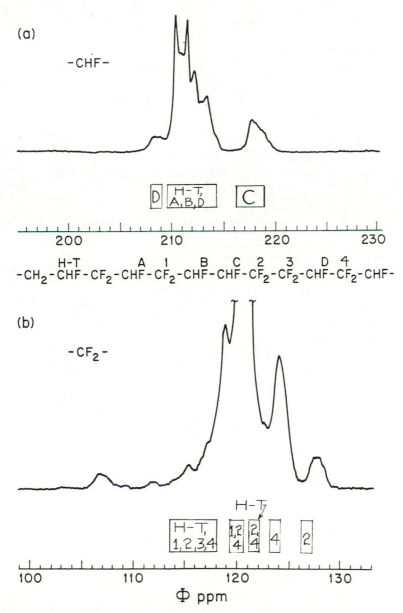

Figure 9. *¹⁹F-NMR spectrum (84.6 MHz) and calculated chemical shifts for PF₃E.*

Acknowledgment

The authors are indebted to Dr. Gordon R. Leader of the Pennwalt Corporation for providing samples of PF_3E.

Literature Cited

1. Tonelli, A. E.; Schilling, F. C. *Acc. Chem. Res.* **1981**, *14*, 223.
2. Tonelli, A. E. *Macromolecules* **1978**, *11*, 565.
3. Tonelli, A. E. *Macromolecules* **1979**, *12*, 83.
4. Schilling, F. C.; Tonelli, A. E. *Macromolecules* **1980**, *13*, 270.
5. Tonelli, A. E. *Macromolecules* **1978**, *11*, 634.
6. Tonelli, A. E. *Macromolecules* **1979**, *12*, 255.
7. Tonelli, A. E.; Schilling, F. C., in preparation.
8. Tonelli, A. E.; Schilling, F. C.; Starnes, W. H., Jr.; Shepherd, L.; Plitz, I. M. *Macromolecules* **1980**, *13*.
9. Tonelli, A. E.; Schilling, F. C. *Macromolecules* **1979**, *12*, 252.
10. Tonelli, A. E. *Macromolecules* **1979**, *12*, 252.
11. Tonelli, A. E. *Macromolecules* **1976**, *9*, 547.
12. Tonelli, A. E. *Macromolecules* **1980**, *13*, 734.
13. Flory, P. J. In "Statistical Mechanics of Chain Molecules"; Wiley: New York, 1969; Chaps. II and IV.
14. Cais, R. E. *Macromolecules* **1980**, *13*, 806.
15. Schilling, F. C. *J. Magn. Reson.* **1982**, *47*, 61.
16. Tonelli, A. E.; Schilling, F. C.; Cais, R. E. *Macromolecules* **1981**, *14*, 560.
17. Tonelli, A. E.; Schilling, F. C.; Cais, R. E. *Macromolecules* **1982**, *15*, 849.
18. Wilson, C. W., III *J. Polym. Sci., Part A* **1963**, *1*, 1305.
19. Wilson, C. W.; Santee, E. R., Jr. *J. Polym. Sci., Part C* **1965**, *8*, 97.
20. Bovey, F. A.; Schilling, F. C.; Kwei, T. K.; Frisch, H. L. *Macromolecules* **1977**, *10*, 559.
21. Tonelli, A. E., unpublished data.
22. Ferguson, R. C.; Brame, E. G., Jr. *J. Phys. Chem.* **1979**, *83*, 1397.

RECEIVED for review January 15, 1982. ACCEPTED February 28, 1982.

Characterization of Molecular Motion in Solid Polymers by Variable Temperature Magic Angle Spinning ^{13}C-NMR Spectroscopy

W. W. FLEMING, J. R. LYERLA, and C. S. YANNONI

IBM Research Laboratory, San Jose, CA 95193

The inclusion of a variable temperature magic-angle spinning capability for solid-state ^{13}C-NMR spectroscopy makes feasible the investigation by ^{13}C-relaxation parameters of structural and motional features of polymers above and below T_g and in temperature regions of secondary relaxations. We report variable temperature (77 K to 323 K) spectral data on polytetrafluoroethylene and polypropylene. Illustrative of the data are the T_1 and $T_{1\rho}$ results for isotactic polypropylene over the temperature range 77–300 K. All carbons in the repeat unit show minima in T_1 and $T_{1\rho}$ that reflect methyl group reorientational motion at the appropriate measuring frequencies (15 MHz and 57 kHz). The $T_{1\rho}$ data for CH and CH_2 carbons indicate the importance of spin–spin as well as spin–lattice pathways in their rotating frame relaxation over much of the temperature interval studied. An interesting spectral observation is the strong motional broadening of the methyl group in the temperature region of the $T_{1\rho}$ minimum. These and other facets of the polypropylene data as well as similar data for other polymers are discussed with respect to their implications for insight into polymer chain dynamics in the solid state.

Aᴌᴛʜᴏᴜɢʜ ᴍᴀɴʏ ᴏꜰ ᴛʜᴇ ᴇᴀʀʟɪᴇsᴛ NMR experiments were performed on solid samples, technical advances in magnetic field homogeneity and pulsed Fourier transform NMR have resulted in NMR being considered a high resolution analytical tool for studying liquid-

0065–2393/83/0203–0455$06.00/0

state samples. Until recently, solid materials could be studied only by so-called broad-line NMR and pulse relaxation NMR measurements on abundant nuclei, usually hydrogen and fluorine. However, the recent advances in the application of magic angle sample spinning, dipolar decoupling, cross-polarization, and multipulse NMR have brought solid samples into the realm of high resolution NMR in which individual resonances of the solid can be resolved. The usefulness of the high resolution experiments is now well established. In particular, high resolution carbon NMR of solids has shown considerable promise, as evidenced by the numerous publications in which magic angle spinning NMR has been used to study polymers and other solid organic and inorganic materials (1−4).

In principle, resolution of individual carbon resonances in bulk polymers allows relaxation experiments to be performed which can be interpreted in terms of main-chain and side-chain motions in the solid. This is a distinct advantage over the more common ^{1}H-NMR relaxation experiments where efficient spin diffusion usually results in the averaging of the relaxation behavior over the ensemble of protons. Thus, a direct interpretation of ^{1}H-relaxation data in terms of unique motions of the solid polymer is often not possible.

Relaxation parameters of interest for the study of polymers include the ^{13}C spin−lattice relaxation time in H_0, T_1; the spin−spin relaxation time, T_2; the nuclear Overhauser enhancement, NOE; the proton and carbon rotating frame relaxation times, $T_{1\rho}^x$; the C−H cross-polarization or cross-relaxation time, T_{CR}; and the proton relaxation time in the dipolar field T_{1D}. Not all of these parameters provide information directly; nonetheless, all the inferred information is important in characterizing motional frequencies and amplitudes in macromolecules.

Although initial studies on solids by cross-polarization (CP), magic angle spinning (MAS), and ^{13}C NMR have been carried out almost exclusively at ambient temperature (5), full exploitation of this spectroscopy requires variable temperature capability. This is particularly true of macromolecules where the accessibility of variable temperature magic angle spinning (VT-MAS) makes feasible the investigation of structural and motional features of polymers above and below the glass transition temperature, T_g, and in temperature regions of secondary relaxations. The spectral data and $T_{1\rho}^c$ measurements by Garroway et al. (6) on epoxy resins over the limited temperature interval from −31 to 51 °C represent the only other VT-MAS ^{13}C NMR results to date. A spinner assembly has been described that is suitable for routine operation over a wide range of temperatures (7). This chapter examines the ^{13}C-relaxation behavior of two common polymers: polytetrafluoroethylene (PTFE) and isotactic polypropylene (PP).

Experimental

The ^{13}C-NMR data at 15.1 MHz were acquired on a modified Nicolet TT-14 NMR system. The features of this spectrometer and of the spinning assembly have been reported previously (4, 7, 8). Samples were machined into the shape of Andrew-type rotors (9) and used directly for the various studies. Temperature variation was achieved by cooling or heating the helium gas used for driving the rotor. The temperature was controlled to ±2 °C with a home-built temperature sensing and heater/feedback network. Spin–lattice relaxation times T_1 were collected using a pulse sequence developed by Torchia (10) that allows CP enhancement of the signals. The $T_{1\rho}$ data were determined at 40 and 58 kHz using $T_{1\rho}$ methodology described elsewhere (5).

Polypropylene, PP, samples were made from compression-molded Pro-Fax PP (Hercules). The material was quenched slowly and had a 70% crystallinity, as determined by ^{1}H NMR, for the 95% isotactic material. The polytetrafluoroethylene, PTFE, was machined from commercial Teflon from du Pont. Data based on an IR band intensity ratio analysis (11) indicated the PTFE sample to be 67% crystalline.

Results and Discussion

The proton-decoupled CP/MAS ^{13}C-NMR spectra of PP as a function of temperature are shown in Figure 1. At ambient temperature, all three carbons of the PP repeat unit are resolved. However, as the temperature is lowered, the methyl resonance begins to broaden significantly and is completely lost in the baseline at about −143 °C. The broadening, which is also seen to occur to a degree for the methine and methylene carbons, arises primarily from incomplete decoupling as the reorientation rate of the methyl group about the C_3-axis becomes comparable to the strength of the decoupling field. Heteronuclear dipolar coupling characterized by correlation frequencies near the decoupling frequency is not decoupled efficiently (12). This phenomenon has been observed (6) for methyl groups in epoxy resins, and is the same mechanism responsible for the motional broadening in crystalline regions of PTFE (13).

The T_1 data for PP over a temperature range from −195 to 24 °C are summarized in Figure 2a. As indicated in the figure, each of the carbons displays individual relaxation rates. The CH and CH_2 carbons have a T_1 minimum at about −113 °C, nearly the same temperature as that reported for the proton T_1 minimum in isotactic PP (14).

If it is assumed that a C−H heteronuclear dipolar relaxation mechanism is operative, the methyl protons probably dominate the relaxation behavior of these carbons over much of the temperature range studied despite the $1/r^6$ dependence of the mechanism. The shorter T_1 for the CH as compared to the CH_2 then arises from the shorter C−H distances. Apparently, the contributions to spectral den-

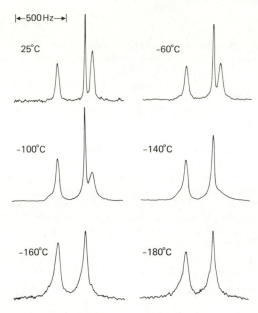

Figure 1. Proton-decoupled CP/MAS ^{13}C-NMR spectra of isotactic PP as a function of temperature from 25 to −180 °C.

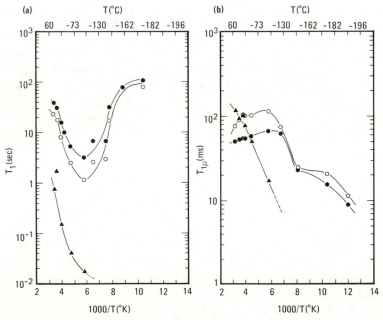

Figure 2. ^{13}C-NMR T_1 relaxation times for the methyl (▲), methylene (●), and methine (○) carbons of PP as a function of temperature at 1.4 T(a), and ^{13}C-NMR $T_{1\rho}$ relaxation times for the same carbons of PP as a function of temperature at 1.4 T (b).

sity in the megahertz region of the frequency spectrum due to backbone motions are minor relative to the side-group motion. The $T_{1\rho}$ data for the CH and CH$_2$ carbons also give an indication of methyl group rotational frequencies (Figure 2b). As the temperature is lowered below -113 °C, the contribution of the methyl protons to megahertz spectral density decreased, yet increased in the kilohertz regime. Consequently, the $T_{1\rho}$ decreases by roughly 10 times between -103 and -195 °C.

The interpretation of carbon $T_{1\rho}$ data is complicated by the fact that spin–spin (cross-relaxation) processes, as well as rotating frame spin–lattice processes, may contribute to the relaxation (15). Only the latter process provides direct information on molecular motion. For the CH and CH$_2$ carbons of PP, the $T_{1\rho}$s do not change greatly over the temperature interval from -113 °C to ambient temperature and, as opposed to the T_1 behavior, the CH$_2$ carbon has a shorter $T_{1\rho}$ than the CH carbon. These results suggest that spin–spin processes dominate the $T_{1\rho}$. However, below -118 °C, the $T_{1\rho}$s for both carbons shorten and tend toward equality. A proton $T_{1\rho}$ minimum (which reflects methyl group reorientation of kilohertz frequencies) has been reported at -180 °C (14). No clear minimum is observed in the ^{13}C-NMR data, perhaps due to an interplay of spin–spin and spin–lattice processes. Nonetheless, it is apparent that the methyl protons are responsible for the spin–lattice portion of the $T_{1\rho}$ relaxation for the CH and CH$_2$ carbons.

Figure 3 shows semilog plots of typical T_1^c and $T_{1\rho}^c$ relaxation data (intensity vs. time) at -2 °C for PTFE. The decays are clearly nonexponential, indicative of multiple relaxation behavior. However, for both plots, the long-time behavior of PTFE can be characterized by one relaxation time, associated with about 65–70% of the total signal intensity as judged by the y-intercept, and in agreement with the IR analysis reported in the experimental section. On this basis and previous ^{19}F-NMR relaxation studies (16), the long-time relaxation component is ascribed to crystalline regions of the polymer. The faster relaxing component of the resonance line is attributed to noncrystalline regions and is not described by a single time constant, but by a distribution of relaxation times in accord with the results on glassy polymers (5).

The ^{13}C-NMR T_1 data for the crystalline component of PTFE are plotted vs. reciprocal temperature in the region from -43 to 47 °C in Figure 4. A shortening of T_1 by two orders of magnitude is observed in

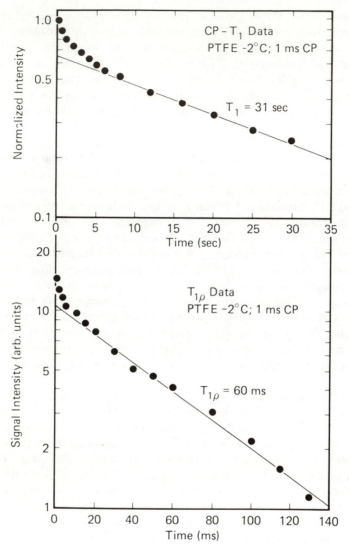

Figure 3. ^{13}C-NMR T_1 *and* $T_{1\rho}$ *relaxation data (intensity vs. time) for PTFE at −2 °C and 1.4 T. The sample had a crystallinity of 67%.*

this temperature interval. The sharp decrease in T_1 between 0 and 20 °C can be attributed to the well-known crystal phase transition at 19 °C (*17*). The increase in specific volume accompanying the unit cell change and slight unwinding of the helix apparently allows rotational motion characterized by correlation times in the range of $10^{-5} - 10^{-8}$ s. The motional broadening of the ^{13}C-NMR resonance line in the tem-

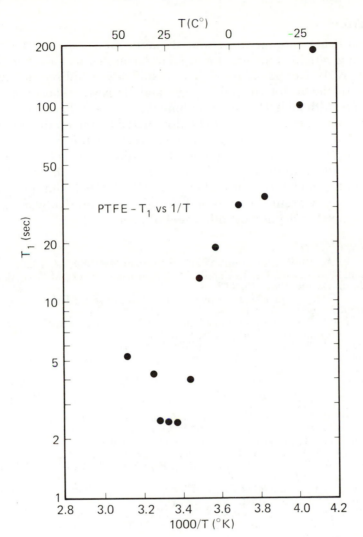

Figure 4. ^{13}*C-NMR* T$_1$ *relaxation times for PTFE as a function of temperature at 1.4 T.*

perature range from 17 to 27 °C has been reported previously (*13*) and also reflects the phase transition.

The ^{13}C-NMR $T_{1\rho}$ relaxation times for PTFE decreased from about 185 ms at −117 °C to 53 ms at 47 °C. In contrast, the ^{19}F-NMR T_1 relaxation times for crystalline PTFE change 20–40 fold over the same temperature range (*16*). The smaller change in the ^{13}C $T_{1\rho}$ is probably due in part to the contribution of the spin−spin interactions to the rotating frame relaxation process.

Conclusion

Significant information about dynamic processes in the solid state using variable temperature dipolar decoupled magic angle spinning ^{13}C-NMR can be obtained. The results are in substantial agreement with the results of pulsed ^{19}F- and ^{1}H-NMR relaxation measurements. Although the polypropylene data suggest that relaxation in methyl-containing polymers can be dominated by the methyl protons, the use of deuterated methyl groups can overcome this contribution if necessary. Possibly the strong methyl contribution to relaxation may be exploited to study intermolecular phenomena such as polymer compatibility and miscibility. Thus, the ^{13}C-NMR technique offers the opportunity to study details of dynamic processes that were previously intractable using most other techniques.

Literature Cited

1. Fyfe, C. A.; Rudin, A.; Tchir, W. *Macromolecules* **1980**, *13*, 1320.
2. Dixon, W. T.; Schaefer, J.; Sefcik, M. D.; Stejskal, E. O.; McKay, R. A. *J. Magn. Reson.* **1981**, *45*, 173.
3. Brown, C. E.; Jones, M. B.; Kovacic, P. *J. Polym. Sci., Polym. Lett. Ed.* **1980**, *18*, 653.
4. Lyerla, J. R. In "Contemporary Topics in Polymer Science"; Shen, M., Ed.; Plenum: New York, 1979; Vol. 3, p. 143.
5. Schaefer, J.; Stejskal, E. O.; Buchdahl, R. *Macromolecules* **1977**, *10*, 384.
6. Garroway, A. N.; Moniz, W. B.; Resing, H. A. *Faraday Discuss. Chem. Soc.* **1979**, *13*, 63.
7. Fyfe, C. A.; Mossburger, H.; Yannoni, C. S. *J. Magn. Reson.* **1979**, *36*, 61.
8. Fyfe, C. A.; Lyerla, J. R.; Volksen, W.; Yannoni, C. S. *Macromolecules* **1979**, *12*, 757.
9. Andrew, E. R. *Int. Rev. Sci.: Phys. Chem., Ser. Two* **1976**, *4*, 1973.
10. Torchia, D. A. *J. Magn. Reson.* **1978**, *30*, 613.
11. Rabolt, J. F., IBM, San Jose, private communication.
12. Rothwell, W. P.; Waugh, J. S. *J. Chem. Phys.* **1981**, *74*, 2721.
13. Fleming, W. W.; Fyfe, C. A.; Lyerla, J. R.; Vanni, H.; Yannoni, C. S. *Macromolecules* **1980**, *13*, 460.
14. McBrierty, V. J.; Douglas, D. C.; Flacone, D. R. *J. Chem. Soc., Faraday Trans. 2* **1972**, *68*, 1051.
15. VanderHart, D. L.; Garroway, A. N. *J. Chem. Phys.* **1979**, *71*, 2773.
16. McCall, D. W.; Douglas, D. C.; Falcone, D. R. *J. Phys. Chem.* **1967**, *71*, 998, and references therein.
17. For example, see Clark, E. S.; Muus, L. T. *Z. Kristallogr.* **1960**, *117*, 119.

RECEIVED for review October 14, 1981. ACCEPTED March 8, 1982.

^{13}C-NMR Studies of Nonordered Regions of Semicrystalline Polymers

L. MANDELKERN

Florida State University, Department of Chemistry and Institute of
Molecular Biophysics, Tallahassee, FL 32306

*Studies of the ^{13}C-NMR relaxation parameters, under
conditions of scalar proton decoupling, are shown to be
an effective method to investigate the properties of the
noncrystalline regions of semicrystalline polymers. Our
work in this area will be reviewed. These works include
the influence of the level of crystallinity and crystalline
morphology on the spin relaxation time, nuclear Overhauser enhancement and resonant line width. The different contribution to the line widths in the molten and
semicrystalline states are assessed by field dependent
and preliminary magic angle spinning experiments. The
influence of temperature and crystallinity on the motion
of branch groups are described. The use of these techniques to study transitions already defined by dynamic
mechanical experiments also is discussed.*

THIS CHAPTER SUMMARIZES ^{13}C-scalar proton-decoupled NMR studies, and their structural implications, that we have carried out. It is presented in the form of a somewhat expanded long abstract (*1*). Studies
of ^{13}C-relaxation parameters can be used in the analysis of the segmental
motions and structure of bulk polymers (*2−4*). Highly sophisticated
and fruitful methods have been developed to study glasses and the
interior structure of the crystalline regions in polymers (*5−8*). However, scalar proton decoupling also has proven capable of yielding
high resolution spectra for amorphous polymers and for the noncrystalline regions of semicrystalline polymers (*3, 4*). By employing this
technique, the influence of the level of crystallinity and crystalline
morphology (supermolecular structure) on the spin−lattice relaxation time (T_1), the nuclear Overhauser enhancement (NOE) and the
line widths ($\nu\frac{1}{2}$) in the noncrystalline region can be assessed. A new
set of techniques can be used to help elucidate the nature of these
complex structures (*9*). In addition, the influence of the glass tempera-

0065−2393/83/0203−0463$06.00/0

ture on these relaxation parameters and the character of other transitions can be studied. In particular, more information about the molecular nature of the secondary relaxation transitions in semicrystalline polymers can be expected. The results obtained by this experimental method, coordinated and correlated with other studies of semicrystalline polymers, are summarized. In the study of semicrystalline polymers, which represents the major endeavor, emphasis is placed on the control and characterization of the level of crystallinity and the morphological form that is developed (10). As described elsewhere, wide ranges in the level of crystallinity and quite different morphological forms can be achieved by using an extensive set of molecular weights and controlled crystallization conditions (11−13). The main work presented here will be restricted to the scalar decoupling technique, with some reference to preliminary studies involving dipolar decoupling and magic angle spinning.

We first consider the influence of crystallinity and crystalline morphology, or the supermolecular structure, on T_1 for pure (undiluted) semicrystalline polymers. If possible, comparison of the T_1 values for a completely amorphous polymer with the same polymer in the semicrystalline state, at the same temperature and pressure is made. Because of favorable crystallization kinetics (14), this comparison can be made with cis-polyisoprene at 0 °C with a degree of crystallinity of about 0.30 (15). This definitive experiment showed that for each of the carbons the T_1 values do not change on crystallization (16).

These experiments were limited to a fixed, relatively low level of crystallinity. Hence, generalizations should not be made solely from these results. However, this conclusion is further enhanced from studies with linear and branched polyethylene (4, 10, 17). As illustrated in Table I, the value of T_1 for both bulk- and solution-crystallized linear polyethylene at ambient temperature is essentially a constant independent of the wide range of crystallinities that can be attained (0.51−0.94), the variety of organized supermolecular structures and random lamellae that can be developed in linear polyethylene, and the mode of crystallization. The crystallites formed from dilute solution, which yield the characteristic platelet habit, yield the same T_1 for the methylene carbon as a bulk-crystallized sample having only a degree of crystallinity of 0.50. Because only scalar decoupling was used in these experiments, the carbon atoms within the crystal do not contribute to the motion. Thus, the fast segmental motions that determine T_1 must be the same in the noncrystalline regions of both types of samples. The constancy of T_1 must then reflect the same type of chain structure in the noncrystalline regions. This result is obviously incompatible with the concept that the crystals formed in dilute solution possess a regularly folded interfacial structure to any significant

Table I. ^{13}C-NMR Spin–Lattice Relaxation Times (ms) for Bulk- and Solution-Crystallized Linear Polyethylene at 45 °C and 67.9 MHz

Degree of Crystallinity	T_1	Morphological Forms
Bulk crystallized		
0.51	369	random lamellae
0.72	358	random lamellae
0.51	355	spherulite
0.57	343	spherulite
0.68	356	spherulite
0.81	348	spherulite
0.94	352	rod
Solution crystallized		
0.80	370	platelet

extent. This conclusion is in accord with those obtained by other physicochemical measurements (*18*). The T_1 values for the backbone carbons of the branched polyethylenes are independent of the type and extent of branching and have the same value as for the linear polymer. Hence, branching does not affect the motion of the methylene backbone carbons.

As shown in Figure 1, both linear and branched bulk-crystallized samples, as well as the solution-crystallized polymers, have the same temperature dependence and are continuous functions from ambient temperature through their respective melting temperatures. Thus, for the polyethylenes, as for *cis*-polyisoprene, T_1 is the same for the completely amorphous polymer and the noncrystalline region of the partially crystalline polymer. Similar results are found for polyethylene oxide and polytrimethylene oxide, where because of the lower melting temperatures more striking examples are given for the continuity of T_1 through the melt (*4*). The finding that T_1 of the noncrystalline region is the same as that of the pure melt (under the same conditions) is, therefore, of general validity. We conclude that the segmental motions manifest in T_1 are independent of all the structural aspects of crystallinity. For semicrystalline polyethylene, the correlation time associated with these motions is essentially the same as that for the interior carbons of the molten n-alkanes (*10*).

The influence of the crystallinity parameters on the nuclear Overhauser enhancement factor (NOEF) has not as yet been studied in great detail. However, some interesting results still remain to be exploited. Linear and branched polyethylenes having degrees of crystallinity of about 0.50 possess full NOEF values at ambient temperature that are the same as that of the pure melt (*10, 19*). As the crystallinity of the linear polyethylene is increased, the NOEF values

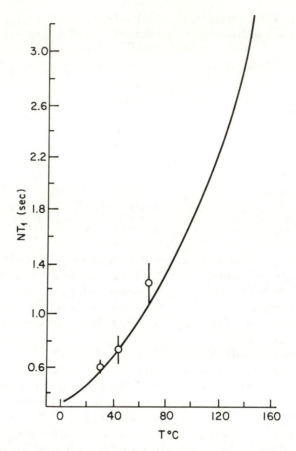

Figure 1. Plot of NT$_1$ vs. T at 67.9 MHz for polyethylenes. Key:—, average values for bulk-crystallized linear and branched samples (4, 9); ○, crystals formed from dilute solution. (Reproduced with permission from Ref. 17. Copyright 1981, John Wiley & Sons, Inc.)

are reduced progressively irrespective of the specifics of the crystalline morphology. These limited data suggest that crystallinity influences this parameter. As the temperature is lowered, the full NOEF is reduced progressively and the limiting theoretical value is approached (20).

In these scalar decoupled experiments, it is appropriate to separate the discussion of the resonant line width in the completely molten, or amorphous, state as contrasted with the semicrystalline state. For the polyethylenes and for polyethylene oxide, the line widths determined above the melting temperature are very dependent on the way the sample is prepared (4, 10, 21). For the same polymer, at a field strength corresponding to 67.9 MHz, the line width can vary from several hundred Hertz in some cases, to just a few in others. The sample must be studied in a continuum form and in as solid a geometric shape as possible. Fissures, cracks, and other such types of geometric irregularities and heterogenieties contribute in a very substantial way to the observed resonant lines and lead to erroneous conclusions (21). When a proper sample is prepared, consistent results can be obtained (21). The T_1 and NOE values for polyethylene above its melting temperature can be described approximately by a single correlation time (10, 21). On this basis, a relatively narrow resonant line, of the order of 0.1 Hz, would be expected at a field strength of 67.9 MHz. Line widths of several Hertz can in fact be observed, which is in accord with these theoretical expectations when allowance is made for possible experimental errors. A consistent set of relaxation parameters is obtained from these data. However, the preparation of proper samples becomes more difficult as the molecular weight is increased (21) so that one can never be sure that the optimum has been achieved, with the lowest possible line width. This problem of the proper sample preparation for pure, undiluted polymers is a major experimental problem and makes difficult any interpretation of the line width in the molten, completely amorphous state.

Upon crystallization, subsequent to cooling, the resonant line widths of linear and branched polyethylene and polyethylene oxide increase by factors of from 10 to 50 at ambient temperatures at a field strength corresponding to 67.9 MHz (10, 21). Substantial increases in line widths upon crystallization also were reported for the *cis*- and *trans*-polyisoprenes (16, 22). The problems described earlier for the molten polymer are not nearly as important under these conditions. For the polyethylenes, the line widths at 30 °C (67.9 MHz) vary from about 400–600 Hz in a very reproducible and systematic manner for levels of crystallinity less than 0.85. The line widths do not depend on the level of crystallinity in this range but are determined solely by the supermolecular structure (4, 21). The lower value of about 400 Hz corresponds to random lamellae (21) irrespective of the level of crystallinity, molecular weight, and branching concentration. On the other hand, the broader lines correspond to a well-developed spherulitic morphology independent of these factors. Only for very high levels of crystallinity (of the order of 0.90) is a direct influence of the extent of crystallization on the line width observed. Thus, the ^{13}C resonant line width of the noncrystalline regions is one of the few observable quantities probing these structures that is influenced by the crystallite organization. Therefore, although the more rapid segmental motions that contribute to T_1 are independent of both the level of crystallinity and the supermolecular structure, the motions governing the line width, and thus T_2, are very dependent on these parameters, which are the principal ones describing the crystallinity of the sample. These conclusions also are applicable to polyethylene oxide (4) and appear to be general for crystalline polymer systems (23).

Because the observations just described offer the possibility of learning more about the noncrystalline regions, it becomes important to identify the major factors that contribute to the resonant line width. The task of resolving the line width into its component parts is not simple because several substantial factors potentially could be contributing (2, 4, 10, 21, 22). Studies designed to establish these major contributors are as yet incomplete, and must obviously involve higher power decoupling and magic angle spinning experiments. Several salient features have evolved, however. The results for a scalar decoupled magic angle spinning experiment are given in Figure 2. A rapid reduction in line width to about 100–150 Hz is observed with the initial increase in spinning frequency. However, at higher spinning frequencies, up to at least 4.5 kHz, the line width remains constant. When the decoupling power is increased, under static nonspinning conditions, the line width is reduced by several hundred Hertz (24). The low-frequency contributions from the chain motions should be removed by this procedure. Magic angle spinning, with dipolar decoupling, yields very similar results to those plotted in Figure 2 (24). The narrowing line width can be attributed to the removal, or sharp reduction, of the contribution from chemical shift anisotropy. An invariant residual line width of 100–150 Hz again is observed. Cross-polarization, magic angle spinning experiments yield amorphous resonances having a similar line width (21). There appears to be a significant, irreducible residual contribution to the amorphous line width of the semicrystalline polyethylene. By more detailed studies of the type outlined previously, a quantitative decomposition of the amorphous line width should be achieved.

Proton-decoupled ^{13}C-NMR relaxation studies of polymers can yield interesting information with respect to both the glass temperature and secondary relaxation processes. We first examine completely amorphous polymers where the major transition is glass formation. For all types of glass-forming substances, the temperature, T_c, at which the backbone carbon resonances collapse, that is, become too broad to be resolved, occurs well above the glass temperature of each material (25, 26). These data are summarized in Table II for a variety of substances whose glass temperatures are well known and noncontroversial. The differences between T_c and T_g range from 30 to 90 °C and, except for polyisobutylene, the ratio T_c/T_g (calculated in Kelvins) for the polymers is in the range 1.2–1.3. This ratio is 1.4 for polyisobutylene, an experimentally significant difference, which can be given

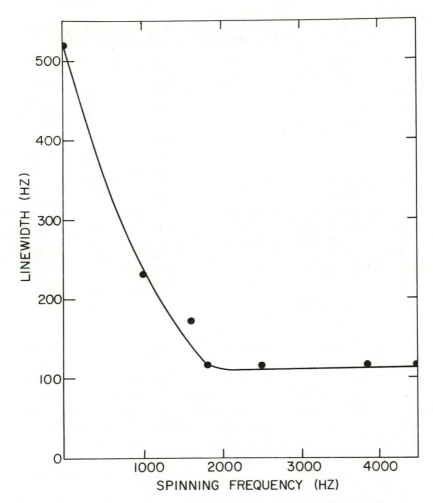

Figure 2. Amorphous phase resonant line width as a function of magic angle spinning frequency. Scalar decoupled ^{13}C Fourier transform NMR spectra obtained at 75.5 MHz for a linear polyethylene sample in powder form; degree of crystallinity 0.65; randomly arranged lamellae. (Reproduced with permission from Ref. 21. Copyright 1979, John Wiley & Sons, Inc.)

a molecular interpretation as discussed later. Apparently, because this ratio appears to be constant, the temperature at which the spectra can no longer be resolved was identified with another amorphous transition previously designated as T_{ll} (27). However, the existence of such a transition was questioned very seriously from a completely different point of view (28, 29). In the present context, the temperature at which a resonance cannot be resolved is related to the line width and thus T_2

Table II. Relation Between Glass Temperature, T_g, and Temperature at Which ^{13}C-NMR Spectra Collapse, T_c, at 67.7 MHz

Substance	$T_c(°C)$	$T_g(°C)$	T_c/T_g
Isotactic polymethyl methacrylate	110	52	1.18
Polyisopropyl acrylate	47	−11	1.22
Poly(n-butyl acrylate)	0	−54	1.25
Atactic polypropylene	30	−20	1.20
$trans$-Polyisoprene	−28	−58	1.14
cis-Polybutadiene	−65	−102	1.22
cis-Polyisoprene	−30	−70	1.20
Polyisobutylene	+10	−70	1.42
Polyvinyl acetate	80	32	1.16
Ethylene−butene copolymer (26 m % butene)	−20	−80	1.33
Ethylene−propylene copolymer (63 m % propylene)	18	−47	1.29
Polyethylene oxide	−38	−70	1.16
Glycerol	−6	−93	1.48

Source: Refs. 25 and 26.

(spin−spin relaxation time) and therefore, the mean correlation time, $\bar{\tau}_c$, for segmental motion.

The values of the correlation times and their temperature dependence can be calculated from the measured T_1 and NOE values. The line broadening so that spectra cannot be resolved corresponds for all cases to correlation times of the order of 10^{-7}s. Simple considerations of the line width in terms of T_2 indicate that correlation times of this order will correspond to the loss of spectra. Thus, if the postulated transition T_{ll} actually exists, it merely corresponds to a correlation time of 10^{-7}s.

However, a natural explanation can be given for the fact that $\bar{\tau}_c$ is of the order of 10^{-7}s at a temperature 1.2−1.3 times greater than T_g. Plots of the correlation time−temperature data yield curves whose shapes, for each polymer, are similar to each other. The curves are displaced, however, along the temperature axis in relation to the respective glass temperatures of the polymers. When plotting the ^{1}H-NMR correlation frequency against temperature, McCall (30) noted that the curvature observed is characteristic of Williams−Landel−Ferry (WLF) (31) behavior. Nevertheless, no further discussion or analysis was made of this point (30). We have noted very similar behavior with the ^{13}C-correlation time−temperature plots (26). Accordingly, we analyzed the data by the WLF theory, taking the known, accepted glass temperature as the reference temperature in their equation. The data for the amorphous polymers obey this theory and thus offer a natural explanation for the essentially universal ratio of the temperature of spectral collapse to the glass temperature. It is not necessary to invoke the existence of a new transition to explain this ratio. Moreover, for those polymers where data are available, so that a direct comparison can be made, the constants, C_1 and C_2, that appear in the WLF theory are the same for both dynamic mechanical and the ^{13}C-NMR studies. The difference between *cis*-polyisoprene and polyisobutylene, which have identical glass temperatures, is due to a large difference in their respective values for C_2. This fact was known for a long time from dynamic mechanical studies (32) and is the reason that the spectral collapse occurs at two different temperatures.

It is tempting to apply the inverse process to obtain the glass temperatures of controversial semicrystalline polymers. This procedure cannot be carried out with the required accuracy because correlation time data cannot be obtained at sufficiently low temperature prior to spectral collapse. However, one can obtain an upper limit to the glass temperature, which is helpful in clarifying the location of the transition and the nature of others. For example, high-resolution ^{13}C-NMR spectra were obtained for linear polyethylene at temperatures as low as -40 °C (25). Based on the correlation described earlier, this observation immediately places a conservative upper limit on the glass temperature of linear polyethylene. This conclusion is supported by the fact that at this temperature the correlation time is of the order of $10^{-7} - 10^{-8}$s, indicating that glass formation must occur at a substantially lower temperature. These results rule out the identification of the glass temperature with the β-transition for at least the linear polymer. They do not allow, however, for a discrimination or selection to be made in the range -150 to -90 °C, as suggested by many investigations (33).

Dynamic mechanical and other methods have established that the branched polyethylenes display a defined, very intense β-transition. This transition, in the vicinity of $-30 - 0$ °C, is very often identified with the glass temperature for this class of polymers. For several representative polymers in this class, resolvable ^{13}C-NMR spectra could be obtained virtually coincident with, or only a few degrees above, the independently determined β-transition (20). The correlation time at the β-transition is of the order of 10^{-9}s. Both these observations demonstrate that for the branched polyethylenes as well, the β-transition cannot be identified with the glass temperature, although a well-defined relaxation clearly exists. The β-transition is well defined so that the underlying molecular basis for this relaxation still remains to be elucidated.

The examples outlined in this chapter illustrate that the scalar decoupled relaxation parameters are useful in describing the segmental motions in amorphous polymers and in the noncrystalline regions of semicrystalline polymers. These parameters are also particularly helpful in understanding secondary transitions and placing upper limits on the glass temperature. They can be further exploited when chemically different carbons can be studied. Complementary studies involving dipolar decoupling and magic angle spinning, which will allow for probing of the crystalline regions, will enable a better understanding to be developed for the molecular basis of these transitions.

Acknowledgments

The author wishes to thank R. A. Komoroski, D. E. Axelson, J. J. Dechter, and A. H. Dekmezian for their major contributions to this work.

Literature Cited

1. Mandelkern, L. *Polym. Prepr. Am. Chem. Soc. Div. Polym. Sci.* **1981**, *22*(1), 276.
2. Schaefer, J. In "Topics in Carbon-13 NMR Spectroscopy"; Levy, G. C., Ed.; Wiley-Interscience: New York, 1974; Vol. 1, 8, 149.
3. Komoroski, R. A.; Mandelkern, L. In "Applications of Polymer Spectroscopy"; Brame, E. G., Jr., Ed.; Academic: New York, 1978.
4. Axelson, D. E.; Mandelkern, L. In "Carbon-13 NMR in Polymer Science"; Pasika, W. M., Ed.; ACS SYMPOSIUM SERIES No. 103; ACS: Washington, DC, 1979; p. 181.
5. Schaefer, J.; Stejskal, E. O.; Buckdahl, R. *Macromolecules* **1974**, *8*, 291.
6. *Ibid.* **1977**, *10*, 384.
7. Earl, W. L.; VanderHart, D. L. *Macromolecules* **1979**, *12*, 762.
8. Fyfe, C. A.; Lyerla, J. R.; Volksen, W.; Yannona, C. S. *Macromolecules* **1979**, *12*, 757.
9. Mandelkern, L. *Trans. Faraday Soc.* **1979**, *68*, 310.
10. Komoroski, R. A.; Maxfield, J.; Sakaguchi, F.; Mandelkern, L. *Macromolecules* **1977**, *10*, 550.
11. Maxfield, J.; Mandelkern, L. *Macromolecules* **1977**, *10*, 1141.
12. Mandelkern, L.; Maxfield, J. *J. Polym. Sci., Polym. Phys. Ed.* **1979**, *17*, 1913.
13. Mandelkern, L.; Glotin, M.; Benson, R. A. *Macromolecules* **1981**, *14*, 22.
14. Wood, L. A.; Bekkedahl, N. *J. Res. Natl. Bur. Stand.* **1946**, *36*, 489.
15. Mandelkern, L. In "Crystallization of Polymers"; McGraw-Hill: New York, 1964.
16. Komoroski, R. A.; Maxfield, J.; Mandelkern, L. *Macromolecules* **1977**, *10*, 545.
17. Dechter, J. J.; Mandelkern, L. *J. Polym. Sci., Polym. Lett. Ed.* **1979**, *17*, 317.
18. Mandelkern, L. *Trans. Faraday Soc.* **1979**, *68*, 454.
19. Inoue, Y.; Nishioka, A.; Chujo, R. *Makromol. Chem.* **1973**, *168*, 163.
20. Dechter, J. J.; Axelson, D. E.; Dekmezian, A.; Glotin, M.; Mandelkern, L. *J. Polym. Sci., Polym. Phys. Ed.* **1982**, *20*, 641.
21. Dechter, J. J.; Komoroski, R. A.; Axelson, D. E.; Mandelkern, L. *J. Polym. Sci., Polym. Phys. Ed.* **1981**, *19*, 631.
22. Schaefer, J. *Macromolecules* **1972**, *5*, 427.
23. Jelinski, L. W.; Schilling, F. C.; Bovey, F. A. *Macromolecules* **1981**, *14*, 581.

24. Axelson, D. E., unpublished data.
25. Axelson, D. E.; Mandelkern, L. *J. Polym. Sci., Polym. Phys. Ed.* **1978**, *16*, 1135.
26. Dekmezian, A.; Axelson, D. E.; Mandelkern, L., in press.
27. Boyer, R. F.; Heeschen, J. P.; Gilham, J. K. *J. Polym. Sci., Polym. Phys. Ed.* **1981**, *19*, 13.
28. Neumann, R. M.; MacKnight, W. J. *J. Polym. Sci., Polym. Phys. Ed.* **1981**, *19*, 369.
29. Neumann, R. M.; MacKnight, W. J. *Polym. Eng. Sci.* **1978**, *18*, 624.
30. McCall, D. W. *NBS Spec. Publ. (U.S.)* **1969**, *310*, 475.
31. Williams, M.; Landel, R. F.; Ferry, J. D. *J. Am. Chem. Soc.* **1955**, *77*, 3701.
32. Ferry, J. D. In "Viscoelastic Properties of Polymers," 2nd ed.; Wiley: New York, 1971.
33. Davis, G. T.; Eby, R. K. *J. Appl. Phys.* **1973**, *44*, 4274.

RECEIVED for review October 14, 1981. ACCEPTED January 21, 1982.

^{13}C-NMR Spectra of Cross-Linked Poly(styrene-co-chloromethylstyrene) Gels

WARREN T. FORD[1] and T. BALAKRISHNAN[2]

Oklahoma State University, Department of Chemistry, Stillwater, OK 74078

^{13}C-NMR spin lattice relaxation times, line widths, nuclear Overhauser effects (NOE), and relative signal areas at 25.2 MHz in CDCl$_3$ at 30 °C were measured for copolymers of styrene with 25 wt % chloromethylstyrenes cross-linked with 0−10% divinylbenzene (DVB). As cross-linking increases, T$_1$ is almost invariant, line widths increase markedly, NOE ratios decrease significantly, the aliphatic signal area decreases markedly, and the aromatic signal area remains constant up to 6% cross-linker and decreases with 10% cross-linker. The results are discussed in terms of distributions of correlation times for polymer motions.

SOLUTIONS OF POLYSTYRENE have been investigated extensively by dynamic ^{1}H- and ^{13}C-NMR spectroscopy to characterize rotational motions of the polymer chains. Relaxation times T_1 and T_2 and nuclear Overhauser enhancement (NOE) factors of ^{13}C require models with broad distributions of correlation times to fit the data (*1−6*). Solid polystyrene also has been investigated by ^{13}C-NMR spectroscopy by using cross-polarization (CP) and magic angle spinning (MAS) to find rotating frame correlation times ($\tau_{1\rho}$), CP relaxation times (τ_{CH}), and NOE values that also are interpreted as due to broad distributions of motional frequencies (*7, 8*). We report here ^{13}C-NMR relaxation measurements on styrene copolymer gels that are insoluble because of divinylbenzene (DVB) cross-linking, but are highly swollen by CDCl$_3$ to give motional freedom more like that of liquids than that of solids. In previous ^{13}C-NMR investigations of polymer gels, high resolution spectra were obtained in good solvents without recourse to CP, dipo-

[1] To whom correspondence should be addressed.
[2] On leave, 1980−81, from University of Madras, India.

lar decoupling, and MAS (9−11). Doskocilova and co-workers (12−14) performed MAS ^{1}H- and ^{13}C-NMR experiments on a solvent-swollen cross-linked polystyrene and interpreted the spectra in terms of narrow line width signals, typical of polymers in solution, superposed on dipolar broadened signals, typical of glassy polymers.

Cross-linked polystyrenes commonly are used as polymeric supports for synthesis and catalysis, ion exchange resins, and size exclusion chromatography column packings. Interest in polymer-supported synthesis and catalysis prompted us to study the motional behavior of the cross-linked styrene−chloromethylstyrene copolymer gels we had prepared for catalyst supports. These samples cover a range of cross-linking from 0.5 to 10% by weight divinylbenzene (DVB). The results are the first systematic data on how degree of cross-linking of gels affects their ^{13}C-NMR spectra.

Experimental

All ^{13}C-NMR spectra were obtained with a Varian XL−100(15) spectrometer equipped with a Nicolet TT−100 PFT unit at 25.2 MHz and ambient temperature (ca. 30 °C) under conventional high resolution conditions. Details of sample preparation and spectral conditions are published elsewhere (15).

Results

All copolymers examined contained 25 wt% chloromethylstyrene, styrene, and 0−10% DVB. Spectra were obtained as CDCl$_3$ solutions or CDCl$_3$-swollen gels. Chloroform is an excellent solvent for the copolymers, giving the largest volume expansion and the narrowest ^{13}C-NMR line widths of any solvent we have found.

The line widths at half-height in spectra of copolymer solutions (Table I) do not depend on concentration up to 25 wt%. The broader lines in spectra of cross-linked polymers are due to the cross-links, not to viscosity changes, as can be seen by a comparison of the data on the 25% solution with the data on the 4% cross-linked gel. Cross-linking increases the breadth of the aromatic region of the spectra dramatically as the *ortho-*, *meta-*, and *para-*carbon resonances merge into a single broad band that even overlaps with the *ipso-*carbon resonance at ≥4% cross-linking. Aliphatic carbon signals also become broader as cross-linking increases until with 10% cross-linker no distinct peaks for the backbone methine and chloromethyl carbon atoms can be seen. The aliphatic band appears to broaden less than the aromatic band as cross-linking increases, but the aliphatic signal area decreases markedly as cross-linking increases.

Data for T_1 and NOE ratios are in Table II. As cross-linking increases, there is an apparent minimum in T_1 of the aromatic carbon

Table I. Line Widths in ^{13}C-NMR Spectra of Cross-Linked Styrene/Chloromethylstyrenes in CDCl$_3$

Sample, % DVB	Wt% Polymer in CDCl$_3$[a]	Line width, Hz			
		o, m	p	Methine	CH$_2$Cl
0 (soluble)	10	20.5	14	15	8
0	19	21.5	15	15	8
0	25	21	15	15.5	9
0.5	11	25.5	20	14	9
1.0	10	40	—[b]	16	11.5
2.0	17	95	—[b]	22.5	15.5
4.0	26		180	36	15
6.0	30		240	65	34
10.0	33		470	—[b]	—[b]

[a] Wt % polymer in solvent-swollen beads excluding solvent in interstitial spaces.
[b] Could not be measured.

atoms at 4% DVB, but the largest and smallest T_1 values in the series are almost within experimental error of one another. The aliphatic carbon T_1 values and the chloromethyl carbon T_1 values decrease slightly as the degree of cross-linking increases. Previous reports of polystyrene ^{13}C T_1 values show a small solvent dependence with the longest T_1 values in the least viscous solvents (16–23). Our T_1 values for soluble copolymers in CDCl$_3$ agree within 10% with earlier data on polystyrene solutions in low viscosity solvents at 30–40 °C when spectrometer frequency differences are taken into account. The NOE ratios in Table II for the soluble polymers agree with literature values at 30–40 °C for atactic polystyrene and for ring-methyl derivatives of

Table II. ^{13}C T_1 and NOE Values of Cross-Linked Styrene/Chloromethylstyrenes in CDCl$_3$

Sample, % DVB	T_1, ms[a]					NOE Ratio[b]	
	o, m	p	CH	CH$_2$	CH$_2$Cl	Aromatic	Aliphatic
0	123	106	97	65	141	2.14	2.25
1.0	112	102	93	50	155	2.10	2.25
2.0	100	100	72	56	132	1.57	1.94
4.0	93		73	51	104	1.58	1.85
6.0	113		78	72	110	1.40	1.88
10.0	—[c]		—[c]	—[c]	—[c]	1.55	1.92

[a] Estimated error limits are ±10% or ±10 ms, whichever is greater.
[b] Signal area fully coupled ÷ signal area gated decoupled. Estimated error limits are ±0.2.
[c] Not determined.

polystyrene to ±0.1 (17−23). The NOE ratios decrease as cross-linking increases, and the decrease is greater for the aromatic signals than for the aliphatic signals.

Peak areas per carbon atom for the aliphatic and aromatic resonances in gated decoupled spectra are compared with the peak area per carbon atom of a polyethylene glycol (PEG) internal standard in Table III. Some of the gated decoupled spectra are in Figure 1. The aromatic/PEG area ratio is effectively 1.0 for samples with up to 6% cross-linking, but the aliphatic/PEG and aliphatic/aromatic area ratios decrease as cross-linking increases to 6%. Both aromatic and aliphatic peak areas are reduced at 10% cross-linking. The results were confirmed for some samples by comparisons of peak areas in fully coupled spectra. Agreement between the gated decoupled and fully coupled aliphatic/aromatic area ratios is remarkably good.

The T_1, line width, and NOE data for soluble polystyrenes cannot be explained by a single rotational correlation time (τ_c) for motion of backbone carbon atoms, or by two correlation times for aromatic carbon atoms due to polymer backbone motion and to liberation of the aromatic rings (1−6). Neither can the cross-linked polystyrene ^{13}C relaxation data be explained by a single τ_c. Several methods are available for calculation of distributions of correlation times in soluble polymers (1−6), but we choose not to apply them and to discuss our results only qualitatively because none of the measurements truly represent the entire sample. This point will be elaborated on later.

As cross-linking increases, T_1 values of backbone and aromatic ^{13}C nuclei change little, line widths increase greatly, and NOE ratios decrease significantly. Cross-linking causes low frequency motions in the polymer that affect T_2 and NOE processes but have little effect on T_1 (7). The small decreases in T_1 of the −CH$_2$Cl group as cross-linking increases indicate that those T_1 values are on the motional narrowing side of the minimum value of T_1, attained at $\tau_c \sim 5 \times 10^{-9}$ s in a 25.2-MHz ^{13}C experiment. The internal rotational motion of the −CH$_2$Cl group makes its T_1 longer than that of any other protonated carbon atom in the polymer in all samples.

The line widths in a ^{13}C-NMR spectrum of solid polystyrene obtained with dipolar decoupling and MAS (24) may be compared with the line widths in the spectrum of a cross-linked styrene copolymer gel. In the former case (Figure 5 of Reference 24), the aromatic and aliphatic line widths are about 300 and 900 Hz, yet in the most nearly comparable spectrum in our work, that of the 6% cross-linked polymer in Figure 1, the aromatic line width is 240 Hz with an additional broad shoulder on its low field side, but the methine and CH_2Cl signals in the aliphatic region still clearly are seen and are only 65 and 28 Hz wide. If we assume that the line widths in the spectrum of solid

Table III. Aromatic and Aliphatic ^{13}C-Signal Areas in Cross-Linked Polystyrenes

Sample[a]	Gated Decoupled Spectra[b]			Fully Coupled Spectra[b]
	Aromatic/ PEG	Aliphatic/ PEG	Aliphatic/ Aromatic	Aliphatic/ Aromatic
Polystyrene	0.98	0.82	0.89	0.89
Uncross-linked copolymer				
10% soln	c	c	0.87	c
19% soln	c	c	0.84	c
25% soln	c	c	0.87	0.91
0.5% DVB	0.96	0.84	0.88	0.84
1.0% DVB	1.12	0.86	0.77	0.77
2.0% DVB	1.06	0.76	0.72	0.60
4.0% DVB	0.93	0.50	0.54	0.58
6.0% DVB	1.04	0.45	0.43[d]	f
10.0% DVB	0.48	0.20	0.42[e]	f

[a] See Table I, footnote a.
[b] Signal areas per carbon atom. PEG, MW 1500, is used as internal standard.
[c] Not determined.
[d] A value of 0.51 in a separate determination with shorter delay between acquisitions.
[e] A value of 0.56 in a separate determination with shorter delay between acquisitions.
[f] Could not be measured.

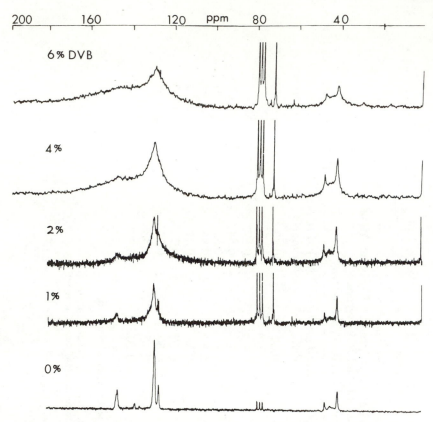

Figure 1. Gated decoupled ^{13}C-NMR spectra of polystyrenes containing 25 wt % chloromethylstyrenes and 0–6% DVB. Spectra of cross-linked samples were taken with a 10-s delay between acquisitions and contain PEG as internal standard. The CDCl$_3$ and PEG peaks have been truncated. The spectrum of soluble polymer was taken with a 1.0-s delay between acquisitions, and the sample contains no PEG. Peak assignments are aromatic C$_1$, 145.3 ppm; aromatic C$_{2356}$, 127.9 ppm; aromatic C$_4$, 125.8 ppm; CH$_2$Cl, 46.4 ppm; backbone methylene, 40–47 ppm; and backbone methine, 40.4 ppm.

polystyrene are due strictly to T_2 processes (all dipolar couplings and chemical shift anisotropies were removed), then comparison with spectra of cross-linked gels indicates that cross-linking limits the aromatic ring motions that affect aromatic ^{13}C T_2 values more than it limits backbone motions responsible for T_2 values of all ^{13}C nuclei. That assumption seems unreasonable. A better explanation for the relative differences in line widths between the spectra of solid polystyrene and of polystyrene gels is that MAS failed to remove all of the dipolar broadening in the aliphatic region of the spectrum of the solid.

A 15-MHz CP/MAS spectrum[3] of our 10% cross-linked copolymer as a dry solid has line widths of about 500 Hz for the protonated aromatic ^{13}C peak and for the aliphatic ^{13}C peak with 2600-Hz spinning. The broader aromatic yet narrower aliphatic line widths in the spectrum of the cross-linked polymer, compared with that of polystyrene in Reference 21, also suggest that 2-KHz MAS failed to remove all of the dipolar broadening in the aliphatic region of polystyrene in the pioneering experiments of Schaefer (*24*).

The most remarkable effect of cross-linking on ^{13}C-NMR spectra of the styrene copolymer gels is the decrease in signal area of the aliphatic resonances relative to both the internal standard and the aromatic signal area. The same effect occurs in gated decoupled spectra that have a pulse repetition time of 11.5 s and in completely coupled spectra that have a pulse repetition time of 0.75 or 1.5 s. Therefore, the signal area losses cannot be due to large increases in T_1 of part of the aliphatic carbon atoms. The relaxation time T_1 increases with correlation time when $\tau_c > 10^8$ s. Previous investigations of fibrous proteins (*25*) and of hydrophilic polymer gels (*10*) also found decreased intensity or absence of some signals in ^{13}C-NMR spectra. The most likely explanation for the disparity in signal areas is that cross-linking limits motion of some of the backbone carbon nuclei so that they are dipolar coupled, and the resulting broad signals are not distinguishable from base line in our spectra. Effectively, the aliphatic/PEG and aromatic/PEG signal area ratios indicate the fractions of aliphatic and aromatic carbon atoms that undergo reorientation at a frequency $>10^6$ s^{-1}. Low frequency motions ($<10^6$ s^{-1}) are present in the backbones of all of the cross-linked polymers. The aromatic carbons maintain rapid reorientation up to a 6% level of cross-linking, but lose signal area with 10% cross-linking. The aromatic carbons can maintain rapid reorientation at higher cross-linking than can the aliphatic carbons because they have an oscillatory motion about the sp^2–sp^3 single bond available to them that the aliphatic carbon atoms lack. This motion is effective in relaxing C_{2356} of the aromatic ring but not C_1 and C_4. Proof for its effect in T_1 processes can be seen in the T_1 data in Table II and in ^{13}C T_1 data on polystyrene in the literature (*16–23*), in which C_{2356} consistently have about a 10% longer value T_1 than does C_4.

The loss of theoretical signal areas with increased cross-linking calls attention to other difficulties in interpretation of NMR relaxation data on polymers. The carbon atoms giving rise to a particular spectral band are structurally heterogeneous due to differences in primary stereochemical sequences and to the many local conformations avail-

---

[3] Obtained by D. O'Donnell, Phillips Petroleum Co.

able to the polymer chains. In the cross-linked styrene copolymers, far greater heterogeneity is found than in homopolymers because seven different monomers (styrene, m-chloromethylstyrene, p-chloromethylstyrene, m-divinylbenzene, p-divinylbenzene, m-ethylstyrene, and p-ethylstyrene) are incorporated into the network, and branches in the network may restrict motional freedom of atoms several bonds away from the branch points. The T_1 values in Table II and most other polymer ^{13}C T_1 values in the literature are based on peak intensities at the maxima of broad spectral bands. These intensities are due to weighed averages of all of the different carbon nuclei giving an NMR signal at the frequency of the band maximum. Those carbon nuclei having the narrowest natural line width contribute most to the intensity at the maximum, and those having the greatest line width contribute least. Dipolar coupled carbon nuclei contribute not at all. All polymer T_1 values measured by peak intensities necessarily weight most heavily the fastest relaxing carbon atoms. We calculated some T_1 values using integrated signal areas instead of intensities at peak maxima, but the results did not change. The parameter T_1 is insensitive to low frequency motions, and is insensitive to rotational correlation times in the experiments of Table II.

Measurements of T_2 suffer from similar difficulties. If the spectral bands are broad, the contribution from field inhomogeneities, T_2^*, to the observed line width is small. The ^{13}C nuclei causing the signal are structurally heterogeneous. Sometimes, individual lines can be resolved for different stereochemical sequences, as in the aromatic C_1 and backbone methylene signals of polystyrene, but when the dispersion of chemical shifts is not great enough to allow resolution, as in the backbone methine signal of polystyrene, one cannot tell how much of the line width at half-height is due to natural T_2 and how much is due to structural heterogeneity. Another problem with the use of line widths to measure T_2 is that they also tend to weight more heavily the carbon atoms having the narrower natural line widths. Carbon nuclei that have longer rotational correlation times give wider bands that contribute less to intensity between the frequencies where intensity is half of its maximum value ($I = I_{max}/2$).

If T_2 was measured by a multipulse method, such as the Carr–Purcell method or one of its improved versions, and peak intensities were used for the T_2 calculation, the results would be biased in favor of the fastest relaxing nuclei. Even T_2 values calculated from integrated signal areas are not true averages if some of the nuclei are dipolar coupled and are not counted in the integrals. NOE measurements, which normally employ integrated signal areas rather than peak intensities, also weight the faster relaxing nuclei more heavily if dipolar coupling causes some of the nuclei not to be counted in the integrals. We conclude that no method can be found to measure relaxation times of heterogeneous polymers, in which some signal is lost by dipolar coupling, that does not bias the measurement in favor of the faster relaxing nuclei.

Even though representative values of T_1, T_2, and NOE in the cross-linked styrene copolymer gels cannot be obtained, the results still enable a qualitative description of the distributions of motions. The signal areas smaller than theory indicate clearly that a fraction of the ^{13}C nuclei are dipolar coupled and have motions that are anisotropic over time periods of $\sim 10^{-6}$ s. The fraction of such nuclei in the polymer backbone increases as cross-linking increases. The loss of signal area is significant in the aromatic band only with 10% cross-linking. In the spectrum of the 10% cross-linked sample, 80% of the aliphatic signal and 58% of the aromatic signal are so broadened that they cannot be distinguished from base line. Increasing line widths and decreasing NOE ratios also indicate that the distributions of correlation times shift to longer values as the degree of cross-linking increases.

Acknowledgments

This research was supported by the U.S. Army Research Office. We thank J. C. Randall, A. A. Jones, and P. G. Schmidt for several helpful discussions.

Literature Cited

1. Wright, D. A.; Axelson, D. E.; Levy, G. C. In "Topics in Carbon-13 NMR Spectroscopy"; Levy, G. C., Ed.; Wiley: New York, 1979; Vol. 3, pp. 104–74.
2. Jones, A. A.; Robinson, G. L.; Gerr, F. E. In "Carbon-13 NMR in Polymer Science"; Pasika, W. M., Ed.; ACS SYMPOSIUM SERIES No. 103, ACS: Washington, D.C., 1979; p. 271.
3. Schaefer, J. *Macromolecules* **1973**, 6, 882.
4. Valeur, B.; Jarry, J. P.; Geny, F.; Monnerie, L. *J. Polym. Sci., Polym. Phys. Ed.* **1975**, *13*, 667, 675, 2251.
5. Jones, A. A.; Stockmayer, W. H. *J. Polym. Sci., Polym. Phys. Ed.* **1977**, *15*, 847.
6. Bendler, J.; Yaris, R. *Macromolecules* **1978**, *11*, 650.
7. Schaefer, J.; Stejskal, E. O. In "Topics in Carbon-13 NMR Spectroscopy"; Levy, G. C., Ed.; Wiley: New York, 1979; Vol. 3, pp. 284–324.
8. Schaefer, J.; Stejskal, E. O.; Buchdahl, R. *Macromolecules* **1977**, *10*, 384.
9. Schaefer, J. *Macromolecules* **1971**, *4*, 110.
10. Yokota, K.; Abe, A.; Hosaka, S.; Sakai, I; Saito, H. *Macromolecules* **1978**, *11*, 95.
11. Manatt, S. L.; Horowitz, D.; Horowitz, R.; Pinnell, R. P. *Anal. Chem.* **1980**, *52*, 1529.
12. Doskocilova, D.; Schneider, B.; Jakes, J. *J. Magn. Reson.* **1978**, *29*, 79.
13. Schneider, B.; Doskocilova, D.; Babka, J.; Ruzicka, Z. *J. Magn. Reson.* **1980**, *37*, 41.

14. Doskocilova, D.; Schneider, B.; Jakes, J. *Polymer* **1980**, *21*, 1185.
15. Ford, W. T.; Balakrishnan, T. *Macromolecules* **1981**, *14*, 284.
16. Allerhand, A.; Hailstone, R. K. *J. Chem. Phys.* **1972**, *56*, 3718.
17. Schaefer, J.; Natusch, D. F. S. *Macromolecules* **1972**, *5*, 416.
18. Heatley, F.; Begum, A. *Polymer* **1976**, *17*, 399.
19. Opella, S. J.; Nelson, D. J.; Jardetzky, O. In "Magnetic Resonance in Colloid and Interface Science"; Resing, H. A.; Wade, C. G., Eds; ACS SYMPOSIUM SERIES No. 34, ACS: Washington, DC, 1976; p. 397.
20. Inoue, Y.; Konno, T.; Chujo, R.; Nishioka, A. *Makromol. Chem.* **1977**, *178*, 2131.
21. Laupetre, F.; Noel, C.; Monnerie, L. *J. Polym. Sci., Polym. Phys. Ed.* **1977**, *15*, 2127.
22. Gronski, W.; Murayama, N. *Makromol. Chem.* **1978**, 179, 1509.
23. Okada, T. *Polym. J.* **1979**, *11*, 843.
24. Schaefer, J.; Stejskal, E. O.; Buchdahl, R. *Macromolecules* **1975**, *8*, 291.
25. Torchia, D. A.; VanderHart, D. L. In "Topics in Carbon-13 NMR Spectroscopy"; Levy, G. C., Ed.; Wiley: New York, 1979; Vol. 3, pp. 325–60.

RECEIVED for review October 14, 1981. ACCEPTED February 22, 1982.

INFRARED SPECTROSCOPY

Probing the Real Structure of Chain Molecules by Vibrational Spectroscopy

GIUSEPPE ZERBI

Istituto Chimica Industriale, Politecnico, Piazza L. Da Vinci, 32, Milan, Italy

The field of vibrational spectroscopy and molecular dynamics is presented and discussed with a review of the basic concepts and results of the more recent techniques developed. Emphasis is placed on the possibilities vibrational spectroscopy offers for providing detailed qualitative and quantitative information on polymer structure, in particular, its use to detect disorder in polymeric materials. The advantages vibrational spectroscopy offers over other physical techniques are mentioned. The motions of the CH_2 group are dealt with primarily. Examples are given of the usefulness of defect calculations for structural analysis. Measurement of the chain length of ordered chains now becomes possible. By selectively deuterating one end of the molecule and by using end-group resonance modes together with CD_2 gap modes, the structural evolution of both ends and tail of a hydrocarbon chain can be followed. Various aspects of vibrational spectroscopy, for example, factor group splitting and the effect of Fermi resonances on Raman spectra are discussed. The state-of-the-art interpretation of the Raman spectrum of polymethylene chains in the all trans *structure is given. Specific phenomena such as the physical transport of matter across crystal boundaries during annealing of polymer solids can be studied.*

C HEMISTS WHO NEED to characterize a given polymer sample or wish to probe the structure of a polymeric material are asked generally to use all the physical or physicochemical techniques presently available that may help in their endeavor. Each technique has advantages and disadvantages because it may provide very detailed information

0065–2393/83/0203–0487$12.25/0
© 1983 American Chemical Society

for some samples in a physical condition, yet it may be useless for other kinds of samples in a different physical state. Often, limitations and advantages are discussed in terms of a technical vocabulary with which the specialists assume everyone is familiar. Thus, a barrier often is built between the community that develops methods and those who may use such methods.

The purpose of this chapter is to focus on the field of vibrational spectroscopy and molecular dynamics of polymers and to try to build a bridge between specialists and users. We purposely shall avoid lengthy mathematical developments and will make an effort to present and discuss the basic concepts and results of the recent techniques that have been developed in the field of vibrational spectroscopy. We hope our efforts may be of help in the structural characterization of polymeric materials.

This study will emphasize the possibilities vibrational spectroscopy offers for the understanding of the structure of real polymers. The word "real" describes the polymeric materials as they actually are prepared in a university or industrial laboratory and which may contain structural as well as chemical disorder directly connected with the chemical or physical treatments that were used during their preparation. Obviously, the reality of a polymeric material as described by a given technique depends on the capability of a given technique to go from overall information to deeper and deeper detail on the molecular scale.

We wish here to (1) analyze how far vibrational spectroscopy and molecular dynamics can go at present in providing detailed qualitative and quantitative structural information on polymers; and (2) illustrate when and how vibrational spectroscopy can be of help in structural analysis when and where other physical techniques fail.

Infinite Perfect Polymers

Polymers as One-Dimensional Crystals. The vibrational analysis of polymer molecules considered as objects with perfect structures has been discussed elsewhere (1–5). We shall point out only a few aspects later in this chapter when dealing with polymers with nonperfect structures.

Let us assume that during the synthesis of the polymer:

1. The chemical reaction performs the desired polymerization with 100% yield, and the linking of the chemical units in the polymer chain (head-to-tail, etc.) occurs without mistakes.

2. If a stereospecific polymerization is taking place, it yields a polymer with perfect tacticity.

3. The intramolecular atom–atom interactions in such a

chemically perfect molecule generate a minimum-energy structure that corresponds to a perfectly regular chain with a given overall shape (planar zig-zag, helical, etc.).

4. The polymer chains pack with a perfect order in a tridimensional lattice. At present, problems of chain folding in single crystals, and other problems are neglected intentionally.

Because intramolecular forces and couplings in organic molecules are much stronger than intermolecular forces, the whole vibrational spectroscopy of polymers first considered the polymer molecule as prepared in Steps 1 through 3 just listed. The decision to study first the polymer molecule as a single chain in an empty space is justified because the intermolecular interactions that occur when the molecule is embedded in a tridimensional lattice are very weak and act as small perturbations on the normal vibrations of the isolated chain where covalent bonding is strongly making up the whole structure.

Such a structurally perfect, infinite, and isolated polymer chain is called a one-dimensional crystal. Indeed, if an operation of roto-translation or screw motion $R(\theta, \delta)$ (where θ is a rotation about the chain axis and δ is a translation along the same axis) is applied to the starting chemical repeating unit, we generate an infinite one-dimensional crystal in which a one-dimensional unit cell with repeat distance d can be located. The one-dimensional unit cell may contain n chemical units organized in a helix with m turns.

Geometrical relationships among the structural parameters θ, δ, and d were fully discussed by structural chemists (6). According to the values of θ and δ, we may find that the one-dimensional crystal is a planar zig-zag chain ($\theta = \pi$, $\delta = d/2$, polyethylene) or a slightly twisted helix ($\theta \approx 24\pi/15$, polytetrafluoroethylene) or a highly twisted helix (e.g., $\theta = \pi/2$, $\delta = d/2$, orthorhombic polyoxymethylene; $\theta = 2\pi/3$, $\delta = d/3$, isotactic polypropylene).

The chemical repeat unit may not coincide with the crystallographic repeat unit even in the one-dimensional crystal. The simplest example is single-chain polyethylene, which consists of a planar zig-zag chain (point group D_{2h}) in which the single CH_2 unit is the chemical repeat unit, and the one-dimensional crystallographic repeat unit is made up of two CH_2 groups in *trans* conformation. A more complex case is that of hexagonal polyoxymethylene, whose one-dimensional crystallographic repeat unit contains nine chemical units coiled in a chain with five turns.

The vibrational analysis of a perfect polymer chain considered as a one-dimensional crystal has to account for the very large number of vibrations that occur in such a system. For an isolated molecule, if N is the number of atoms then $3N - 6$ degrees of vibrational freedom or normal modes must be described. Analogously, for an infinite polymer chain, an infinite number of normal modes that occur simultaneously must be accounted for. As in a small finite molecule, the overall vibrational motion of the polymer chain can be described as the superposition of an infinite number of normal modes. The fact that the observed IR and Raman spectra are generally very simple suggests that very few of these innumerous normal modes contribute to the absorption of IR light or to the Raman scattering process, even if physically all the other modes do exist and perform their duties in making the whole molecule vibrate and in helping to determine several physical properties (7).

Well-established theories of molecular dynamics (8) show that if the assumption that atoms vibrate in a harmonic potential is made, the equation of motion can be solved ending with the solution of an eigenvalue equation:

$$G \, F \, L = L \, \Lambda \tag{1}$$

The vibrational coordinates chosen to set up Equation 1 are the traditional vibrational internal displacement coordinates (8).

In Equation 1, the matrix G contains all the information on the geometry of the molecule and the mass of the atoms, F contains the information on the intramolecular potential in the harmonic approximation, and Λ is a diagonal matrix giving the vibrational frequencies λ_K.

The solution of the secular determinant

$$|G \, F - E \, \Lambda| = 0 \tag{2}$$

provides the vibrational frequencies $\nu_K(\text{cm}^{-1}) = [\lambda_K/(4\pi^2 c^2)]^{1/2}$ (where c is the speed of light) and finally, with some more calculations, one can determine the vibrational displacements, L_i^K, of each atom, thus being able to know the shape of each of the $3N - 6$ normal vibrations (8).

The solution of Equations 1 or 2 requires knowledge of the geometry of the molecule and of the vibrational potential. Although the geometrical parameters may be derived from other experiments, the vibrational force field is derived from the analysis of the vibrational spectra of similar molecules (9, 10). However, a discussion on this important point is outside the scope of this study. The reader is referred to the relevant literature on molecular potentials (8–11) for deriving a critical evaluation of the reliability of the various intramolecular potentials proposed in the literature. As a working hypothesis, we assume for the time being that for polymers the vibrational potential is known to such an extent as to allow reliable calculations to be performed (9, 12−15).

For the calculation of the vibrational normal modes, Q_K, of a polymer chain, one must be able to treat the vibrations of an infinite one-dimensional crystal that can generate an infinite number of normal modes. Let N be the number of atoms in the chemical repeat unit and P the number of units, P being very large.

To make mathematics more directly understandable, the physics of the phenomenon of the vibrations of an infinite polymer chain are examined briefly. Taking the nine normal vibrations of the $-CH_2-$ unit (three masses, nine vibrational degrees of freedom), let us focus on the CH_2 symmetric stretching mode, ν_s (CH_2). If a long chain of CH_2 units is constructed with a well-defined and translationally regular geometry of the chain, by exciting the ν_s (CH_2) mode of the first unit such a motion is transmitted to the next CH_2 and to the next one, and so on along the chain (16). By imposing a certain phase relationship, ϕ, between the vibrational displacements of the first unit with respect to the nearest neighbor and then to the second nearest one and so on, a cooperative normal mode, $Q^\phi_{CH_2s}$, is established throughout the whole chain. Such a mode is a standing wave-like mode with wavelength $\lambda = 2(p/j)d$ (where $j = 0, \frac{1}{2} \ldots P^{-1}$) during which all atoms move around their equilibrium position with the same frequency, $\nu(\phi)$, such as in any normal mode of a small molecule (2, 3). The main variable in this problem is the phase difference, ϕ, between neighboring units; ϕ can take any value from 0 to 2π. The corresponding frequencies, $\nu_{CH_2s}(\phi)$, of the CH_2 symmetric stretching mode, taken before as an example, will vary smoothly depending on the intramolecular coupling between oscillators. Such a coupling is described again by the G and F matrices as for small molecules. The plot of ν vs. ϕ collects the infinite normal modes of ν_s (CH_2) into a curve called a dispersion curve.

Because nine degrees of vibrational freedom exist for the single CH_2 group, in a similar manner nine dispersion curves will be constructed throughout the same range of values of ϕ.

The actual calculation of the dispersion curves can be made by using an eigenvalue equation similar to that of Equation 1, where each of the quantities now becomes phase dependent, that is, they become modulated by the phase factor ϕ (14−17).

$$G\,(\phi)\, F\,(\phi)\, L\,(\phi) = L\,(\phi)\, \Lambda\,(\phi) \tag{3}$$

In this case

$$G\,(\phi) = G^\circ + \sum^s G^s\, e^{is\phi} + \sum^s \tilde{G}^s\, e^{-is\phi} \tag{4a}$$

$$F\,(\phi) = F^\circ + \sum^s F^s\, e^{is\phi} + \sum^s \tilde{F}^s\, e^{-is\phi} \tag{4b}$$

in Equations 4a and 4b, matrices $G°$ and $F°$ correspond to those of the isolated $-CH_2-$ unit, and G^s and F^s describe the interactions of each CH_2 with its neighbor s units away on both sides.

Dispersion curves were calculated for several one-dimensional perfect polymer chains (14, 15, 18, 19) and they provided basic information on the coupling of intramolecular vibrations. In turn, the experimental knowledge of the dispersion curves provides the information necessary for the precise definition of the intramolecular mechanics of these types of molecules.[1]

In Equations 4a and 4b, the index, s, accounts for the distance of intramolecular interactions between chemical repeating units. The precise knowledge of s so far has been neglected generally because for most of the covalent simple and classical polymers considered, s does not extend further than the second nearest neighbor (12). This phenomenon means that the intramolecular interatomic interactions for most of the atoms organized in a network of covalent bonds do not extend too far, but fall off very quickly, as expected. This behavior is not the case with more recently prepared organic polymers, such as polyacetylene, that become conducting or semiconducting when suitably doped with different substances.

Experiments show that the delocalization of the π-electrons in these systems extends very far along the chain and that small perturbations at one site of the chain are felt at large distances further along the molecule. For practical reasons, both spectroscopy (21–24) and quantum mechanics (25) in their calculations must truncate their interactions at some value of s. For these systems, the actual extent of interaction is not yet known; its knowledge should be very relevant to the understanding of the electrical properties of these systems and to the related materials in physics and biology.

To be more precise in the definition of the dispersion curves for a one-dimensional crystal (2), one can state that if N is the number of atoms per chemical repeat unit, $3N$ roots of the secular determinant are calculated for each value of ϕ using Equation 3. Then, $3N$ branches can be drawn in the dispersion curve when plotted as a function of the phase shift. Two of these branches tend to $\nu = 0$ cm^{-1} when ϕ tends to 0. Depending on the geometry of the polymer chain, one of these branches may again reach $\nu = 0$ for $-\pi < \phi < \pi$. The dispersion curves are periodic functions of ϕ with period 2π. Hence, it is only necessary to consider the dispersion curves in the interval $-\pi < \phi < \pi$.

Approximately half of these solutions are enough and are plotted

---

[1]Experimental dispersion curves generally are derived from neutron scattering experiments; discussion of this technique is outside the scope of this study.

generally in the interval $0 < \phi < \pi$ *(14, 15, 18)*. The two branches in the dispersion relation that reach zero at $\phi = 0$ are called acoustical branches because at $\phi = 0$ they describe a rigid motion of the chain with the center of mass of the unit cell moving in phase with all the others.

For the normal modes (generally called phonons) along the $3N - 2$ optical branches, the center of mass remains undisplaced for all values of ϕ, including $\phi = 0$. The origin of the name "optical branches" is because some phonons may interact with the electromagnetic wave thus absorbing and/or scattering light of proper frequency.

Once the infinite normal frequencies (phonon frequencies) of the one-dimensional crystal are organized along the many branches of the dispersion relation, one must find which of these phonons interact with an electromagnetic field thus absorbing IR light or scattering visible light in Raman experiments. Selection rules were worked out *(16)* and state that if θ is the angle with which units are rotated in the generation of the whole chain by the screw motion $R(\theta, \delta)$, then the normal modes with phase difference $\phi = 0$ and $\phi = \theta$ are IR-active; motions with $\phi = 0, \theta$, and 2θ are Raman-active. Using the language of solid-state physics, normal modes corresponding to $\phi = 0, \theta$, and 2θ in the most classical cases possess wave vector $k = 0$ in reciprocal space, ($k = \phi/d$). For brevity, these normal modes will be called "$k = 0$ phonons."

If a stretch-oriented sample is examined with polarized IR light, phonons with $\phi = 0$ have a transition moment parallel to the drawing direction that generally coincides with the chain direction. Phonons with $\phi = \theta$ show a transition moment generally perpendicular to the chain axis *(1-2)*. Normally by extrusion or by stretching, molecular chains are oriented predominantly parallel to the chain axis but are oriented randomly in a plane orthogonal to the chain axis. Full orientation along the three crystallographic planes for IR studies only has been achieved on samples of syndiotactic polypropylene *(26)*.

A fully tridimensionally oriented specimen of single crystal mats of polyethylene with a very small mosaic spread was reported *(27)*, but this finding has not been repeated by others.

Oriented specimens as extruded rods also were examined by Raman spectroscopy in different scattering geometries *(28-30)*. By using these experimental techniques, detailed experimental information on the symmetry species of the spectroscopically active phonons can be obtained. The classification into symmetry species makes possible an improvement in the matching between frequencies predicted by theories and those observed in actual experiments.

IR and Raman experiments and calculations have been performed by numerous investigators in the last twenty years on many polymeric

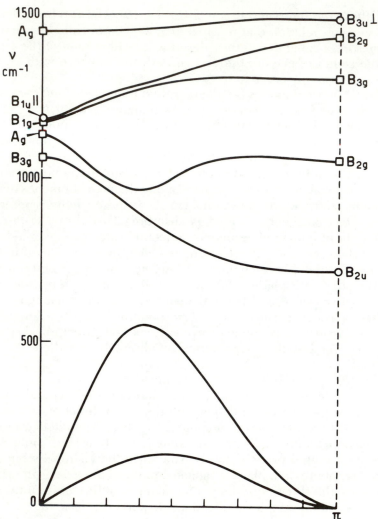

Figure 1. Dispersion curves of single-chain polyethylene in the 1500–
0-cm⁻¹ range, calculated with the force field of Ref. 88. Spectroscopic activity for k = 0 modes (φ = 0, π) is indicated. Direction of
transition moments for IR-active modes in a stretch-oriented sample
are given. The symbols ‖ and ⊥ mean transition moments parallel and
perpendicular to the stretching directions, respectively. Key: ○, IR;
and □, Raman.

systems considered to be perfect, one-dimensional crystals. Calculations and experiments were improved along with the technological
development of computers and of IR and Raman spectrometers (2, 31)
(Figure 1).

A comment on the relevance of such types of studies in polymer science seems appropriate. In particular, we would like to examine briefly the types of information derived from the study of the vibrational spectra of polymers as perfect, one-dimensional crystals and compare them with the information derived from other physical techniques. We neglect in this discussion the relevance of these kinds of studies to the specialized basic field of molecular and crystal dynamics but will analyze their relevance for polymer chemists.

The first observation of practical importance is that theories of polymer dynamics provided the way to interpret in detail the observed spectra of polymers that otherwise could not be interpreted in terms of the well-known spectroscopic correlations (*32, 33*). Indeed, as seen before, the few spectroscopically active phonons along the dispersion branches may occur in spectral regions that are away from those predicted by spectral correlations. Their frequencies depend on the extent of intramolecular couplings, which are, in turn, related to the geometrical shape of the polymer chain.

Hence, those spectroscopically active phonons generate bands in IR and/or Raman spectra that are very characteristic of a polymer chain coiled in a specific conformational structure, which is supposed to extend for a considerable length in space.

The IR and/or Raman bands observed thus provide direct information that the chain has taken up a conformationally regular structure in space (one-dimensional translation periodicity) (*34*) as generated by a rototranslational operator with well-defined values for θ and δ. The so-called bands due to $k = 0$ phonons that appear in the spectrum on careful slow solidification from melt or crystallization from solution are called "regularity bands" (*34*) because they refer to a specific operator $R(\theta,\delta)$. In addition, $k = 0$ modes of stretch-oriented samples can provide symmetry species and, possibly, information on δ and θ, hence, on the shape of the molecule. On melting, these bands disappear almost completely because most of the conformational regularity collapses (*34, 35*). If some weak bands still remain, it means that sections of chains in the melt (or solution) are still helical with the same $R(\theta, \delta)$ still operating on a short range (*36*). Determination of the length of short regular segments in a liquid polymer (or in solution) is still under investigation.

On melting or dissolving, Step 3 in the formation of the polymer is removed and the spectrum of a substance formed during Steps 1 and 2 remains. This spectrum provides information on chemical and stereochemical structure. In such a case, classical spectroscopic correlation tables again may be of help. A full analysis of this problem has been carried out recently for two samples of isotactic and syndiotactic polypropylene in the melt and solid state (*30, 37*).

The use of regularity bands (or $k = 0$ modes) for a quick determination of the geometry of a polymer chain in the one-dimensional state is not straightforward and easy. The choice of a model that will be consistent with the observed spectra must be based on a careful comparison of experiments and calculations that use carefully selected vibrational potentials. x-Ray diffraction studies provided more detailed information on the structure of many polymers (31). In a few cases, however, the shape of the molecular chain first was determined from vibrational studies (38).

The polymorphic form of polyvinylidene fluoride was detected by IR spectroscopy, and the shapes of the two modifications first were determined from IR spectra (39). The most recent example of the application of IR spectroscopy to the structural studies of polymers is that of polyacetylene either in the *cis* or *trans* form. The structures of both modifications were determined entirely from IR and Raman studies (40); however, x-ray diffraction data are hindered greatly by the unusual nature of the sample.

Although x-rays may be used to measure interatomic distances more easily, vibrational spectroscopy finds shapes of molecules that are consistent with the experimental spectrum. Calculations of normal modes may come later based on some assumed geometry. If the overall shape of the molecule has a certain evidence, refinement of force field and geometrical parameters to fit the experimental spectrum may provide somewhat better geometrical parameters.

Polymers in Three Dimensions. Step 4 in the building of the perfect polymer material consists of packing the molecular chains into a tridimensional lattice. When organic molecules crystallize, the vibrational spectrum shows new bands in the IR or Raman spectra that are related directly to the arrangements of the molecules in the crystal. The considerations previously made for one-dimensional crystals must now be extended to three dimensions.

Tridimensional periodicity is reached by minimization of intermolecular interactions, and the molecules will adjust to form the crystal generating a tridimensional pattern that can be described by its symmetry properties in space and by the number of molecules in the repeating unit cell. Polymer molecules, by controlled experiments, can crystallize in tridimensional networks thus adding to the already discussed one-dimensional periodicity along the chain and the periodicity across chains.

The vibrational motions propagating along the chain will then be perturbed slightly by the weak intermolecular forces and new vibrations (phonons) will be generated transverse to the chain axis, with molecules bouncing or sliding rigidly against each other. Dispersion curves still can be calculated for phonons propagating within the crystal along all possible directions (2, 41).

Again, only a few phonons will be spectroscopically active according to the symmetry of the crystals and according to the number of molecules in the tridimensional crystallographic unit cell. The concepts developed for small molecules fully apply to crystalline polymers (Figure 2). Let $k = 0$ modes of the one-dimensional crystal be considered, in a first approximation, internal modes of the molecule.

Because of intermolecular (interchain) coupling, each internal mode splits into multiplets, with the maximum degree of multiplicity being equal to the number of chains per unit cell. The symmetry of the crystal may reduce the multiplicity of the corresponding bands observed in the spectra either because some transitions may be symmetry forbidden or because vibrations may be degenerate (*42, 43*). Such a splitting is generally called "factor group splitting," with the extent of splitting being due to the strength of the intermolecular interactions (*43*). These intermolecular interactions are different for different normal modes because the extent of interaction depends on the vibrational amplitude for each normal mode. The larger the amplitude (atoms move nearer to each other), the larger the splitting. Because intermolecular forces are weak, splittings will be generally small and, for the reasons just mentioned, may be unobserved for many normal modes that do not bring atoms of one chain in close contact with atoms of the neighboring chains.[2] "External modes," owing to the motions of the rigid chains (or quasi-rigid chains) one against the other, give rise to external lattice modes that occur in the very low frequency region of the spectrum (*42, 43*).

Because the concept of factor group splitting is of great help in understanding the physical properties of polymethylene chains (discussed later), a description of the same phenomenon from another viewpoint is given.

If two energy levels of proper symmetry and described by the corresponding wavefunctions accidentally coincide or occur very near in energy, the wavefunctions may mix and the levels may repel each other. The extent of repulsion depends on the extent of mixing of the starting wavefunctions. The extent of mixing (hence of the splitting) slowly decreases and disappears when energies no longer match and move away from each other. Factor group splitting originates from precisely this situation.

The vibrational energy level for a given normal mode for one isolated molecule (natural frequency) coincides in energy with that of the neighboring unit(s) in the unit cell. If symmetry allows, repulsion takes place and a doublet (or multiplet) occurs. The coupling of the

---

[2] As an example, in crystalline orthorhombic polyethylene, only CH_2 bending and rocking show a nonnegligible splitting (~ 25 cm^{-1}).

SINGLE CHAIN   ORTHORHOMBIC CRYSTAL

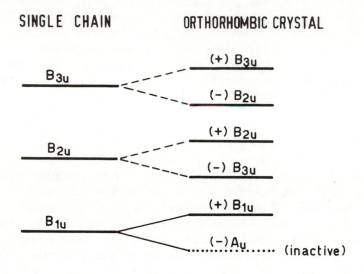

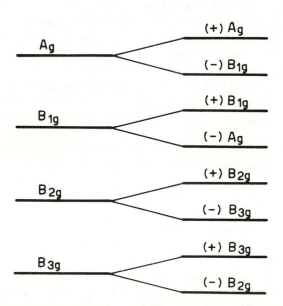

Figure 2. Scheme for factor group splitting in poluethylene. Key: top, IR; and bottom, Raman spectra. (See Refs. 28 and 29 for definition of reference axes.)

levels is related to the dynamical coupling (F and G matrices of Equation 3) within the crystal.

One of the ways to check whether an observed splitting originates from intermolecular interactions in crystals is to look at the spectra of mixed crystals of isotopic species. Generally, mixed crystals of hydrogenated and deuterated materials were studied because they are more easily available (42, 43). Thus, by isotopic dilution preparation of crystals is possible in which the molecule under study can become a guest in a lattice of the isotopic derivative.

The interatomic potential in isotopic derivatives is the same as that of the parent molecule (44); hence, changes in frequencies are related only to changes of the masses of the atoms (G matrix of Equation 1). For the isotopically diluted molecule, a given energy level (natural frequency) of the guest molecule does not match with that of the molecules in the host lattice and thus remains unperturbed. A single band will then appear somewhere in the middle of the observed doublet (or multiplet), its intensity being proportional to the amount of isolated guests in the host lattice. At very large dilutions, the doublet (multiplet) disappears because only isolated islands in the host lattice exist. The method of isotopic dilution has been used widely in solid-state vibrational spectroscopy of simple organic molecules (42, 43) and has received a great deal of attention in recent years in the field of polymers, mainly by Krimm and coworkers (45, 46). The concept of guest and host molecules will be taken up again later in a somewhat different context.

In the field of polymer spectroscopy, factor group splitting and external lattice modes are of particular importance for the determination of the crystallinity of a polymer sample. The determination of the crystallinity of polymers has been reviewed extensively, and a reanalysis of the problem is outside the scope of this work (2, 34).

The only important aspect is that vibrational spectroscopy allows the differentiation between one-dimensional and tridimensional order. In fact, $k = 0$ modes from an isolated chain (i.e., regularity bands) inform us that the polymer chain has coiled up in a given conformation. When regularity bands split into a multiplet and, possibly, lattice modes are observed (crystallinity bands), we are informed that a tridimensional periodicity has been formed. The concept of regularity and crystallinity bands is not new (34), but it seems not yet well accepted in practical polymer chemistry and physics where often regularity bands still are used to study and measure crystallinity. A particular feature of the solid state of a polymeric material has to be pointed out. Researchers report (47) that by suitable thermal treatments polymer solids do not immediately pack in tridimensional lattices but may form aggregates of polymer chains with one-dimensional longitudinal order but with no lateral periodicity. This phase generally has been named "smectic." According to the previous discussion, the spectrum of the smectic form of a polymer clearly will only show regularity bands while crystallinity bands will not occur. By proper annealing of the material, chains move inside the crystal and adjust to form a tridimensional arrangement. At this stage, crystallinity bands (factor group splitting and lattice modes) will occur.

Experimentally, for most of the polymers studied, regularity bands have been observed. A textbook case for vibrational spectroscopy of polymers is that of isotactic polypropylene (Figure 3). For such a polymer, Steps 1 through 4 in the formation of the polymer can be followed separately. Tacticity bands are observed for the spectrum in melt and solution and allow for differentiation between the syndiotactic derivative (37).

The coiling of the chain in a three-fold helix isolated from the neighboring chain is followed by observing the appearance of regularity bands when a liquid sample is quenched into the smectic form (47). Factor group splittings testifying to the formation of an ordered tridimensional arrangement recently were observed in IR and Raman spectra after suitable annealing of the smectic sample (48) (Figure 3).

Disordered Polymers

The vibrational structural analysis of polymers, when considered as an infinite and perfect system, showed quickly its weakness because all polymers in their real structure contain a large amount of matter in a state of disorder. An understanding of the erroneously called "amorphous" bands then becomes essential. Even single crystals of polymers contain an intrinsic disorder because the very long molecular chain has to fold back and forth to crystallize (7). Whatever may be the type of folding (49), a conformational disorder certainly must be considered because the molecule in a single crystal is different from the regular one generated by the operation $R(\theta,\delta)$. This symmetry operation still acts in the region of the crystallites where chains can still be considered as one-dimensional crystals.

As discussed previously (50–54), all four steps (1 through 4) in the make-up of a polymeric material generally fail in performing their duties. In reference to the four steps discussed earlier, the consequences of a nonideal polymerization are the following:

1. Inversion of type of linking occurs.
2. Stereoregularity is affected.
3. Conformational disorder is introduced.
4. Stacking faults, dislocations, etc., easily occur.

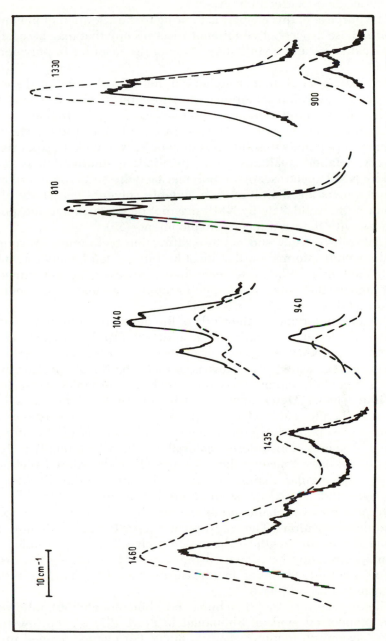

Figure 3. *Raman spectrum of isotatic polypropylene from smectic (one-dimensional) to crystalline (tridimensional) orders. Raman regularity bands split into doublets due to crystallinity.*

Numerous studies of these types of errors were reported using different physicochemical techniques.

The purpose of this chapter is to bring to the attention of polymer chemists those aspects of vibrational spectroscopy that may be useful in the qualitative and quantitative study of the disorder in polymeric materials.

The main goal of the researchers in the field of vibrational spectroscopy is to develop vibrational spectroscopy in such a way as to provide information that other techniques cannot give. Indeed, competition, for example, with NMR spectroscopy in the study of stereoregularity of polymers in solution is unnecessary. Quick, inexpensive, and very detailed qualitative and quantitative information is provided by NMR spectroscopy on polymer tacticity and type of linking when polymers are dissolved in a suitable solvent. The study of a polymer in the solid state by NMR may not be so quick and inexpensive even with the most recent developments (55).

Analogously, x-ray and neutron diffraction techniques on solids provide overall information that must be interpreted by suitable molecular modeling (49). These techniques, however, do not provide direct signals that arise from specific atomic groups in a specific structure (49).

Of more importance is the study of the detailed conformation of a polymeric material, especially in the solid state. In this case, the contributions from NMR, x-ray, and/or neutron scattering are not as specific. Therefore, we wish to show here that vibrational spectroscopy can provide detailed information on a much smaller molecular scale.

Finite Chains: Determination of Chain Length. All models that try to describe the reality of a solid polymer consider the existence of segments of supposedly perfect chains and domains of disordered material. Domains of disordered material may be of different type, but generally one may conceive the existence (1) of disordered sections that interrupt regular chains; and (2) of domains where disordered matter surrounds crystallites with ordered structure (7).

Ordered polymer chains are interrupted by small or large sections of disordered material. The question we wish to answer is how to measure the chain length of the ordered chains. Recent studies of Raman spectroscopy have opened the way to determining the existence of such finite chains and to measuring their length either in the solid or in the liquid phase.

The original idea by Mizushima and Shimanouchi (56), followed by the pioneering work of Shimanouchi et al. (57) on the low frequency "accordion motion" of finite molecular chains, opened up a new field of Raman spectroscopy. The simplest case of polymethylene chains that comprise all normal hydrocarbons with increasing

length and that end with polyethylene will be considered first. The
second to the last acoustical branch (ω_5) in the dispersion curve of
single-chain polyethylene (*14, 15, 58*) refers to the normal modes that
are related to the cooperative stretching of the C−C bonds and the
bending of the CCC angles. For finite chains of various lengths, the
normal mode with only one node located in the middle of the chain
produces the lowest frequency lying on this branch (Figure 4). The
plot of the corresponding eigenvectors shows that this mode is just
that of an accordion that extends and compresses the molecule along
the chain axis leaving its center unshifted. It is then a totally symmetric Raman-active mode.

If the model of the additivity of bond polarizabilities is accepted
(*59, 60*), the sum of the contributions from each C−C bond predicts
that the change in polarizability, hence the Raman intensity, must be
very large. This prediction is experimentally verified (*57, 58*). Indeed,
for polymethylene chains such a low-frequency mode scatters light
more than any other normal mode.

Because the slope of the acoustical branch is directly related to
the longitudinal elastic coefficient of the system, Shimanouchi (*58*)
compares the normal hydrocarbon chain to an elastic rod, with Young's
modulus E, uniform density ρ, and length L. The frequency of the
longitudinal acoustic mode (LAM) is then related to its length by the
following relationship:

$$\omega \ (\text{cm}^{-1}) = \frac{1}{2L}\sqrt{\frac{E}{\rho}}$$

The observation of such a strong Raman line at low frequencies for
polymethylene chains allows the direct determination of their length,
L, once reasonable values for E and ρ are assumed. Such a discovery
has made LAM spectroscopy a very powerful tool for the determination of chain lengths in molecules or polymers of polymethylene.

Comparison between chain lengths from LAM spectroscopy and
small angle x-ray or neutron scattering allows the measurement of the
angle of tilt of the molecular chains with respect to the lamella surface
in single crystals (*61*). The direct use of the straightforward formula by
Shimanouchi provided important information in morphological studies of polyethylene (*62, 63*). Moreover, study of the profile of the
Raman line made possible the determination of the chain-length distribution in samples of polyethylene (*64*).

The extension of Shimanouchi's method from *n*-hydrocarbons to
polyethylene implies that the disordered matter at both ends of the
perfect stems in polyethylene crystals does not affect both frequencies
and intensities of the LAM mode. This problem is not yet fully solved.
Polymethylene chains may be terminated by different kinds of atomic

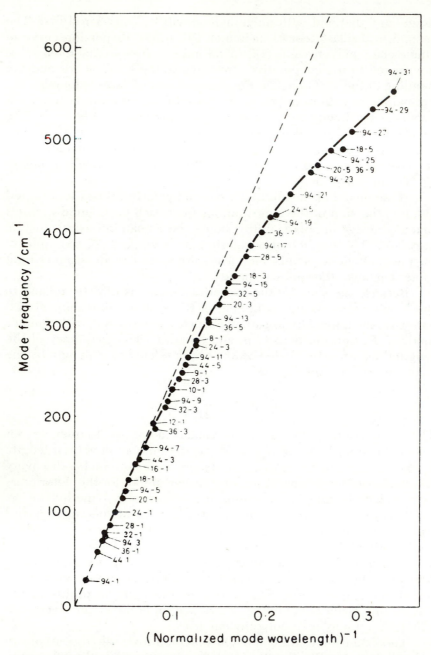

Figure 4. Lower frequency ω₅ dispersion branch of single-chain poly-ethylene and location of LAM modes for n-alkanes (from Ref. 57). Each point is labeled with two numbers: the first refers to the number of carbon atoms in the chain, the second to the order of the LAM mode. (Reproduced with permission from Ref. 57. Copyright 1967.)

groups for which mass and conformation may be relevant in the perturbation of LAM modes. For solid polyethylene or liquid or amorphous substances with a polymethylene residue, conformational irregularities at both ends may again perturb the LAM mode.

Following the initial idea of Strobl (65), Krimm and coworkers (66–68) considered the problem of the LAM mode of an elastic rod whose ends were perturbed symmetrically by masses and/or forces. The effect of conformational disorder on a realistic molecular model was first considered by Zerbi et al. (69) who showed that the introduction of conformational defects at the end of the polymethylene chain introduces a transversal component in the longitudinal motion, thus affecting frequencies and intensities. Consequently, more than one normal mode in the low-frequency region acquires character of LAM thus giving intensities to other modes lying nearby. This issue has been restated and expanded by Snyder et al. who actually calculated the intensity of LAM modes in liquid n-alkanes and determined the concentrations of rotamers in the same liquid compounds (70).

The problem of the perturbation by the end-groups is not yet completely solved because the contribution of mass and force constant defect at the ends of polymethylene chains are not understood fully. Rabolt discussed long-chain alcohols (71) and recently Minoni and Zerbi (72–74) discussed, in a more general way, fatty acids and the case where molecules join at both ends with some sort of bond, such as in phospholipids and membranes. LAM spectroscopy mainly has been restricted to n-alkanes because these systems are easier to handle. However, LAM modes were observed for isotatic polypropylene (75) and polyoxymethylene (76). The theory of LAM of helical polymers still, however, is undeveloped.

Conformational Disorder

General Concepts. The discussion primarily will focus on the possibility of detecting the conformational disorder in a polymer chain using vibrational spectroscopy. Use of the spectrum implies detailed knowledge, as far as possible, of the extent to which the introduction of some kind of conformational disorder perturbs the normal vibrations of the polymer. The first duty is then to discuss the dynamics of conformationally disordered chains and then to predict, if possible, the observations expected in the IR and/or Raman spectrum.

Just the same concepts, however, can be applied to the study of chemical (77), stereospecific (78, 79), and mass defects (80). As stated earlier, however, chemical and stereospecific defects are studied more easily by NMR methods because these two structural characteristics of the polymer chain also are maintained in solution, thus making NMR an easier and cheaper method of analysis. IR and Raman spectroscopy again become very useful when polymers cannot be dissolved, thus making NMR analysis less effective. Mass defects are particularly interesting when hydrogen atoms are replaced by deuterium atoms in a preassigned way and site in the molecule. (This case will be treated in detail later.)

From a theoretical viewpoint, several investigators tackled the various problems and tried to calculate the shape of the normal modes and the vibrational frequencies of conformationally disordered polymers. Several kinds of conformational disorder are possible, such as the existence of a few well-characterized kinks (jogs, folds, etc.) (7) embedded in a host lattice of translationally regular polymer chain or instead, a distribution of conformational disorder with varying geometry, concentration, and distribution along the chain (7).

The discussion that follows again will center on the simplest case of polymethylene chains that comprises a large class of compounds that has been the subject of extensive research (81). Attempts to extend such kinds of studies to other polymers are still few and very tentative. We first will consider a few, well-isolated conformational defects in a host lattice of translationally regular one-dimensional crystals.

Typical kinks such as $G(T)_nG'$ or $GGTGG$ (where G and T refer to *gauche* and *trans*, respectively), for example, are energetically likely to occur in a polymer (82). The former introduces only a local distortion in the chain, and if n is odd, the trajectory of the chain continues to be straight in the same direction as before, with the chain only lying on a different, but parallel, plane. The latter defect describes the trajectory for a tight fold re-entry of the polyethylene chain along the [200] crystallographic plane.

The perturbations by these kinds of conformational distortions in an otherwise regular one-dimensional lattice were treated mainly in the following ways:

1. By numerical techniques (50–54, 78–80, 83–85).
2. By application of Green's function techniques (86, 87).
3. By studying the spectra of short model compounds (88).

Details of the techniques are discussed in various specialized articles and were reviewed recently (81).

We feel that the numerical technique is best suited for treating realistic cases of real polymers that may contain different types of disorder with any type of distribution and concentration. Our efforts were made along these lines and we shall report here on the basic concepts of these studies. A very long, one-dimensional lattice that happens to host a defect consisting of a few well-characterized conformational defects will be considered. Each defect has its own natural frequency (or frequencies). If such frequency (ies) happen to occur within the band of frequencies of the host lattice (i.e., they occur within the frequency band spanned by the dispersion curves of the perfect, one-dimensional crystal) strong or weak coupling with the other modes may occur, with large or small perturbations of frequencies and displacements. Each type of vibrational mode thus formed is called a "resonance mode" and the corresponding frequency is called a "resonance frequency" because the proper frequency of the defect is resonating with those of the perfect lattice that lie nearby in energy. The resulting atomic displacements of the defect almost may be washed out by the phonons of the host lattice (Figures 5a and 5b) thus introducing a very small perturbation or may instead generate normal modes in which the atomic displacements mostly are clustered around the defect, leaving the motions of the neighboring unit almost unperturbed. Resonance modes (or in-band modes) then can be distinguished from quasi-localized, or pseudoresonance modes. Their occurrence in the spectrum depends on the intrinsic spectroscopic intensity associated with these normal modes. The problem of the intensities in IR and Raman spectroscopy will be dealt with in detail later. If the proper or natural frequency of the defect occurs in an energy region of the spectrum where no other modes lie nearby with which to couple, the vibrational mode of the defect cannot be transmitted by the lattice, and its displacements are damped immediately.

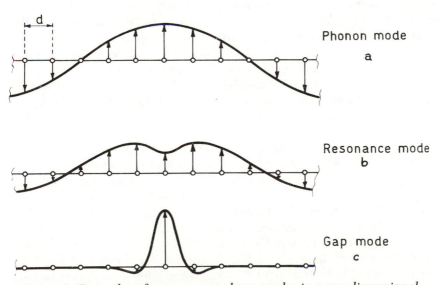

Figure 5. Examples of resonance and gap modes in a one-dimensional lattice modeling a simple, single-chain polymer. Displacements are orthogonal to the chain axis, but the phonon propagates along the axis.

It then follows that the mode is highly localized at the defect (Figure 5c), and the natural frequency is not changed greatly from the hypothetical case of an isolated state. In such a case, one deals with a gap mode or an out-of-band mode because the frequency occurs either in the gap between two dispersion curves in the dispersion relation or above the first dispersion curve of the perfect host lattice.

These concepts are not new because they are derived from solid-state physics (89). The only difference is that solid-state physics generally has dealt with very simple types of impurities that could be treated easily. Organic polymers instead usually contain defects with each one contributing many characteristic frequencies, and the number and distribution of defects may be such that nonnegligible dynamical perturbations and coupling between defects may occur throughout the host lattice.

For this reason, we think that the understanding of real polymers requires the use of a numerical method that is capable of treating realistic cases making analytical methods elegant, but unusable. Indeed, the chemical reality of a polymer must be kept without introducing into the calculations chemically and physically unacceptable simplifications. For this reason, we believe that the method of the negative eigenvalue theorem (NET) proposed by Dean (90) and heavily used by our group is best suited for the study of the spectra and structure of disordered polymers. Details of the method are discussed elsewhere (52, 53).

NET just solves secular equations of very large sizes, thus allowing treatment of disordered polymers as very large molecules. Basically, the same secular equation as in Equation 2 is solved. The extent of intradefect or interdefect coupling is just determined by the shape of the defect and the masses of the atoms (G matrix), and by the intramolecular potential (F matrix). The shape of the normal modes (hence the extent of localization on the defect) is judged after calculation of the vibrational displacements, which can be made by suitable numerical methods (91, 92).

Detailed Conformational Information. We report here a few detailed examples of frequencies due to specific conformational defects in a *trans*-polymethylene chain. This information may be of general use for all molecules that contain polymethylene residues.

RESONANCE MODES. GG *defects: CH_2 wagging.* The simplest case of a defect mode that occurs within the frequency band spanned by the wagging motions of the *trans* host lattice, but which shows a highly localized character, is the CH_2 wagging motion of the CH_2 group isolated between two GG structures. The localization can be understood by the fact that the GG structure introduces a strong deformation of the polymethylene chain thus decoupling the wagging of the *trans* sections from the one localized on the defect. This mode represents the typical case of a pseudolocalized mode generated mostly by molecular geometry. These cases are considered as typical situations of organic solid-state physics. The frequency calculated with the force field (88) is at 1359 cm^{-1}.

GTG *defects: CH$_2$ wagging.* This defect introduces two wagging modes because two CH$_2$ groups are joined by a bond in *trans* conformation and are decoupled partly from the host lattice by *gauche* conformations. The intradefect coupling generates out- and in-phase modes calculated near 1375 and 1309 cm^{-1}, respectively. The calculated values depend slightly on the force field used (93). Based on the force field used by Snyder (88), their values are calculated at 1375 and 1309 cm^{-1}, respectively when the defect is embedded in a long *trans* lattice.

GGTGG *defects: CH$_2$ wagging.* This defect, as already mentioned, is the one that may exist on the surface of the single crystal of polyethylene if the chain re-enters in the crystal with a tight fold re-entry along the [200] crystallographic plane (7, 49). Zerbi et al. have tried to predict where this defect, if it exists, may absorb in the IR or scatter in the Raman. These modes may help eventually in the study of the structure of the surface of single crystal polyethylene which is, at present, a matter of strong debate (49).

Calculations were made and several characteristic resonance modes were described. In the CH$_2$ wagging region, the calculated modes are practically those calculated for the GG and GTG defects when combined in the GGTGG defect by taking the sum and differences (symmetry combinations) with respect to the local two-fold axis crossing the C$\overset{T}{-}$C bond (94).

Other resonance modes for such a defect were calculated in other regions of the spectrum, but they were not as characteristic, with the exception of the CH$_2$ rocking, which will be discussed in the next section.

The experimental test of the calculated numbers of this section can only be made if one studies a molecular system that is known for certain, from other independent techniques, to contain the desired defects rigidly held by the structure of the molecule. x-Ray diffraction work demonstrated that the cyclic molecule C$_{34}$H$_{68}$ contains two GGTGG defects joining two *trans* planar sections (95). The IR spectrum shows absorption bands just near where calculations predict the wagging modes of the GGTGG defect to occur.

Frequency matching, however, may not be such good physical evidence because polymer molecules generally are so complicated that a band can be found in the spectrum wherever one wishes. Application of recent theories to the prediction of IR intensities has provided the prediction of an intensity pattern in very good agreement with the observed one (96) thus giving more selective information and adding more faith in these kinds of calculations.

GAP MODES. *CH$_2$ rocking.* According to the previous discussion, gap modes occur in a frequency region of the spectrum where no other vibrational modes of the perfect host lattice occur. Conformational and configurational gap modes were calculated and observed by Rubcic et al. for configurationally and conformationally disordered polyvinyl chloride in the frequency gap below the CCl stretching (77–79). This calculation was performed to help understand the complex spectrum in this frequency range, which has been the subject of great interest among various groups.

Because this chapter deals mainly with the motions of CH$_2$ groups, the typical gap mode calculated when a head-to-head defect is introduced in a polyvinyl chloride single chain will be mentioned. When the sequence $-CH_2-CH_2-$ between two heavy boundaries occurs in a molecule, a rocking mode is calculated near 850 cm^{-1} (77). This information, however, is not new, because the study of copolymers of ethylene and propylene already empirically showed the existence of an absorption owing to such a group (97).

GGTGG *defects: CH$_2$ rocking.* More interesting for structural diagnosis are the somewhat complex vibrational motions that can be described primarily as CH$_2$ rocking of several CH$_2$ groups on and near the GGTGG mode. Such a vibrational mode is calculated to occur at 714 cm^{-1}, just below the limiting $k = 0$ of the CH$_2$ rocking mode of the all *trans* sequence, in the gap between CH$_2$ rocking and C$-$C bending branches (ω_9 and ω_5), respectively (94). This mode was observed in the spectrum of the cyclic hydrocarbon C$_{34}$H$_{68}$ mentioned earlier for the CH$_2$ wagging modes (94).

As an example of the usefulness of defect calculations for structural diagnosis, we quote the case of such cyclic molecules. The simultaneous occurrence of the resonance CH$_2$ wagging modes near 1350 cm^{-1} and of the gap mode near 714 cm^{-1} reportedly could be taken as a strong indication of a GGTGG defect (94). Their relative intensity should depend on the number of CH$_2$ groups in the *trans* planar host lattice. This fact was verified for a group of cyclic paraffins (98).

CD$_2$ Rocking Modes as Probes for the Local Conformation of Polymethylene Chains. The principle of the technique may be explained as follows. If in a polymethylene chain one CH$_2$ is replaced by a CD$_2$ group, the CD$_2$ rocking mode occurs near 620 cm^{-1} in the gap between the CH$_2$ rocking and the top of the ω_5 branch. Calculations show that the motion is highly localized on the CD$_2$ and that the vibration extends slightly only to the nearest CH$_2$ group on both sides. Moreover, this mode is strongly conformation dependent.

The first calculation and application of this concept were presented by Snyder (99), who tried to understand the structure of partially deuterated polyethylene. We repeated the calculation for long hydrocarbon chains (100) and long-chain fatty acids (72–74) where CD_2 groups were selectively located at different sites along the polymethylene chain.

Calculations show that the CD_2 rocking frequency strongly depends on the conformation about the C–C bond to which it is attached on both sides. Table I shows the results of calculations for n-nonadecane where the CD_2 group has been placed in position C-2 (thus labeling the head of the molecule) and in position C-10 (thus labeling the middle of the molecule.

For $^{T}CD_2{}^{T}$, $^{G}CD_2{}^{T}$, and $^{G}CD_2{}^{G}$, frequencies are predicted to occur near 620, 650, and 670 cm^{-1}, respectively where no other absorption occurs. Observations on n-nonadecane fully support the theoretical predictions (100). The GG structure was not observed because of its small population, according to the rotational isomeric state model.

Using these labels, the evolution of the conformation of the ends and of the inner part of the n-nonadecane molecule could then be followed from the solid state through phase transition and melting. A

Table I. Calculated CD2 Rocking Gap Frequencies for a Polymethylene Molecule as a Function of Local Conformation Used as Markers for Conformational Mapping of Polymethylene Chains

Chain				*Frequency (cm⁻¹)*
$\{-CH_2-CH_2-CD_2-CH_3\}$				
T		T		615
G		T		658
T		G		658
$\{-CH_2-CH_2-CD_2-CH_2-CH_2-\}$				
T	(T	T)	T	
G	(T	T)	T	
G	(T	T)	G	~615
G'	(T	T)	G'	
T	(T	G)	T	
G	(T	G)	T	
G	(T	G)	G	~650
G'	(T	G)	T	
T	(G	G)	T	
T	(G	G)	G	~677
G	(G	G)	G	

mechanism for phase transition and melting was proposed on the basis of the observation of these gap modes together with the occurrence in IR and/or Raman of other resonance modes (100). With n-hydrocarbons with a limited chain length, such as n-nonadecane, no evidence was found for the existence or formation of conformational kinks like those claimed by others to be the origin of the melting of the crystal.

CD$_2$ rocking modes selectively placed in a polymethylene chain thus become very useful probes for mapping the conformational topology of such chains in various physical states. We believe that this method is uniquely important because it can be applied to molecules in the solid as well as the liquid or solution states.

We recently applied this method to the mapping of the conformational topology of some fatty acids in the solid and solution state (72–74). Although CD$_2$ groups placed at sites situated away from the carboxyl group behave in the same way as hydrocarbons, some other efforts must be made when, by approaching the carboxyl group, CD$_2$ rocking may couple with the vibrational motion localized on the COOH group (72–74).

Resonance Modes as Markers of Conformational Mobility of Polymethylene Chains

Structurally useful defect modes were calculated and observed that can be used for the study of the conformational mobility of long alkyl residues such as are found in fatty acids, phospholipids, membrane systems, and other systems. We now will focus on a few internal modes of the CH$_3$ group, namely the internal deformation (umbrella motion, CH$_3$ U) with natural frequency near 1375 cm^{-1} and the external deformation, δ CH$_3$, with natural frequency near 890 cm^{-1} (Figure 6). The former mode shows a well-known characteristic medium intensity band in the IR spectrum; the latter shows a clear peak in the Raman spectrum. The natural frequencies of both these modes occur just in the middle of the frequency band spanned by two dispersion curves. Coupling thus is expected to occur. From the viewpoint of dynamics, these modes actually should be considered end-group modes (or surface modes of one-dimensional crystals) that were shown to be damped quickly along the crystal extending for a few CH$_2$ groups along the chain (100).

If the geometry of the chain changes near the CH$_3$ group, the coupling changes and the frequencies as well as the vibrational amplitudes change. This result is exactly what was predicted from calculations and tested experimentally on n-hydrocarbons (100, 101). If a structure such as —CH$_2$^{T}CH$_2$^{G}CH$_2$–CH$_3$ is generated in the system, modes are generated at 1342 and 866 cm^{-1}. If a structure such as— CH$_2$^{G}CH$_2$^{T}CH$_2$–CH$_3$ occurs, new modes are generated at 1355 and

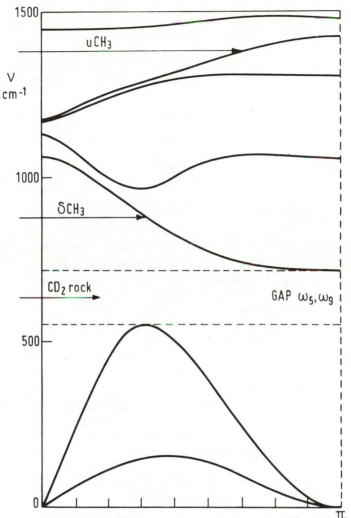

Figure 6. Dispersion curve of single-chain polyethylene in the 1500−0-cm⁻¹ range. Location of CD₂ rocking gap modes and CH₃ deformation resonance modes. Arrows for the latter two modes indicate which range of phonons of the host lattice are likely to couple with them.

844 cm⁻¹. Other resonance modes are generated in the spectrum as indicated in Table II. Calculations were tested on *n*-hydrocarbons in the α-phase and in the melt (*100*).

Of particular importance are the new modes that occur on the low frequency side of δ CH₃ and that are easily observable in the Raman spectrum. Although the absorptions in the IR region may be overlapped partially by absorptions of other defects, the Raman-active

Table II. Calculated Resonance Frequencies as Markers for Conformational Mobility of Chain Ends in Molecules with Polymethylene Residues

	Frequency (cm⁻¹)	
Chain End	Calculated	Observed
$\begin{array}{cc} T & G \end{array}$ $-CH_2-CH_2-CH_2-CH_3$	1348 874 769	1342 (IR) 866 (R^a, IR) 766 (IR)
$\begin{array}{cc} G & T \end{array}$ $-CH_2-CH_2-CH_2-CH_3$	1367 851	~1355 (IR) 844 (IR, R^a)

Note: Observed data are from *n*-nonadecane (*100, 101*).
^{a} R refers to Raman.

modes near the δ CH_3 are isolated and may be used as highly selective markers in structural studies of complex molecules (*101*).

These signals clearly were observed in *n*-nonadecane through its phase transition and melting thus allowing attention to focus, on a molecular scale, on the structural evolution of chain ends with temperature (*100*). Using intensity spectroscopy (*102, 103*) (*see* later discussion), quantitative determinations also were attempted. Independent information on chain ends was obtained from the study of CD_2 rocking gap modes that were placed selectively in position C-2 in the molecule of *n*-nonadecane. The information was in full qualitative and quantitative agreement. By such selective deuteration on one end of the molecule, and by using end-group resonance modes together with CD_2 gap modes, the structural evolution of both ends and tail of a hydrocarbon chain can be followed.

The Raman Spectrum as a Sensitive Probe for the Overall Conformation of a Polymethylene Chain

Because knowledge of the structure and structural evolution of the polymethylene chain in several polymeric systems or biological molecules is of great importance, many attempts were made in the past few years to correlate the observed changes of the Raman spectrum with the conformational changes of the molecules (*104*).

Disagreements can be found among various researchers who mainly base their interpretation on qualitative intuitions on the vibrational assignment of the Raman spectrum. The Raman spectrum of these systems can be obtained easily from samples in awkward physical states such as an aqueous solution, molecules spread on a surface as thin layers, and other states. Thus, the Raman spectrum has become an essential source of data on these complex organic and biological systems.

A substantial improvement in the interpretation of the observed spectra in terms of molecular structure has come about when the existence of Fermi resonances between the vibrational levels of CH_2 sequences were pointed out. The historical development of these new ideas are presented here. Our discussion is limited to the Raman spectrum, even if the IR spectrum also is affected. First Snyder et al. presented the idea that both the CH stretching and bending regions of the Raman spectrum are affected by Fermi resonances (*105*). The evidence they gave was still qualitative, but was supported by some experiments as well as by the results of previous dynamic calculations on perfect polymethylene chains.

The main concepts are the following (*see* Figure 7). For an all *trans* polymethylene chain, the vibrational frequencies lying on the lower portion of the dispersion branch of the CH_2 rocking (Figure 7) have their first overtone levels ($\sim720 \times 2 = \sim1440$ cm^{-1}) just falling close to the fundamental levels of the CH_2 bending motions. On their turn, the first overtones of the levels along the CH_2 dispersion curve occur just in the frequency region where the fundamental CH stretchings occur ($2 \times \sim1440 = \sim2880$ cm^{-1}). When symmetry allows, Fermi resonance can take place (in such a case many Fermi resonances) and levels repel each other with mutual borrowing of intensity.

The issue is complicated further by the fact that polymethylene chains normally crystallize in an orthorhombic lattice. Intermolecular interactions in the unit cell generate new levels thus providing more levels that may participate in the Fermi resonances.

These concepts were translated in a more qualitative way by a mathematical treatment for the Fermi resonance interactions in *n*-alkanes restricted to the Fermi interactions of the first overtones of the CH_2 bending modes with the CH stretching fundamentals of a single chain (*106*).

At the same time, a very similar and parallel dynamic treatment that included the interactions of CH_2 rocking with bending and bending with CH_2 stretching for single chain as well as for chains organized in a tridimensional lattice was reported. Moreover, intensity calculations also were included (*107, 108*).

Of particular interest to the latter group was not just the interpretation of the spectrum originating from the perfect, crystalline, all *trans* structure of the polymethylene chain but was the fact that new spectral features could be predicted when the Fermi resonance scheme of the *trans* structure is modified by the introduction of *gauche* conformers. This information may be relevant for all researchers interested in the study of the structure of actual or real polymethylene systems with a partially ordered and a partially conformationally disordered structure.

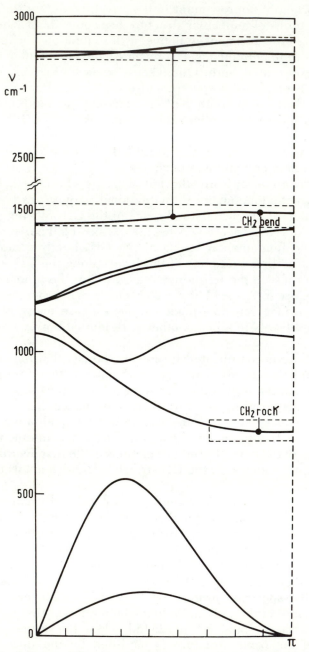

Figure 7. Scheme of Fermi resonances occurring in trans *single-chain polyethylene.*

The investigation by Abbate et al. (*108*) currently is being followed by others (*109, 110*).

Without going into the mathematical details, we will briefly sketch the state of the art in the interpretation of the Raman spectrum of polymethylene chains in the all *trans* structure.

1. We will first describe a polymethylene chain in an orthorhombic lattice with two molecules per crystallographic unit cell. Not only the IR-active doublet of the B_{2u} and B_{3u} species of the CH_2 rocking modes, but also all the IR-inactive phonons along the dispersion branches, ω_9, generate two phonon states of proper symmetry and phase to enter in Fermi resonance with the CH_2 bending motions near 1400 cm^{-1}. Figures 8a and 8b

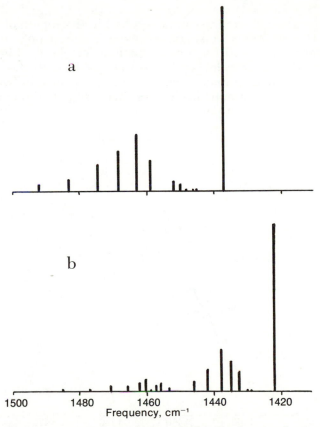

Figure 8. Fermi resonance between the fundamental of δ CH_2 of species A_g and the first overtone of CH_2 rockings, (a), and Fermi resonances between the fundamental δ CH_2 of species B_{1g} and the first overtones of CH_2 rockings (b).

show the calculated sequence of A_g and B_{1g} modes arising from the CH_2 bending fundamentals that entered in
Fermi resonance with the overtones of the CH_2 rockings,
lending intensities to them (spectroscopically, in the
double harmonic approximation, intensities of overtone
and combination levels should be zero). Figure 9 shows
the calculated Raman spectrum obtained by convolution
of Lorenzian bands of both sequences weighed by the
intensity factor predicted by intensity calculations (108).
The calculated spectrum agrees with the observed one.
The peaks near 1420 and 1440 cm^{-1} form the factor group
doublet due to the orthorhombic lattice while the broad
scattering near 1460 cm^{-1} is only due to the superposition
of two phonon levels of a *trans* chain that have borrowed
intensity from the two fundamentals near 1420 and 1440
cm^{-1}.

2. The same reasoning can be applied to the region of the
   CH stretchings. A single all *trans* isolated polymethylene chain will be considered first for clarity. Figure
   10a shows the calculated spectrum (with intensities included) when intramolecular Fermi resonance between
   various CH_2 groups has not yet been switched on. The

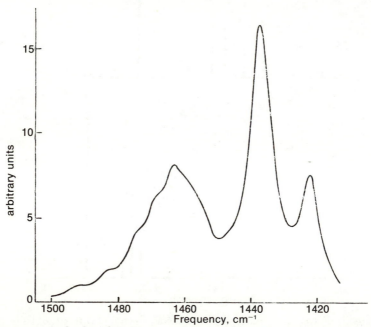

*Figure 9. Calculated Raman spectrum in the 1450-cm^{-1} range obtained
as convolution with Lorenzian bands of the transitions of Figures 8a
and 8b.*

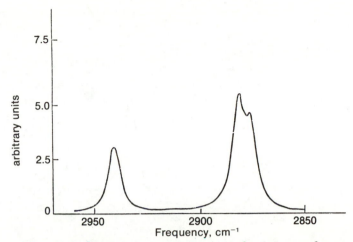

Figure 10a. Calculated Raman spectrum in the CH stretching region taking into account the Fermi resonance only within an isolated CH_2 group.

only Fermi resonance is that which exists within the single and isolated CH_2 group for which the first overtone of the CH_2 bending (A_1) interacts with the symmetric CH_2 stretching (A_1), while the antisymmetric CH_2 stretching (B_1) remains unaffected. This resonance was first pointed out by Lavallay and Sheppard in 1,1,1,3,3,3-hexadeuteropropane (*111*). If the intramolecular CH_2 Fermi resonance is switched on, the manifold of overtones from CH_2 bending modes occurs and spreads just above the CH_2 symmetric stretching. A strong repulsion between the overtone levels and the CH_2 fundamental takes place (*see* Figure 10b). The fundamental moves toward 2850 cm^{-1} and lends intensity to the manifold of overtone levels thus decreasing its apparent intensity. The background scattering is raised substantially from 2850 to 2990 cm^{-1}. The B_{1g} CH_2 antisymmetric stretch remains unperturbed and floats on top of the A_g background scattering.

3. If a tridimensional orthorhombic lattice now is considered, new overtone levels are generated and the density of overtone states in the CH stretching region increases, all possibly in Fermi resonance. Calculations show [in agreement with the original idea of Snyder et al. (*105*)] that these levels cluster mainly below and between the B_{1g} peak near 2880 cm^{-1} and the A_g peak near 2850 cm^{-1} thus again mainly affecting the intensity and frequency of the A_g fundamental.

4. We will now move backward from the perfect structure

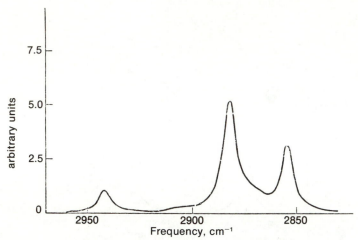

Figure 10b. Calculated Raman spectrum in the CH stretching region for the all trans-*polymethylene chain when Fermi resonances are switched on and inserted in the resonance scheme of Figure 10a.*

toward a more disordered structure. First, the intermolecular interactions of the orthorhombic lattice will be removed by placing the molecule: a, as a guest in a host lattice of perdeuterated material (*104*); and b, in a cage of a clathrate (*112*). The tridimensional order can be destroyed by making some sort of smectic structure as in the α-phase observed for *n*-nonadecane (*100*).

We expect the valley between the peaks near 2850 and 2880 cm^{-1} to become deeper or emptier because intermolecular Fermi resonances were switched off. The A_g 2850-cm^{-1} band is expected eventually to shift slightly toward higher frequencies and to regain the intensity that it lent to the levels that were removed. The much questioned lateral interactions proposed by Gaber and Peticolas for phospholipids and measured by the same group by taking the ratio of the intensities in the CH stretching region by changing the environment of the *trans* chains (*104*) finds in the facts just presented a possible explanation.

5. The long *trans* sequences will now be interrupted by *gauche* distortions. Because both the CH$_2$ rocking and bending modes in *gauche* conformations occur at somewhat different frequencies, the population of vibrational levels participating in the *trans* interaction scheme decreases and the spectral distribution both in the CH$_2$ bending and CH$_2$ stretching region is modified. Qualitatively, the vibrational spectrum should change from that of a full, all *trans* single-chain molecule toward that of a

fully decoupled structure, thus tending toward the situation described by the spectrum of Figure 10. Of particular importance is that the 2850-cm^{-1} peak should move
toward higher frequencies close to the always unper-
turbed band near 2880 cm^{-1} and the $2 \times CH_2 \delta$ peak near
2930 cm^{-1} should gain more intensity because it tends to
acquire its localized identity and does not share with the
two phonon manifold the intensity borrowed from the
CH_2 symmetric stretching. If the predicted behavior actually occurs, the peak near 2930 cm^{-1} should increase
its intensity when *gauche* conformations are generated
with respect to the band near 2850 cm^{-1}, which represents the all *trans* segments. The relative ratio I_{2850}/I_{2940}
then is expected to change, decreasing by increasing the
concentration of conformationally distorted non-*trans*
C—C bonds.

These predictions were verified in a series of experiments by
Wunder and Murajver on several model systems. Liquid *n*-hexadec-
ane at different temperatures (*109*), solutions of hexadecane at various
concentrations and temperatures (*109*), polyethylene from the solid to
the melt at various temperatures (*110*), and solid samples of polyethylene with varying content of an amorphous substance (*108*) all show
the same kind of predicted behavior. The same fact also was observed
for liquid fatty acids at different temperatures (*113*). Wunder and coworkers (*109, 110*) point out that such a ratio is very sensitive to small
changes in the *trans*—*gauche* ratio. It becomes then possible to follow
the thermal conformational evolution of a polymethylenic system. A
parallel, but not yet quantitatively followed, behavior is observed for
the Raman line near 1460 cm^{-1}, which again is characteristic of long
sequences of *trans* structures and decreases when the population of
trans conformers decreases.

One of the advantages of the method just proposed is that it can be
adopted for numerous polymethylene systems in all physical conditions and phases. Because CH stretching modes are strong scatterers
and scatter in a region generally free from overlapping, the method
can be adopted for a large variety of substances in extreme experimental conditions.

From just the quick observation of a spectrum of a liquid *n*-alkane
or a phospholipid system, we learn that a fraction of the substance is in
the *trans* conformation and another is in a distorted structure whose
relative ratio changes according to temperature, solvent, chemical or
physical treatment, and for other reasons. The problem still unsolved
is the precise accounting of the Raman intensities for an absolute or
relative quantitative determination of the populations of conformers.
An interesting attempt was made (*109*) to relate the observed ratio of

the intensities I_{2930}/I_{2840} with the *trans/gauche* population as determined from the statistical isomeric state model.

Because our proposed interpretation of the spectral changes observed in the Raman spectra as just discussed may find some immediate applications in various fields, a word of caution needs to be said on these matters. The observed changes of the Raman spectra in the CH stretching region of systems containing *n*-alkane residues in going from ordered to disordered phases may be the result of a combination of several phenomena, one of which may be the simple Fermi resonance interactions just discussed. More complicated phenomena may be envisaged (such as intermolecular effects, band shape changes, changes in anharmonicity of the CH bonds, changes in harmonic diagonal or/end off diagonal force constants with torsional angles or with the environment, etc.). To our knowledge, neither theory nor experiments have shown whether these additional phenomena concomitantly exist and which one plays the dominant role (if any) in making up the observed spectrum. More work is needed on this important issue.

Lattices with Tridimensional Disorder

The appearance and disappearance of a tridimensional order in crystalline materials and specifically of crystalline polymethylene molecules received little attention in the past, although, as shown here, most of the attention was focused on intramolecular order and disorder. For polymers, the existence of the factor group splitting (noted earlier) helps in the precise definition of crystallinity bands and informs us that molecular chains are packed with a tridimensional order. The study of the extent of intramolecular coupling in polymethylene chains was carried out extensively by Krimm and coworkers in an attempt to understand the organization of molecular chains in single crystals grown from solution or from the melt (45, 49). The spectra of isotopically diluted crystals were used as a tool and very small frequency changes on isotopic dilution were rationalized in terms of the organization and regular clustering of chain stems within the crystal (49). This issue is at present a key problem in polymer morphology and polymer solid-state physics (49).

As already discussed, the factor group splitting arises because of the coupling of the natural frequency of the oscillators of the isolated chain with the identical ones of the chains in the surrounding crystal.

The possibility of removing the intermolecular coupling of a given vibrational level, while leaving some others practically unperturbed recently was considered (114, 115). In a mixed crystal of *n*-alkanes where the two components differ slightly in length, full miscibility in solids without any phase separation was shown to occur for

n-alkanes of the type $C_n + C_{n\pm2}$ (*116*). If n is sufficiently large, the sequences of normal modes labeled with the index j along the dispersion curves of the infinite polymethylene chain generally show different frequencies for the two molecules. The sequences of the CH_2 rocking modes that show the largest dispersion (all lying approximately on the ω_9 branch of polyethylene) now will be considered.

Although for large j values the frequency differences are large for the two molecules (tending toward small j and to the limiting $j = 0$ modes), the frequency mismatch of the two frequencies decreases (Figure 11). For chains longer than ~5 CH_2 units, the in-phase $j = 0$ mode always occurs near 720 cm^{-1} for all *trans* planar n-hydrocarbons.

In a rough approximation, one may first expect that in a mixed crystal of n-alkanes of the type just mentioned, where molecules can pack with high intermolecular order, the splitting of the CH_2 normal modes depends on the frequency mismatch. Singlets will appear for modes with many nodes (large j), and splittings may appear for normal modes with a small number of nodes (small j) tending to the well-known splitting for the limiting in-phase motion ($j = 0$) near 720 cm^{-1}. These expectations were verified in the IR spectra of several mixed crystals of n-alkanes with lengths slightly differing (*114, 115*), where practically only the limiting rocking mode near 725 cm^{-1} shows splitting.

The IR spectrum of mixed crystals of n-alkanes with different lengths then can become a tool for the study of molecular motions in the solid state. Indeed, the concepts just described and the experiments that were derived allowed detailed information to be gained on the capability of hydrocarbon molecules to migrate within a crystal and from one crystal to another when placed in physical contact but separated by well-defined crystal boundaries. The phenomenon first reported by Ungar and Keller (*117*) can be followed in detail (*114, 115, 118*) with the spectroscopic techniques described in this chapter.

In a mechanical mixture of crystalline compound A + crystalline compound B (A and B being two n-alkanes of general formula $A = C_n H_{2n+2}$ and $B = C_{n\pm2} H_{2(n\pm2)+2}$, n either even or odd); n must be some integer such that the structure will be orthorhombic or monoclinic (*119*). The spectrum of the mechanical mixture $A + B$ shows throughout the whole spectrum the factor group splitting of each compound independent from the other, each one having its own orthorhombic structure, thus allowing normal in- and out-of-phase coupling of each vibrational level. By annealing at different temperatures below, up to, and above the solid–solid phase transition, each doublet of the modes at large j becomes a singlet, although the region near 720 cm^{-1} still shows the normal doublet. The disappearance of the doublet on annealing of the mechanical mixture occurs because molecules of type A

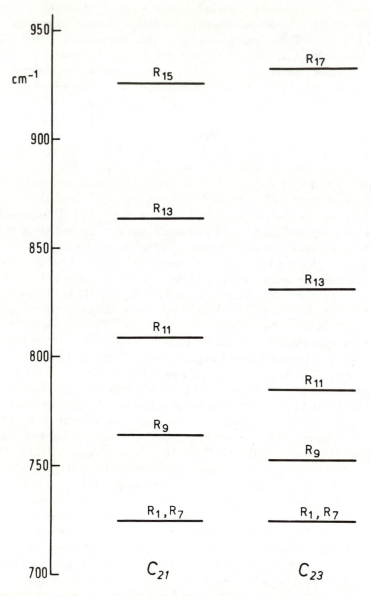

Figure 11. Scheme of the observed frequency mismatch for a few components of CH_2 rocking for $C_{21}H_{44}$ and $C_{23}H_{48}$. Factor group splitting was removed from the figure for simplicity. [Values of $j + 1$ are reported here because fixed boundary conditions are imposed for both ends (see Ref. 3).]

migrate from their own crystal into the crystal of Compound B and vice versa. Islands of A into B and of B into A then are formed. At the end of the process, the compound obtained is just that which can be obtained independently by co-crystallization of A and B from a suitable solvent.

The rate of migration and the energetics of the phenomenon were studied (*114, 115, 118*) using the vibrational spectrum and calorimetric measurements. Moreover, using many of the spectroscopic signals derived from the study of the dynamics of disordered materials discussed previously, one can then state that molecules migrate at larger distances in proportion to their length and at different rates according to the temperature and time of annealing. Molecules migrate substantially as rigid bodies without introducing noticeable conformational defects. When mixed crystals are formed, the geometrical mismatch of chain lengths within the crystal forces a few molecules to tilt their heads with *GT* or *TG* structure at either end.

The phenomenon, however, is not restricted to molecules with unequal length. Self-diffusion or self-mixing between molecules of the same length was observed and studied by using solid crystalline $C_{36}H_{74}$ and its perdeuteroderivative $C_{36}D_{74}$. Details of the phenomenon as studied from spectroscopy (*114, 115*) and a discussion on its possible origin are given elsewhere (*118*).

With the techniques presented in this chapter, it becomes possible to study a phenomenon that gives a positive answer to the still open question (*7*) of whether during annealing of polymeric solids a physical transport of matter across crystal boundaries occurs. At least for polymethylene chains in the solid state, our studies seem to prove that matter is transported across crystal boundaries on annealing and that chain migration can take place in solids also without the existence of amorphous or conformationally irregular domains in the material through which chains should move (or do move) supposedly with a snake-like motion (*120*). The most relevant information derived from these studies is that hydrocarbon chains move within and across crystallites substantially as rigid bodies. Snake-like hopping from site to site within the bulk should not occur in this case because molecules are crystalline and no sizable conformational deformations were yet detected spectroscopically at temperatures just below the melting point where migration is very fast.

Vibrational Intensities: A New Future for IR and Raman Spectroscopy of Polymers

Recent basic studies of vibrational spectroscopy tackled the problem of predicting the absorption coefficients in the IR region or the scattering power in the Raman region for various molecular species (*121*). Until a few years ago, the study of vibrational intensities was conducted by a few specialists of molecular spectroscopy who dealt with the basic measurements on several model molecules and discussed the various theoretical techniques for accounting for the observed spectrum.

Recently, efforts have been made, and are being made, to apply these techniques to molecules of practical importance for deriving quantitative information on molecular species or molecular conformations of more general interest to chemists. The attempts made by a few researchers in the field of vibrational (IR and Raman) intensities just parallel the efforts made by numerous groups during the last thirty years to calculate the vibrational frequencies from force constants.

Vibrational intensities are related to the electrical properties of molecules (change in the total electric dipole moment or polarizability during the vibrational motion). If a model of such electrical properties can be found, it can be made parametric using the experimentally observed intensities. Such parameters (called intensity parameters) may be used to predict the unknown intensities of another similar molecule.

Among the various models proposed and tested (122, 123), the valence electrooptical scheme (124) seems well suited for intensity calculations. The corresponding parameters are called IR or Raman electrooptical parameters (eop). Considerable effort has been made recently to develop such a detailed model and to carry out a least-squares fitting of the eop values to the experimental intensities of many molecules.

Unfortunately, the number of molecules for which reliable experimental measurements are available is still very limited, being acceptable for IR spectroscopy, but extremely limited for Raman spectroscopy. For n-hydrocarbons, IR measurements are available for methane, ethane, propane, and several deutero derivatives in the gas phase. Data on liquid n-butane and n-pentane were collected recently (125). Solid n-hydrocarbons from n-butane to n-heptane were measured by Snyder (126).

Most of these experimental data were included in least-squares fittings (127) of a set of eop values after the necessary analyses of the data in terms of the various theoretical technicalities were carried out. These eop values accurately predict the spectra of n-decane (128), of polyethylene, and of perdeuteropolyethylene (127, 129).

Analogously, the Raman spectra of polyethylene and perdeuteropolyethylene were predicted accurately with the eop values derived from smaller n-hydrocarbons (28, 29, 129). Attempts to jump ahead and use the eop values obtained for the prediction of the absorption coefficients of defects in polymethylene molecules are still tentative, but nevertheless, are of exploratory relevance. An example of the application of intensity calculations in conformationally disordered poly-methylene chains is the prediction of the intensity pattern for the group of resonance modes near 1350 cm^{-1} generated by the *GGTGG* defect in a cyclic $C_{34}H_{68}$ molecule as discussed (*96*) previously. Figure 12 shows the pattern calculated from group intensities compared with the experimental spectrum. The agreement is very good and shows again the usefulness of these kinds of work.

For the vibrational spectra of perfect, single-chain polyethylene, intensity calculations have helped in the detailed interpretation of the Raman spectrum in the CH_2 bending and stretching regions where Fermi resonances occur (*108*) as discussed earlier.

Finally, for the Raman spectrum of perdeuteropolyethylene, intensity calculations were decisive in solving the controversy of the vibrational pattern near 980 cm^{-1}. The pattern near 980 cm^{-1} can be described (*28, 29*) as a strong doublet with a very small and almost unnoticeable shoulder on the high-frequency side of the strongest peak. The peak near 970 cm^{-1} was assigned to the CD_2 deformation mode, and the peak near 990 cm^{-1} was taken to be the CD_2 rocking mode. Calculations indicate that the CD_2 rocking mode contributes an

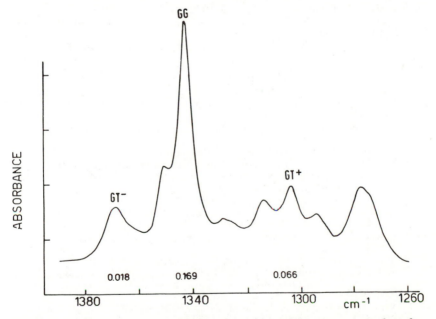

Figure 12. Comparison between the observed IR spectrum of cyclic $C_{34}H_{68}$ solid in the 1260–1380-cm^{-1} region and the intensities calculated for the CH_2 wagging resonance modes of the GGTGG defect. Numbers on the bottom refer to calculated intensities. No simulation of the spectrum was made because the spectrum is overlapped with other bands.

extremely weak scattering in this region, and the relative integrated area of the strong doublet approaches the value calculated for the CD_2 bending motion. As already suggested by others (130), the whole doublet (989–970 cm^{-1}) must then be assigned to a factor group splitting of the CD_2 deformation mode while the very weak shoulder is instead the scattering due to the CD_2 rocking mode.

Generally, intensity studies also can help greatly in the characterization of polymeric materials in the perfect or disordered state if the spectral intensities of many more model molecules are measured carefully. We hope to convince more people to carry out such measurements.

Literature Cited

1. Krimm, S. *Adv. Polym. Sci.* **1960**, 2, 51.
2. Zerbi, G. In "Applied Spectroscopy Reviews"; Brame, E. G., Ed.; Dekker: New York, 1969; Vol. 1.
3. Zbinden, R. "Infrared Spectroscopy of High Polymers"; Academic: New York, 1964.
4. Koenig, J. L. In "Applied Spectroscopy Reviews"; Brame, E. G., Ed.; Dekker: New York, 1971; Vol. 4.
5. Koenig, J. L. *J. Polym. Sci., Part D* **1972**, 59.
6. De Santis, P.; Giglio, E.; Liquori, A. M.; Ripamonti, A. *J. Polym. Sci., Part A* **1963**, 1, 1383.
7. Wünderlich, B. "Macromolecular Physics"; Academic: New York, 1976, Vols. 1 and 2.
8. Wilson, E. B.; Decius, J. C.; Cross, P. C. "Molecular Vibrations"; McGraw-Hill: New York, 1955.
9. Zerbi, G. In "Modern Methods in Vibrational Spectroscopy"; Orville-Thomas, W. J.; Barnes, A. J., Eds.; Elsevier: Amsterdam, 1977.
10. Shimanouchi, T. In "Physical Chemistry, An Advanced Treatise"; Eyring, H.; Anderson, D.; Jost, W.; Eds.; Academic: New York, 1970; Vol. 4.
11. Califano, S. "Vibrational States"; Wiley: London, 1976.
12. Schachtschneider, J. H.; Snyder, R. G. *Spectrochim. Acta* **1963**, 19, 17.
13. Snyder, R. G.; Zerbi, G. *Spectrochim. Acta* **1967**, 23A, 391.
14. Piseri, L.; Zerbi, G. *J. Chem. Phys.* **1968**, 48, 3561.
15. Zerbi, G.; Piseri, L. *J. Chem. Phys.* **1968**, 49, 3840.
16. Higgs, P. W. *Proc. R. Soc. (London)* **1953**, A220, 472.
17. Piseri, L.; Zerbi, G. *J. Mol. Spectrosc.* **1968**, 26, 254.
18. Tasumi, M.; Shimanouchi, T.; Miyazawa, T. *J. Mol. Spectrosc.* **1962**, 9, 261.
19. Piseri, L.; Powell, B. M.; Dolling, G. *J. Chem. Phys.* **1973**, 57, 158.
20. Chiang, C. K.; Drug, M. A.; Gau, S. C.; Heeger, A. J.; Shirakawa, H.; Louis, E. J.; Mac Diarmid, A. G.; Park, Y. W. *J. Am. Chem. Soc.* **1978**, 100, 1019.
21. Gavin, R. M.; Rice, S. A. *J. Chem. Phys.* **1971**, 55, 2675.
22. Kakitani, T. *Prog. Theor. Phys.* **1974**, 51, 656.
23. Zerbi, G.; Zannoni, G. *Chem. Phys. Lett.* **1982**, 1982, 50.
24. Zannoni, G.; Zerbi, G. *Chem. Phys. Lett.* **1982**, 1982, 55.
25. Suhay, S. *J. Chem. Phys.* **1980**, 73, 3843.
26. Peraldo, M.; Cambini, M. *Spectrochim. Acta* **1965**, 21, 1509.
27. White, J. W. In "Structural Studies of Macromolecules by Spectroscopic Methods"; Irvin, K. J., Ed.; Wiley: London, 1976.

28. Masetti, G.; Abbate, S.; Gussoni, M.; Zerbi, G. *J. Chem. Phys.* **1980**, *73*, 4671.
29. Abbate, S.; Gussoni, M.; Zerbi, G. *J. Chem. Phys.* **1980**, *78*, 4680.
30. Zerbi, G.; Masetti, G. In "Analytical Raman Spectroscopy"; Kiefer, W., Ed.; Wiley: New York, in press.
31. Tadokoro, H. "Structure of Crystalline Polymers"; Wiley: New York, 1979.
32. Jones, R. N. In "Techniques of Organic Chemistry"; Weissberger, A., Ed.; Wiley-Interscience: New York, 1956; Vol. 9.
33. Bellamy, L. J. "The Infrared Spectra of Complex Molecules"; Methuen: London, 1958.
34. Zerbi, G.; Ciampelli, F.; Zamboni, V. *J. Polym. Sci., Part C* **1964**, 141.
35. Tosi, G.; Zerbi, G. *Chim. Ind.* **1973**, *55*, 334.
36. Zerbi, G.; Gussoni, M.; Ciampelli, F. *Spectrochim. Acta* **1967**, *23A*, 301.
37. Masetti, G.; Cabassi, F.; Zerbi, G. *Polymer* **1980**, *21*, 143.
38. Schachtschneider, J. H. *J. Polym. Sci.* **1964**.
39. Cortili, G.; Zerbi, G. *Spectrochim. Acta* **1967**, *23A*, 285.
40. Shirakawa, H.; Ikeda, S. *Polym. J.* **1971**, *2*, 231.
41. Tasumi, M.; Shimanouchi, T. *J. Chem. Phys.* **1965**, *43*, 1245.
42. Dows, D. In "Physics and Chemistry of Organic Solid State"; Fox, D.; Labes, M. M.; Weissbarger, A., Eds.; Wiley-Interscience: New York, 1963; Vol. 1.
43. Vedder, W.; Hornig, D. F. *Adv. Spectrosc.* **1962**, *2*, 189.
44. Herzberg, G. "Infrared and Raman Spectra of Polyatomic Molecules"; Van Nostrand: Princeton, NJ, 1959.
45. Bank, M. I.; Krimm, S. *J. Appl. Phys.* **1969**, *10*, 248.
46. Krimm, S.; Cheam, T. C. *Faraday Disc. Chem. Soc.* **1979**, *68*, 244.
47. Natta, G.; Danusso, F. "Stereoregular Polymers and Stereospecific Polymerisation"; Pergamon: New York, 1967; Vols. 1 and 2.
48. Masetti, G.; Zerbi, G., unpublished data.
49. *Faraday Disc. Chem. Soc.* **1979**, *68*, 244.
50. Zerbi, G. *Pure Appl. Chem.* **1971**, *26*, 501.
51. *Ibid.*, **1973**, *36*, 35.
52. Zerbi, G. In "Lattice Dynamics and Intermolecular Forces"; Califano, S., Ed.; Academic: New York, 1975.
53. Zerbi, G. In "Advances in the Preparation and Properties of Stereospecific Polymers"; Ciardelli, F.; Lenz, R., Eds.; Reidel: Dordrecht, The Netherlands, 1979.
54. Zerbi, G. *Makromol. Chem. Suppl.* **1981**, *5*, 253.
55. Lyerla, J. R. "Contemporary Topics in Polymer Science"; Shen, M., Ed.; Plenum: New York, 1979; Vol. 3.
56. Mizushima, S.; Shimanouchi, T. *J. Am. Chem. Soc.* **1949**, *71*, 1320.
57. Schaufele, R. F.; Shimanouchi, T. *J. Chem. Phys.* **1967**, *47*, 3605.
58. Shimanouchi, T. In "Structural Studies of Macromolecules by Spectroscopic Methods"; Ivin, K. J., Ed.; Wiley: London, 1976.
59. Shimanouchi, T.; Tasumi, M. *Indian J. Pure Appl. Phys.* **1971**, *9*, 958.
60. Gussoni, M.; Abbate, S.; Zerbi, G. In "Raman Spectroscopy, Linear and Nonlinear"; Lascombe, J.; Huong, P. V., Eds.; Wiley: New York, 1982; p. 24.
61. Fraser, G. V. *Indian J. Pure Appl. Phys.* **1978**, *16*, 344.
62. Dluglosz, J.; Fraser, G. V.; Grubb, D.; Keller, A.; Odell, J. A.; Goggin, P. L. *Polymer* **1976**, *17*, 471.
63. Farrell, C. J.; Keller, A. *J. Mater. Sci.* **1977**, *12*, 966.
64. Snyder, R. G.; Krause, S. J.; Scherer, D. R. *J. Polym. Sci., Polym. Phys. Ed.* **1978**, *16*, 1953.
65. Strobl, G. R.; Eckel, R. *J. Polym. Sci.* **1976**, *A2*, *14*, 913.
66. Hsu, S. L.; Krimm, S. *J. Appl. Phys.* **1976**, *47*, 4265.
67. Hsu, S. K.; Ford, G. V.; Krimm, S. *J. Polym. Sci., Polym. Phys. Ed.* **1977**, *15*, 1769.

68. Krimm, S. *Indian J. Pure Appl. Phys.* **1978**, *16*, 335.
69. Zerbi, G.; Gussoni, M. *Bull. Am. Phys. Soc.* **1979**, *24*, 347.
70. Scherer, J. R.; Snyder, R. G. *J. Chem. Phys.* **1980**, *72*, 5798.
71. Rabolt, J. F. *J. Polym. Sci., Polym. Phys. Ed.* **1979**, *17*, 1457.
72. Minoni, G.; Zerbi, G. *J. Phys. Chem.*, in press.
73. Minoni, G.; Zerbi, G. *Polym. Prepr., Am. Chem. Soc., Div. Polym. Chem.* **1982**.
74. Minoni, G.; Zerbi, G.; Tulloch, M. *J. Chem. Phys.*, in press.
75. Hsu, S. L.; Krimm, S.; Krause, S.; Yeh, G. S. Y. *J. Polym. Lett.* **1976**, *14*, 195.
76. Hendra, P. J.; Majid, H. A. *Polymer* **1977**, *18*, 573.
77. Rubcic, A.; Zerbi, G. *Chem. Phys. Lett.* **1975**, *34*, 343.
78. Rubcic, A.; Zerbi, G. *Macromolecules* **1974**, *7*, 754.
79. *Ibid.*, 759.
80. Tasumi, M.; Zerbi, G. *J. Chem. Phys.* **1968**, *48*, 3813.
81. Fanconi, B. *Annu. Rev. Phys. Chem.* **1980**, *31*, 265.
82. McCullough, R. L. *J. Macromol. Sci. Phys.* **1974**, *B9*, 97.
83. Zerbi, G.; Piseri, L.; Cabassi, F. *Mol. Phys.* **1971**, *22*, 241.
84. Zerbi, G.; Sacchi, M. *Macromolecules* **1973**, *6*, 692.
85. Masetti, G.; Cabassi, F.; Morelli, G.; Zerbi, G. *Macromolecules* **1973**, *6*, 700.
86. Hölzl, K.; Schmid, C.; Hägele, P. C. *J. Phys. Chem.* **1978**, *11*, 9.
87. Kozyrenko, V. B.; Kumpanenko, I. V.; Mikhailov, I. D. *J. Polym. Sci., Polym. Phys. Ed.* **1977**, *15*, 1721.
88. Snyder, R. G. *J. Chem. Phys.* **1967**, *47*, 1316.
89. Genzel, L. In "Optical Properties of Solids"; Nudelman, S.; Mitra, S. S., Eds.; Plenum: New York, 1969.
90. Dean, P. *Rev. Mod. Phys.* **1972**, *44*, 127.
91. Wilkinson, J. H. *Comput. J.* **1958**, *1*, 90.
92. Gussoni, M.; Zerbi, G. *J. Chem. Phys.* **1974**, *60*, 4862.
93. Shimanouchi, T. *J. Phys. Chem. Ref. Data* **1973**, *2*, 121.
94. Zerbi, G.; Gussoni, M. *Polymer* **1980**, *21*, 1127.
95. Newman, B. A.; Kay, H. F. *J. Appl. Phys.* **1967**, *38*, 4105.
96. Zerbi, G.; Gussoni, M.; Abbate, S.; Jona, P. In "Intensities in Infrared and Raman Spectroscopy"; Person, W.; Zerbi, G., Eds.; Elsevier: Amsterdam, 1982.
97. A group of articles on these matters in Ref. 34.
98. Trzebiatowski, T., Thesis, Univ. of Mainz, Mainz, Federal Republic of Germany, 1980.
99. Snyder, R. G.; Poore, M. W. *Macromolecules* **1973**, *6*, 709.
100. Zerbi, G.; Magni, R.; Gussoni, M.; Holland Moritz, K.; Bigotto, A.; Dirlikov, S. *J. Chem. Phys.* **1981**, *75*, 3175.
101. Zerbi, G.; Magni, L.; Gussoni, M. *J. Mol. Struct.* **1981**, *73*, 235.
102. Zerbi, G.; Gussoni, M. *Bull. Am. Phys. Soc.* **1981**, *26*, 327.
103. Colombo, L.; Gussoni, M.; Zerbi, G., presented at the National Meeting of the Italian Association of Physical Chemistry, Albano, Italy, 1981.
104. Gaber, B. P.; Peticolas, W. L. *Biochim. Biophys. Acta* **1977**, *465*, 260.
105. Snyder, R. G.; Hsu, S. L.; Krimm, S. *Spectrochim. Acta* **1978**, *34A*, 395.
106. Snyder, R. G.; Scherer, J. R. *J. Chem. Phys.* **1979**, *71*, 3221.
107. Zerbi, G.; Abbate, S. *Chem. Phys. Lett.* **1981**, *80*, 455.
108. Abbate, D.; Zerbi, G.; Wunder, S. L. *J. Phys. Chem.* **1982**, *86*, 3140.
109. Wunder, S. L. *Macromolecules*, in press.
110. Wunder, S. L.; Merajver, S. D. *J. Chem. Phys.* **1981**, *74*, 5841.
111. Lavallay, J.; Sheppard, N. *Spectrochim. Acta* **1972**, *28A*, 845.
112. Snyder, R. G.; Scherer, J. R.; Gaber, P. *Biochim. Biophys. Acta* **1980**, *601*, 47.
113. Pison, P., Thesis, Univ. of Trieste, Trieste, Italy, 1981.

114. Piazza, R.; Zerbi, G., presented at the National Meeting of the Italian Association of Physical Chemistry, Albano, Italy, 1981.
115. Piazza, R., Thesis, Univ. of Milano, Milan, Italy, 1981.
116. Mazee, W. M. *Anal. Chim. Acta* **1957**, *17*, 97.
117. Ungar, G.; Keller, A. *Colloid Polym. Sci.* **1979**, *90*, 257.
118. Zerbi, G.; Piazza, R.; Holland Moritz, K. *Polymer*, in press.
119. Kitaigorodskij, A. I. "Organic Chemical Crystallography"; Consultant Bureau, New York, p. 181.
120. De Gennes, P. G. *J. Chem. Phys.* **1971**, *55*, 572.
121. "Vibrational Intensities in Infrared and Raman Spectroscopy"; Person, W. B.; Zerbi, G., Eds.; Elsevier: Amsterdam, 1982.
122. Person, W. B.; Steele, D. "Molecular Spectroscopy"; Spec. Publ. Chem. Soc. (London) **1974**, *2*, 357.
123. Gussoni, M. In "Advances in Infrared and Raman Spectroscopy"; Clark, R. J. H.; Hester, R. E., Eds.; Heyden: London, 1979; Vol. 6, p. 61.
124. Gribov, L. A. "Intensity Theory for Infrared Spectra of Polyatomic Molecules"; Consultant Bureau: New York, 1964.
125. Zerbi, G.; Gussoni, M.; Colombo, L. *Bull. Am. Phys. Soc.* **1980**, *25*, 401.
126. Snyder, R. G. *J. Chem. Phys.* **1965**, *42*, 1744.
127. Gussoni, M.; Abbate, S.; Zerbi, G. *J. Chem. Phys.* **1979**, *71*, 3428.
128. Gussoni, M.; Abbate, S.; Dragoni, B.; Zerbi, G. *J. Mol. Struct.* **1980**, *61*, 355.
129. Abbate, S.; Gussoni, M.; Masetti, G.; Zerbi, G. *J. Chem. Phys.* **1977**, *67*, 1519.
130. Boerio, F. J.; Koenig, L. *J. Chem. Phys.* **1970**, *52*, 3425.

RECEIVED for review October 14, 1981. ACCEPTED January 26, 1982.

Deformation Studies of Polymers by Time-Resolved Fourier Transform IR Spectroscopy

D. J. BURCHELL and S. L. HSU[1]

University of Massachusetts, Department of Polymer Science and
Engineering, Amherst, MA 01003

*The construction of a miniature closed-loop servo-con-
trolled hydraulic tester interfaced to a Fourier transform
IR spectrometer is described. Macroscopic information
in the form of stress−strain curves can be collected along
with the microstructural changes in the form of IR data.
Time-resolved spectroscopy was applied to measure
spectroscopic changes in measurement times as short
as 500 μs.*

TRADITIONALLY, EXTERNAL STRESS- OR STRAIN-INDUCED microstructural changes in polymers, such as chain segment orientation rate or
unit cell distortion, have been measured by a combination of dynamic
x-ray, light scattering, and birefringence techniques (*1−6*). However,
a direct and profitable approach to studying the response of crystalline
or amorphous chain segments is vibrational spectroscopy. In many
polymers the IR spectra exhibit absorption bands that can be assigned
to vibrations of chain segments in the crystalline or noncrystalline
phase only. Furthermore, these vibrational bands may be characteristic of the details of local conformation. If band assignments are established and if the directions of the transition moments with respect to
the chain axis are known, vibrational spectroscopy offers the advantage that stress-induced changes in orientation, conformation, and
packing can all be measured.

The sensitivity of the spectroscopic technique in detecting structural changes associated with polymers deformed by external stress
has been illustrated (*7−12*). Changes in frequency and intensity have
been related to molecular stress (*7, 10*), chain axis orientation (*8, 11*),
viscoelastic behavior (*12*), and conformational changes (*7, 10, 12*).

[1] To whom correspondence should be addressed.

0065−2393/83/0203−0533$06.00/0

Most of the experiments in vibrational spectroscopic studies, however, suffer because the short time-dependent phenomena cannot be measured conveniently. Fourier transform IR spectroscopy (FTIR) has proven to be a powerful tool in polymer characterization. However, it has been applied only to the observation of events that are stationary in time, or at least stationary with respect to the measurement time. The multiplex characteristic (the ability to measure all spectral elements) of the interferometer together with the high energy throughput, provide FTIR with a substantial gain in signal to noise ratio for a given measurement time, as compared to a dispersive instrument. Therefore, the use of FTIR to study time-dependence phenomena is feasible. Measurements of band position, shape, and relative intensity are retained.

It is not practical, if not impossible, for the interferometer to follow mechanically the rapid structural changes that can occur in a millisecond time scale. Furthermore, fast microstructural changes in polymers can be masked due to coadding scans to improve signal-to-noise ratio (10). Therefore, we have developed a time resolved FTIR to follow the rapid structural changes. Our experimental apparatus and preliminary results are reported here.

Experimental

Stretching Apparatus. Our laboratory has developed a computer-controlled hydraulic stretching device that can be used for stress relaxation, creep, and harmonic deformation experiments in conjunction with a rapid scanning FTIR. This hydraulic system is schematically outlined in Figure 1 and incorporates a closed-loop DC servo-controller that drives a hydraulic servo-valve. A solid state DC to DC displacement transducer (LVDT) and a load cell supply information regarding displacement and load applied to the sample, respectively. Inputs to the controller can also be attached to a sweep function generator, enabling a predefined periodic oscillatory strain to be applied to a prestretched sample (to prevent slackening, but well below the yield point). In addition, both the load and displacement values applied to the sample are available in digitized form and are accessible by our minicomputer. This extremely accurate macroscopic mechanical information in the form of stress–strain data is collected along with microscopic information in the form of IR spectra. Through the feedback loop to the controller, both the stroke and load are programmable.

Time-Resolved Fourier Transform IR Spectroscopy. In these experiments, repetitive oscillatory strains are applied to prestretched polymers and many cycles of data are collected and sorted in time space. The interferogram for any particular segment of time will be reconstructed and transformed into the frequency space. The principles representing several different approaches to this type of time-resolved FTIR are discussed elsewhere (13–16). Each approach incorporates some form of ordered sampling technique that generally results in the collection of a large amount of data that must later be sorted to obtain properly time ordered interferograms.

Our interferometer is a continuous scanning type and uses a HeNe laser to determine accurately the changing position of a moving mirror relative to

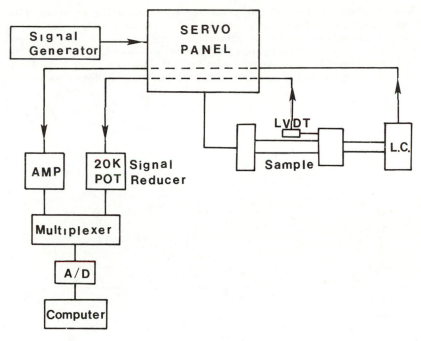

Figure 1. Schematic drawing of the stretching apparatus.

the fixed mirror. The transform of a monochromatic source (6328 Å of the HeNe laser) is sinusoidal in nature. The zero crossings of this curve define the coordinates in the retardation space, x. The moving mirror at constant velocity therefore intercepts these zero crossings at well-defined time intervals, Δt, which can be calculated as

$$\Delta t = \frac{X_m}{vN} \tag{1}$$

where X_m is the maximum optical path difference, v is the velocity, and N is the number of data points. For a time evolving event, a single scan of the interferometer represents a collection of interferogram elements, $I(x, t_n)$, where x is the optical path difference and t_n is the time at nth data point. It is shown in Equation 2.

$$t_n = t_0 + \frac{x}{v} \tag{2}$$

A constant time interval, t_0, is measured relative to a white light central burst that is approximately 300 data points before the IR central burst. In most of these experiments t_0 was taken to be zero.

The first event is triggered by the white light crossing. In subsequent events, triggers are delayed by multiples of Δt. This is continued until N events have been collected and subsequently provide a definite relationship between the retardation of the moving mirror and the time into the event. Thus the set of the initial interferogram elements of each of these N spectra represents the interferogram of the sample at time t_0 (strain equals zero), and is shown in Figure 2. The interferogram corresponding to the structure of the sample at a time of $n\Delta t$ into the deformation are the set of interferogram elements that correspond to the nth data point of each of the N time averaged interferograms. With our instrumentation, the theoretical maximum time resolution achievable is 15 μs (time/data point). Our computer software is based on a commercial package generally sold with the Nicolet instruments.

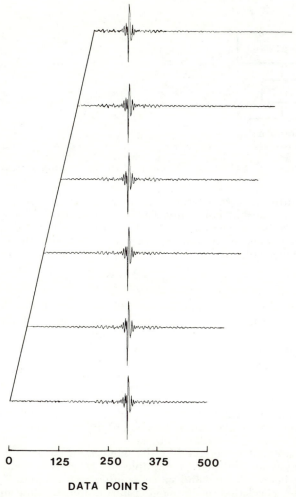

0   125   250   375   500

DATA POINTS

Figure 2. The six successive interferograms collected under the time-resolved mode. The line represents the interferogram elements for t = 0.

Results

We were interested in applying the new spectroscopic technique to elucidate the structural response rate of the subunits in the ion-containing ethylene methacrylic acid copolymers. These polymers have crystalline and amorphous lamellar phases of polyethylene, and ion domains of submicroscopic dimensions (10 Å–1 μm). However, neither the aggregate structure nor size has been defined completely (*17–20*). Although the enhancement of properties of these ionomers is well known, a number of fundamental problems remain in understanding the structure–property relationships.

First, it is important to establish and calibrate the changes in vibrational band width (shape), intensity, frequency, and dichroism as a function of strain amplitude and strain rate. In the case of several polymers, particularly polyethylene, the polarized IR method has the advantage that the transition moment's directions (with respect to the chain axis) are known to a good approximation for many IR absorption bands. The dichroic ratios measured for several bands are shown in Figure 3. For strain amplitudes as small as 2–4%, measurable changes

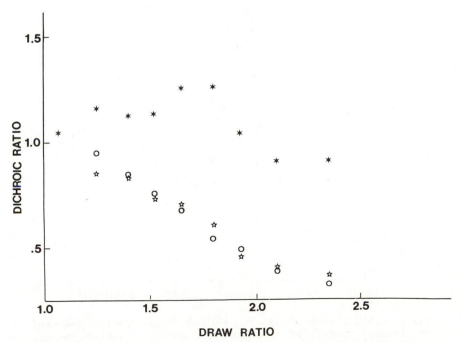

*Figure 3. Dichroic ratio of selected IR bands in ethylene methacrylic acid copolymer. Key: ○, 1700 cm^{-1}; *, 935 cm^{-1}; and ☆, 720 cm^{-1}.*

occur in the dichroic ratios of several well-defined bands, such as the CH₂ group frequencies (CH_2 stretching, bending, rocking, etc.), $C=O$ stretching vibration, and $O-H$ stretching of the acid group. Second, several bands also exhibit frequency changes for spectra taken with incident polarization parallel to the applied stress. This is shown in Figure 4.

The experimental conditions for the initial time-resolved experiment were number of data points, 2048 (8-cm⁻¹ resolution); time resolution, 500 μs; period of external stress, 50,000 μs (i.e., 20-Hz defor-

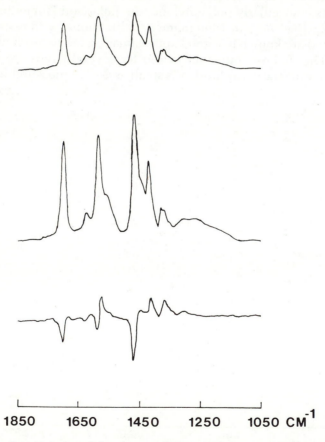

Figure 4. Strain-induced frequency shifts in ethylene methacrylic acid copolymer containing Zn^{2+} ions. Key: top, parallel polarization; middle, perpendicular polarization; and bottom, the ratio of parallel to perpendicular.

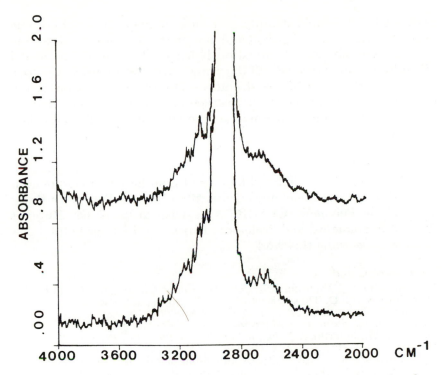

Figure 5. The intensity change associated with the 2673-cm⁻¹ band. The lower curve is at t = 0; *the upper curve is obtained 500 μs later.*

mation); strain amplitude, 3%; velocity, 1.43 cm/s; and measurement time, ~50 min. Each spectrum obtained in the time-resolved mode without straining the sample yields identical spectra as a function of time. However, when external strain is applied, the successive time-resolved spectra exhibit features that follow the strain amplitude. One such region is shown in Figure 5. We observed the dichroic ratio of the 2673-cm⁻¹ parallel band to change as a function of strain in times as short as 500 μs. This appears to be the first application of the time-resolved technique to following the structural changes of polymeric solids.

The number of times the sample is deformed in a time-resolved spectroscopic experiment is proportional to the number of coadded scans needed to achieve a respectable signal-to-noise ratio, and to the number of data points needed to reconstruct the interferogram to achieve the resolution needed. The main difficulty we encountered is associated with sample stability. For each sample (~10 μm thick) one needs to determine the period and amplitude of the external driving force for reproducible deformation behavior. Despite experimental difficulties, time-resolved FTIR can contribute significantly to a better understanding of the structure–property relationship. A detailed comparison between microstructural response in low density polyethylene and ethylene–methacrylic ionomers will be published elsewhere.

Acknowledgments

This research is supported by a grant from Research Corporation and one from the Cooperative University of Massachusetts-Industry Research on Polymers (CUMIRP). We wish to thank Dr. Huppler (Nicolet Instruments) and Professors Farris and Stein for helpful discussions concerning this work.

Literature Cited

1. Wilkes, G. L. *J. Macromol. Sci. Rev. Macromol. Chem.* **1974**, *10*, 149.
2. Suhiro, S., Ph.D. Thesis, Kyoto Univ., Kyoto, Japan, 1978.
3. Oda, T.; Stein, R. S. *J. Polym. Sci., Part A* **1972**, *2(10)*, 685.
4. Read, B. E.; Stein, R. S. *Macromolecules* **1968**, *1*, 116.
5. Glenz, W.; Peterlin, A. *J. Polym. Sci., Polym. Phys. Ed.* **1971**, *9*, 1191.
6. Read, B. E.; Hughes, D. A. *Polymer* **1972**, *13*, 495.
7. Wool, R. P.; Statten, W. O. In "Applications of Polymer Spectroscopy"; Brame, E. G., Jr., Ed.; Academic: New York, 1978; Chap. 12.
8. Uemura, Y.; Stein, R. S. *J. Polym. Sci., Part A* **1972**, *2(10)*, 1691.
9. Reynolds, J.; Sternstein, S. S. *J. Chem. Phys.* **1964**, *41*, 47.
10. Siesler, H. W. *Makromol. Chem.* **1979**, *180*, 2261.
11. Gotoh, R.; Takenaka, T.; Hayama, N. *Kolloid-Z.* **1965**, *205*, 18.
12. Davies, G. R.; Smith, T.; Ward, I. M. *Polymer* **1980**, *21*, 221.
13. Sakai, H.; Murphy, R. E. *Appl. Opt.* **1978**, *17*, 1342.
14. Murphy, R. E.; Cook, F. H.; Sakai, H. *J. Opt. Soc. Am.* **1975**, *65*, 600.
15. Mantz, A. W. *Appl. Opt.* **1978**, *17*, 1347.
16. Mantz, A. W. *Appl. Spectrosc.* **1976**, *30*, 459.
17. Longworth, R.; Vaughan, D. J. *Polym. Prepr., Am. Chem. Soc., Div. Polym. Chem.* **1968**, *9*, 525.
18. Marx, C. L.; Caulfield, D. F.; Cooper, S. L. *Macromolecules* **1973**, *6*, 344.
19. Binsberger, F. L.; Kroon, G. F. *Macromolecules* **1973**, *6*, 145.
20. MacKnight, W. J.; Taggart, W. P.; Stein, R. S. *J. Polym. Sci., Polym. Symp.* **1974**, *45*, 113.

RECEIVED for review October 14, 1981. ACCEPTED February 8, 1982.

IR Spectra of Polymers and Coupling Agents Adsorbed onto Oxidized Aluminum

F. J. BOERIO and C. A. GOSSELIN

University of Cincinnati, Department of Materials Science and Metallurgical Engineering, Cincinnati, OH 45221

The principles of external reflection IR spectroscopy are considered and the design of experiments for obtaining IR spectra of thin films on metal mirrors is described. The IR reflection technique is then used to determine the structure of thin films on mechanically polished 2024 aluminum alloy mirrors. The air-formed oxide on such mirrors is characterized by an absorption band near 960 cm^{-1} and is identified as amorphous Al_2O_3. This oxide is converted to pseudoboehmite during immersion in distilled water at temperatures up to 100 °C. The structure of films formed by adsorption of an organosilane coupling agent, γ-aminopropyltriethoxysilane (γ-APS), onto freshly polished mirrors is a strong function of pH and adsorption time. Films formed by adsorption from solutions acidified by addition of HCl are composed of polysiloxanes in which the amino groups are protonated. Films obtained during 1-min adsorption at pH 10.4 are similar to those previously obtained on iron and are composed of polysiloxanes. When γ-APS is adsorbed for 15 min at pH 10.4, the air-formed oxide is dissolved and composite films of γ-APS and an aluminum oxide are deposited on the surface. The structure of thin films formed by epoxy polymers adsorbed onto aluminum mirrors also is considered.

T HIN FILMS ON METAL SURFACES are important in areas such as adhesion, corrosion, corrosion inhibition, catalysis, and lubrication. As a result, there has been substantial interest in developing surface analysis techniques such as x-ray photoelectron (XPS), Auger electron (AES), and secondary ion mass spectroscopy (SIMS) for determining

0065–2393/83/0203–0541$06.00/0

the structure of such films. External reflection IR spectroscopy is potentially an extremely useful technique for determining the structure of thin films on metal surfaces. However, few such applications have been reported because of the relatively low sensitivity of IR spectroscopy for all but the most carefully designed experiments.

Any attempt to obtain the IR spectrum of a thin film on the surface of a metal necessarily involves a reflection experiment. The most important aspect of this experiment concerns the standing wave created by the combined incident and reflected beams. Under most circumstances, this wave has nearly zero electric field amplitude at the surface of the metal, and little interaction with a surface film is obtained. However, when large, almost grazing, angles of incidence and radiation polarized parallel to the plane of incidence are used, the electric field amplitude of the standing wave at the surface of the metal is greatest and maximum interaction with a surface film is obtained.

The size or depth of an absorption band in external reflection IR spectroscopy may be obtained using the formalism given by Heavens (1). Consider a metal substrate with refractive index $\tilde{n}_2 = n_2 - ik_2$ covered by a thin film with refractive index $\tilde{n}_1 = n_1 - ik_1$ as shown in Figure 1 and let the refractive index of the surrounding medium be n_0. When parallel polarized radiation is reflected from the filmed substrate as shown in Figure 1, the total electric field amplitude reflected back into the surrounding medium is given by

$$r = (r_1 + r_2 e^{-2i\delta_1})/(1 + r_1 r_2 e^{-2i\delta_1}) \tag{1}$$

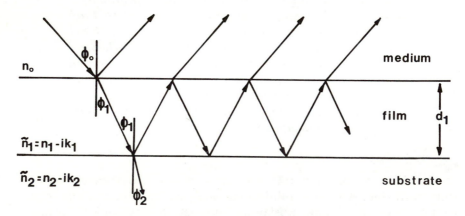

Figure 1. Model for reflection–absorption IR spectroscopy from a thin film on a reflecting substrate.

where

$$r_1 = \frac{n_0 \cos \phi_1 - \tilde{n}_1 \cos \phi_0}{n_0 \cos \phi_1 + \tilde{n}_1 \cos \phi_0} = \rho_1 e^{i\beta_1} \tag{2}$$

$$r_2 = \frac{\tilde{n}_1 \cos \phi_2 - \tilde{n}_2 \cos \phi_1}{\tilde{n}_1 \cos \phi_2 + \tilde{n}_2 \cos \phi_1} = \rho_2 e^{i\beta_2} \tag{3}$$

and

$$\delta_1 = (2\pi \tilde{n}_1 d_1 \cos \phi_1)/\lambda \tag{4}$$

where δ_1 is the phase thickness of the film and β_1 and β_2 are the phase shifts for radiation reflected at the air/film and film/substrate interfaces, respectively. The quantities ϕ_1 and ϕ_2 can be obtained from ϕ_0, n_0, $\tilde{n}_1$, and $\tilde{n}_2$ using Snell's law:

$$n_0 \sin \phi_0 = \tilde{n}_1 \sin \phi_1 \tag{5}$$

$$\tilde{n}_1 \sin \phi_1 = \tilde{n}_2 \sin \phi_2 \tag{6}$$

However, because $\tilde{n}_1$ and $\tilde{n}_2$ are complex, ϕ_1 and ϕ_2 are also complex and do not represent angles of refraction. The intensity of the reflected beam (reflectivity) is given by $R = rr^*$ where r^* is the complex conjugate of r. The depth of an absorption band then is obtained as

$$A = (R - R_0)/R_0 \tag{7}$$

where R_0 and R are the reflectivities for $k_1 = 0$ and for $k_1 \neq 0$.

When radiation polarized perpendicular to the plane of incidence is used, the analysis remains exactly the same but r_1 and r_2 are obtained as

$$r_1 = \frac{n_0 \cos \phi_0 - \tilde{n}_1 \cos \phi_1}{n_0 \cos \phi_0 + \tilde{n}_1 \cos \phi_1} = \rho_1' e^{i\beta_1'} \tag{8}$$

$$r_2 = \frac{\tilde{n}_1 \cos \phi_1 - \tilde{n}_2 \cos \phi_2}{\tilde{n}_1 \cos \phi_1 + \tilde{n}_2 \cos \phi_2} = \rho_2' e^{i\beta_2'} \tag{9}$$

These equations are rather difficult to solve by hand but are programmed easily for solution by computer techniques. The results obtained indicate that the band depth A is negligible for perpendicular polarized radiation for all angles of incidence. The band depth for parallel polarized radiation is also negligible for angles of incidence less than about 50° but then rises sharply, reaching a maximum near 88°. An example is shown in Figure 2 where the band depth is shown as a function of angle of incidence for parallel polarized radiation and for $n_0 = 1.0$, $\tilde{n}_1 = 1.3 - 0.1i$, $\tilde{n}_2 = 3.0 - 30.0i$, $d_1 = 1.50$ nm, and $\lambda = 7.5\,\mu m$. These optical constants represent a moderately strong IR absorption band in a thin film on a highly reflecting metallic substrate (2).

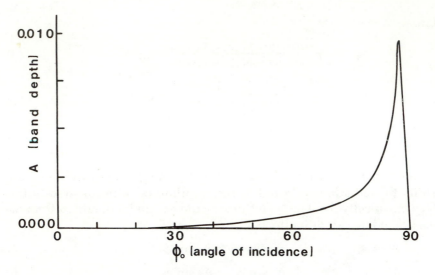

Figure 2. Band depth as a function of the angle of incidence for a thin film on a metallic substrate; $n_0 = 1.0$, $\tilde{n}_1 = 1.3 - 0.1i$, $\tilde{n}_2 = 3.0 - 30.0i$, $d = 1.50$ nm, and $\lambda = 7.5$ μm.

The optimum angle of incidence of approximately 88° is difficult to obtain in practice. As a result, the sensitivity obtained in most experiments is somewhat less than the theoretical maximum. To obtain usable spectra from extremely thin films (e.g., monomolecular) it is usually essential to enhance the band depth by making multiple reflections or by using a computer-controlled spectrophotometer to average several scans of the same spectral region. The number of reflections that can be used is related to the reflectivity of the substrate. The sensitivity of the technique is, therefore, greatest for good reflectors such as copper, gold, silver, and aluminum and is somewhat less for moderate reflectors such as iron and titanium.

The purpose of this chapter is to demonstrate the use of IR spectroscopy for surface analysis by determining the structure of thin films formed by the adsorption of polymers and organofunctional silanes onto the oxidized surfaces of aluminum mirrors. Silanes such as γ-aminopropyltriethoxysilane (γ-APS) are widely used as coupling agents to improve the wet strength of glass fiber reinforced composites. The use of silanes as primers to improve the hydrothermal stability of aluminum/epoxy adhesive bonds also has been described (3, 4). However, very little information has been reported concerning the molecular structure of thin films formed by the adsorption of silanes onto aluminum. Similarly, epoxy polymers are widely used for adhesive bonding of aluminum but little is known about the molecular structure of such polymers adsorbed onto aluminum.

Experimental

Aluminum sample mirrors for IR spectroscopy were prepared by mechanically polishing 2024−T3 alloy with magnesium oxide polishing compound
on kitten ear cloth. The mirrors then were rinsed repeatedly in distilled water
and blown dry using a strong stream of nitrogen.

Films were formed on the freshly polished mirrors using the techniques
described here and the mirrors were then mounted in an external reflection
accessory (Harrick Scientific Co.) configured to provide two reflections at 78°
as shown in Figure 3. IR spectra of the surface films then were obtained using
a Perkin−Elmer Model 180 IR spectrophotometer. A silver bromide wire grid
polarizer was placed in front of the entrance slit to the monochromator and
oriented so as to transmit radiation polarized parallel to the plane of incidence
at the sample mirrors. In most cases, spectra were recorded differentially by
mounting a pair of polished, but unfilmed, mirrors in a second reflection
accessory in the reference beam of the spectrophotometer. This procedure
allowed elimination of sloping baselines due to changes in the substrate reflectivity and elimination of absorption bands due to surface oxides on the

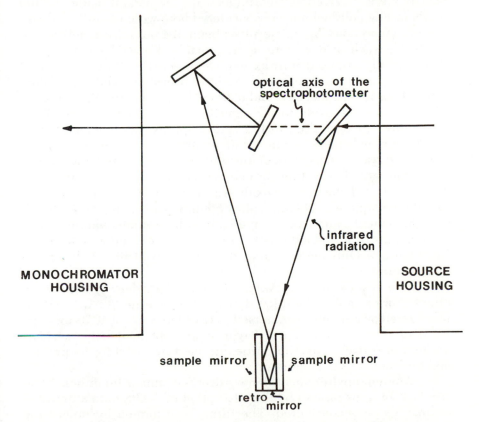

*Figure 3. Sampling arrangement for external reflection IR spectros-
copy using any number of reflections at a 78° angle of incidence.*

substrate. However, in some cases the reflection accessory was removed from the reference beam so that the structure of the oxide could be determined. The sample mirrors were usually examined with an ellipsometer (Rudolph Research Model 436) before and after the adsorption of organic compounds, enabling the thickness of the oxide and the adsorbed films to be estimated.

Results and Discussion

Characterization of Oxides on Aluminum Mirrors. Previous results showed that IR spectroscopy was very useful for characterizing the oxides formed on electropolished copper substrates (5). Accordingly, a similar investigation was carried out for mechanically polished aluminum substrates (6). The IR spectra obtained from freshly polished aluminum mirrors were characterized by a single strong band near 960 cm^{-1} (see Figure 4A). The spectra of anodic oxides on aluminum are characterized by a similar band (7). Moreover, the thermal oxides obtained on aluminum at temperatures less than about 400 °C usually are considered as amorphous (8). The natural oxides on the mechanically polished mirrors considered here were identified, therefore, as amorphous Al_2O_3. Results obtained from ellipsometry indicated that such oxides were approximately 30–50 Å in thickness. When freshly polished mirrors were immersed in boiling distilled water for 5 min, spectra characterized by absorption bands near 3399, 3099, 1640, 1375, 1080, 847, and 650 cm^{-1} were obtained (see Figure 4B), enabling the oxide to be identified as pseudoboehmite (7, 9) having the structure $Al_2O_3 \cdot XH_2O$ with X approximately equal to two.

Freshly polished aluminum mirrors then were immersed in distilled water at various temperatures below 100 °C and the rate at which the natural oxide transformed to pseudoboehmite was observed (6). At 63 °C, only the natural oxide was obtained after 5 min, but only pseudoboehmite was obtained after 60 min (see Figure 5). At 40 °C, only the natural oxide was observed after 1 h. Pseudoboehmite began to form after 16 h, and after 24 h only pseudoboehmite was observed (see Figure 6). Only the natural oxide was observed after 22 h at room temperature.

These results are somewhat different than those obtained by others. For example, Alwitt (9) showed that bayerite (β-$Al_2O_3 \cdot 3H_2O$) was formed on aluminum immersed in distilled water at 40 °C for 24 h. No evidence for formation of bayerite was obtained here, perhaps because of surface-active impurities in the water used for hydrothermal treatment of the aluminum.

γ-Aminopropyltriethoxysilane Adsorbed onto Aluminum. The structure of films formed by the adsorption of γ-APS onto aluminum mirrors also was considered. The films were formed by immersing freshly polished mirrors into 1% aqueous solutions of γ-APS for appropriate times and then blowing the excess solution off the mirrors

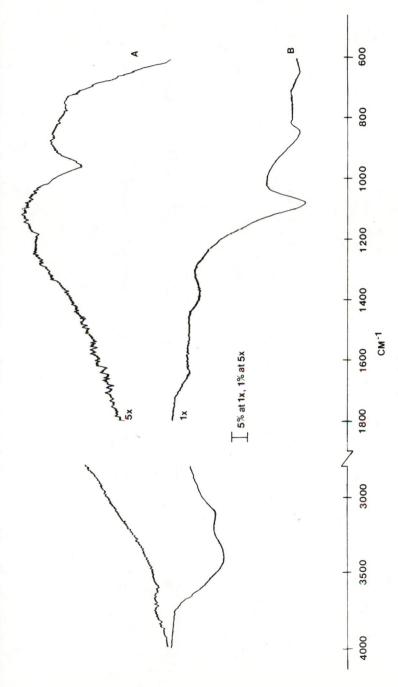

Figure 4. IR spectra of oxides formed on mechanically polished 2024 aluminum alloy: A, natural oxide; B, oxide obtained after immersion in boiling water for 5 min. (Reproduced with permission from Ref. 6. Copyright 1981, Gordon and Breach Science Publishers, Inc.)

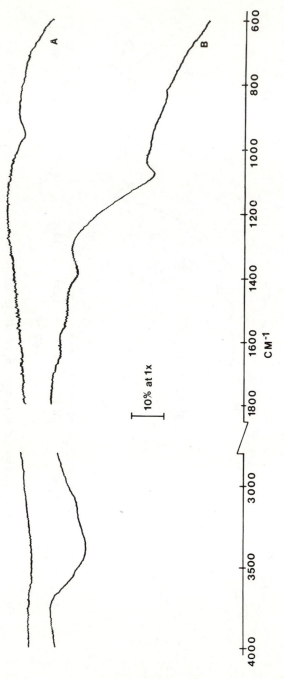

Figure 5. IR spectra of oxides formed on mechanically polished 2024 aluminum alloy during immersion in water at 63 °C for 5 min (A) and 60 min (B).

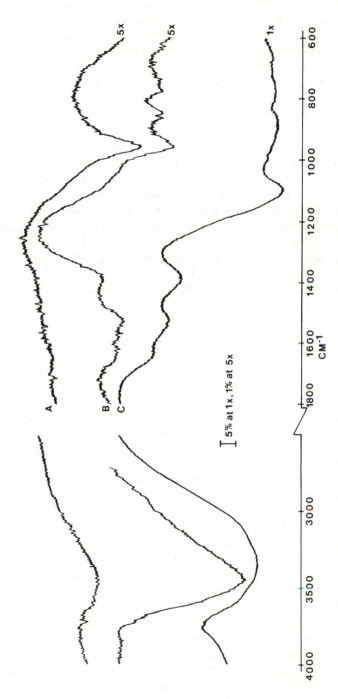

Figure 6. IR spectra of oxides formed on mechanically polished 2024 aluminum alloy during immersion in water at 40 °C for 1 h (A), 16 h (B), and 24 h (C). (Reproduced with permission from Ref. 6. Copyright 1981, Gordon and Breach Science Publishers, Inc.)

using a strong stream of nitrogen. Films formed by adsorption from solutions acidified to pH 6.8 by adding HCl were characterized by absorption bands near 3000, 1600, 1500, and 1110 cm^{-1} (*see* Figure 7). The bands near 1600 and 1500 cm^{-1} were assigned to asymmetric and symmetric deformation modes of protonated amino groups (*10*), indicating that γ-APS was adsorbed at pH 6.8 to form amine hydrochlorides. The strong, broad band near 3000 cm^{-1} was assigned to the stretching vibrations of NH$_3^+$ groups (*10*). The stretching vibrations of the ethoxy groups in the γ-APS monomer were assigned to intense bands near 1105, 1080, and 960 cm^{-1} (*11*). These bands were not observed for γ-APS adsorbed onto aluminum at pH 6.8, indicating that the adsorbed silane was highly hydrolyzed. However, the strong band near 1110 cm^{-1} (*see* Figure 7) was assigned to an SiOSi asymmetric stretching vibration, indicating that the adsorbed silane had polymerized on the aluminum surface to form siloxane polymers. Overall, the IR spectra of γ-APS adsorbed onto aluminum at pH 6.8 were similar to those of γ-APS adsorbed onto iron at similar pH values (*12*), indicating that the structures were also similar.

Films formed by adsorption of γ-APS onto aluminum at pH 8.5 (*6*) were similar to those formed at pH 6.8 and were characterized by absorption bands near 3300, 3000, 1600, 1500, and 1080 cm^{-1} (*see* Figure 8A). The bands near 3000, 1600, and 1500 cm^{-1} were assigned again to the stretching and deformation modes of NH$_3^+$ groups. The strong band near 1080 cm^{-1} was assigned again to an SiOSi asymmetric stretching vibration. However, in an earlier investigation of γ-APS adsorbed onto iron mirrors at comparable pH values, the SiOSi asymmetric stretching mode was observed near 1130 cm^{-1} (*12*). The low frequency observed for the aluminum substrates was considered to indicate that γ-APS was adsorbed onto aluminum as low molecular weight oligomers, perhaps including some SiOAl bonds. The broad band near 3300 cm^{-1} for γ-APS adsorbed onto aluminum at pH 8.5 (*see* Figure 8A) was assigned to the stretching vibrations of residual silanol groups.

The structure of films formed by the adsorption of γ-APS onto aluminum mirrors from 1% aqueous solutions at pH 10.4 depended on the adsorption time (*6*). Films formed during 1-min immersion were similar to those previously observed for the adsorption of γ-APS onto iron at pH 10.4 (*12*) and were characterized by absorption bands near 2920, 1630, 1570, 1470, 1330, and 1100 cm^{-1} (*see* Figure 8B). The band near 2920 cm^{-1} was assigned to a methylene stretching vibration in the propylamine groups and the band near 1100 cm^{-1} was assigned to an SiO asymmetric stretching vibration. Bands were observed near 1570, 1470, and 1300 cm^{-1} for γ-APS adsorbed onto iron at pH 10.4 and were the subject of considerable discussion (*11*, *12*). However, Boerio

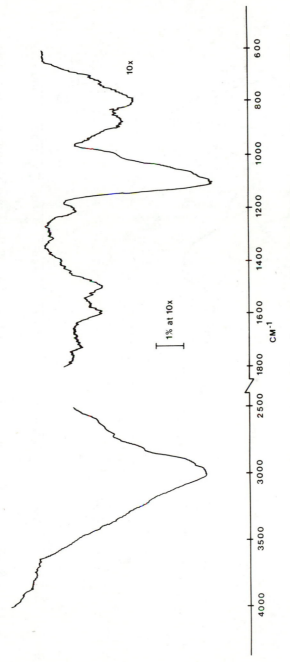

Figure 7. IR spectrum of thin film formed by adsorption of γ-APS onto aluminum mirrors from 1% aqueous solution of γ-APS at pH 6.8.

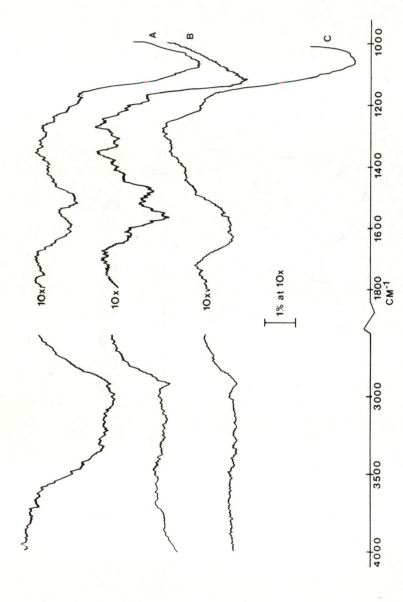

Figure 8. IR spectra of films formed by the adsorption of γ-APS films onto aluminum mirrors. Key: A, 15-min adsorption at pH 8.5; B, 1-min adsorption at pH 10.4; and C, 15-min adsorption at pH 10.4. (Reproduced with permission from Ref. 6. Copyright 1981, Gordon and Breach Science Publishers, Inc.)

and Williams showed that these bands are not observed for γ-APS adsorbed onto iron in the absence of CO_2 and assigned them to an amine bicarbonate species (13). They also observed bands near 1640, 1570, and 1470 cm^{-1} for certain aliphatic amines such as n-dodecyl-amine that were exposed to air for long periods of time but not for the same amines recrystallized in the absence of CO_2 (13). Accordingly, the bands near 1640, 1570, 1470, and 1330 cm^{-1} for γ-APS adsorbed onto aluminum at pH 10.4 for 1 min also were assigned to an amine bicarbonate species.

Films formed during 15-min adsorption at pH 10.4 were characterized by a weak band near 2920 cm^{-1}, a broad weak band extending from 1650−1500 cm^{-1}, and a strong band near 1080 cm^{-1} (*see* Figure 8C). The band near 2920 cm^{-1} was assigned to a methylene stretching mode as indicated earlier. The band near 1080 cm^{-1} similarly was assigned to an SiOSi asymmetric stretching mode. As noted already, the corresponding vibration for γ-APS adsorbed onto iron at pH 10.4 is near 1130 cm^{-1} and the relatively low frequency for aluminum substrates may indicate a strong interaction between γ-APS and the oxide (6). However, aluminum oxides are soluble at pH 10.4 (9) and the strong band observed near 1080 cm^{-1} for γ-APS adsorbed onto aluminum for 15 min at pH 10.4 may be related in part to an oxide that was precipitated on the surface following dissolution of the native oxide. In fact, when the reflection accessory was removed from the reference beam, the oxide absorption band near 960 cm^{-1} (*see* Figure 9A) was observed to decrease in intensity during 15-min adsorption of γ-APS at pH 10.4 (*see* Figure 9B), indicating dissolution of the native oxide and, perhaps, precipitation of a different oxide on the surface. The broad band extending from 1650−1500 cm^{-1} was assigned to the deformation mode of the amino groups but also may indicate hydration of the precipitated oxide.

The amine bicarbonate species formed by the adsorption of γ-APS from aqueous solutions at pH 10.4 onto iron and copper substrates was shown to be unstable (12). During atmospheric exposure, the band near 1470 cm^{-1} gradually decreased in intensity while the band near 1570 cm^{-1} gradually increased in frequency and finally approached 1590 cm^{-1}. Boerio and Williams used x-ray photoelectron spectroscopy to show that Cu(I) ions found in γ-APS films adsorbed onto copper at pH 10.4 were oxidized to Cu(II) ions during exposure to atmospheric moisture and assigned the band observed near 1590 cm^{-1} to the deformation mode of amino groups coordinated to Cu(II) ions (14). Somewhat similar results may be obtained for iron substrates. By comparison, films formed by the adsorption of γ-APS onto aluminum for 15 min at pH 10.4 were stable during atmospheric exposure. Few changes were observed in the IR spectra of such films even after

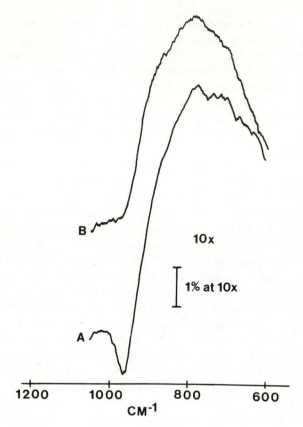

Figure 9. IR spectra of the characteristic amorphous Al$_2$O$_3$ band near 960 cm^{-1} for freshly polished aluminum mirrors (A) and aluminum mirrors following 15-min immersion in 1% aqueous solution of γ-APS at pH 10.4 (B).

exposure times as long as 7 days (*see* Figure 10), indicating that such films may be composed at least partly of a reprecipitated oxide.

Epoxy Polymers Adsorbed onto Aluminum.   We also investigated the structure of films formed by epoxy polymers adsorbed onto aluminum mirrors. The spectrum shown in Figure 11 was obtained from mirrors that were spin-coated at 3,000 revolutions per minute (rpm) with a 0.1% solution of an epoxy polymer (Epon 1007, Shell Chemical Co.) in methyl ethyl ketone to provide a film approximately 40 Å in thickness. The spectrum was characterized by strong bands near 1508, 1240, 1179, and 827 cm^{-1} and by weaker bands near 1600 and 1030 cm^{-1}. The strong band near 1508 cm^{-1} and the weak band near 1600 cm^{-1} were assigned to stretching modes of the benzene rings. The strong band near 1240 cm^{-1} and the weak band near 1030 cm^{-1} were assigned

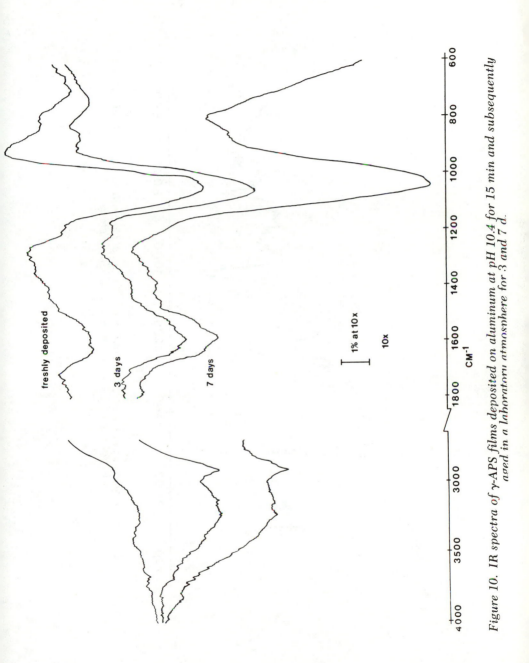

Figure 10. IR spectra of γ-APS films deposited on aluminum at pH 10.4 for 15 min and subsequently aged in a laboratory atmosphere for 3 and 7 d.

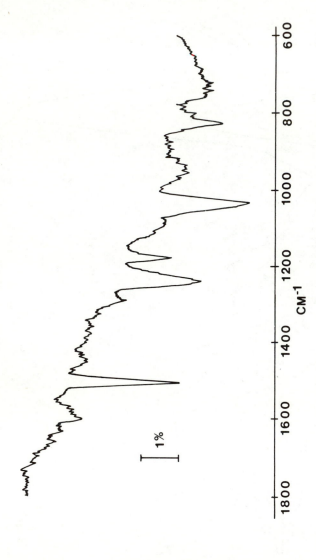

Figure 11. IR spectrum of epoxy polymer (Epon 1007, Shell Chemical Co.) spin-coated onto aluminum mirrors from 0.1% solution in methyl ethyl ketone.

to phenyl ether and aliphatic ether stretching modes, respectively. In-plane and out-of-plane deformation vibrations of ring hydrogen atoms were assigned near 1179 and 827 cm^{-1}, respectively.

The electric vector in external reflection IR spectroscopy is perpendicular to the surface of the sample mirrors. As a result, vibrations having transition moments perpendicular to the surface appear more strongly in reflection spectra than do vibrations having transition moments parallel to the surface. This polarization effect can be used to determine the orientation of adsorbed species that form ordered structures on the surface of metals. For example, the relative intensities of the symmetric and asymmetric carboxylate stretching modes were used to determine the orientation of lauric acid adsorbed onto iron mirrors (*15*). Such orientation effects can sometimes be observed in reflection spectra of epoxy polymers adsorbed onto aluminum. However, the results are poorly reproducible, perhaps because of the heterogeneous nature of most commercial epoxy polymers. More reproducible results may be obtained using materials with narrower molecular weight distribution.

Conclusions

The results just described indicate that external reflection IR spectroscopy is an effective technique for some types of surface analysis. IR spectra of films only a monomolecular layer in thickness can be obtained. Subtle changes in the structure of surface oxides, chemical reactions occurring in adsorbed species, and the orientation of ordered adsorbed species can be determined.

The results described also indicate that the adsorption of γ-aminopropyltriethoxysilane onto aluminum from aqueous solutions is a complex process that is very dependent on pH. At high pH values, the natural oxide is dissolved readily. A new oxide is precipitated and γ-APS also is adsorbed. At intermediate pH values, the dissolution of the oxide is considerably slower and γ-APS is adsorbed onto the natural oxide to form siloxane polymers in which the amino groups are protonated.

Acknowledgments

This research was supported in part by grants from the Office of Naval Research and the National Science Foundation Polymers Program.

Literature Cited

1. Heavens, O. S. In "Optical Properties of Thin Solid Films"; Butterworths: London, 1955; Chap. 4.
2. Greenler, R. G. *J. Chem. Phys.* **1966**, *44*, 310.

3. Patrick, R. L.; Brown, J. A.; Cameron, N. M.; Gehman, W. G. *Appl. Polym. Symp.* **1971**, *16*, 87.
4. Boerio, F. J.; Gosselin, C. A. *Proc. 36th Annu. Conf.—SPI Reinf. Plast./ Compos. Inst. Soc. Plast. Ind.*, Sec. 2G, 1981.
5. Boerio, F. J.; Armogan, L. *Appl. Spectrosc.* **1978**, *32*, 509.
6. Boerio, F. J.; Gosselin, C. A.; Dillingham, R. G.; Liu, H. W. *J. Adhes.* **1981**, *13*, 159.
7. Vedder, W.; Vermilyea, D. A. *Trans. Faraday Soc.* **1969**, *65*, 561.
8. Dignam, M. J.; Fawcett, W. R.; Bohni, H. *J. Electrochem. Soc.* **1966**, *113*, 656.
9. Alwitt, R. S. In "Oxides and Oxide Films"; Diggle, J. W.; Vijh, A. K., Eds.; Dekker: New York, 1976; Vol. 4, Chap. 3.
10. Colthup, N. B.; Daly, L. H.; Wiberley, S. E. In "Introduction to Infrared and Raman Spectroscopy," 2nd ed.; Academic: New York, 1975; pp. 324–25.
11. Boerio, F. J.; Schoenlein, L. H.; Grievenkamp, J. E. *J. Appl. Polym. Sci.* **1978**, *22*, 203.
12. Boerio, F. J.; Cheng, S. Y.; Armogan, L. *J. Colloid Interface Sci.* **1980**, *73*, 416.
13. Boerio, F. J.; Williams, J. W. *Proc. 36th Annu. Conf., SPI Reinf. Plast./ Compos. Inst. Soc. Plast. Ind.*, Sec. 2F, 1981.
14. Boerio, F. J.; Williams, J. W., unpublished data.
15. Boerio, F. J.; Chen, S. L. *J. Colloid Interface Sci.* **1980**, *73*, 176.

Received for review January 4, 1982. Accepted May 11, 1982.

Fourier Transform IR (FTIR) Studies of the Degradation of Polyacrylonitrile Copolymers

M. M. COLEMAN and G. T. SIVY

Pennsylvania State University, Materials Science and Engineering Department, Polymer Science Section, University Park, PA 16802

Fourier transform IR spectroscopy (FTIR) offers considerable potential for studying the complex reactions occurring in the degradation of polyacrylonitrile (PAN) copolymers. Results obtained from studies of PAN copolymers containing methacrylic acid, acrylamide, and vinyl acetate are reviewed and results of a terpolymer of PAN containing vinyl acetate and itaconic acid are presented.

FOURIER TRANSFORM IR (FTIR) results on the degradation of polyacrylonitrile (PAN) and its α-deuterated analogue were presented previously (1, 2). From the results of these studies we concluded that the mechanism of the degradation of PAN under reduced pressure at 200 °C involved cyclization to yield an imine followed by tautomerism to the enamine. The enamine is subsequently oxidized to yield a final pyridone structure. The chemical structure of the cyclic pyridone has several implications concerning the formation of carbon fibers from PAN precursors. The ladder structure imparts rigidity along the chain in the form of sequences of cyclized groups. Additionally, we believe that extensive interchain hydrogen bonding between the polymer chains through the C=O and N—H groups will have a major role in maintaining the structure of the degraded material as it is pyrolized finally to carbon fibers. If the number of cyclized sequences in the initial stages of degradation can be maximized and the number of oxidative side reactions minimized, superior high-performance carbon fibers could be obtained. The degradation of PAN at 200 °C under reduced pressure is a relatively slow process; presumably, initiation is by anionic impurities or initial degradation products. However, inclusion of specific comonomers into the PAN chain, such as vinyl acetate (VAc), methacrylic acid (MAA), acrylamide (AM), acrylic acid (AA),

0065–2393/83/0203–0559$06.00/0

and itaconic acid (IA), markedly affects the rate of degradation. The pyrolysis of PAN copolymers containing the comonomers AA, VAc MAA, IA, and AM separately has been studied using primarily thermal analysis techniques (3, 4). Heating rates in all cases were 10 °C/min and temperatures to 500 °C were considered. However, we are concerned primarily with the initial reactions occurring in the degradation of PAN copolymers at temperatures below the onset of the major exothermic reaction (i.e., <200 °C), and under reduced pressure. At these temperatures significant weight loss does not occur and the major reaction is the cyclization of the acrylonitrile units (3).

We have discussed previously the results of FTIR studies of the degradation of three acrylonitrile (AN) copolymers containing approximately 4 wt % of MAA, AM, and VAc (5–8). This chapter summarizes the major conclusions of our previous work and presents preliminary results obtained from a terpolymer of AN, VAc, and IA.

Experimental

The PAN homopolymer has been described earlier (1). Copolymers of AN containing approximately 4 wt % of MAA, AM, and VAc were synthesized by a continuous free-radical polymerization technique and have reported weight-average molecular weights in the range of 100,000 ± 20,000. Additionally, a terpolymer of AN containing 2.2 and 3.8 wt % of VAc and IA were supplied. Thin films (<0.5 mil) employed for the FTIR degradation studies were prepared by evaporation of a 3% dimethyl sulfoxide (DMSO) solution. The solvent was removed by stirring in water for several days and subsequently dried under high vacuum. The films were then transferred to KBr discs and placed in a specially designed cell contained in the spectrometer that allows continuous monitoring while under reduced pressure at elevated temperatures. The IR spectra were recorded on a Digilab model FTS–15/B FTIR spectrometer. Each spectrum was recorded at a resolution of 2 cm^{-1} utilizing 128 scans. The frequency scale is accurate to 0.02 cm^{-1}. Particular care was made to ensure that the films were sufficiently thin to be in a range where the Beer–Lambert law is obeyed (9).

Results and Discussion

Room temperature absorbance IR spectra of undegraded PAN, PAN/VAc, PAN/AM, and PAN/MAA are presented in Figure 1. Of particular interest are the major IR bands characteristic of the acrylonitrile unit in the polymer chain at 2240 (C≡N stretching frequency), 1451 (CH$_2$ bending frequency), 1360, 1250, and 1070 cm^{-1} (mixed modes). Major IR bands characteristic of the comonomers present in the three copolymers studied are as follows: VAc and MAA at 1735 cm^{-1} (carbonyl stretching vibrations), and AM at 1735, 1685, and 1605 cm^{-1} (mixed modes containing contributions from C=O, C=N stretching, and N–H bending vibrations).

Figure 2 shows the initial absorbance IR spectra of the four polymers obtained as soon as the samples had attained a temperature of

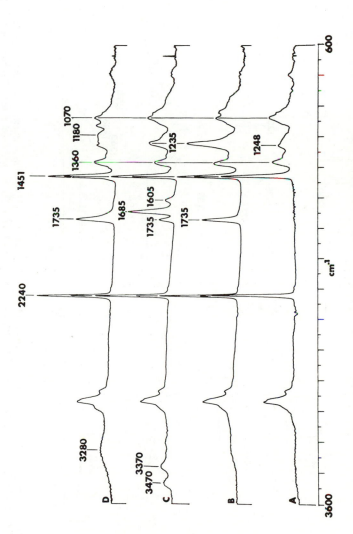

Figure 1. FTIR spectra recorded at room temperature. Key: A, PAN homopolymer; B, PAN/VAC; C, PAN/AM; and D, PAN/MAA. (Reproduced with permission from Ref. 5. Copyright 1981, Pergamon Press.)

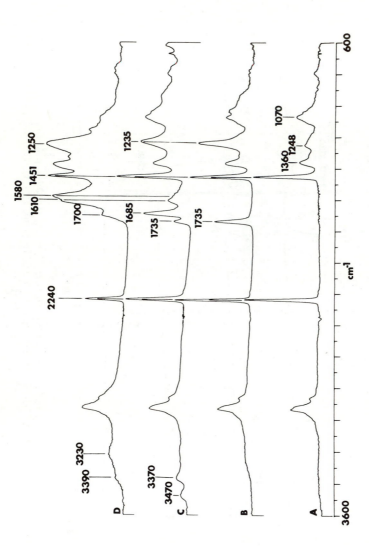

Figure 2. Initial FTIR spectra recorded immediately at a sample temperature of 200 °C. Key is the same as in Figure 1. (Reproduced with permission from Ref. 5). Copyright 1981, Pergamon Press.)

200 °C under a reduced pressure of 5×10^{-2} torr (approximately 10 min). A comparison of the room temperature spectra and the initial spectra shows that the PAN and PAN/VAc polymers show no significant degradation (IR bands at 1610 and 1580 cm^{-1} are characteristics of the cyclized pyridone structure) (*1, 2*). The spectrum of the PAN/AM copolymer does indicate that minor degradation has occurred. In contrast the PAN/MAA copolymer obviously has degraded quite extensively (note the presence of the relatively strong 1610- and 1580-cm^{-1} bands which are not present in the room temperature spectrum; the presence of a new band at 1700 cm^{-1} and the reduction in intensity of the C≡N stretching frequency at 2240 cm^{-1}).

After 4 h at 200 °C (Figure 3) one can readily detect the cyclized pyridone structure in the PAN homopolymer by the appearance of the 1610/1580-cm^{-1} bands. However, the degradation is not extensive as indicated by the relatively strong C≡N stretching frequency at 2240 cm^{-1}. The PAN/VAc copolymer definitely has degraded to a greater extent than pure PAN but there is still evidence of a significant amount of undegraded material present. Conversely, both the PAN/MAA and PAN/AM copolymers have almost completely degraded as evidenced by the very weak contribution of the C≡N stretching frequency.

To compare the relative rates of degradation of PAN and the three copolymers at 200 °C under a reduced pressure of 5×10^{-2} torr we plotted a graph of the normalized absorbance of the C≡N band at 2240 cm^{-1} as a function of time (Figure 4). The assumptions made in the preparation of this graph deserve further comment. Normalized absorbances for the C≡N stretching frequency band at 2240 cm^{-1} are readily obtained for the PAN and PAN/VAc polymers from measurement of the peak height or area of this band as a function of degradation time at 200 °C. However, in the case of the PAN/MAA and PAN/AM samples, a reduction of the absorbance of the C≡N band has occurred by the time the sample has attained 200 °C due to significant degradation. Normalization was achieved using the approximation that the absorbance of the C≡N band of the PAN and PAN/VAc polymers was determined on the same film at room temperature (Figure 1) and on immediately attaining a temperature of 200 °C (Figure 2). Band broadening and a reduction of intensity occurs at elevated temperatures, and it was determined that a decrease in peak intensity by a factor of 0.85 occurs at 200 °C. No evidence of degradation was observed in the initial 200 °C spectra of the PAN or PAN/VAc polymers (*see* Figure 2). For the PAN/AM and PAN/MAA films the absorbance of the C≡N band was measured at room temperature and multiplied by 0.85 to yield an estimation of the undegraded absorbance at 200 °C. Normalized intensities were then calculated and plotted as shown in

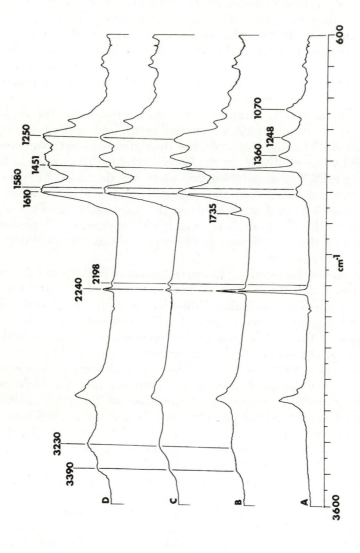

Figure 3. FTIR spectra recorded after 4 h at 200 °C. Key is the same as in Figure 1. (Reproduced with permission from Ref. 5. Copyright 1981, Pergamon Press.)

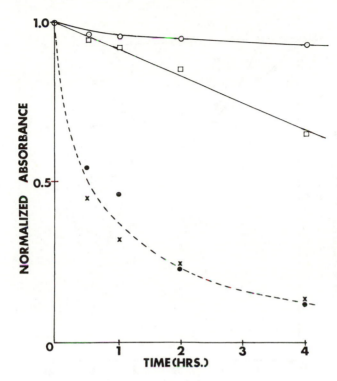

Figure 4. Normalized absorbance measurements of the C≡N band as a function of time at 200 °C. Key: ○, PAN; □, PAN/VAc; ●, PAN/MAA; and ×, PAN/AM. (Reproduced with permission from Ref. 5. Copyright 1981, Pergamon Press.)

Figure 4. Accurate absolute rates of degradation at 200 °C cannot be obtained for the PAN/AM and PAN/MAA polymers but this approximation at least demonstrates significant relative rate differences in the degradation of the four polymers.

Similar rate studies were also performed at 160 and 130 °C under reduced pressure and are summarized in Figure 5. No significant degradation was detected for the PAN homopolymer or the PAN/VAc copolymer at these temperatures. However, at 160 °C degradation of the PAN/MAA and PAN/AM copolymers occurs at a reasonable rate. These studies indicate that in the case of the PAN/MAA copolymer, the cyclization reaction is initiated almost immediately. Conversely, an induction period of approximately 1 h is necessary for the PAN/AM copolymer, after which cyclization occurs at a faster rate and is more extensive, at times exceeding 2 h, when compared to the PAN/MAA copolymer. This finding is consistent, although not as readily detectable, with the results obtained at 200 °C. The initial spectra obtained at

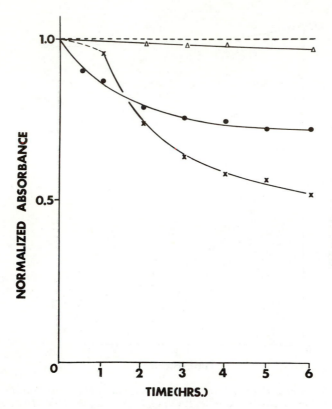

Figure 5. Normalized absorbance measurements of the C≡N stretching frequency at 2240 cm⁻¹ as a function of time and temperature. Key: ---, PAN/AM copolymer at 130 °C; △, PAN/MAA copolymer at 130 °C; ●, PAN/MAA copolymer at 160 °C; and ×, PAN/AM copolymer at 160 °C. (Reproduced with permission from Ref. 5. Copyright 1981, Pergamon Press.)

200 °C (Figure 2) show that the PAN/MAA copolymer has degraded to a greater extent than the PAN/AM copolymer. However, after 4 h there is little difference in the extent of degradation of either polymer. This result implies that initiation of the degradation of the PAN/MAA occurs more rapidly, but this is offset by a more rapid degradation of the PAN/AM copolymer upon initiation. Further support for this hypothesis is found from a consideration of the data obtained at 130 °C. The PAN/MAA copolymer still degrades at a relatively slow rate at this temperature. In the case of the PAN/AM copolymer, however, we were unable to detect any significant degradation for time periods up to 4 h.

In summary, the inclusion of MAA or AM comonomers in the PAN chain has the greatest effect on the rate of degradation. Detailed spectral results and mechanistic implications of studies of the degradation of the PAN/MAA and PAN/AM copolymers at 130 and 160 °C, respectively, are presented elsewhere (6, 8).

Finally, to further illustrate the potential of FTIR spectroscopy for degradation studies, preliminary results obtained for a terpolymer of PAN containing both VAc and IA are presented. Frankly, it would have been more desirable to study a copolymer of PAN and IA because the presence of the VAc increases the complexity of the spectra and their interpretation. The terpolymer was readily available, however, and the preliminary results sufficiently interesting to report here. Figure 6 shows the IR spectra from 1500 to 2000 cm^{-1} of the terpolymer at room temperature (A), as soon as the sample had attained a temperature of 130 °C (B), and after 15, 30, 120, and 360 min at 130 °C (C–F, respectively). At room temperature, a single band at 1740 cm^{-1} is observed and represents a composite carbonyl stretching vibration associated with both VAc and IA units. Upon heating to 130 °C, the most striking difference is the appearance of the sharp rotational bands throughout

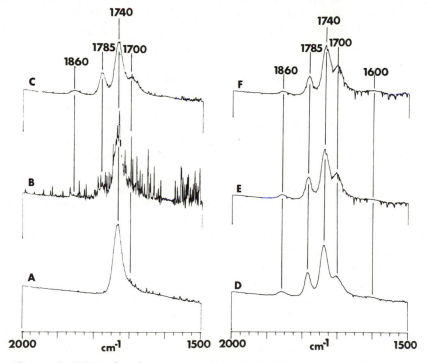

Figure 6. FTIR absorbance spectra of the PAN/IA/VAc terpolymer in the range 1500–2000 cm^{-1} as a function of temperature or time under a reduced pressure of 5 × 10^{-2} torr. Key: A, room temperature spectrum; B, initial spectrum at 130 °C; C, 15 min at 130 °C; D, 30 min at 130 °C; E, 120 min at 130 °C; and F, 360 min at 130 °C.

this spectral region that are associated with water. On close examination a weak band at 1785 cm⁻¹ is also discernible. This band, together with a weak counterpart at 1860 cm⁻¹, is more obvious after 15 min at 130 °C (Spectrum C). The bands attributable to water have now almost completely disappeared, which is to be expected because the system is under vacuum. The two bands at 1785 and 1860 cm⁻¹ are consistent with the formation of a cyclic anhydride. Thus it appears that at 130 °C, the IA unit condenses to yield the following:

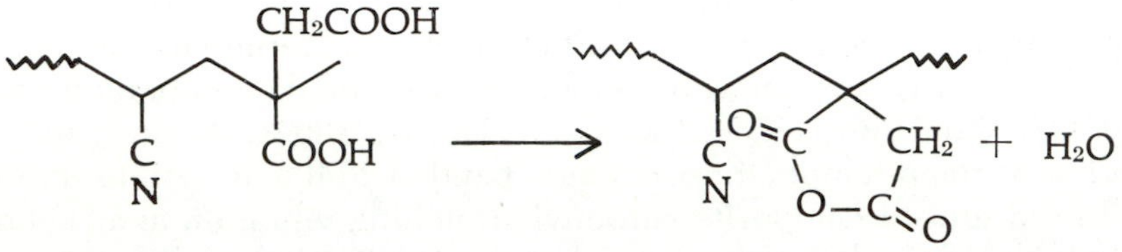

An additional band at 1700 cm⁻¹ is also observed as a shoulder on the 1740-cm⁻¹ band. This band, along with a small reduction of the nitrile absorbance, and the weak absorbance near 1600 cm⁻¹, is consistent with the formation of the pyridone-type ring structure similar to that discussed for the PAN/MAA copolymers (5). Little change is observed in the spectra between 1 and 6 h at 130 °C (Spectrum F). The 1700- and 1600-cm⁻¹ bands are slightly more prominent and the cyclic anhydride is still present. At 130 °C the VAc units are assumed to play no role in the degradation (7).

After 6 h at 130 °C the sample was heated to 160 °C. Spectra recorded as soon as the sample had attained 160 °C (A) and after 15 and 30 min at this temperature (B and C, respectively) are shown in Figure 7. Comparing these three spectra demonstrates that as a function of time at 160 °C the concentration of the cyclic anhydride is decreasing while the amount of cyclized pyridones are concurrently increasing. Raising the temperature further to 200 °C, the rate of cyclization of the PAN units increases rapidly to form pyridone-type structures as seen in Figure 8. There is no evidence of cyclic anhydrides at this temperature. At 200 °C we would also anticipate that the VAc units will now serve to initiate PAN cyclization and the spectra are consistent with this assumption (7).

This study illustrates the type of information that one may obtain from FTIR studies of the thermal degradation of a rather complex terpolymer system. We emphasize that the results are preliminary and we have yet to determine the relevance of the results on the overall degradation mechanism of this system.

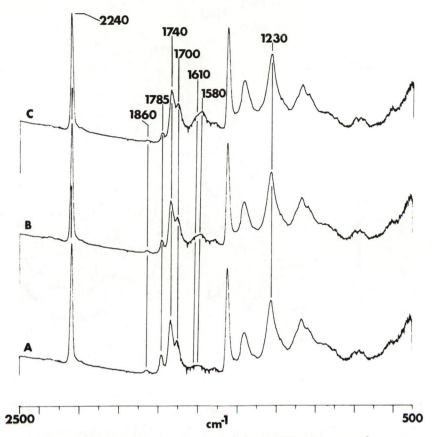

Figure 7. FTIR absorbance spectra of the PAN/IA/VAc terpolymer in the range 500–2500 cm⁻¹ as a function of time at 160 °C under a reduced pressure of 5 × 10⁻² torr. Key: A, initial spectrum at 160 °C after 6 h at 130 °C; B, 15 min at 160 °C; and C, 30 min at 160 °C.

Acknowledgments

This research was supported by the Applied Research Laboratory of The Pennsylvania State University under contract with the Naval Sea Systems Command.

The polymers used in this study kindly were supplied by S. Olive of the Monsanto Company.

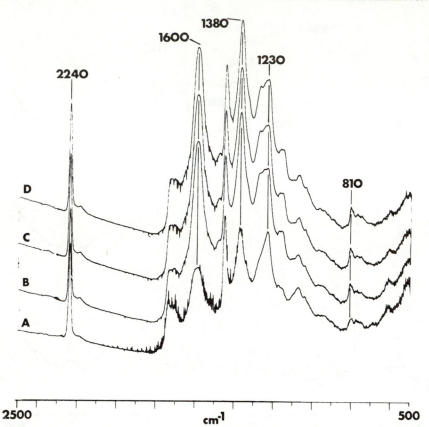

Figure 8. *FTIR absorbance spectra of the PAN/IA/VAC terpolymer in the range 500–2500 cm⁻¹ as a function of time at 200 °C under a reduced pressure of 5 × 10⁻² torr. Key: A, initial spectrum at 200 °C after 6 h at 130 °C and 30 min at 160 °C; B, 15 min at 200 °C; C, 30 min at 200 °C, and D, 1 h at 200 °C.*

Literature Cited

1. Coleman, M. M.; Petcavich, R. J. *J. Polym. Sci., Polym. Phys. Ed.* **1978,** *16,* 821.
2. Petcavich, R. J.; Painter, P. C.; Coleman, M. M. *J. Polym. Sci., Polym. Lett. Ed.* **1979,** *17,* 165.
3. Grassie, N.; McGuchan, R. *Eur. Polym. J.* **1972,** *8,* 257.
4. Grassie, N.; McGuchan, R. *Eur. Polym. J.* **1973,** *9,* 113.
5. Coleman, M. M.; Sivy, G. T. *Carbon* **1981,** *19,* 123.
6. Sivy, G. T.; Coleman, M. M. *Carbon* **1981,** *19,* 127.
7. Coleman, M. M.; Sivy, G. T. *Carbon* **1981,** *19,* 133.
8. Sivy, G. T.; Coleman, M. M. *Carbon* **1981,** *19,* 137.
9. Coleman, M. M.; Painter, P. C. *J. Macromol. Sci., Rev. Macromol. Chem.* **1977,** *16,* 197.

RECEIVED for review October 14, 1981. ACCEPTED February 10, 1982.

Fourier Transform IR Spectroscopy for the Study of Polymer Degradation

Thermal and Thermooxidative Degradation of Polyethylene Terephthalate

ELI M. PEARCE, BERNARD J. BULKIN,[1] and MO YEEN NG

Polytechnic Institute of New York, Department of Chemistry, Brooklyn, NY 11201

The thermal and thermooxidative degradation of polyethylene terephthalate (PET) were studied using IR spectroscopy. Spectra as a function of time and temperature are presented for the thermooxidative reaction and analyzed in terms of the products formed, rate of product formation, and relative stability of functional groups in the polymer. The degradations of PET samples with different properties are also compared. Thermal degradation is studied from the viewpoint of the IR spectra of the gases evolved from the sample, under a nitrogen stream, when the temperature is changed at a constant rate. For both thermal and thermooxidative degradation, the chemical reactions involved in generating the observed products are detailed.

V̇IBRATIONAL SPECTROSCOPY is being used extensively to characterize polymers, changes in polymer structure, and chemical modification of polymers. Among the important problems in polymer chemistry is stabilization against the chemical reactions caused by heat, oxygen, moisture, UV radiation, and other environmental factors.

In this chapter we illustrate the application of Fourier transform IR (FTIR) spectroscopy to study the thermal and thermooxidative degradation of polyethylene terephthalate.

[1] To whom correspondence should be addressed.

0065−2393/83/0203−0571$06.75/0

Background

The application of FTIR techniques to problems of polymer aging is really the application of Beer's law and our accumulated knowledge of group frequencies. Using group frequencies, we assign vibrational modes to functional groups. With Beer's law and spectral subtraction, the manner in which these functional groups change in concentration with time is evaluated.

To assess functional group stability, the relative functional group stability approach is used (1). This approach is a way of viewing Fourier transform (FT) spectral data acquired as a function of time. In this approach, x and y designate functional groups absorbing at v_x and v_y. If two spectra (1 and 2, undegraded and degraded polymer, respectively) are obtained, then an absorbance (A) difference spectrum can be computed as follows:

$$\Delta A_x = A_2^{v_x} - KA_1^{v_x}$$

$$\Delta A_y = A_2^{v_y} - KA_1^{v_y}$$

In a usual absorbance subtraction experiment, sample thickness and extinction coefficient are assumed to be constant throughout the measurement. Setting $K = 1$ thus gives the change in concentration of the functional group absorbing at each wavenumber. However, another way to view these data is to vary K to force the absorbance difference at one wavenumber to be zero, for example:

$$\Delta A_x = A_2^{v_x} - KA_1^{v_x} = 0 \quad \text{or} \quad K = A_2^{v_x}/A_1^{v_x}$$

Then with this value of K, ΔA_y is computed. One finds that ΔA_y is either greater than, equal to, or less than zero, indicating that functional group y is either more stable than, as stable as, or less stable than functional group x (respectively). This simple approach to absorbance subtraction, the relative functional group stability approach, is quite useful for analyzing data from a polymer being degraded. The calculation may be applied to functional groups of species formed as reaction products as well. This application will be illustrated with reaction products of polyethylene terephthalate (PET).

Experimental: Isothermal Degradation

All of the spectra and computations on spectra in this study were carried out on a Digilab Model FTS20B, run at 2-cm^{-1} resolution. Spectra were obtained in the external reflection–absorption mode at near normal incidence. A film of PET was cast on a polished aluminum plate, either from the melt or from phenol solution. The plate was then mounted on a horizontal heated stage in a specular reflectance attachment (Wilks Model 22 or Barnes Engineering Model 134). The sampling techniques are described in detail elsewhere (2).

Results: Isothermal Oxidative Degradation of PET

Figure 1 shows the difference spectra obtained in the thermooxidative degradation of PET for various times and temperatures. Each trace in Figure 1 is the initial spectrum subtracted from the spectrum at the indicated time. At 210 °C, it is clear that significant changes do not take place after 2 h. When the temperature is raised to 230 °C several differences, both positive and negative, appear after 2 h. These are explained as being due to annealing of *trans−gauche* conformers (3). An increase in temperature to 250 °C changes the chemis-

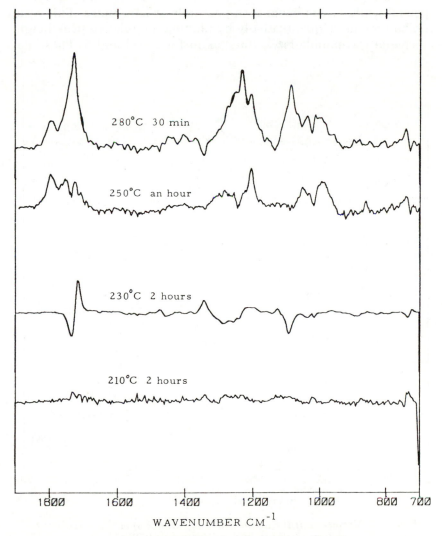

280°C 30 min

250°C an hour

230°C 2 hours

210°C 2 hours

1800   1600   1400   1200   1000   800   700

WAVENUMBER CM^{-1}

*Figure 1. Absorbance difference spectra (time = t indicated in figure
− time = 0) for PET at several temperatures.*

try markedly. The difference spectrum in Figure 1 now shows the growth of many new bands even after 1 h. This growth is accelerated by further temperature increases, illustrated by the several strong positive deviations after 30 min at 280 °C. These changes are irreversible and represent the thermooxidative degradation of PET.

Figures 2 and 3 show the evolution of these difference spectra with time at 250 °C and 280 °C, respectively. The time evolution immediately demonstrates that degradation, as revealed by these traces, is not simply the progressive growth of new bands. Rather, some bands are seen to increase first then decrease in the difference spectra, while others begin to appear only at later times. This result can be seen more quantitatively by plotting the relative peak heights at several wavenumbers vs. time, as in Figures 4 and 5. These data

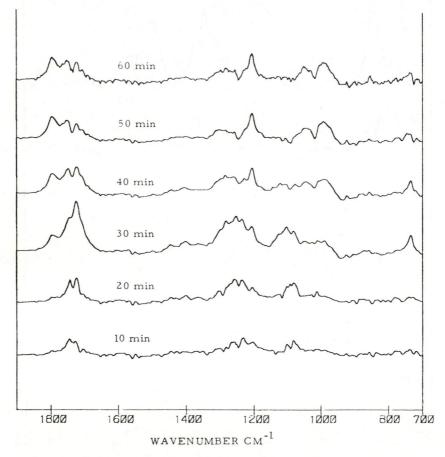

Figure 2. *Variation of the absorbance difference spectrum (time = t − time − 0) for PET with time at 250 °C.*

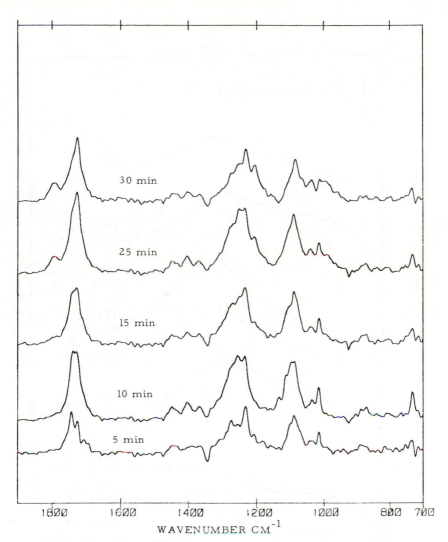

30 min

25 min

15 min

10 min

5 min

WAVENUMBER CM⁻¹

Figure 3. Variation of the absorbance difference spectrum (time = t − time = 0) for PET with time at 280 °C.

show the growth of primary degradation products that are consumed in subsequent reactions. The actual kinetic profiles at different wavenumbers can be quite complex, as seen, for example, in the 1730-cm⁻¹ plot at 280 °C. This complexity results from overlapping absorption bands arising from several different carbonyl degradation species.

Although the relative functional group stability approach can be applied to the bands in the original PET spectrum, it is interesting to

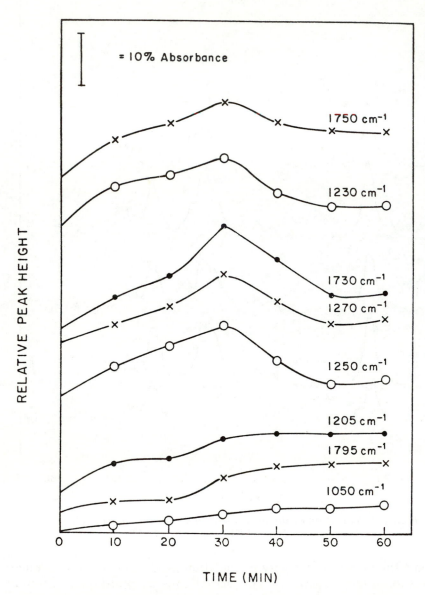

Figure 4. Kinetic profiles (absorbance vs. time) for several IR absorption bands observed in the thermooxidative degradation of PET at 250 °C. The profiles are arbitrarily displaced along the ordinate.

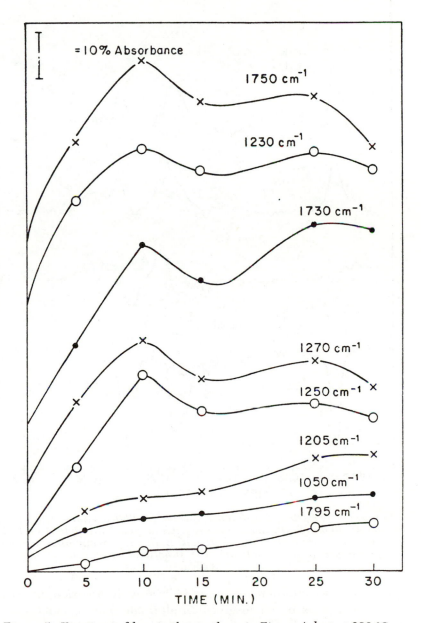

Figure 5. Kinetic profiles similar to those in Figure 4, but at 280 °C.

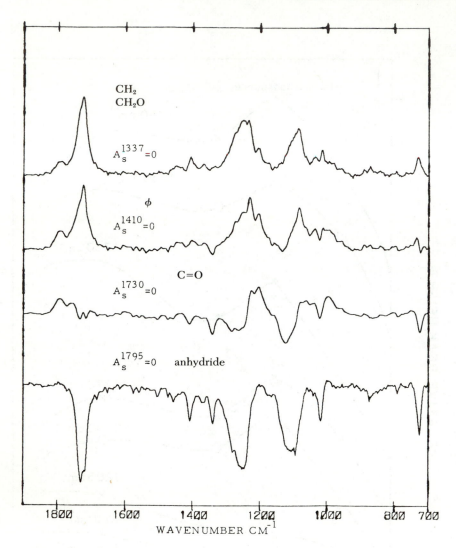

Figure 6. Absorbance difference spectra for PET samples (degraded at 280 °C, 30 min for initial spectrum) calculated according to the relative functional group stability approach. Both the wavenumber at which absorbance is forced = 0 and the functional group assignment are indicated.

apply it to the bands of reaction products as well. An example of this is shown in Figure 6, which gives an alternate way of looking at the degradation data. In Figure 6 an anhydride product absorbing at 1795 cm^{-1} is seen in small amount even in the initial sample. When the absorbance difference at 1795 cm^{-1} is forced to be equal to zero, all other bands appear to be negative. Conversely, the methylene and CH$_2$O linkage are degraded in the early stages of the reaction. Thus, when an absorption band characteristic of these functional groups (at 1337 cm^{-1}) is forced to be equal to zero, many positive features appear. Functional groups manifesting themselves in the bands at 1410 and 1730 cm^{-1} are clearly intermediate cases.

Thus far a series of results on a single sample of PET has been discussed. But PET, as other polymers, may vary in many ways. Some of these are common to many polymer systems, such as variation in molecular weight or intrinsic viscosity. Others are unique to PET, such as a variation in the concentration of diethylene glycol (DEG) linkages —OCH$_2$CH$_2$OCH$_2$CH$_2$O— in the polymer. The concentration of COOH end groups may also be a useful parameter of characterizing samples.

Four samples of PET were compared and their properties are shown in Table I. Results discussed previously in this chapter are for Sample 1. Comparable results are now shown for Samples 2—4. Figure 7 shows kinetic profiles for several samples. Samples 1 and 2 have similar DEG contents and molecular weights (MW), but differ greatly in COOH content. The degradation patterns of these two samples are rather similar. The main observable difference is that the anhydride concentration increases more rapidly in Sample 1. This formation of anhydride may be associated with the presence of the COOH groups.

Samples 1 and 3 differ mainly in the DEG concentration. Data show that the lowering of the DEG concentration reduces the rate of thermal degradation. This result is further accentuated in Sample 4, where the COOH content and the DEG content are both relatively low. In this sample the formation of anhydride is very small even after relatively long times.

The effect of MW on the degradation rate is not great, and can be understood as being due to a lower melt viscosity in the samples with lower molecular weight. Work on additional samples controlling these parameters in other ways would be necessary for clear separation of the effect of molecular weight on thermal degradation rate from the other effects discussed here.

Discussion: Isothermal Degradation

The IR absorption spectrum of PET has been addressed by several investigators (3—7). By combining these data, results of previous mechanistic studies of the degradation (8, 9), and knowledge of IR group frequencies (10, 11), most of the bands observed in these experiments can be assigned. These assignments are collected in Table II.

Table I. Characteristics of PET Samples Studied

Sample	COOH	DEG	IV	M_w
5	16	1.96	0.71	6.1×10^4
11	70	1.96	0.61	4.9×10^4
6	16	0.94	0.74	6.4×10^4
1	8	1.0	0.47	3.5×10^4

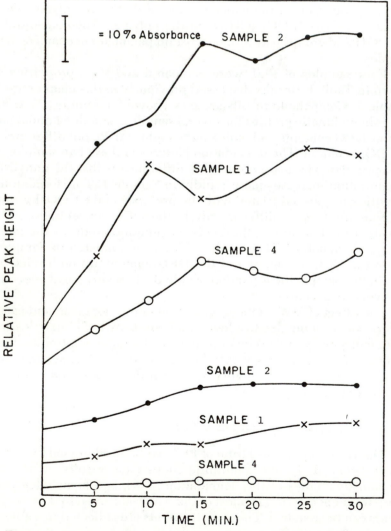

Figure 7. Kinetic profiles for several PET samples degraded at 280 °C.
The top three curves are 1730 cm^{-1} and the bottom curves are at
1795 cm^{-1}.

Table II. Tentative IR Absorption Assignments for PET Before and After Degradation

Wavenumber (cm⁻¹)	PET B.D.	PET A.D.	Functional Group	Stretching Frequency	Vibration Mode
3081	X	X	C–H	ν_{C-H}	ring C–H
3067	X	X	C–H	ν_{C-H}	ring C–H
2962	X	– X	C–H	ν_{C-H}	methylene
2889	X	– X	C–H	ν_{C-H}	methylene
1790–1800		+ X	(O=)C–O–C(=O)	$\nu_{C=O}$	asymmetric ester[a]
~1770		+ X	C=O	$\nu_{C=O}$	ester[a]
1740–1750		+ X	(O=)C–O–C(=O)	$\nu_{C=O}$	symmetry "carbonyl"
1740–1750		+ X	C=O	$\nu_{C=O}$	ester
1730	X		C=O	$\nu_{C=O}$	"carbonyl"
1730		+ X	C=O	$\nu_{C=O}$	benzoic acid
1700	X	+ X	C=O	$\nu_{C=O}$	ring C–C
1582	X	X	C–C	ν_{C-C}	ring C–H, in plane
1504	X	X	C–H	δ_{C-H}	trans-ethylene glycol
1475	X	– X	H–C–H	δ_{H-C-H}	gauche-ethylene glycol
1453	X	– X	H–C–H	δ_{H-C-H}	ring C–H, in plane
1410	X	X	C–H	δ_{C-H}	ring C–C
1370	X	– X	C–H	ω_{C-H}	gauche-ethylene glycol
1337	X	– X	C–H	ω_{C-H}	trans-ethylene glycol
1240–1290	X	– X	O–C–H	δ_{O-C-H}	trans-ethylene glycol
1240–1290	X	– X	C–C–H	δ_{C-C-H}	trans-ethylene glycol
1240–1290	X	X	C–C	ν_{C-C}	ring-ester
1240–1290	X	+ X	C–O	ν_{C-O}	C(=O)O
1240–1290	X	+ X	C=O	$\delta_{C=O}$	in plane
1240–1290	X	– X	C–H	τ_{CH_2}	methylene
1270	X	+ X	C–C–C	ν_{C-C-C}	ring C–C–C
1250	X	+ X	C–O	ν_{C-O}	C(=O)O
1230	X	+ X	(O=)C–O–C(=O)	ν_{C-O}	in plane

Continued on next page

Table II—Continued

Wavenumber (cm⁻¹)	PET		Functional Group	Stretching Frequency	Vibration Mode
	B.D.	A.D.			
1205	X	+ X	(O=)C-O-C(=O)	$\delta_{C=O}$	(or ν_{C-O-C})
1126	X	+ X	C-O	ν_{C-O}	ester
	X	X	C-C-C	δ_{C-C-C}	ring
	X	+ X	C=O	$\delta_{C=O}$	
	X	- X	C-O	ν_{C-O}	O-CH₂
1109	X	X	C-H	δ_{C-H}	ring, in plane
	X	X	C-C	ν_{C-C}	ring
1096	X	+ X	C-O	ν_{C-O}	C(=O)O
	X	- X	C-O	ν_{C-O}	O-CH₂
1050	X	- X	(O=)C-O-C(=O)	ν_{C-O-C}	
1018	X	X	C-C-C	ν_{C-C-C}	ring
	X	X	C-C	ν_{C-C}	ring
	X	- X	C-H	δ_{C-H}	ring, in plane
973	X	+ X	C-O	ν_{C-O}	O-CH₂
	X	- X	C-O	ν_{C-O}	C(=O)O
895	X	X	C-H	ρ_{C-H}	gauche-ethylene glycol
875	X	X	C-H	δ_{C-H}	ring, out of plane
875	X	X	C-C	δ_{C-C}	ring-ester
875	X	+ X	C=O	$\delta_{C=O}$	out of plane
845	X	- X	C-H	ρ_{C-H}	trans-ethylene glycol
727	X	+ X	C=O	$\delta_{C=O}$	out of plane
	X	X	C-H	δ_{C-H}	ring, out of plane
502	X	X	C-C	ν_{C-C}	ring-ester
	X	X	C=O	$\delta_{C=O}$	in plane

[a] Assignments derived principally from Ref. 17.
Key: B.D., before degradation; A.D., after degradation; X, absorption present; +, absorbance increase; −, absorbance decrease; ν, stretching; δ, bending; ω, wagging; ρ, rocking; τ, twist; and ester[a], $-C_6H_5C(=O)OCH_2C(=O)OCH_2CH_2OC(=O)C_6H_5-$.

The complex series of changes in the 1700–1750-cm^{-1} region, where several bands are present and change in intensity with time, is a result of many carbonyl species. These include aldehydes, carboxylic acids, and eventually one of the anhydride carbonyl modes (the other is at 1795 cm^{-1}).

Discussion: Thermooxidative Degradation Schemes

Directly following from these results are some conclusions concerning degradation mechanism: (1) the weak point toward thermooxidative degradation in PET is the CH$_2$ group; (2) the CH$_2$ group located in DEG is much more reactive than that normally in PET; (3) the fluctuating increase of the carbonyl group indicates that some products such as COOH, CHO, and others may form, with multiple consecutive reactions taking place; and (4) the almost constant value for the anhydride group implies that an equilibrium or a steady state may be reached between the anhydride group and other groups.

The α-position of ethers is attacked easily by oxygen even at room temperature. A simple ether can autoxidize with atmospheric oxygen to form peroxide. This α-position vulnerability is the reason that the CH$_2$ group in DEG is more reactive, there being two α-positions in DEG. A suggestion for an autoxidation process is therefore reasonable.

The normal thermooxidative mechanism (involving the ethylene glycol linkage) is illustrated in Scheme I (8, 9).

The autoxidative mechanism (involving the diethylene glycol linkage) is illustrated in Scheme II.

The reactions between the anhydride and the other groups are illustrated in Scheme III.

From the Schemes I–III, the products containing the carbonyl species are $C_6H_5COOCH_2COOH$, CH_3CHO, $C_6H_5COOCH_2CHO$, $C_6H_5COOCH_2CH_2OCHO$, and $C_6H_5COOCH_3$. As a result, the band at 1730 cm^{-1} is actually an overlapped band; this is why it shifts in frequency during the degradation. Some of these products can be further oxidized, the gaseous products such as CH_3CHO can escape, and the carbonyl group intensity is therefore expected to fluctuate with time. The production and escape of the gaseous products such as CH_3CHO, CO_2, and H_2O explain the decrease in the thickness of the sample film after degradation.

The double bond product forms during this degradation; however, a band characteristic of a double bond was not evident in our difference spectra. Possibly, the double bond products react rapidly to a graft cross-linking by a free radical mechanism, thus making their concentrations too low to be detected, especially considering the weak IR activity of double bonds.

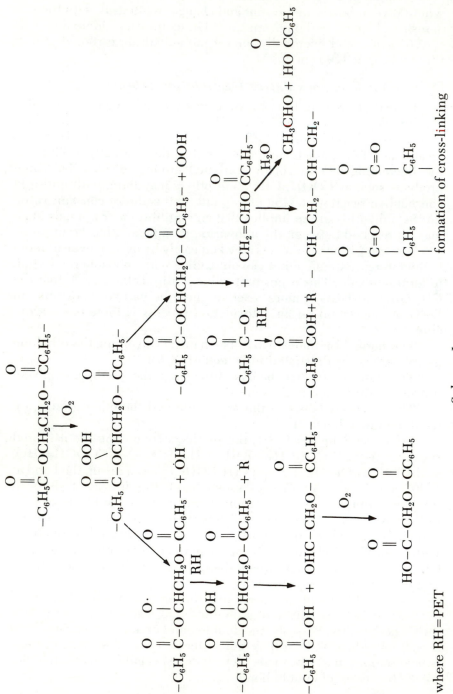

Scheme I

where RH=PET

Also, a band for the intermediate peroxide group $(-OOH)$ was not found. This may be due to interference or its low concentration combined with a low extinction coefficient. The band characteristic of peroxide is about 885 cm^{-1} and a band exists at 895 cm^{-1}, which is characteristic for the CH$_2$ rocking of ethylene glycol (CH$_2$ decreases during degradation). These two bands are too close together and may interfere.

Experimental: Evolved Gas Analysis of PET Under Nitrogen

In the preceding section we described the IR spectroscopic results on degradation of PET viewed from the residual polymer. To complement such studies, the spectra of volatile products also must be examined.

One approach to this examination is polymer aging under isothermal conditions in a sealed chamber. This approach allows the measurement of both residual and volatile products. The chamber is used as a gas cell. Results of such an experiment are described elsewhere (1, 2).

Another approach is measurement of evolved gases in a stream of flowing carrier gas as a function of temperature. This method has been described in detail (12, 13).

Results: Evolved Gas Analysis of PET Under Nitrogen Atmosphere

A sample of PET was heated at 10 °C/min. The sample was held at 55 °C for several minutes to remove loosely bound water and then heated to 550 °C while spectra were taken continuously. A total of 125 spectra were taken during this period. Each spectrum represents an average of the evolved gases in a 4 °C temperature range. A nitrogen atmosphere was used throughout.

Quick examination of the spectral files allows immediate identification of CO, CO$_2$, and H$_2$O among the products of the decomposition. From a frequency characteristic of each of these gases, a plot of absorbance vs. temperature can be made (Figures 8–10). In this decomposition, the evolution of CO and CO$_2$ shows nearly identical temperature profiles. Water evolution at low temperatures is probably not significant, and the higher temperature water evolution again follows a similar profile to CO and CO$_2$. Although we have not calibrated these intensities to determine relative concentrations, the quantity of water probably is considerably less than that of CO$_2$.

Aldehyde, particularly acetaldehyde, is clearly present in the evolved gas analysis spectra. By 400 °C, all of the bands characteristic of acetaldehyde in the gas phase (14) can be seen clearly. This can be illustrated by plotting a temperature profile of the absorbance at 2700 cm^{-1}, shown in Figure 11. This low wavenumber C–H stretch is characteristic of aldehydes. Similar temperature profiles are seen for the carbonyl stretch, and other bands in the fingerprint region.

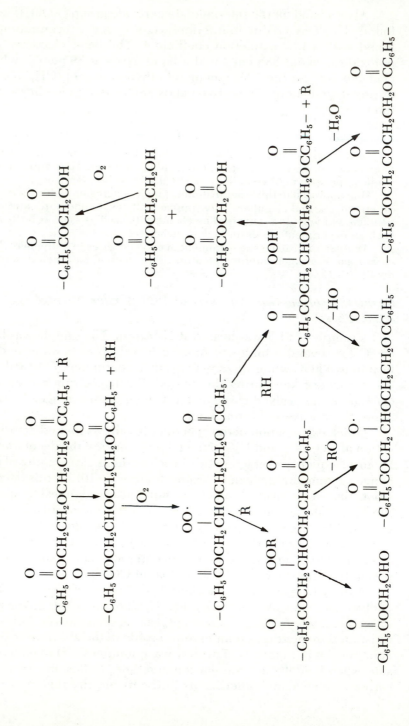

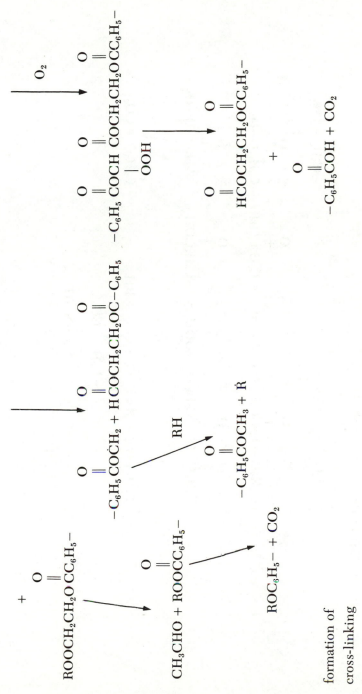

Scheme II

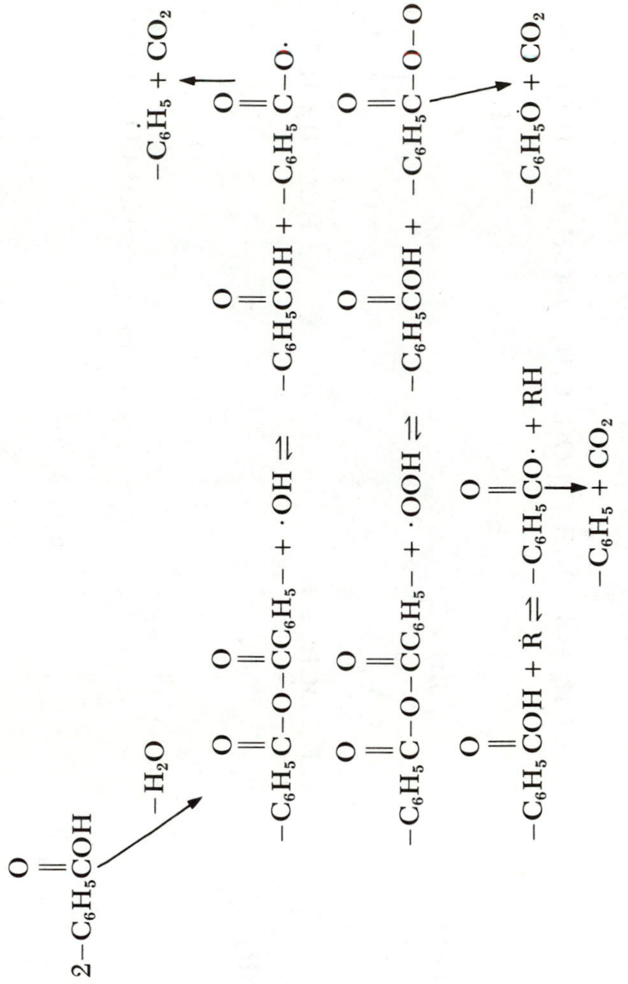

$$2 - C_6H_5\overset{O}{\overset{\|}{C}}OH \xrightarrow{-H_2O} -C_6H_5\overset{O}{\overset{\|}{C}}-O-\overset{O}{\overset{\|}{C}}C_6H_5- + \cdot OH \;\rightleftharpoons\; -C_6H_5\overset{O}{\overset{\|}{C}}OH + -C_6H_5\overset{O}{\overset{\|}{C}}-O\cdot \;\longrightarrow\; -C_6\dot{H}_5 + CO_2$$

$$-C_6H_5\overset{O}{\overset{\|}{C}}-O-\overset{O}{\overset{\|}{C}}C_6H_5- + \cdot OOH \;\rightleftharpoons\; -C_6H_5\overset{O}{\overset{\|}{C}}OH + -C_6H_5\overset{O}{\overset{\|}{C}}-O-O\cdot \;\longrightarrow\; -C_6H_5\dot{O} + CO_2$$

$$-C_6H_5\overset{O}{\overset{\|}{C}}OH + \dot{R} \;\rightleftharpoons\; -C_6H_5\overset{O}{\overset{\|}{C}}\cdot + RH \;\longrightarrow\; -C_6\dot{H}_5 + CO_2$$

Scheme III

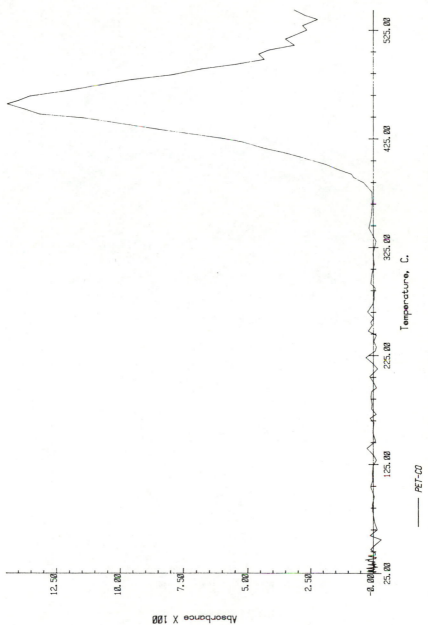

Figure 8. Absorbance characteristic of evolved CO (relative to blank region of spectrum) vs. temperature for PET samples.

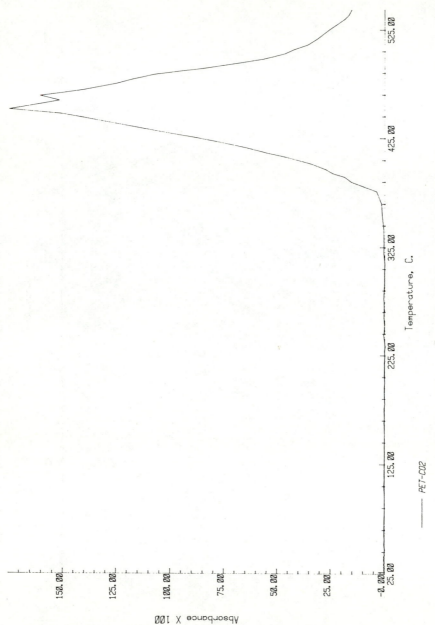

Figure 9. Absorbance characteristic of evolved CO₂ from PET vs. temperature.

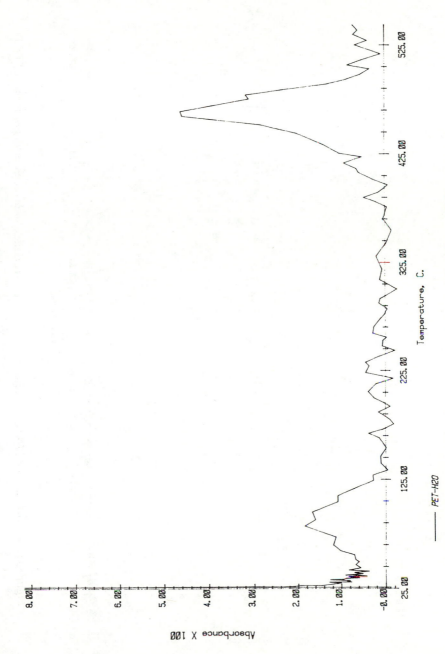

Figure 10. Absorbance characteristic of evolved H₂O from PET vs. temperature.

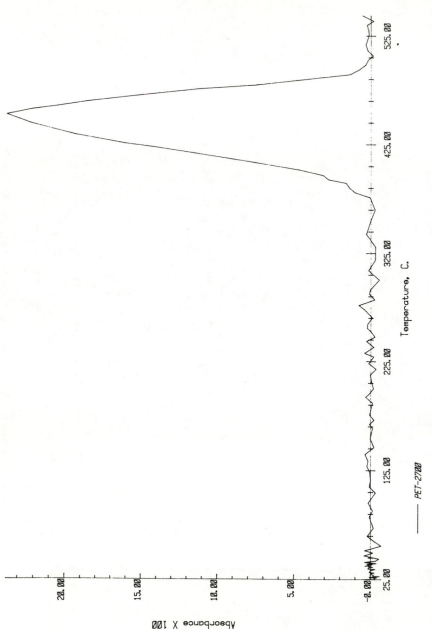

*Figure 11. Absorbance at 2700 cm⁻¹, characteristic of evolved aldehyde (seen in degradation of PET)
vs. temperature.*

In addition to the acetaldehyde band, a number of other bands appear in the spectrum, increasing in intensity above 375 °C, peaking near 450 °C, and then decreasing. These bands, at 3590, 3080, 1760 (overlapped), 1350 (overlapped), 1270, 1180, 1085, and 711 (over-lapped) cm^{-1} all have identical temperature profiles. They are the spectrum of benzoic acid (*14*).

Discussion: Evolved Gas Analysis

That PET should decompose into benzoic acid and acetaldehyde is not surprising. Under a nitrogen atmosphere the decomposition of this polymer is thus quite simple (*15, 16*).

Products such as CO, CO_2, and H_2O are undoubtedly formed from the primary decomposition products, as well as through other reactions. The data would not permit detection of minor products, particularly given the large number of bands present from the species already mentioned.

Acknowledgment

Evolved gas analysis was carried out at the Philip Morris Research Center in Richmond, VA, through the courtesy of John Lephardt. We are pleased to acknowledge his assistance.

Literature Cited

1. Lin, S. C.; Bulkin, B. J.; Pearce, E. M. *J. Polym. Sci. Polym. Chem. Ed.* **1979**, *17*, 3121.
2. Bulkin, B. J.; Pearce, E. M.; Chen, C. S. *Appl. Spectrosc.*, in press.
3. D'Esposito, L.; Koenig, J. L. *J. Polym. Sci. Polym. Phys. Ed.* **1976**, *14*, 1731.
4. Ward, I. M.; Wilding, M. A. *Polymer* **1977**, *18*, 327.
5. Boerio, F. J.; Bahl, S. K. *J. Polym. Sci. Polym. Phys. Ed.* **1976**, *14*, 1029.
6. Boerio, F. J.; Bahl, S. K. *Spectrochim. Acta* **1976**, *32A*, 987.
7. Cunningham, A.; Ward, I. M.; Willis, H. A.; Zichy, V. *Polymer* **1974**, *15*, 749.
8. Buxbaum, L. H. *Angew. Chem. Int. Ed. Eng.* **1968**, *7*, 182.
9. Buxbaum, L. H. *Polym. Prepr. Am. Chem. Soc. Div. Polym. Chem.* **1967**, *8*, 552.
10. Bellamy, L. J. "The Infrared Spectra of Complex Molecules," 3rd ed.; Chapman and Hall: London, 1975; Vol. 1.
11. *Ibid.*, 2nd ed.; 1980 Vol. 2.
12. Lephardt, J. O.; Fenner, R. A. *Appl. Spectrosc.* **1980**, *34*.
13. *Ibid.* **1981**, *35*, 95.
14. Welti, D. "Infrared Vapour Spectra"; Heyden: London, 1970.
15. Pohl, H. A. *J. Am. Chem. Soc.* **1951**, *73*, 5660.
16. DePuy, C. H.; Kong, R. W. *Chem. Rev.* **1960**, *60*, 431.
17. Boerio, F. J.; Bahl, S. K.; McGraw, G. E. *J. Polym. Sci., Polym. Lett. Ed.* **1974**, *12*, 13.

RECEIVED for review October 14, 1981. ACCEPTED August 2, 1982.

Photoacoustic Fourier Transform IR Spectroscopy and Its Application to Polymer Analysis

D. WARREN VIDRINE and S. R. LOWRY

Nicolet Instrument Corporation, Madison, WI 53711

Photoacoustic spectroscopy (PAS) in the UV-visible region, although a promising technique, is limited in its application by the prevalence of PA saturation and by the paucity of structural information in that spectral region (relative to the mid-IR). The recent development of Fourier transform IR photoacoustic spectroscopy (FTIR/PAS) has made mid-IR PA detection practical for solid and liquid samples. Because of the relatively low volume absorptivities characteristic of vibrational absorptions, PA signal saturation is generally not a problem except for some inorganic materials. The applications of FTIR/PAS to polymer analysis have been in three directions so far: (1) obtaining IR spectra of samples that are intractable to most traditional sample preparation methods; (2) identifying or ruling out the existence of sample preparation artifacts in IR spectra obtained by traditional methods; and (3) surface and depth profiling studies of sample surfaces. Applications of FTIR/PAS to the analysis of polymers and composite materials are presented, along with comparisons of PAS with spectra obtained (on the same materials) using other IR measurement techniques.

THE PHOTOACOUSTIC EFFECT WAS DISCOVERED by Alexander Graham Bell (*1*) who found that many materials gave off an audible sound when illuminated by modulated light. As shown in Figure 1, the heating at intervals of the sample surface by the modulated light causes the air adjacent to the sample to be heated at intervals. The

0065–2393/83/0203–0595$06.00/0
© 1983 American Chemical Society

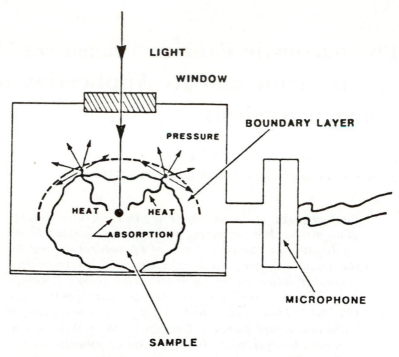

Figure 1. Idealized photoacoustic cell for solid samples.

intermittent expansion of this air causes an alternating pressure wave—sound (2). So, photoacoustic detection directly measures the absorption of light energy at the sample surface. In 1975, UV-visible photoacoustic spectroscopy (PAS) became a commercial reality, and immediately found application in solids analysis. Two factors have previously limited its application: the phenomenon of photoacoustic (PA) saturation (analogous to a completely absorbing band) that limits quantitative application, and the relative paucity of structural information for many polymers in the UV region. The weaker sources and lower photon energy of the mid-IR region at first prohibited IR-PAS; however, when PA detection was demonstrated (3) with a Michelson interferometer spectrometer (in the visible region), the solution of the mid-IR sensitivity problem [by using interferometric Fourier transform IR (FTIR) spectrometers] was inevitable. FTIR spectrometers permit more effective use of measurement time by allowing measurement of all IR frequencies at all times during each scan (allowing much signal-averaging of noise), and have characteristically higher light throughputs than dispersive spectrometers (no slits are required because there is no monochromator). These sensitivity advantages offset the characteristically low thermal efficiency of the untuned air-mediated PA microphone cell and permit useful spectra to be obtained in reasonable measurement times.

The PAS Experiment

Although the basic PAS experiment is quite simple, the analog signal measured at the microphone is a result of a complex transfer of energy from the IR source to the microphone. Although PAS is fundamentally a surface technique, the actual thermal response is a rather complicated function of modulation frequency, surface morphology, sample absorptivity, and thermal diffusivity. Attempts to model the experiment have been partially successful, but a general theory is not available currently.

A major point to be made concerning PAS is the direct relation of IR absorbance to the signal measured with the microphone. Unlike normal transmittance or reflectance techniques where one measures the loss of signal caused by the sample absorbing radiation, PAS measures the true magnitude of the interaction. This ability to have maximum signal when the sample is present reduces the dynamic range problems sometimes encountered in FTIR.

Figure 2 shows the PAS interferogram from a sample of carbon black. Because carbon black is a strong broad range absorber the signal is quite similar to a normal transmitting spectrum, even though the measuring procedure is quite different. Figure 3 shows the PAS interferogram from a vapor phase sample of methanol and carbon disulfide (CS_2). This sample has a few very strong absorption lines, and the interferogram shows the beat pattern between these frequencies. As stated previously, the magnitude of the interferogram is related directly to the sample. If more sample were in the cell, the signal would be proportionally larger.

An important effect in PAS is the huge signal enhancement observed when the sample is in the vapor phase. This is often a problem when a sample contains adsorbed water. After the sample is sealed in the PAS cell, a small amount of desorbed water can make a large contribution to the final spectrum. The problem can be reduced by purging the sample before sealing the cell or by performing a spectral subtraction with a water reference spectrum.

Comparison of Methods

Crucial to understanding the uses of FTIR/PAS is the consideration and comparison of alternative IR methods of measuring polymer surface properties. The most direct methods are physical microtoming and include surface grinding, etching, and slicing (classical microtoming). In these methods the IR spectra and their properties are

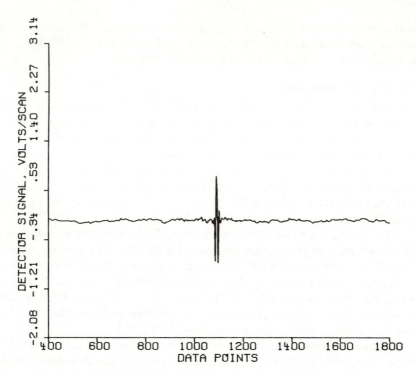

Figure 2. PAS interferogram from a carbon black sample.

not usually the point of controversy; they are essentially simple transmission spectra. Rather, the trick of these techniques lies in the microtoming process itself. The following section attempts a description of the operational characteristics of the spectroscopic microtoming, or surface methods in use.

Reflectance Methods.   Because most surface methods are reflection methods, it seems appropriate to define some descriptive terms in order to insert a modicum of order (and a minimum of further confusion) into this field. "Plane specular reflection" is the reflection from a polished flat surface, such as a common first-surface mirror. "Scattered specular reflection," or "diffused specular reflection" occurs when there are single reflections from a rough surface, such as a sandblasted gold surface. "Diffuse reflection" occurs when there are multiple reflections and transmissions of each ray within a more or less transparent scattering sample, such as powdered KBr. Although the latter two often have the same spatial characteristics, i.e., great or total diffusion of reflected light, the nature and photometry of these measurements is quite different. The term "f/5" refers to the common focusing and collection angle found in most commercial IR spectrometers, and "f/1" refers to the practical maximum focus angle and collection angle. This is usually nominally near f/1, but the useful f/number is higher. By using these terms, we may list and describe the major reflectance methods in use.

"NORMAL" SPECULAR REFLECTANCE.   This term commonly refers to high-angle specular reflection from a plane surface. In practice, an f/5 beam, with an incidence angle between 45° and 80° to the surface plane, or a Christiansen spectrum of optically thick nonscattering samples, is most often used to measure coatings on metal surfaces. In this use, the reflection off the metal surface is not of interest, and we are measuring actually the double-pass transmission spectrum of the coating. This method suffers in sensitivity for extremely thin films because the electric vector of the light approaches zero very near the metal surface.

GRAZING ANGLE REFLECTANCE.   Grazing angle reflectance (GAR), or ellipsometry is a true surface technique and refers to the low-angle reflection from a plane surface. In practice, either a collimated beam or an f/5 beam is used with the nominal incidence angle usually being 5° to 10° from the surface. The angular dispersion of an f/5 beam is

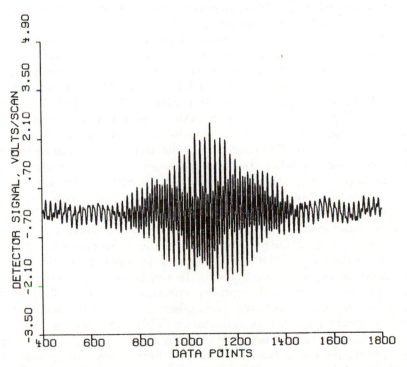

Figure 3. PAS interferogram from a vapor phase sample of methanol and carbon disulfide.

significant here and will influence the resulting photometry. Near normal incidence, the spectrum is simply a weak Christiansen, or refractive-index spectrum of the material. At grazing angles, the spectrum of the perpendicular polarization is a reasonable simulacrum of the absorption spectrum of the surface. GAR requires a smooth, highly polished surface of fairly large area as highly focused low f/number optics increase the incidence angle dispersion. The method is useful for both polished, optically thick samples and for very thin (monolayer) films on metal. Spectral features appear only for light of polarization, so a polarizer is often used for the measurements. The equivalent depth penetration is on the order of 1 μm.

MULTIPLE EXTERNAL REFLECTANCE. This method is used most commonly as a rough but sensitive version of grazing angle reflectance. In practice, an f/1 to f/5 beam is bounced into the space between two polished plates (for background spectra, these are mirrors; otherwise one or both are sample), and exits after multiple reflections. Because the beam is not collimated, and the samples are often not completely flat, considerable incidence angle dispersion commonly exists, and the technique is most often used where sensitivity rather than photometric definability is paramount, as in the measurement of submonomolecular layers on metal.

DIFFUSE REFLECTANCE. Diffuse reflectance (DRF or DRIFT) is the reflectance of a sample associated with light being reflectively scattered multiple times within the optically scattering sample. Photometrically, the sample absorptivity is related to the reflectivity by the Kubelka–Munk equation. In practice, f/1 collection optics are used to collect as much (20% or so) of the diffusely reflected light as is practical. Two forms of DRF accessory are available: those that eliminate the central f/5 portion of the collected beam, and those that attempt to collect the whole beam. The former is valuable when a scattering sample has a smooth polished surface with its specular Christiansen contribution. For most samples, including powders, the specular component is also diffused by surface roughness and cannot be eliminated in this way. Nonscattering rough-surfaced samples are best measured in a DRF accessory, but the reflectance is diffused specular reflectance (DSR, not DRF), and does not follow the Kubelka–Munk absorptivity relationship. In practical DRF spectroscopy, the reflectivity of high-absorptivity spectral regions is often dominated by specular reflectivity, even when the rest of the spectrum is predominantly true diffuse reflectance. This can cause a specious, antinomial, poltergeisterhaft effect known as band reversal, where the absorptivity peak of a band actually has higher reflectivity than the band wings. In general, DRF is useful for two classes of polymer samples: with powders or other highly scattering samples, and with nonscattering samples having small, scattering blemishes on their surfaces. The ease of use of DRF makes it a very desirable method for repetitive measurements when the sample's optical properties and analytical requirements permit.

ATTENUATED TOTAL REFLECTANCE. Attenuated total reflectance (ATR), or multiple internal reflectance (MIR), is the most commonly used technique of surface measurement in IR spectroscopy. Light traveling inside a high refractive index, transparent crystal is reflected against the ATR crystal surface from the inside. Because the reflection is below (more grazing than) the critical angle total dielectric reflection takes place, and the light is not attenuated. However, the electric vector of the light extends beyond the crystal surface, and an optically absorbing, low refractive index substance placed in contact with the crystal surface will attenuate the reflected light. Because this electric vector, or evanescent wave, is of significant intensity within only a few micrometers of the surface, only the surface spectral absorptivity of the sample is measured.

In practice, an f/1 or f/5 beam is introduced into the crystal. The high refractive index of the crystal reduces the angular dispersion of the beam within the crystal. Crystals of various refractive indices are used. For relatively deep penetration, low refractive index crystals are used. Common crystals are KRS-5 (thallium bromoiodide, refractive index 2.2), and zinc selenide (2.42). For shallower penetrations, silicon (3.4) or germanium (4.0) is used. Penetration may also be decreased by reflecting the light at a lower, more grazing angle. Reflection at a high angle increases penetration, but angles near the critical angle (limit of total reflection) produce spectra with serious band asymmetries. Samples must make good, intimate contact with the crystal surface. For quantitative work, this requires liquids or very flat or deformable solids. Fortunately, for qualitative work or relative quantitation less perfect sample contact is permissible and even textured surfaces and powders may be suitable.

Nonreflection Surface Methods. EMISSION SPECTROSCOPY. This nonreflection method uses the IR light emitted by the sample itself as the source. The sample is placed on a thermostatted stage at a preinterferometer f/1 beam focus. The beam then originates at the sample, is modulated by the interferometer and then detected (the detector must be at a different temperature than the sample, and liquid N_2-cooled detectors are usually used). The spectral zero and 100% points are defined by substituting a gold surface or a carbon black (or platinum black) surface in place of the sample. Results are expressed as percent emissivity or as watts per steradian. There are three general classes of samples, each analogous to the type of high-angle reflectance spectroscopy. Optically thick, nonscattering samples give featureless gray body emission spectra, with a small Christiansen component similar to that obtained with specular reflectance. Thin films on metal give inverted-transmissionlike spectra, again similar to the specular reflectance case. Scattering samples can give spectra similar to inverted diffuse reflectance spectra. In general, for the majority of sample types in the mid-IR, similar results can be obtained with reflectance methods, without paying the energy and signal-to-noise penalty of a cooler source (reflectance methods use a 1500-K Globar source, while emission samples are usually 300−600 K).

PHOTOACOUSTIC DETECTION. Photoacoustic detection (PA or PAS) is the subject of this chapter and only information for comparison will be given here. An f/1 or f/5 beam is focused onto the sample stage of the PA microphone cell. The sample cavity in commercial cells is commonly 4−12 mm in diameter and a few millimeters deep, but larger samples can sometimes be accommodated. For reasons outlined elsewhere (4), the detectivity of the PA microphone cell is less than other thermal detectors (such as TGS or DTGS), and measurement times are, therefore, longer (for equivalent spectral signal-to-noise) than with other methods. The spectrum resembles an inverted transmission spectrum in appearance and approximate photometry. Scattering effects that sometimes cause such baseline tilt and curvature in other methods usually do not visibly affect PA baselines. High absorptivity organic IR chromophores rarely give PA signals that are more than about 80% of saturation, and even strong inorganic bands, e.g., Si−O stretch, are incompletely saturated. Sample surface morphology exerts a mild effect on spectrum intensity, and a smaller effect on photometry as compared to other methods.

General Applications

Four uses for FTIR/PAS suggest themselves. First, PAS offers a way of avoiding interfering Christiansen effects on band shape and location. These distortions are particularly noticeable with highly absorbing samples (see Figure 4). For instance, the technique is being used to obtain spectra of inorganic salts that are free of Christiansen contributions (5). Diffuse reflectance (DRF) spectroscopy of very fine powders and ATR spectroscopy with high-index plates are other techniques capable of minimizing Christiansen contributions; but both these techniques impose strict limitations on sample morphology.

As a second use, PAS offers a way of looking at undisturbed surfaces. ATR, the common alternative for surface investigations, requires a flat or deformable surface for good sample contact, and requires sample contact. If these requirements can be met, of course, ATR may offer a much less time-consuming way of obtaining the information. Similarly, if the native sample happens to be a fine, scat-

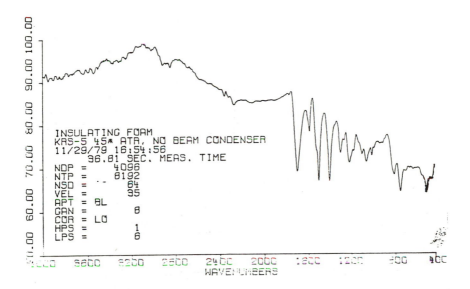

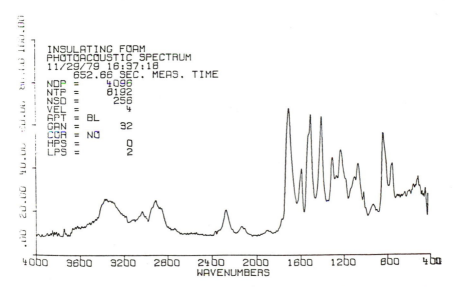

Figure 4. FTIR/PAS spectra that illustrate interfering Christiansen effects on band shape and location.

tering powder without important bands of very high absorptivity, and information about the bulk composition is desired, DRF may easily be the method of choice. But, when spectra that are independent of sample surface morphology are desired, PAS has some intrinsic advantages. Figure 5 illustrates the spectral differences seen when a series

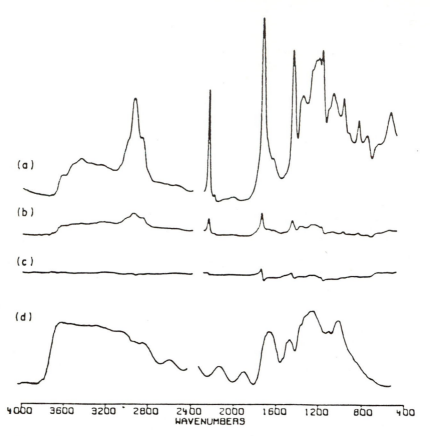

Figure 5. Diffuse reflectance spectra of the same samples of a nitrile-containing resin. Key: a, powder; b, sawn surface; c, smooth surface; and d, pellets.

of nitrile resin samples that differ only in surface morphology are measured. For comparison, note that the same samples measured by PAS show identifiably similar spectra (*see* Figure 6). Of course, if the resin samples were finely ground and mixed with KBr, very similar spectra would be obtained by DRF or KBr pellet in a shorter time. To illustrate this, note that the DRF spectra of Figure 5 required only 10 s of measurement time apiece, while the PA spectra of Figure 6 required 5 min apiece.

Thirdly, mechanically intractable or noxious samples may be analyzed without resorting to messy sample preparation techniques. In this case, one is essentially trading preparation time or effort for a longer measurement time.

Finally, PAS can offer a relatively simple way of calibrating mea-

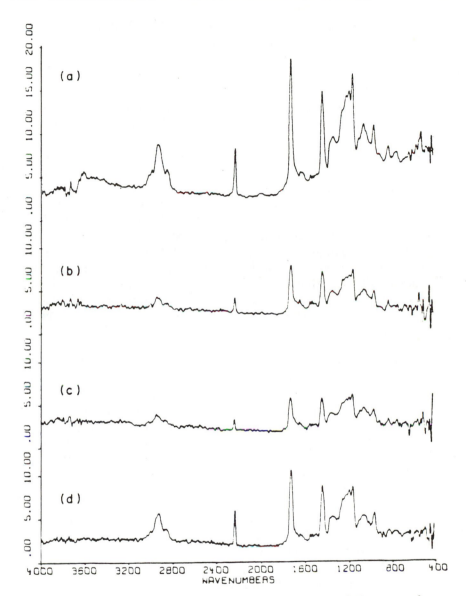

Figure 6. FTIR/PAS of a nitrile-containing resin in different surface morphologies. Key is the same as in Figure 5.

surements obtained by other methods. For instance, the curve relating the reflective attenuation of a band by ATR, and the band absorptivity can make the calibration of an ATR analysis easier. Similarly, a reference PA spectrum of an undisturbed surface compared with a spectrum obtained by a faster technique can make analytical interferences due to the faster technique's artifacts of preparation obvious, and thus easily circumvented.

Relevant Parameters

There are four interrelated factors that affect FTIR/PAS spectra: sample properties, the physical constants of the gas in the cell, the light modulation frequency, and the cell characteristics.

The prime determinants of the effective depth of penetration or measurement depth are the thermal properties of the sample. Most organic polymers have similar thermal conductivity and heat capacity; therefore, for bulk resins and composites, the effective measurement depths are reasonably similar. Fine powders, finely foamed samples, and extremely thin samples tend to give PAS of higher intensity, and the photometric scale of such spectra should be viewed with caution. The photometric theory of PAS has been presented elsewhere (6) in some detail. Although there are some problems with the work, the basic concepts are useful for the mid-IR. Also, it is a useful generalization that scattering and Christiansen contributions to PA spectra are very small, but these contributions do exist and must be considered for critical work.

The working medium used in PA cells designed for solid samples is generally air at atmospheric pressure. Helium has been shown to double the signal obtained, with a corresponding increase in the signal-to-noise ratio obtained. Theory and some experiments suggest that gases with high molecular weights give poor results. Apparently, no experiments have been done that investigate the effect of pressure, although it would be an unlikely coincidence if atmospheric pressure were optimum. In any case, the sensitivity enhancement obtained by changing the working gas must be judged against the convenience of using ambient air.

The modulation frequency of the light also affects the effective measurement depth. In FTIR spectrometers, this modulation is produced by the Michelson interferometer. The modulation frequency is proportional to the wavenumber position in the spectrum, and proportional to interferometer mirror velocity. The effective measurement depth is proportional to the square root of the inverse of the modulation frequency, according to Rosencwaig and others, although for practical samples this may not be an exact relationship.

Standard gas-microphone photoacoustic cells have a frequency response that varies inversely with frequency at frequencies below the cell's residual resonance frequency. Figure 7 illustrates the frequency response of a typical PA cell. By using a higher mirror velocity to achieve a smaller effective measurement depth, a severe signal-to-

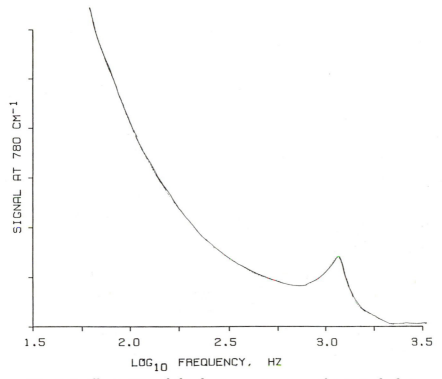

Figure 7. Illustration of the frequency response of a typical photo-acoustic cell.

noise penalty is carried if modulation frequencies above the cell resonance frequency are used. This fact imposes a lower limit of a few micrometers on attainable effective penetration depths. Within these limits, PAS can be used to profile composition as a function of depth. Figure 8 is such a series of PA spectra obtained from a sample of poisoned catalyst. The numbers 0–24 represent increasing mirror velocity (and thus decreasing effective measurement depth) on a logarithmic scale. For instance, velocity 10 is twice velocity 0, and velocity 20 is twice velocity 10. Several features in the spectra suggest a change in composition with depth. In particular, the changes in relative heights of the bands around 700 cm^{-1} indicate differences in the aromatic substitutional patterns for the different depths.

A Study of a Polymer Surface

Considerable research has gone into evaluating the performance of perfluorinated sulfonic acid polymers as separators in the electrolytic cells used in the production of chlorine and sodium hydroxide.

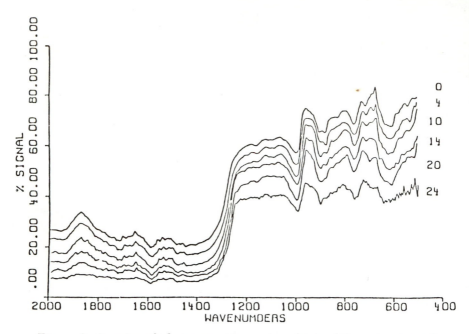

Figure 8. A series of photoacoustic spectra obtained from a poisoned catalyst. The spectra can be used to profile the composition as a function of depth.

One specific polymer used in these cells is a du Pont material called Nafion. The structure of Nafion is shown here:

$$(CF_2 \ CF_2)_n \ (CF \ CF_2)_m$$

$$O$$

$$CF_3-C-O \ CF_2 \ CF_2-X$$

$$H$$

where X = SO$_2$F, SO$_3^-$ H$^+$, SO$_3^-$NA$^+$ and SO$_2$N−

The sulfonamide groups (−SO$_2$NH−) were created by reaction of the polymer in the sulfonyl fluoride form with the corresponding amine. A significant improvement in membrane performance was observed when a modified sulfonamide layer was formed with ethylene diamine (EDA).

A large number of successful ATR experiments were performed on various forms of Nafion produced for research investigations.

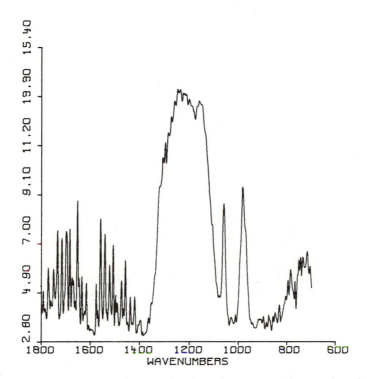

*Figure 9. PAS spectrum of a sample of production membrane placed in
the PAS cell with the untreated side face up.*

However, in order to improve the mechanical strength of the membranes in actual production cells the Nafion membrane is reinforced
with a cross screen of Teflon fibers. The Teflon fibers create a rough
membrane surface where most of the actual Nafion is located. ATR
experiments on the production membrane were quite difficult and the
resulting spectra were frequently dominated by features due to the
Teflon. Transmission experiments on these samples failed because the
C−F stretching mode blanked the entire fingerprint region.

The problem of analyzing the treated and untreated sides of a
polymer membrane where good surface contact is impossible seemed
well suited for PAS. A sample of production membrane was placed in
the PAS cell with the untreated side ($-SO_3^-Na^+$) up. Figure 9 shows the
resulting PAS spectrum. This spectrum has been ratioed to a carbon
black spectrum. The strong water peaks between 1800 and 1400 cm^{-1}
indicate that water has been desorbed from the hydrophilic surface of
the membrane. No attempt was made to remove the water. This spectrum allows a comparison to be made between the desorption prop-

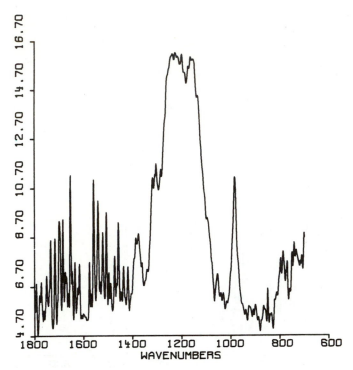

Figure 10. PAS spectrum of a sample of production membrane placed in the PAS cell with the treated side face up.

erties of the two sides of the membrane. The major peak in the spectrum at 1200 cm^{-1} is due to the C–F stretch. The flat top of this peak indicates clearly that this band is saturating. The major peak of interest occurs at about 1050 cm^{-1}. This peak corresponds to the symmetric SO$_3^-$ stretching mode. The asymmetric mode occurs under the large C–F peak and the peak below 1000 cm^{-1} seems to be an overlap of a C–S–O vibrational mode and a C–F mode. Figure 10 shows the spectrum of the treated side of the membrane. The key features here are the almost complete loss of the SO$_3^-$ peak and the formation of a peak at approximately 1375 cm^{-1}. This can be attributed to the SO$_2$N asymmetric stretch. Figure 11 shows the spectral subtraction data for this study. The basis of this subtraction was the nulling of the large C–F peak and is shown in more detail in Figure 12. Great care must be taken when interpreting the results of PAS subtractions. Currently, there is not a good function or theory that provides a linear response that can be applied generally to PAS data. The problem becomes particularly difficult when dealing with saturated peaks; however, in this case the subtraction results appear very meaningful. The nonsubtracted data clearly show the $-SO_3^-Na$ symmetric stretch and the $-SO_2N-$ asymmetric stretch. The other two bands revealed in the subtraction are both visible as shoulders and the locations correspond well to $-SO_3^-Na^+$ asymmetric stretch and the sulfonamide symmetric stretch observed in other compounds. A final observation that can be made from this spectrum is the good subtraction of the water vapor. This indicates that no preferential desorption is observed between the two sides of the membrane.

This study provided information that would be extremely difficult or impossible to obtain by other techniques. The subtraction worked

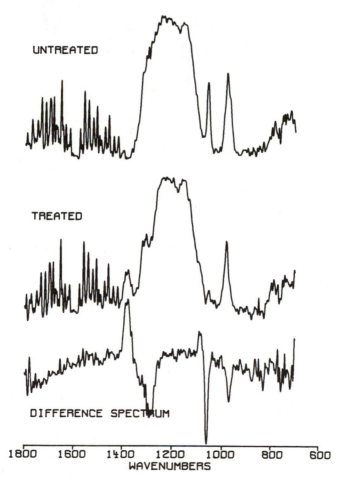

Figure 11. Spectral subtraction data for the PAS spectra discussed in Figures 9 and 10.

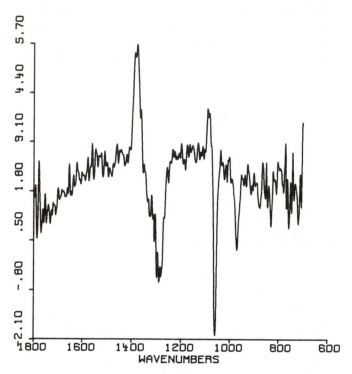

Figure 12. Detail of the nulling of the large C−F peak from the subtraction data shown in Figure 11.

quite well and revealed the two peaks that correspond to the other two S−O stretches.

Summary and Prospectus

FTIR/PAS is a new method of obtaining IR spectra of solid samples. It has some unique advantages in several sampling situations where: (1) spectra free of Christiansen contributions are desired, (2) surface measurements of undisturbed surfaces are required, (3) mechanically intractable or noxious samples must be measured, or (4) sapient analytical design would be facilitated by reference spectra with PAS characteristics. The application of FTIR/PAS to polymer problems is still in its infancy, and therefore, this is not an overview of a well-used method, but rather an introduction to the principles of FTIR/PAS, and a summary of existing work as it relates to possible polymer applications. The amount of literature on polymer applications is still quite small. Several papers of polymer interest were presented at the 1981 International FTIR Conference (4), and extended abstracts are available for these. The Literature Cited section will hopefully serve as a guide to the groups and authors currently active in this field.

Literature Cited

1. Bell, A. G. *Am. J. Sci.* **1880**, *20*, 305.
2. Christiansen, C. *Ann. Phys. (Leipzig)* **1884**, *23*, 298.
3. Farrow, M. M.; Burnham, R. K.; Eyring, E. M. *Appl. Phys. Lett.* **1978**, *33*, 735.
4. Mehicic, M.; Kollar, R.; Grasselli, J. G. *Proc. 1981 Intl. Conf. FTIR Spectrosc.* Columbia, S.C.; Sakai, H., Ed.; Vol. 289, S.P.I.E., Bellingham, WA.
5. Laufer, G.; Huneke, J. T.; Royce, B. S. H.; Teng, Y. C. *Appl. Phys. Lett.* **1980**, *37*, 517.
6. Rosencwaig, A. In "Optoacoustic Spectroscopy and Detection"; Pao, Y.-H., Ed.; Academic: New York, 1977.
7. Rockley, M. G.; Devlin, J. P. *Appl. Spectrosc.* **1980**, *34*, 407.
8. Teng, Y. C.; Royce, B. S. H. *J. Opt. Soc. Am.* **1981**, in press.
9. Vidrine, D. W. In "Fourier Transform Infrared Spectroscopy: Application to Chemical Systems"; Ferraro; Basile, Eds.; Academic: New York, 1981; Vol. 3, Chap. 4.
10. Vidrine, D. W. *Appl. Spectrosc.* **1980**, *34*, 314.

RECEIVED for review December 30, 1981. ACCEPTED October 22, 1982.

ANALYTICAL PYROLYSIS/ GAS CHROMATOGRAPHY/ MASS SPECTROMETRY

Pyrolysis Gas Chromatography, Mass Spectrometry, and Fourier Transform IR Spectroscopy

Introductory Material

S. A. LIEBMAN and E. J. LEVY
Chemical Data Systems, Inc., Oxford, PA 19363

Despite successful pyrolysis gas chromatographic (PGC) applications in laboratories throughout the world (*1*), diverse experimentation has made standardization of the method difficult. However, modern pyrolyzers coupled with high-performance chromatographic and spectroscopic detection systems now provide the means to study polymers, in varied sample forms and amounts, in a reproducible and informative manner. Currently, ASTM committees are assessing a standardization method for pyrolysis using the molecular thermometer concept (*2*). Also, statistical evaluation of the reproducibility of the method, as well as an aid in data interpretation, has been applied to key systems. This statistical or chemometrics approach permits handling of complex chromatographic and spectral results in an efficient, informative way by using pattern recognition, multivariate, or factor analysis techniques (*3*). Features of unique pyrolysis experimentation and interpretive aids to define microstructure and degradation of macromolecular systems have been emphasized by many researchers. Much of the information obtained by pyrolysis gas chromatography (GC)/mass spectroscopy (MS) or Fourier transform IR spectroscopy (FTIR) would be difficult to obtain by traditional spectroscopic or chromatographic techniques alone. If a single word were needed to describe the major attribute of analytical pyrolysis, it would be versatility—the ability to analyze diverse types of synthetic, biopolymers, and geopolymers in a realistic, informative manner.

Pyrolysis Basics

Experimentalists have used a wide range of homemade and commercial pyrolysis units for the analysis of complex materials (*4*). However, the most successful systems in modern laboratories are those that provide the needed control of pyrolysis temperature, heating rate, and other parameters. The factors affecting filament pyrolysis are as follows:

1. Current, voltage, or final temperature
2. Heating rate (dependent on current, filament mass and geometry, and rate of heat dissipation)
3. Velocity of carrier gas
4. Duration of pyrolysis process
5. Sampling method, and amount and form of sample
6. Variations in coil/ribbon geometry
7. Pressure variations in pyrolysis unit

The pyroprobe (Chemical Data Systems, Inc.), which uses resistive heating and temperature sensing with a platinum filament or ribbon, specialty microfurnaces, lasers, and Curie point pyrolyzer arrangements, are used with varying degrees of sophistication. A large majority of analytical problem solving may be performed with rather moderate instrumentation, such as interfacing the pyrolysis unit to a GC capable of modern high-performance operation (i.e., capillary column separation, multiple detectors, and close temperature and flow control). A typical polymer fingerprint (or pyrogram) of the pyrolysis fragments separated in such a GC analysis provides microstructural information when studies are conducted with proper experimentation. For example, the pyrograms resulting from the pulse pyrolysis with a Pyroprobe at 750 °C of atactic and isotactic polypropylenes (PP) are easily distinguishable patterns (Figure 1). The full interpretive value of analytical pyrolysis data for microstructural studies of PP systems has been reported (5), and is discussed regarding other polymers in this volume.

Extended Applications

Residual Volatiles. Another aspect of polymer characterization involves the analysis of trace residuals—solvents, monomers, and additives. With minor experimental modifications, static headspace analysis may be conducted by isolating the Pyroprobe interface from the GC flow stream and heating the sample for the desired time/temperature interval (e.g., 10 min at 150 °C) below the degradation temperature. The volatiles are then swept from the interface by the He carrier stream into the appropriate GC column held at ambient or subambient temperature to trap the evolved gases. Temperature programming of the column in the normal manner provides a direct analysis of the sample. Figure 2 shows this comparative analysis of volatiles of two PP samples obtained from different manufacturers. The residual volatile contents are significantly different. Similarly,

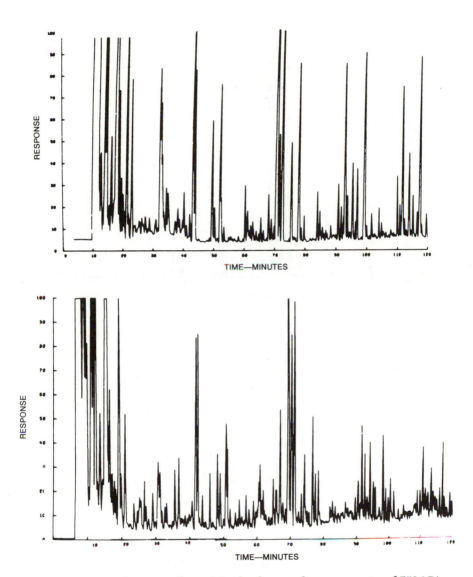

*Figure 1. Pyrolysis capillary GC of polypropylenes at a rate of 750 °C/
10 s. Key: top, isotactic; and bottom, atactic.*

the analytical pyrolysis experiment may be easily configured to simulate some process or in-use exposure condition (e.g., heating in air at
200 °C for 2 min). The same sample may then be subjected to flash
pyrolysis conditions to provide full characterization data. In this manner, minor variations or residuals from polymerization or purification
schemes do not complicate the microstructural analysis. More impor-

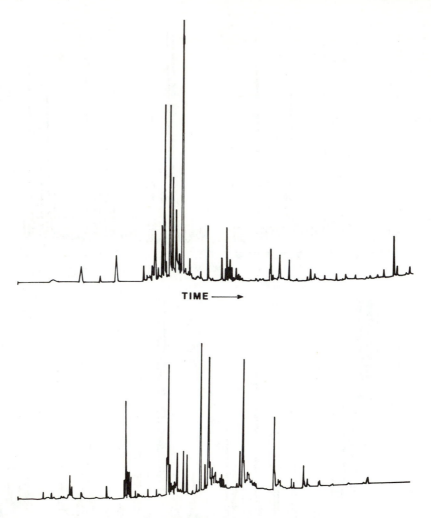

Figure 2. Volatiles in polypropylenes.

tantly, performance data may be correlated to the presence or absence of some components that may be critical for polymerization or processing conditions.

Programmed Pyrolysis and FTIR. In addition to the flash pyrolysis mode, analysts may use a programmed pyrolysis or time-resolved PGC configuration. This approach was needed to relate degradation fragments produced from conventional thermogravimetric analyses (TGA) to products produced by pyrolysis. This result was achieved by programming the Pyroprobe coil at a heating rate of 2, 5, 10, etc., °C/min to correspond to the rate used in the TGA experiments

(6). Interfacing the Pyroprobe to an automatic GC sampling device, or FTIR (7) provided the necessary information. PGC data were obtained for vinyl chloride—propylene copolymers and complex composites in microstructural and degradation studies.

Lephardt reported previously (8, 9) that evolved gas analysis with an FTIR to serve as a sensitive spectral monitor for chosen volatiles. Figures 3–5 present the information from this type of experimentation. Lignin and cellulose samples were heated linearly and the released volatiles were removed continuously by the N_2 carrier gas, and then passed through a heated gas cell mounted in the spectrometer. During the sample heating, IR spectra of the evolving species were collected repetitively. Spectral absorbances at frequencies known to be due to the chosen chemical species were subsequently plotted vs. temperature to obtain evolution profiles. These evolution profiles are shown for the above systems. The Kraft pine lignin samples provided an evolution profile of phenolic species (Figure 3). The spectral subtraction technique permitted separation of two phenols (A and B) and patterns for each were obtained (Figure 4). Additionally, SO_2 detection gave information on the cross-linking system. A cellulose sam-

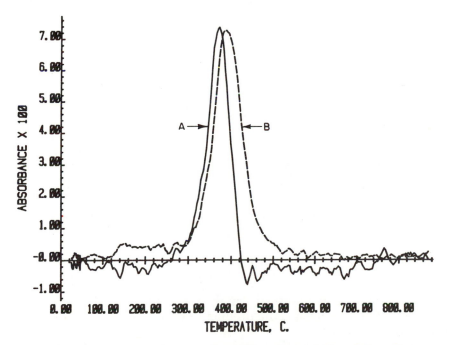

Figure 3. FTIR-evolved gas analysis. Phenol A and Phenol B evolution profiles from lignin. (Reproduced from Ref. 8. Copyright 1977, American Chemical Society.)

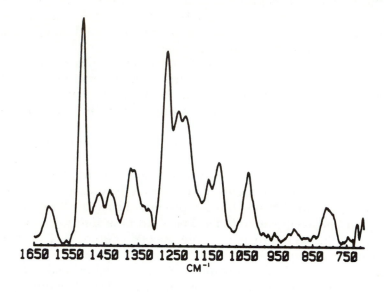

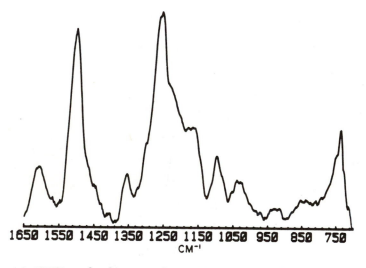

Figure 4. FTIR-evolved gas analysis. IR spectra (1650–750 cm⁻¹) of Phenol A (top) and Phenol B (bottom).

ple was examined in a thermal depolymerization mode by following an apparent end-group specific product (formic acid) vs. a chain product (H_2O) (Figure 5).

The evolved gas FTIR experiments provide invaluable insight regarding kinetics and mechanisms in thermal degradation studies. The ability to optimize product formulations with respect to specialty additives (e.g., smoke suppressants, flame retardants, promoters) has been achieved with such simulation studies for the material sciences (*10*). The particular advantage of FTIR is evident over the experimentally limited, but perhaps more quantitative approach, of automatic GC sampling and analysis.

Pyrolysis MS and GC/MS

Synthetic Polymers. The growing field of pyrolysis MS and GC/MS of synthetic polymers has benefited from the contributions of many researchers using Curie point, Pyroprobe, laser, or other direct insertion pyrolyzers. Some of the outstanding developments documented deal with polyamides and nylon degradation mechanisms (*11, 12*). Other work has involved catalytic effects on polystyrene thermal degradation at the initiation stage. The catalytic effect of small amounts of Lewis acids, salts, or bases was studied with condensation polymers to assess the sensitivity of the initial degradation temperature of the system and production of oligomers and monomers. Also, new extensions of laser microprobe mass analysis (LAMMA) are reported by Hercules et al. in applications to synthetic polymers.

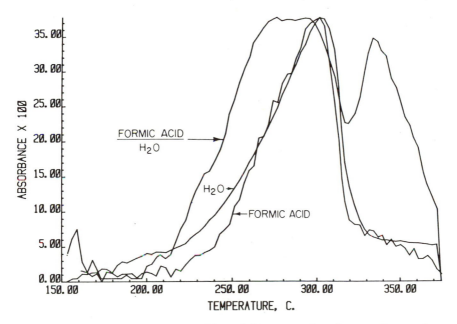

Figure 5. FTIR-evolution profiles of formic acid and water from cellulose.

Natural Polymers.   Unique information provided from temperature programming the thermal device and leading degradation products of natural polymers directly into the MS has been reviewed (*13*). The experimental arrangements needed for such studies have been specialized to provide limited but significant data on a variety of biopolymers. An overview discussed in this volume highlights the advances made in the biochemical and biomedical field, not only with pyrolysis MS, but with the full analytical pyrolysis input. From Reiner's initial report in 1965 (*14*) on fingerprinting microorganisms using flash pyrolysis GC, to the current high sensitivity, sophisticated thermal-programming MS analysis, one can appreciate the accomplishments in this field. Additionally, to be aware that advanced pyrolytic techniques have been applied to both synthetic and natural polymers with directly analogous methodologies is a focal point of this volume.

Interpretive Aids

For the most demanding of analytical pyrolysis applications— complex composites, microstructural details of synthetic biopolymers and geopolymers—the input of chemometrics is proving to be invaluable. Pattern recognition and other statistical or computer-assisted techniques are being applied to aid interpretation of pyrograms at an increasing rate with many successful applications (*1, 3*). Effective data management is mandatory for the massive amount of information generated in typical pyrolysis GC/MS or FTIR experiments. Illustration of such experimental designs using the chemometrics approach is discussed elsewhere in this volume.

Literature Cited

1. "Selected Bibliography of Analytical Pyrolysis (1973–1980)," Applications Laboratory, Chemical Data Systems, Inc.
2. Levy, E. J.; Walker, J. G. *J. Chromatogr. Sci.*, in press.
3. "Chemometrics: Theory and Applications"; Kowalski, B. R., Ed.; ACS SYMPOSIUM SERIES No. 52, ACS: Washington, D.C., 1977.
4. "Selected Bibliography of Solids Pyrolysis (1960–1973)," Applications Laboratory, Chemical Data Systems, Inc.
5. Sugimura, Y.; Nagaya, T.; Tsuge, S.; Murate, T.; Takeda, T. *Macromolecules* **1980**, *13* (4), 928.
6. Liebman, S. A.; Ahlstrom, D. H.; Foltz, C. R. *J. Polym. Sci., Polym. Chem. Ed.* **1978**, *16*, 3139.
7. Liebman, S. A.; Ahlstrom, D. H.; Griffiths, P. R. *J. Appl. Spectrosc.* **1976**, *30*, 335.
8. Fenner, R. A.; Lephardt, J. O. *J. Agric. Food Chem.* **1981**, *29*, 846.
9. Lephardt, J. O. et al. *J. Appl. Spectrosc.* **1980**, *34* (2), 174.
10. Liebman, S. A. et al., Presented at the ACS 172nd Natl. Meet., Org. Chem. & Plastics Div., San Francisco, CA, 1976.
11. Luderwald, I. *Angew. Makromol. Chem.* **1978**, *74*, 165.
12. *Ibid.* **1978**, *67*, 193.
13. Risby, T. H.; Yergey, A. L. *Anal. Chem.* **1978**, *50*, 327A.
14. Reiner, E. *Nature (London)* **1965**, *206*, 1272.

Accepted October 27, 1981

Microstructural Characterization of Copolymers by Pyrolysis–Glass Capillary Gas Chromatography

S. TSUGE, Y. SUGIMURA, T. KOBAYASHI, T. NAGAYA, and
H. OHTANI

Nagoya University, Faculty of Engineering, Department of Synthetic
Chemistry, Nagoya 464 Japan

*Fundamental conditions to obtain characteristic and re-
producible high-resolution pyrograms of polymer samples were first studied using pyrolysis–glass capillary
gas chromatography [P(GC)²]. The effect of thermal
conductivity of carrier gas and splitting conditions were
examined and discussed together with the reproducibility of the results, the secondary thermal reactions of the
degradation products, and column contamination by
less volatile products. Then, the established P(GC)²
technique was applied for the microstructural investigation of tapered block copolymers of styrene (S) and
isoprene (I). The resulting high-resolution pyrograms of
the copolymers were interpreted in terms of the amounts
of the junctions of the different monomer units along the
copolymer chain.*

AMONG VARIOUS THERMAL ANALYSIS TECHNIQUES, pyrolysis–gas chromatography (PGC) has a relatively short history. However, owing
to recent developments such as highly efficient glass capillary columns, specific identification of the peaks in the pyrograms by gas
chromatography (GC)–mass spectroscopy (MS) systems, and highly
specific pyrolysis devices, PGC has made great strides toward being a
powerful tool in structural characterization of high polymers. The
structural information obtained by PGC is sometimes unique and complementary to that obtained by the conventional spectroscopic methods such as IR and NMR.

In this work, some fundamental studies on the effect of thermal
conductivity of the carrier gas and splitting conditions were first made

to obtain highly reproducible pyrograms using pyrolysis–glass capillary gas chromatography [P(GC)²]. Then, the established P(GC)² technique was applied successfully for characterization of microstructures in tapered block copolymers of styrene (S) and isoprene (I) [TB−P(S−co−I)].

Experimental

The schematic flow diagram of the P(GC)² system is shown in Figure 1. A vertical microfurnace-type pyrolyzer (1, 2) was attached directly to a gas chromatograph with a glass capillary column (ID 0.3 mm × 50 m, OV−101). The glass splitter with a split ratio of 1:55 was modified to be heated independently to the desired temperatures. The dead volume of the splitter was packed with 5% of OV−101 on Diasolid H (80−100 mesh) and maintained at 250 °C. Samples between 0.1 and 0.2 mg were pyrolyzed mostly at 510 °C under a flow of carrier gas (55 mL/min). The column temperature was programmed from 50 to 250 °C at a rate of 4 °C/min. The peak identification was carried out using a directly coupled quadrupole mass spectrometer, JMS−Q10A from Jeolco, which was operated in a chemical ionization mode.

Results and Discussion

Fundamental P(GC)² Conditions. The splitless mode is not always best for P(GC)² because the very low velocity of the carrier gas in the pyrolyzer sometimes causes undesirable secondary reactions of the characteristic degradation products (2, 3). Another problem for P(GC)² is column deterioration by less volatile tarry products that may form during the thermal degradation of high polymers. Furthermore, in the splitting mode operation, the product composition actually entering the capillary column sometimes differs from the original one, depending on the volatility of each component and the temperature of the splitter. This behavior causes serious problems in reproducibility and quantification of the results because the degradation products of polymer samples usually consist of complex mixtures with a wide range of volatilities.

To overcome these problems, a modified splitting device was used (2). As shown in Figure 1 (D), the dead volume of the splitter was packed with an ordinary packing material with the same liquid phase as that for the capillary column. The splitter was maintained independently at the maximum temperature of the column (250 °C). This arrangement enabled smooth and reproducible splitting of the degradation products with a wide range of volatility. This splitter also provided column protection from less volatile tarry products without lowering the resolution of the resulting pyrograms.

Another serious problem in PGC of polymers often is caused by ghost peaks during repeated runs. To prevent this phenomenon, as shown in Figure 1, small amounts of carrier gas were fed through a

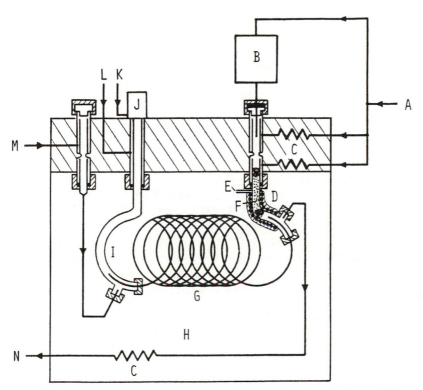

Figure 1. Schematic flow diagram for P(GC)². Key: A, carrier gas (N₂ or He); B, microfurnace-type pyrolyzer; C, flow resistance; D, splitter; E, heater for splitter; F, packing material; G, glass capillary column; H, column oven; I, outlet glass tube; J, flame ionization detector (FID); K, air; L, hydrogen; M, make-up gas; and N, vent.

resistance (C) to the injection and inlet tubes that otherwise would be dead ends and cause ghost peaks of the degradation products.

Thermal conductivity of carrier gas plays a very important role in PGC (4) because exchange of heat energy associated with the thermal degradation of polymer samples is done mostly through the carrier gas. In the furnace-type pyrolyzer, for example, heat energy necessary for the thermal degradation of samples is supplied through the carrier gas from the outer furnace. The excess heat energy retained by the resulting degradation products has to be released as soon as possible through the carrier gas as the products are transferred to the cooler zone (either the connection tube between the pyrolyzer and the separation column, or the separation column itself). Heat-energy transfer through the carrier gas should change depending on the thermal conductivity of the carrier gas. Thus, reproducibility of the results and the degree of undesirable secondary reactions might be affected by the nature of the carrier gas utilized.

Table I summarizes the reproducibility data obtained for three kinds of methyl methacrylate (MMA)−styrene (S) copolymers [P(MMA-co-S)]s using He and N_2 as carrier gases, whose thermal conductivities at 100 °C are 40.8×10^{-2} and 7.3×10^{-2} cal cm^{-1} s^{-1} deg^{-1}, respectively. For each copolymer sample, ten repeated runs were made using about a 0.1-mg sample. As was expected, significantly better reproducibilities can be obtained when He is used as carrier gas than when N_2 is used. This difference is mostly attributable to the big difference in their thermal conductivities (a factor of six). Almost the same trends were observed for various kinds of polymer samples. These data suggest that He is the best inert carrier gas for PGC.

Recombination reactions during the pyrolysis of the sample sometimes cause noncharacteristic peaks on the resulting pyrograms. These reactions cannot be neglected when large amounts of sample are used because the transient concentration of the degradation products in the carrier gas at the hot-zone in the pyrolyzer changes as a function of sample size under given PGC conditions. For example, when about 1 mg of a physical blend of polystyrene (PS) and polymethyl methacrylate (PMMA) (PS/PMMA = 25 wt/75 wt) is pyrolyzed under a flow of N_2 carrier gas, a typical recombination product of the hybrid dimer, S−MMA, can be observed on the pyrogram. However, when the carrier gas is changed to He or a smaller sample size (less than 0.2 mg) is used the recombination peak becomes indiscernible; with N_2 as carrier gas the same results are observed. Therefore, the limiting sample size for a polymer material should be examined empirically under given experimental conditions.

The other important experimental factor is the cleanliness of the sample holder. Carbide- or nitride-type residues on the sample holder sometimes cause fairly large catalytic effects on the pyrolysis of poly-

Table I. Effect of Carrier Gas on the Reproducibility of the Pyrograms of Various P(MMA–co–S)s

Sample Composition (S Mol Fraction)	Coefficient of Variance (%) for Observed Relative Peak Intensities (S/MMA)[a]	
	He	N_2
0.182	0.53	2.25
0.411	0.29	2.22
0.843	1.00	3.00

[a] Coefficient of variance for 10 repeated runs using about 100 μg of the copolymer sample.

mers. For example, when PS is pyrolyzed in a contaminated sample holder after repeated use with an acrylonitrile—styrene copolymer (which yields fairly large amounts of carbonaceous residues), the observed peak ratio of monomer, dimer, and trimer on the pyrogram significantly differs from that obtained with a clean sample holder. Furthermore, on the former pyrogram, a fairly strong α-methylstyrene peak is observed. This compound might be formed through the catalytic secondary thermal reactions on the surface of the uncleaned sample holder. Therefore, the surface of the sample holder should always be cleaned as well as possible after each run.

Microstructural Characterization of Tapered Copolymers. The morphology and many physical properties of copolymers are affected strongly by sequence distributions of the component monomer units. In this section, the P(GC)² technique was used to estimate the number of junctions in the TB−P(S−co−I)s. Various kinds of TB−P(S−co−I)s, synthesized through a simultaneous "living" anionic copolymerization, were supplied by H. Kawai at Kyoto University (5). Figure 2 shows a hypothetical primary structure of a typical copolymer together with the transmission electron micrograph of a thin film of about 30-nm thickness after staining with osmium tetroxide solution. Here, the dark and the bright stripes and islands represent polyisoprene and polystyrene phases, respectively. The image contrast of the two phases for the tapered block copolymer is slightly different from that for the A−B-type ideal block copolymer.

The number of junctions is a good measure of the degree of mixing of the associated two monomer units in the polymer chain. However, estimation of the number from the ¹H-NMR spectra of the copolymer is very difficult. Figure 3 shows a typical pyrogram of the tapered copolymer (S/I = 50/50). Assignments of the peaks in the pyrogram are listed in Table II. The identification was carried out primarily by a directly coupled GC−MS system. Preliminary evidence indicated that hybrid peaks such as IS, ISS, and SIS were not observed on the pyrogram of a physical blend of PS and PI under the utilized pyrolysis conditions. However, on the pyrogram of the TB−P(S−co−I) (Figure 3) we can see various kinds of hybrid peaks, which reflect the existence of junctions of different monomer units.

Table III summarizes the observed results from the pyrograms of the various copolymers. Estimation of the minimum junctions percent (MJ%) was calculated using the following equation:

$$MJ\% = \tfrac{1}{2}\, SI + \tfrac{1}{3}\, (ISS + SIS) \qquad (1)$$

where IS, ISS, and SIS are the relative yields of the hybrid peaks. The factors one-half and one-third are associated with the minimum number of junctions in the corresponding hybrid products. The estimated MJ% suggests that the tapered copolymers such as TB−1 to TB−3 have almost comparable numbers of junctions such as ⌇⌇S−I⌇⌇ along the polymer chain (at least about 4 junctions within 100 monomer units). The value 0.17 for the ideal block copolymer (TB−5) is very close to the theoretical junction percent of 0.18 in the bracket calculated from the average molecular weight of 49,000. On the other hand, TB−4 with a molecular weight of 97,000 should have junctions about 1.4 times that of the ideal block copolymer.

As discussed in this chapter, the specific high-resolution pyrograms of polymer samples obtained by P(GC)² are quite effective in obtaining microstructural information of polymers provided that the PGC conditions are controlled properly.

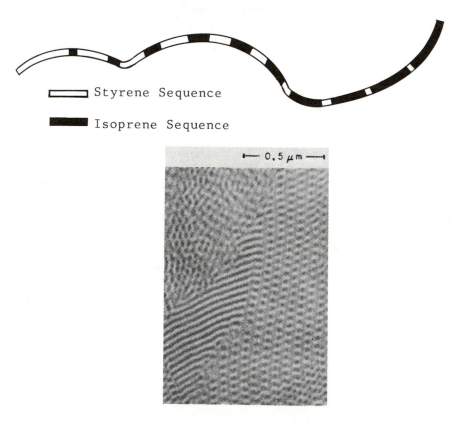

Figure 2. Structure of tapered block copolymer of styrene (S) and isoprene (I); the lower picture is a transmission electron micrograph of TB–2 (in Table III).

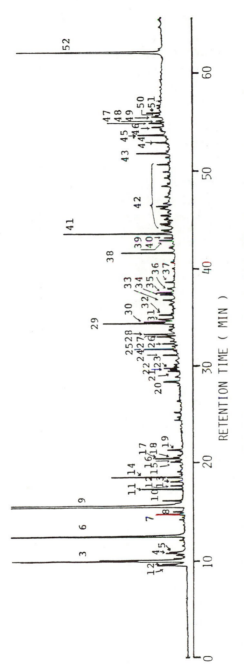

Figure 3. Typical pyrogram of TB–P (S–co–I). Sample: TB–2 (Table III); the peak numbers are the same as those in Table II.

Table II. Peak Assignments of the Pyrogram Shown in Figure 3

Peak No.	Structure	Related Sequence[a]
1	C=C–C–C	I
2	C=C–C=C	I

3	C=C–C=C
	  $\mid$
	  C
		I

4,5	C_6H_8	I
6	Ph–C	S
7	Ph–C–C	S
8	C_8H_{10}	II
9	Ph–C=C	S
10	C_9H_{14}	II
11	Ph–C–C=C	S
12	Ph–C–C–C	S
13	$C_{10}H_{16}$	II

14	Ph–C=C
	   $\mid$
	   C
		S

15	Ph–C=C–C	S
16–18	$C_{10}H_{16}$	II II
19	Ph–C=C–C–C	IS
20	$C_{12}H_{16}$	IS
21	$C_{12}H_{14}$	IS
22	$C_{12}H_{16}$	IS
23	$C_{12}H_{14}$	IS
24–28	$C_{13}H_{16}$	IS
29,30	$C_{14}H_{18}$	IS
31	$C_{14}H_{16}$	IS
32,33	$C_{14}H_{18}$	IS

34	Ph–C=C
	   $\mid$
	   Ph
		SS

35	Ph–C–C–Ph	SS
36	$C_{14}H_{16}$	IS

37	Ph–C–C
	   $\mid$
	   Ph
		SS

38	Ph–C–C–C–Ph	SS
39	cis Ph–C=C–Ph	SS
40	trans Ph–C=C–Ph	SS

41	Ph–C–C–C=C
	        $\mid$
	        Ph
		SS

42	mostly SS	SS
43,44	$C_{21}H_{24}$	SIS or ISS
45	$C_{20}H_{22}$	SIS or ISS
46–51	$C_{21}H_{24}$	SIS or ISS

52	Ph–C–C–C–C–C=C
	       $\mid$ $\mid$
	       Ph Ph
		SSS

[a] I, isoprene unit; and S, styrene unit.

Table III. Estimated Minimum Junctions in TB–P(S–co–I)s

Sample No.	Composition[a] (Nature)	Molecular Weight ($M_n \times 10^{-4}$)	Fragment Distribution (Monomer Unit Ratio %)							Estimated[b] Minimum Junction (%)
			I	S	II	IS	SS	ISS +SIS	SSS	
TB–1	S/I(50/50) (tapered)	4.3	23.0	57.5	1.9	5.7	4.0	3.4	4.6	4.0
TB–2	S/I(50/50) (tapered)	6.3	26.1	55.0	2.6	5.9	3.3	3.1	4.0	4.0
TB–3	S/I(50/50) (tapered)	10.6	28.1	54.8	2.2	5.5	3.3	2.8	3.3	3.7
TB–4	S–S/I(25–25/50) (tapered)	9.7	18.6	67.7	1.2	0.33	3.0	0.23	9.0	0.24
TB–5	S–I(50/50) (ideal block)	4.9	18.0	68.2	1.3	0.34	2.9	—	9.2	0.17(0.18)

[a] S, Styrene unit; and I, isoprene unit (molar ratio).
[b] Calculated using Eq. 1; the theoretical value is 0.18.

Acknowledgment

The authors are indebted to H. Kawai at Kyoto University for providing the TB−P(S−co−I) samples.

Literature Cited

1. Tsuge, S.; Takeuchi, T. *Anal. Chem.* **1977,** *49,* 348.
2. Sugimura, Y.; Tsuge, S. *Anal. Chem.* **1978,** *50,* 1968.
3. Schomburg, G.; Dielman, R.; Husman, H.; Weeke, F. *J. Chromatogr.* **1976,** *122,* 55.
4. Tsuge, S.; Sugimura, Y.; Nagaya, T. *J. Anal. Appl. Pyrolysis* **1980,** *1,* 221.
5. Tsukahara, Y.; Nakamura, N.; Hashimoto, T.; Kawai, M.; Nagaya, T.; Sugimura, Y.; Tsuge, T. *Polym. J.* **1980,** *12,* 455.

RECEIVED for review October 14, 1981. ACCEPTED December 22, 1981.

Structural Analysis of Polymeric Materials by Laser Desorption Mass Spectrometry

JOSEPH A. GARDELLA, JR., SUSAN W. GRAHAM, and
DAVID M. HERCULES[1]

University of Pittsburgh, Department of Chemistry, Pittsburgh, PA 15260

The application of laser desorption mass spectrometry (LDMS) to the analysis of various polymers is discussed. Development of this laser approach as a means of structure determination was accomplished by investigation of various operational parameters of the laser microprobe, development of an interpretative model, application and later testing of the model through analysis of results obtained from technical polymers, exploration of the analytical capabilities of the model, and application of the method to the study of complex biopolymer systems. Minimization of the laser power density was found to be important in the generation of structure-related fragments. The results presented support the model of spectral analysis involving the backbone side chain method. This method proved effective in differentiating each of the methacrylate polymers studied. The LDMS technique provides fragmentation patterns not readily available by other methods.

T HE FIELD OF MASS SPECTROMETRY (MS) recently has expanded to encompass solid samples with low volatility, made possible by the development of several new vaporization methods (1–19). Obtaining mass spectra of solids for many researchers has been limited by the relative newness of the methods. The use of many of these methods is by no means routine. The present work is part of an overall program to develop the analytical capabilities of laser desorption mass spectrometry (LDMS). LDMS utilizes a focused laser as a vaporization/ionization source for a time-of-flight mass (TOF) spectrometer (20).

---

[1] To whom correspondence should be sent.

0065–2393/83/0203–0635$11.50/0

This method originally was applied as a microprobe for biomedical research, but its use as a solid state mass spectrometer shows promise in the effort to provide a routine, reproducible means of solid state structure determination by MS.

Table I lists several methods that exist for generation of ions from solids for subsequent mass analysis. Some of these methods use a separate ion source, the common electron impact (EI) or chemical ionization (CI) cell, and utilize the high energy source field for vaporization only. The mechanisms involved in ion production for these methods are not well known. Some work has shown that all processes involved are similar (21) but, at this stage of development, many controversies exist about mechanisms that dominate in vaporization and ion production.

All of the work reported in this chapter has been published elsewhere (1–5). This chapter summarizes our research on mass spectra of polymers using LDMS, a relatively recent development for structural characterization of polymers.

Each method described in Table I has distinct advantages and disadvantages, and none is universal in its application. The pyrolysis methods are coupled easily and commonly through a gas chromatography (GC) column to the mass spectrometer, thus adding the separation power of the GC instrument. They are well known and relatively inexpensive but may be applied only to relatively volatile solids. Field desorption (or field ionization) is a very sensitive but not a very reproducible method. Fission fragment induced desorption (FFID), using the plasma from ^{252}Cf, shows unique capabilities to generate high mass ions but requires long analysis and data collection times. Secondary ion mass spectrometry (SIMS) is inherently a surface analytical method. Although high sensitivity to surface structure can be advantageous, it also can cause significant problems for interpretation of surface/bulk effects. The neutral beam [fast atom bombardment (FAB)] methods are at a very early stage of development but show great promise. Results indicate that the FAB methods produce results equivalent to SIMS and FFID.

The laser approach to volatilization/ionization, and its application to a very important class of solid materials, polymers, is treated here. The polymer analysis field is one where mass spectrometric structure determination is very uncommon. Polymers are the typical example of a low volatility solid, creating difficulties for ion formation. References listed in Table I applying these methods to polymers are only the work done by this group (1–5, 17) and the laser pyrolysis study (14), which are examples where detailed polymer structure analysis by MS was attempted for more than one polymer.

The goals of the present work are to develop LDMS as a means of

Table I. Volatilization/Ionization Methods Used in Solid State MS

Inductive Source	Volatilization Method Title	Further Ionization	References to Polymer Analysis
Heat	pyrolysis	none	7
		EI[a]	10, 11
		CI[b]	10
Electric field	field desorption (FD) or field ionization (FI)	none	12, 13
Laser	laser desorption (LD) or laser pyrolysis	EI,CI	14, 15
		none	14
	laser ionization (LI)	none	1–6
Charged particle beam	fission fragment induced desorption (^{252}Cf) (FFID)	none	16
	secondary ion mass spectrometry (SIMS)	none	17
Neutral particle beam	atom ion emission	none	18
	fast atom bombardment (FAB)	none	19

[a]EI is electron impact ionization of gas-phase molecules.
[b]CI is chemical ionization of gas-phase molecules.

structure determination by (1) investigation of various operational phenomena of the laser microprobe, (2) development of an interpretation model for results obtained, (3) application of the model and subsequent testing through analysis of results obtained from technical polymers, (4) discussion of analytical capabilities of the method, and (5) applications of LDMS to complex biopolymers.

Results will be presented to support the model of spectral analysis involving the backbone/side chain method. Ions are explained as being generated from the polymer backbone or from a pendant side chain group.

Experimental

The polymers analyzed in these studies are listed in Tables II and III. Most of the polymers are commercially available (Scientific Polymer Products, Inc.). Biomer (Ethicon) and Avcothane (Avco-Everett) are biocompatible polymers. Some polymers (other than Biomer and Avcothane) contained a siloxane impurity and were extracted with hexane to remove the impurity to below the level of detectability.

The polymers were mounted on electron microscope grids under a quartz glass cover slip and placed on the spectrometer sample holder for vacuum pumpdown and subsequent analysis. Laser desorption (LD) spectra were recorded by an Inficon Leybold Heraeus, Inc. LAMMA 500 spectrometer. The output of a frequency quadrupled Q-switched Nd–YAG laser (λ = 265 nm) is focused through microscope optics (10×, 32×, 100×) and an oil immersion lens to an ultimate spot size of <0.5 μm on either a thin section or a powder. Changes in laser spot size have little effect on the mass spectra. A set of filters is used to adjust the laser pulse density, which may range from 10^7 to 10^{10} W/cm² over a pulse width of about 15–20 ns. Ions are accelerated at 3 keV into the drift tube of a time of flight mass spectrometer. The mass spectrometer has a nominal range of 0–2000 amu, with a resolution ($m/\Delta m$) of 850 at 1000 amu. The output from a 17-stage electron multiplier is coupled to a transient recorder after amplification. The transient recorder functions as a storage buffer for selected portions of the mass spectrum. Its timing sequence is triggered by a photodiode signal from the laser pulse, and in all cases mass spectra were obtained from single laser shots. Spectra are displayed on an oscilloscope; a strip-chart recorder is used for hard-copy output. Mass spectra are linear in time, giving a mass scale proportional to $(m/z)^{1/2}$. Mass assignments were determined by measurement along the chart paper and hand calculations. The operational parameters of the LAMMA 500 have been presented in greater detail elsewhere (20, 22, 23).

Method Development

An advantage of a laser source for vaporization/ionization in MS of solids is the variability of the source characteristics (24). Unlike most sources used for solid state MS, the laser gives (1) a wide range of sampling parameters to be controlled, (2) microprobe capabilities, and (3) the possibility of controlled repetitive excitation. In the LAMMA 500, most operational parameters can be varied. However, in practice

Table II. Polymers Analyzed

Sample Name	Polymer Name	Structure	Characteristics
PE	polyethylene	$-(CH_2-CH_2-)_n$	high density beads
PTFE	polytetrafluoroethylene	$-(CF_2-CF_2-)_n$	powder
PVC	polyvinyl chloride	$-(CH_2-CH-)_n$, Cl	low mol wt powder
PP	polypropylene	$-(CH_2-CH-)_x$, CH_3	isotactic fine powder
PS	polystyrene	$-(CH_2-CH-)_x$, C_6H_5	glassy chunks
PAA	polyacrylamide	$-(CH_2-CH-)_x$, $C=O$, NH_2	fine powder
N6	polycaprolactam	$-(N-(CH_2)_5-C-)_x$, H, CH_3, $=O$	beads
PDMS	polydimethylsiloxane	$-(Si-O-)_x$, CH_3, CH_3	liquid
Biomer	polyether/polyurethane	*see* Figure 1	film
Avcothane	polyether/polyurethane/ polydimethylsiloxane	*see* Figure 2	molded sheets

Table III. Poly(R-methacrylate) Polymer Series

Sample ID	R	Characteristics
PMMA	methyl	fine beads/secondary std.
PEMA	ethyl	fine beads/secondary std.
PIPMA	isopropyl	glassy/siloxane impurity
PNBMA	n-butyl	large beads/easily oxidized secondary std.
PSBMBA	sec-butyl	granular
PIBMA	isobutyl	granular/secondary std.
PTBMA	tert-butyl	crystalline
PCHMA	cyclohexyl	powder
PPMA	phenyl	granular powder
PBZMA	benzyl	granular beads/siloxane impurity

$$\left(-CH_2 - \underset{\underset{O \diagdown O-R}{\overset{|}{C}}}{\overset{\overset{CH_3}{\overset{|}{\underset{|}{C}}}}{}} - \right)_x$$

the pulse width and wavelength are constant, while laser power density and focused spot size are varied to yield a wide variety of ionization conditions.

In the process of method development, the effects of laser power density on the informational content of polymer spectra were investigated. This work amplifies the initial results of Unsold et al. (6) where the epoxy resin EPON 812 was analyzed at various laser power densities. At higher power densities, atomic ions were the dominant species in the spectrum; at lower power densities, more structure-related molecular fragments were seen. Because the structure of the epoxy studied by Unsold et al. is not easily definable, a better example of this phenomenon would be to perform a similar analysis on a polymer with known structure.

In particular, data for polyisobutyl methacrylate (PIBMA) are shown in Table IV. Increased rearrangement peaks and backbone atomic and molecular ions (C_n^-, C_nH^-) are evident at high laser power ($\sim 10^{10}$ W/cm^2). Ions with structures corresponding to simple bond breaking can be seen at lower laser power densities ($m/z = 41, 55$, and 73 especially).

This effect is illustrated by the decreasing relative intensity of the

Table IV. LD Results for Polyisobutyl Methacrylate Negative Ion Power Density Dependence

Mass (amu)	Structure	Relative Intensity[a]		
		Low Power	Medium Power	High Power
36, 37	C_3^-, C_3H^- $\diagdown CH_3$	nd	med	med
41	$-C \diagup{CH_2} \diagdown{CH_2}$	med	med	wk
45	$HCOO^-$	med	med	wk
48	C_4^-	med	stg	vstg
49	C_4H^-	med	stg	vstg
55	$-CH=C \diagup CH_3 \diagdown CH_3$	med	med	med
73	$-OCH_2CH(CH_3)_2$ $CH_2=$	stg	med	med
85	$CH_3-C-COO^-$	med	stg	vstg
149		med	med	wk
165		med	wk	wk
181		med	wk	wk

[a] Relative to base peak in spectrum. Low power is $\sim 10^7$ W/cm²; high power is $\sim 10^{10}$ W/cm².
Key to abbreviations: nd, not detected; wk, weak; med, medium; stg, strong; and vstg, very strong.

m/z 73 peak in the negative ion spectrum at high power, assignable to the $^-$OR ion (side chain), and the increased relative intensity of the m/z 49 and 85 amu peaks, which are assigned to backbone structures (Table IV). Because the m/z 85 peak is of much higher intensity than all other C_n^-, C_nH^- peaks, it is assigned to the $(M-R)^-$ fragment (M represents monomer). When the experiment is performed at the highest possible laser power densities, the negative ion spectrum reduces to the C_n^-, C_nH^- pattern from stripping and rearrangement to polyacetylenes, which was observed for polyethylene (6) and other straight chain polymers.

The laser power density dependence also seems to be related to polymer morphology. Because a wide variety of sample forms was investigated (but not for a single sample), different absolute intensity dependence was observed. No firm conclusions could be drawn without a different series of forms for a single sample. Matrix effects affected relative and absolute ion intensities in previous work (25). Thus, this dependence is expected.

The final operational parameter to be discussed in the high intensity negative ion spectra in LDMS (2). The utility of the negative spectra, from an informational standpoint, is realized when results for polymers with functionality likely to yield a stable negative ion are compared to results from other polymers without such a functionality (1). The use of the negative ion spectra will be discussed further for individual polymers.

Development and Application of the Model

The backbone/side chain model of spectral interpretation, for LDMS results, involves assigning peaks to structures related to the polymer backbone or a pendant side chain group, assuming a minimum of cross-linking between backbone chains. The following typical backbones will be characterized: straight chain hydrocarbon, fluorinated hydrocarbon, secondary amide and siloxane, and various side chains. Additionally, the results from the study of complex biopolymers will be presented. The ability of LDMS to distinguish very small differences in side chain functionality and to distinguish between very similar biopolymers will be illustrated.

Straight Chain Carbon Backbones. Polyethylene (PE) shows typical straight chain hydrocarbon polymer patterns in both the positive and negative ion spectra (Figure 1). The positive ion spectrum shows clusters of ions, each approximately 12 amu apart. Each of these mass groupings can be assigned to three to six different ions, usually C_n^+, C_nH^+, $C_nH_2^+$, $C_nH_3^+$, and other ions. Besides the peaks assignable to the carbon chain, peaks are assigned to common processing contaminants: Na$^+$ (23), Al$^+$ (27), K$^+$ (39, 41), and Co$^+$ (59). Small peaks are

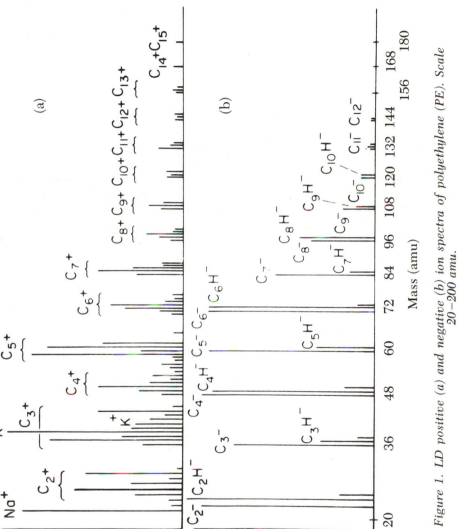

Figure 1. LD positive (a) and negative (b) ion spectra of polyethylene (PE). Scale 20–200 amu.

noted at high laser powers between $C_n H_m^+$ clusters. Most likely, these peaks are due to ions present from rearrangement during the ionization process. Rearrangement would be expected, but these events are minor compared to ejection and ionization of molecular fragments directly related to the structure. The highest mass ion that was detected was a C_{15} group.

The negative ion spectrum of PE shows unusual intensity, because most techniques (outside of chemical ionization) do not produce negative ion spectra of equal intensity to the positive ion spectra. The major peaks parallel the positive ion pattern but show a pattern of C_n^- molecular clusters, where only C_n^-, $C_n H^-$, and $C_n H_2^-$ appear. The $C_n H_2^-$ peak usually is seen only at high laser powers. For C_n^-, where n is an odd number, the peak assignable to $C_n H^-$ is of very low intensity, relative to the C_n^-. The even n cluster has $C_n H^-$ intensity nearly equal to or greater than the C_n^- intensity. This effect was explained as a result of the stability of polyacetylenes having the same structure. In general, for the polymers studied, the backbone pattern relative intensities of $C_n H_m^{+,-}$ clusters increase to $n = 3, 4, 5$, or 6 after which a general decrease in intensity occurs. For PE, where the monomer is a C_2 group, $C_2 H_n^{+/-}$ (related to the monomer, M $\pm$ H) is the most intense peak in the negative ion spectrum. The highest mass negative ion detected was at m/z 144 (C_{12}^-).

Polytetrafluoroethylene (PTFE), shown in Figure 2, offers a check of the conclusions drawn from PE. Besides common contaminants mentioned already, the positive spectrum exhibits very few fragments, but those seen are very characteristic. The peaks assignable to C_n^+ fragments have much lower intensity than the $C_n F_m^+$ fragments. In particular, the $C_3 F_m^+$ series is the most intense, showing C_3^+, $C_3 F^+$, $C_3 F_2^+$, $C_3 F_3^+$, and $C_3 F_5^+$ at m/z 36, 55, 74, 93, and 131, respectively. Besides the base peak, CF^+ (31 amu), the second highest relative intensity is 69 amu corresponding to CF_3^+, which is due to branching and/or rearrangement in the plasma.

The negative ion spectrum exhibits much the same pattern as PE; F^- at m/z 19 is the base peak for the negative ion spectrum. The next most intense peaks are assignable to the C_n^- ion series (m/z 12, 24, 36, etc.). The $C_n F^-$ molecular ions can be assigned to peaks at m/z 31, 43, 55, 67, 79, 91, and 103. However, the pattern of alternating intensities for odd and even n values is not evident for $C_n F^-$. These intensities follow the pattern of a general decrease with increasing values of n. Some organic contaminants present are seen as masses assignable to $C_n H^-$ peaks, but they have low relative intensity. The odd mass of F (19 amu) leads to a unique fingerprint for Teflon in both positive and negative ion spectra. In both spectra, the highest mass fragment detected is at 131 amu.

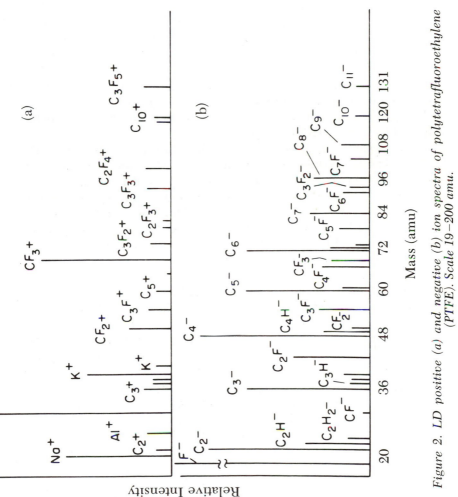

Figure 2. LD positive (a) and negative (b) ion spectra of polytetrafluoroethylene (PTFE). Scale 19–200 amu.

Heteroatom-Containing Backbones. The presence of a new functionality incorporated into the backbone of the polymer presents separate problems for interpretation. For polycaprolactam (Nylon N6), the backbone is assumed to be a C—C backbone punctuated by carbonyl and amine functions. The separation of ions owing to the backbone and side chain may be tested by comparing the secondary amide function to the results from the primary amide, polyacrylamide (PAA).

Figure 3 shows the spectral results for N6. Besides the normal $C_n H_m^+$ backbone, the positive ion spectrum yields a few key peaks. Of prime importance is the observation of $(M + H)^+$, which was detected easily at m/z 114. Other important peaks are at m/z 18 (NH_4^+), and the series of peaks at 30, 44, and 56. The structures assigned to these peaks, given in Table V, provide important confirmation of the secondary amide functionality, along with the peak of 84 amu, which is assigned to a $(M - CH_3NH_3)^+$ structure.

The negative ion spectrum has four interesting features. The C—C backbone produces the typical C_n^-, $C_n H^-$ pattern up to $n = 6$. Also, unlike the PAA spectrum, more $C_n N^-$, $C_n NH^-$ peaks are assigned to odd n structures because of stabilities discussed in earlier work. These peaks are listed in Table V. Monomer-related peaks are observed at 112 and 113 amu [$(M - H)^-$ and M^-, respectively] albeit at very low intensity. Finally, four peaks are observed at m/z 118, 134, 150, and 166 but are probably impurity peaks.

Two important aspects of the results for the polydimethylsiloxane (PDMS) polymer, given in Figure 4, will be mentioned. First, the positive ion mass spectrum can provide confirmation of assignments of ions to PDMS fragments in earlier work (4). These assignments were based on work from pyrolysis GC/MS on PDMS (11). Second, we can describe a new pattern from the siloxane backbone, which should be quite different from the hydrocarbon-based polymers just reported.

The positive ion spectrum yields some important features. In addition to the expected Si^+, $SiOH^+$, and $SiCH_x^+$ ions, Si_2^+ from rearrangement and the $Si(CH_3)_3^+$ peak at 73 are present based on the assignment in earlier work (4, 5, 11). Additionally, a $(M + H)^+$ peak is not present at m/z 75. Other structures for positive ions indicate a good deal of rearrangement. The last important feature is peaks in the higher mass region at m/z 175, 191, and 207. Cyclic structures were assigned to these peaks by earlier investigators (4, 5, 11).

The negative ion spectrum yields some familiar C_n^-, $C_n H^-$ fragments, from rearrangement, along with a pattern of $(SiO)_x C_n H_m^-$ fragments. The pattern of rearrangement was not observed in other LD work with PDMS in a mixture (4, 5), which leads to the speculation that the high mass pattern is of low intensity because of rearrangement.

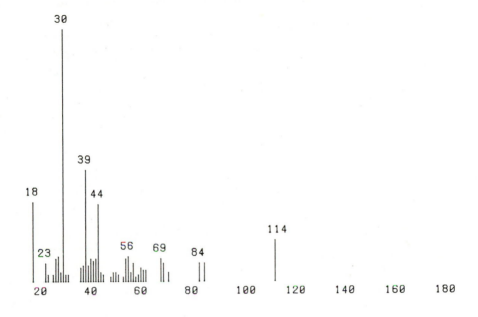

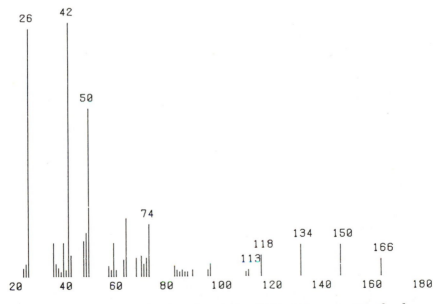

Figure 3. LD positive (top) and negative (bottom) ion spectra of poly-caprolactam (N6). Scale, 18–200 amu.

Table V. Structures of Ions for Polycaprolactam LD Results

Positive Ions		Negative Ions	
Structure	Mass	Structure	Mass
$CH_2=NH_2^+$	30	$-C\equiv N$	26
$O=C=NH_2^+$	44	$-C\equiv C-C\equiv N$	50
$CH\equiv CHCH_2NH_3^+$	56	$-C\equiv C-C\equiv C-C\equiv N$	74
$CH\equiv C-CH_2CONH_3^+$	84	$-C\equiv C-C\equiv C-C\equiv C-C\equiv N$	98
$CH_2=CH-CH_2CONH_3$	86		

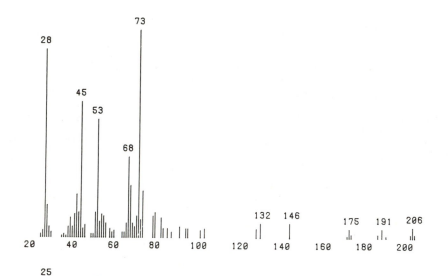

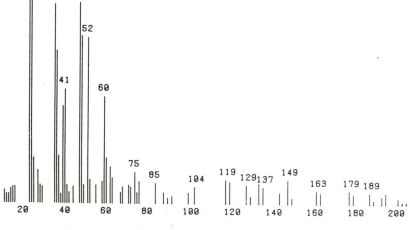

Figure 4. LD positive (bottom) and negative (top) ion spectra of poly-(dimethylsiloxane) (PDMS). Scale 20–220 amu.

In the siloxane data, a large amount of rearrangement is seen in both positive and negative ion spectra. Structures proposed are based on those suggested for positive ions by earlier research (*11*). Matrix effects may play an important role here because siloxane is a volatile liquid. More detailed studies at lower intensities are warranted to evaluate the role of rearrangement. However, siloxane fragments with the pattern $(S_1O)_xC_nH_m^+$ support the concept of structure-related fragmentation.

Simple Side Chain Functionality.   Results are presented from polymers with simple side chain functionalities to illustrate the basic approach to separating assignments between the pendant group and the backbone for interpreting LD spectra.

An initial view is given by considering results in Figure 5 for polyvinyl chloride (PVC). The positive ion spectrum shows a molecular fragment ion cluster pattern similar to PE, that is, the groups based on each $C_nH_m^+$, with increasing relative intensity to C_3-C_5, and then a gradual decrease to a C_{22} fragment at *m/z* 264. Peaks assignable to a Cl-containing molecular ion $(C_3H_2Cl^+)$ are observed at *m/z* 73, 75. No other Cl-containing positive ions were detected. Matched with the negative ion results (the C_n^-, C_nH^- peaks with alternating C_nH^- intensities up to C_{17} at *m/z* 204 and Cl$^-$ at *m/z* 35, 37) we are provided with a simple fingerprint for PVC.

The negative ion PVC spectrum is the initial result where a negative ion may be expected, from the polymer structure, albeit a simple chloride ion. Indeed it is quite diagnostic, the negative ion spectrum is of a backbone with a single chlorine attached, because no other ions having chlorine-containing structures are assigned.

Comparing the polypropylene (PP) spectra in Figure 6 to the results just given indicates that PP is essentially a methyl-substituted PE. Thus, we expect a similar backbone pattern, along with peaks that are due to the methyl group. The positive ion spectrum displays several key peaks that allow differentiation between PE and PP. The CH_x relative intensity in the PP spectrum is much higher than in the PE spectrum. High relative intensity of the C_3H_x group, especially the 43-amu peak $(M + H)^+$ and the 41-amu peak $(M - H)^+$ is seen along with higher intensity in the $C_4H_x^+$ and $C_5H_x^+$ clusters. Higher relative intensity also is seen for the $2M^+$ region of *m/z* 84−87. These important features are related to the structure of PP. No impurities are detected in this sample (Na^+ or K^+). The highest mass ions detected were C_{15}^+ at *m/z* 180 and C_{12}^- at 120.

Results for a more complex side chain, polystyrene (PS), are given in Figure 7. The first feature that can be seen is the increased intensity for all clusters in the positive spectrum. Most of these ions are attributed easily to the backbone of PS, and this feature may be a matrix

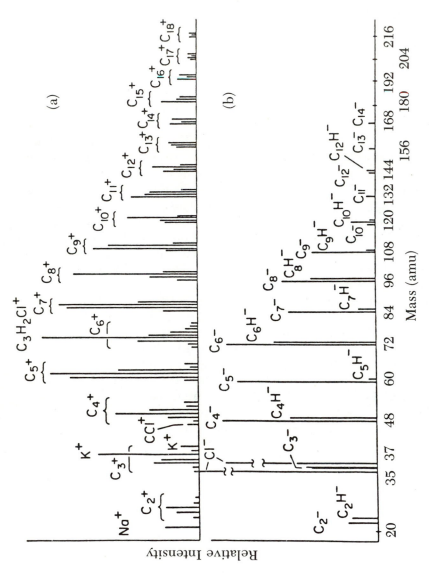

Figure 5. LD positive (a) and negative (b) ion spectra of polyvinyl chloride (PVC). Scale 20–220 amu.

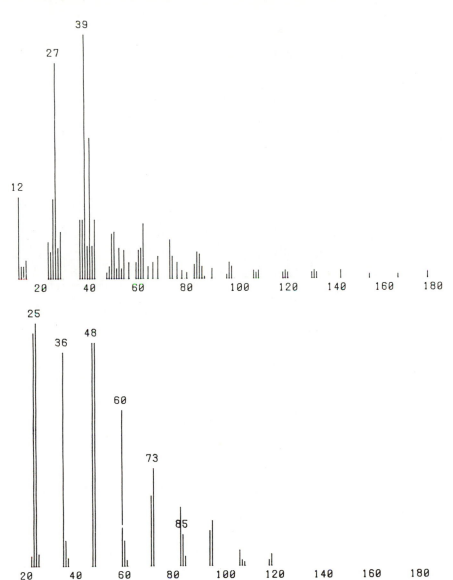

Figure 6. LD positive (top) and negative (bottom) ion spectra of poly-propylene (PP). Scale 10–200 amu.

effect, because the PS is not a powder as are the other polymers, but it is received as glassy chunks. Aside from the backbone structure, the key peaks that dominate (due to the side chain) are peaks assignable to cyclic ions at m/z 77 and 91 ($C_6H_5^+$ and $C_7H_7^+$). Also of high relative intensity are ions due to $(M + H)^+$ (105), $(M - H)^+$ (103), and $(M +$

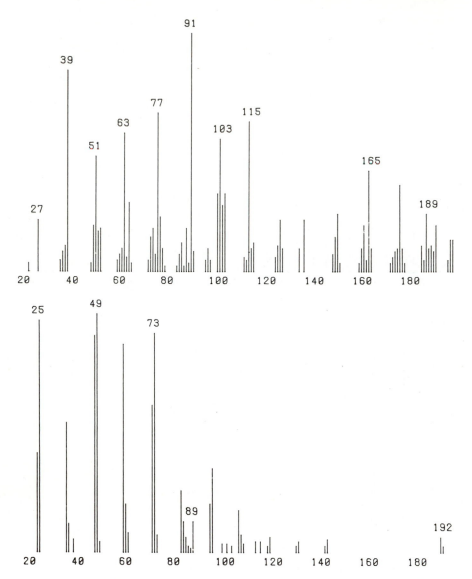

*Figure 7. LD positive (top) and negative (bottom) ion spectra of poly-
styrene (PS). Scale 20–200 amu.*

CPh)$^+$ at 193. Finally, the peak at m/z 115 can be attributed to a $C_9H_7^+$
ion. The highest mass ion seen was at m/z 216. Structures of these ions
are summarized in Table VI for both the positive and negative ion
spectra. In the negative ion spectrum, other than the C_n^-, C_nH^-
backbone peak, only two peaks are dominant: m/z 89 (CPh)$^-$, which is
analogous to CH$^-$ and the highest mass peak seen at 192 amu,

(CPhCh$_2$CPh$^-$). Again, the structure of the polymer provides a unique signature when both positive and negative ion spectra are used.

The study of simple side chains will be concluded by considering results for PAA. PAA results in Figure 8 illustrate still another phenomenon of interest observed in the laser ionization/vaporization process. Besides the usual backbone pattern, the positive ion spectrum indicates a good deal of sodium and potassium impurities. Coupled with the amide functionality, one might expect to observe cationization, which was reported for various organics (25) analyzed by LD. This method is utilized extensively as an ionization enhancer for obtaining mass spectra of involatile solids by other methods (1−18). In these results the amide function seems particularly sus-

Table VI. Structures of Ions for Polystyrene LD Results

Positive Ions		Negative Ions	
Structure	*Mass*	*Structure*	*Mass*
[benzene ring] +	77	C$^-$ [phenyl]	89
[⊕ ring]	91		
CH$_2$=C$^+$ [phenyl]	103		
CH$_3$−CH$^+$ [phenyl]	105	$^-$C—CH=CH [two phenyl]	192
[fused bicyclic]	115		
HC=C=CH$_2$$^+$ [two phenyl]	193		

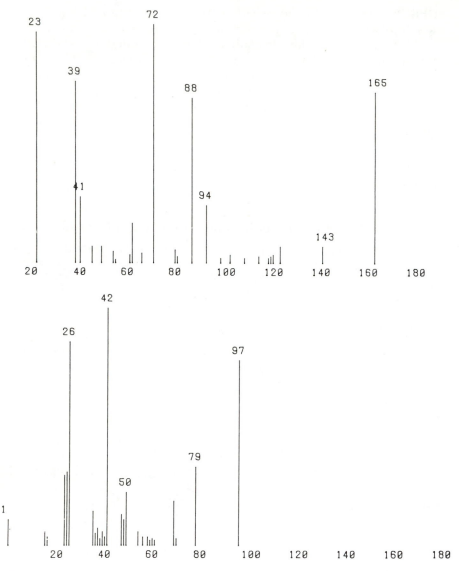

Figure 8. LD positive (top) and negative (bottom) ion spectra of polyacrylamide (PAA). Scale 20–200 amu.

ceptible to cationization by sodium, as the peaks at m/z 94 (M + Na)$^+$
and 165 (2M + Na)$^+$ are the most intense peaks in the spectrum.
Protonation is seen, to a lesser degree, with peaks at 72 (M + H)$^+$
and 143 amu (2 M +H)$^+$ among the high intensity peaks. Cationization
by K$^+$ is not detected, even though large peaks at m/z 39 and 41 are
observed. This phenomenon of selectivity in cationization also was

observed in earlier work (26). Few ions are seen in the negative spectrum with m/z 26, 42, and 50 peaks assigned to CN^-, OCN^-, and C_3N^- (related to the polyacetylene series in stability). Two other peaks are seen at low intensity (m/z 79, 97), which are probably pyridine-like ions formed by rearrangement.

Comparison of results from PAA (primary amide) and the secondary amide (N6), shows more cationization in the side chain pendant PAA; in fact, none is observed in the N6 spectra. In comparison with the negative-ion spectra, PAA generates only one nitrogen-containing polyacetylene ion structure (C_3N^- at m/z 50), while the presence of the nitrogen in the backbone of N6 can be deduced from the several high mass fragments similar to earlier LD work (1−3).

Detailed Studies of Side Chain Differences

In this section, results are considered from the series of polymethacrylate polymers. This series illustrates examples of small differences in side chain functionality, as each member of the series is considered.

Backbone Characterization. In this analysis, characteristic LD positive and negative ion backbone spectra can be developed by considering that all members of the series form common clusters of molecular ions. The remaining peaks in each spectrum then are assigned to side chain-related structures.

Table VII lists the methacrylate backbone assignments for both the positive and negative LD spectra. In the positive ion spectrum, a series of $C_nH_m^+$ ions in clusters is observed up to $n = 22$. Differentiation between the methacrylate backbone and that of PE is possible because the highest intensity methacrylate backbone peak is due to a $C_3H_m^+$ structure (typically $C_3H_3^+$ at m/z 43), while PE shows the base peak in the C_2 region (27 amu, $C_2H_3^+$). The methacrylate backbone is

Table VII. LD Results, Methacrylate Backbone

Positive Ion Spectrum

m/e	Formula	Relative Intensity[a]	m/e	Formula	Relative Intensity
24	C_2^+	vwk	65	$C_5H_5^+$	wk
25	C_2H^+	vwk	66	$C_5H_6^+$	vwk
26	$C_2H_2^+$	wk	67	$C_5H_7^+$	wk
27	$C_2H_3^+$	stg			
28	$C_2H_4^+$	wk	69	$C_4H_5O^+$	med
29	$C_2H_5^+$	med			
			72	C_6^+	vwk
31	CH_3O^+	wk	73	C_6H^+	wk
			74	$C_6H_2^+$	med
36	C_3^+	vwk	75	$C_6H_3^+$	med
37	C_3H^+	med	76	$C_6H_4^+$	vwk

Continued on next page

Table VII. Continued

m/e	Formula	Relative Intensity[a]	m/e	Formula	Relative Intensity
38	$C_3H_2^+$	med	77	$C_6H_5^+$	wk
39	$C_3H_3^+$	base			
40	$C_3H_4^+$	med	79	$C_6H_7^+$	wk
41	$C_3H_5^+$	stg	81	$C_6H_9^+$	vwk
42	$C_3H_6^+$	wk	83	$C_6H_{11}^+$	vwk
43	$C_3H_7^+$, $C_2H_3O^+$	med			
			85	C_7H^+	med
45	$C_2H_5O^+$, $COOH^+$	med	86	$C_7H_2^+$	wk
			87	$C_7H_3^+$	med
48	C_4^+	vwk	88	$C_7H_4^+$	vwk
49	C_4H^+	vwk	89	$C_7H_5^+$	vwk
50	$C_4H_2^+$	med			
51	$C_4H_3^+$	med	91	$C_7H_7^+$, $C_6H_3O^+$	wk
52	$C_4H_4^+$	wk			
53	$C_4H_5^+$	med	95	C_8H^+	wk
54	$C_4H_6^+$	vwk	96	$C_8H_2^+$	vwk
55	$C_4H_7^+$	med	97	$C_8H_3^+$	wk
60	C_5^+	wk	104	C_9^+	vwk
61	C_5H^+	med	105	C_9H^+	vwk
62	$C_5H_2^+$	med	106	$C_9H_2^+$	wk
63	$C_5H_3^+$	med	107	$C_9H_3^+$	wk
64	$C_5H_4^+$	wk	108	$C_9H_4^+$	wk

Negative Ion Spectrum

m/e	Formula	Relative Intensity	m/e	Formula	Relative Intensity
16	O^-	vwk	72	C_6^-	med
17	OH^-	vwk	73	C_6H^-	med
24	C_2^-	med	74	$C_6H_2^-$	med
25	C_2H^-	stg	84	C_7^-	med
26	$C_2H_2^-$	med	85	C_7H^-	stg
			86	$C_7H_2^-$	wk
36	C_3^-	base			
37	C_3H^-	stg	96	C_8^-	med
38	$C_3H_2^-$	med	97	C_3H^-	med
41	C_2HO^-	med	108	C_9^-	wk
			109	C_9H^-	wk
43	$C_2H_3O^-$	med	110	$C_9H_2^-$	wk
45	$C_2H_5O^-$, $COOH^-$	med	120	C_{10}^-	wk
			121	$C_{10}H^-$	wk
48	C_4^-	stg	122	$C_{10}H_2^-$	wk
49	C_4H^-	stg			
50	$C_4H_2^-$	med	132	C_{11}^-	vwk
51	$C_4H_3^-$	wk	144	C_{12}^-	vwk
53	C_3HO^-	wk	156	C_{13}^-	vwk
60	C_5^-	med	168	C_{14}^-	vwk
61	C_5H^-	med			
62	$C_5H_2^-$	med	180	C_{15}^-	vwk

Note: Other fragments present at very weak intensity are $C_nH_m^+$, n = 10, 11, 12, 13, 14, 15, 16, and 17.

[a] Relative intensity—intensity ratioed to the base peak ($C_3H_3^-$ = 39 amu, C_3^- = 36 amu).

Key to abbreviations: vwk, very weak; wk, weak; med, medium; stg, strong; and base, most intense peak in spectrum.

very similar to the spectrum of PP, which also has a base peak at 39 amu. However, PP can be differentiated from the methacrylates by the presence in the methacrylate spectrum of the peak at 45 amu. Thus, the positive ion spectrum allows characterization of the methacrylate backbone, while showing distinct similarities to previous straight chain hydrocarbon polymers with similar structures.

Few methacrylate-specific backbone peaks are seen in the negative ion spectrum listed in Table VII. Generally, the same C_n^-, C_nH^- pattern of ions up to $n = 20$ observed for most straight chain hydrocarbons is dominant, particularly at higher laser power densities. Few other peaks were detected characteristic of the backbone.

Side Chain Characterization. The structure-related ions assigned to peaks that differentiate each member of the polymer series will be studied. For purposes of discussion, these ions will be considered in three groups: short chain alkyls [polymethyl methacrylate (PMMA), polyethyl methacrylate (PEMA), and polyisopropyl methacrylate (PIPMA)], isomeric butyls [poly(n-butyl) methacrylate (PNBMA), poly(sec-butyl) methacrylate (PSBMA), PIBMA, and poly(t-butyl) methacrylate (PTBMA)], and cyclic groups [polycyclohexyl methacrylate (PCHMA), polyphenyl methacrylate (PPMA), and polybenzyl methacrylate (PBZMA)]. This discussion will provide further evidence that the backbone spectrum can be separated easily from ions originating from the side chain functionality.

SHORT CHAIN ALKYLS. Figures 9–11 show spectra for PMMA, PEMA, and PIPMA, respectively. Results for PMMA can be interpreted with regard to the backbone spectrum described earlier. Peaks at 12–15 amu, assigned to CH_x in the positive ion spectrum, provide direct evidence for the methyl side chain, because they are related to the ester group positive ion (R^+) and are not present in the backbone spectrum. Further evidence of simple bond scission along the side chain is provided by peaks at m/z 31 (OCH_3^+), 59 ($CO_2CH_3^+$), 101 (M + H)$^+$ (M is the monomer unit), and 144 (M + H + C_3H_7)$^+$. These peaks provide important keys in elucidating structure, because they involve the R or M structures directly. In the negative ion spectrum, peaks assigned to simple side chain structures are at m/z 31 ($^-OCH_3$) and 43 ($COCH_3^-$). More complex monomer-related peaks are those at m/z 141 (M + C_3H_5)$^-$ and 185 (2M − R)$^-$.

PEMA (Figure 12) shows similarities to the PMMA spectra. In the positive ion spectrum, peaks in the R^+ region (m/z 25–29) ($C_2H_x^+$) are of greater relative intensity than for the backbone spectrum. Peaks at 57 (COR^+), 73 (CO_2R^+), and 115 amu (M + H)$^+$ provide distinction from other members of the series in the same manner as for PMMA. Negative ion spectra show the ^-OR peak at m/z 45 ($^-OCH_2CH_3$), of greater relative intensity than from the backbone peak, which is normally very weak.

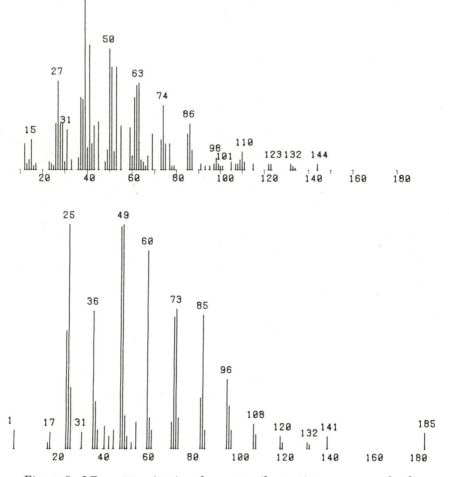

*Figure 9. LD positive (top) and negative (bottom) ion spectra of poly-
(methyl methacrylate) (PMMA). Scale 0–200 amu.*

PIPMA (Figure 11) shows many more distinctive peaks than
PMMA and PEMA. This result may be due to the glassy sample matrix, which is more difficult to vaporize/ionize, because of the larger
effective sample thickness and therefore lower effective laser power.
In the positive ion spectrum, peaks at R^+, m/z 45 $[CH(CH_3)_2^+]$; OR^+,
m/z 59 $[OCH(CH_3)_2^+]$; and CO_2R^+, 87 $[CO_2CH(CH_3)_2^+]$ follow the ex-

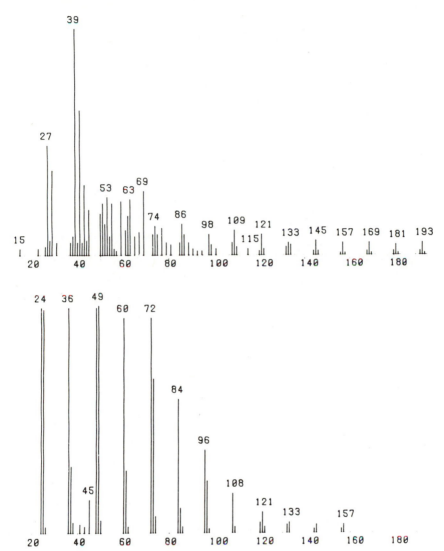

Figure 10. LD positive (top) and negative (bottom) ion spectra of poly-(ethyl methacrylate) (PEMA). Scale 10–200 amu.

pected results from generalizations based on PMMA and PEMA. The negative ion spectrum shows strong peaks at $^-$OR, m/z 59 [$^-$OCH(CH$_3$)$_2$]; COR$^-$, m/z 71 [COCH(CH$_3$)$_2$$^-$]; and m/z 127 (M − H)$^-$, which provide useful structural evidence.

ISOMERIC BUTYL METHACRYLATES.   Results for these polymers are given in Figure 12. The general approach involved examination of the

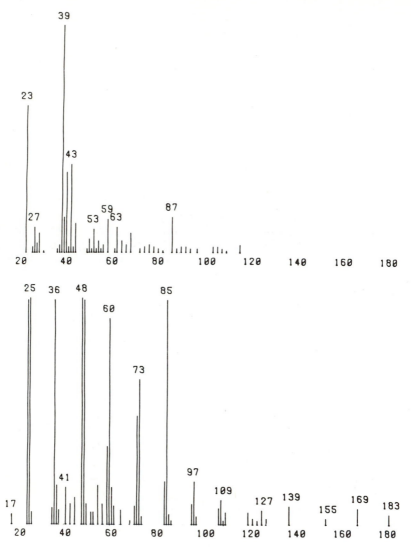

Figure 11. LD positive (bottom) and negative (top) ion spectra of poly-
(isopropyl methacrylate) (PIPMA). Scale 0–200 amu.

R⁺ and ⁻OR regions at *m/z* 57 and 73 in the positive and negative ion
spectra, respectively. The spectra exhibited other peaks assigned to
positive ions at *m/z* 85 (COR⁺) and 101 (CO₂R⁺) and the negative ion at
m/z 141 (M − H)⁻. The peaks used for differentiation are shown in
Figure 13. The important relationship to be seen is the increasing
relative intensity of R⁺ at *m/z* 57 (relative to the backbone base peak at
m/z 39) as primary, secondary, and tertiary carbonium ions are gener-

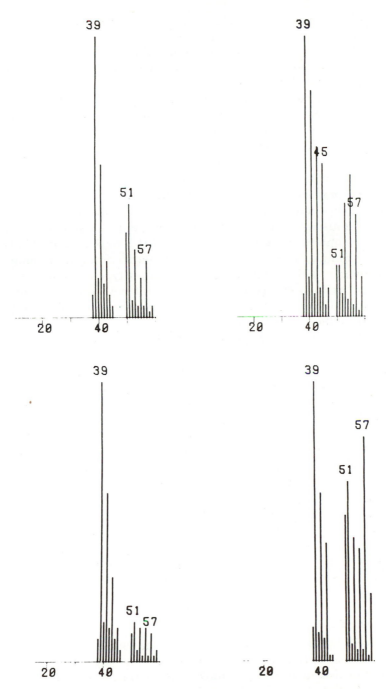

Figure 12. LD positive ion spectra of poly(butyl methacrylates). Key: upper left, n-butyl; upper right, sec-butyl; lower left, isobutyl; and lower right, tert-butyl. Scale as shown.

ated. Similarly, the ratio of peaks at m/z 73/85 $[^-OR/(M - R)^-]$ in the negative ion spectrum increases from n-butyl → sec-butyl/isobutyl → t-butyl side chains. These results follow the stability series of the carbonium ions and thus allow for differentiation based on a well-known chemical stability.

CYCLIC METHACRYLATES. Results from PPMA and PBZMA are shown in Figures 13 and 14. An important observation is the continuance of the R^+, ^-OR pattern established by the discussion just given. Specifically, the ^-OPh assignment at m/z 93 in the negative ion PPMA spectrum illustrates the pattern expected. The peak is so intense that it dominates the spectrum. Although m/z 77 (Ph^+) is a strong peak in the positive ion spectrum, fulfilling the R^+ expectation, the peak at m/z 69 that is assigned to $(M - OR)^+$ is of major interest. The peak is of unusually high relative intensity and is assignable to the fragment complementary to the ^-OPh ion; this assignment makes it seem as if the pair of ions is generated together from the monomer. This phenomenon has been observed in other types of samples (26) and it has been called "pair production." Interestingly, this mechanism seems to occur only when negative ion formation is strong, as illustrated by the results for PBZMA.

The PBZMA polymer yields spectra that have dominant ions in the positive ion spectrum. The $C_7H_7^+$ ion, m/z 91 (R^+), is characteristic of the benzyl moiety. No pair production was seen in this case; the negative spectrum shows only the backbone $C_n^{-\cdot}$, C_nH^- pattern.

The results for PCHMA showed only two important peaks, one at m/z 55 $(C_4H_5^+)$ in the positive ion spectrum and the other at 42 $(C_2H_2O^-)$ in the negative ion spectrum. Although not a strict example of the pair production, the pairing of these two peaks, which together form the ^-OR fragment, is supported by the well-known decomposition reactions seen in EI/MS of saturated cyclics, to yield $C_4H_x^+$ and C_2H_2.

The results from each of the three polymers in this group show similarities to well-documented ion–molecule reactions in other MS methods, follow known ion stabilities well, and suggest the new mechanism of pair production, which is observed because of the ability of LD to produce high intensity negative ion spectra.

Biopolymers

Thus far LD studies of polymers have shown that it is possible to distinguish not only between the backbone and the side chain of a polymer but also between very similar side chains. The goal of the LD biopolymer study is to identify similarities and differences in the LD spectra of Biomer and Avcothane, which are more complex than the

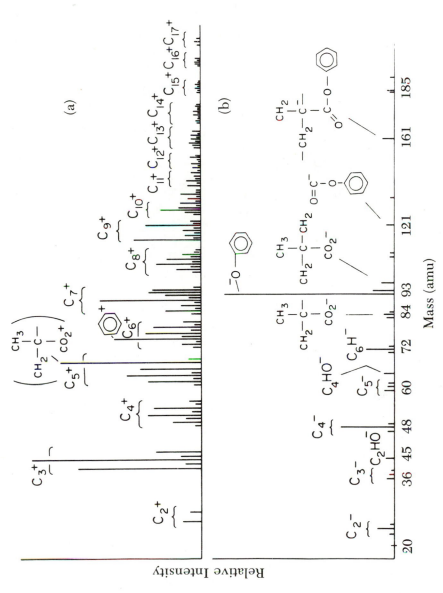

Figure 13. LD positive (a) and negative (b) ion spectra of poly(phenyl methacrylate) (PPMA). Scale 20–200 amu.

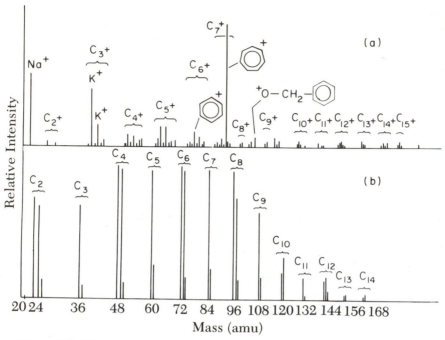

Figure 14. LD positive (a) and negative (b) ion spectra of poly(benzyl methacrylate) (PBZMA). Scale 20–200 amu.

others, i.e., polyether/polyurethanes. The structures of the two bio-polymers are shown in Figures 15 and 16. The polyethers of Biomer and Avcothane are similar, but not identical, and are assumed to be the backbone of the biopolymers, while the urethane fragment is the same for both Biomer and Avcothane. Additionally, Avcothane con-tains 10% by weight PDMS, and LD spectra of Avcothane should con-tain peaks attributable to PDMS.

Biomer. Positive and negative ion LD mass spectra were ob-tained for both Biomer and Avcothane; the Biomer spectra are shown in Figure 17. The positive ion LD spectrum of Biomer (Figure 17) shows Na^+ and K^+, which are typical contaminants seen in many ma-terials. The other major peaks in the low mass region of the positive ion spectrum are at m/z 43, 55, 71, and 106. Tetramethylene oxide is the monomer for the polyether segment of Biomer, and the peaks at 43 and 55 amu correspond to $C_2H_3O^+$ and $C_3H_3O^+$, which can be inter-preted as fragments of the monomer unit. The m/z 71 peak corresponds to $C_4H_7O^+$, which can be accounted for by either a cyclic (**1**) or linear (**2**) structure resulting from α-cleavage at adjacent ether linkages, accompanied by hydrogen transfer.

Cleavage α or β to the ether linkage and hydrogen transfer can account for the m/z 43 and 55 peaks giving a rational progression for

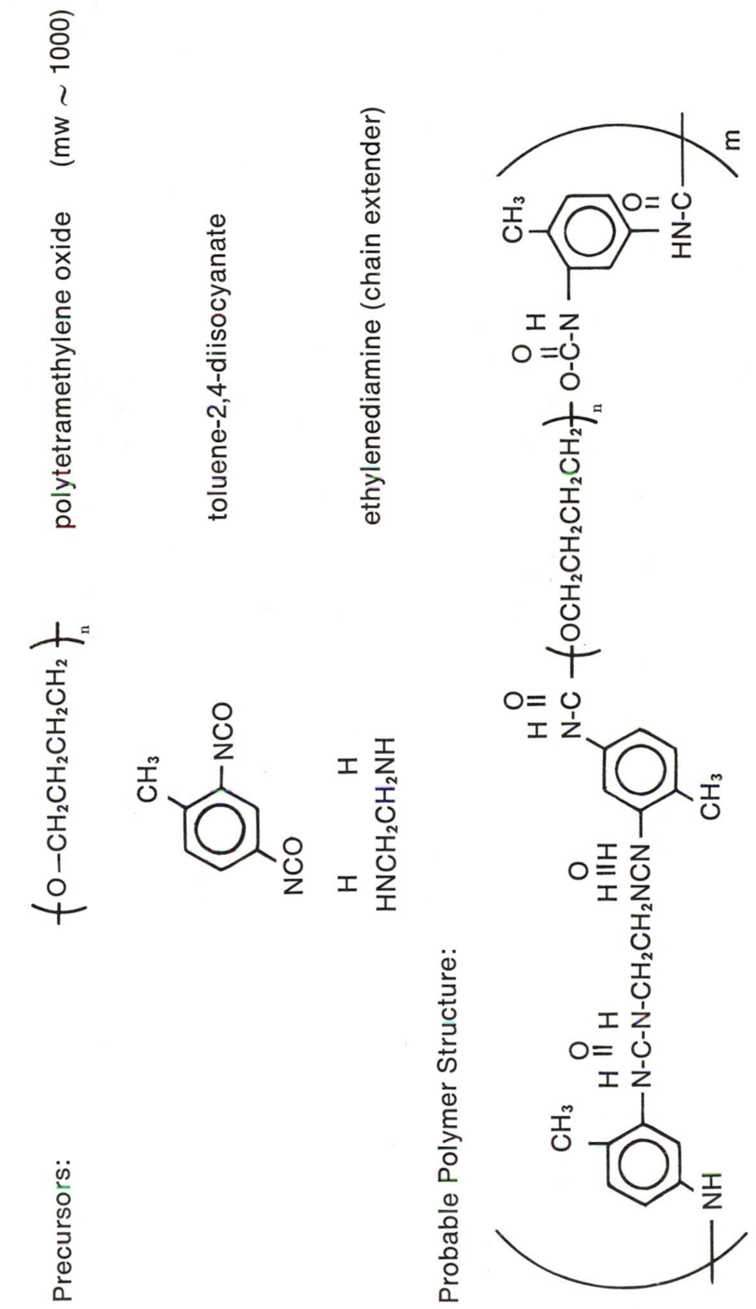

Precursors:

polytetramethylene oxide   (mw ~ 1000)

toluene-2,4-diisocyanate

ethylenediamine (chain extender)

Probable Polymer Structure:

Figure 15. Precursors and probable polymer structure of Biomer.

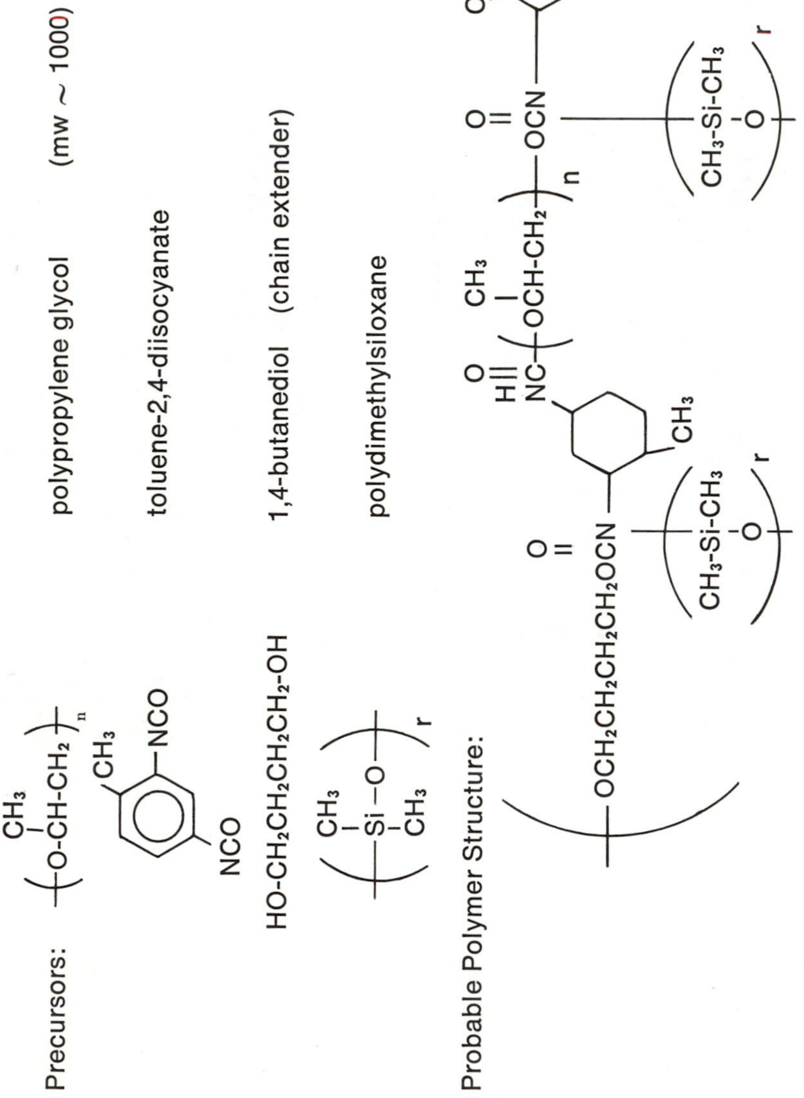

Figure 16. Precursors and probable polymer structure of Avcothane.

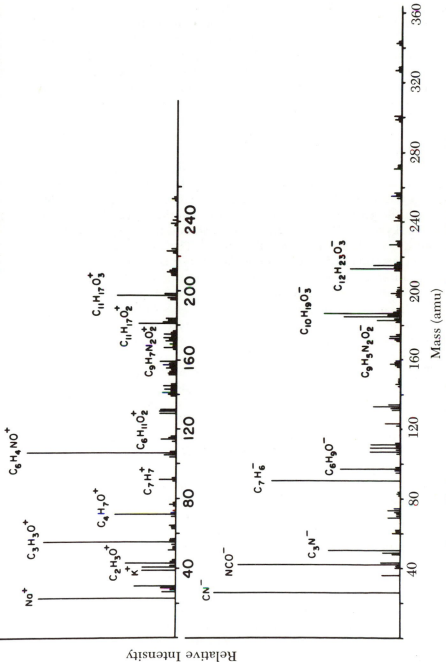

Figure 17. Positive (top) and negative (bottom) ion LAMMA spectra of Biomer.

the m/z 43, 55, and 71 ions. A more logical mass progression might be m/z 43, 57, and 71. The centroid of the second cluster is at m/z 57; however, the peak at m/z 55 is much more intense than the one at 57. Although previous LD spectra of polymers showed primarily even-electron ions (2, 3), α- and β-cleavage in the polyether fragments would give rise to odd-electron ions producing LD spectra similar to EI spectra of polyethers. Linear structures for m/z 55 and 71 also can result from cleavage α or β to the ether linkage. However, the only reasonable linear structure for m/z 43 is $CH_3-C\equiv O^+$, requiring cleavage at an $O-C$ bond, which, in traditional positive ion MS, is considered unlikely (27).

The peaks at m/z 43 and 55 could possibly be interpreted as arising from $C_3H_7^+$ and $C_4H_7^+$ as in the PE spectra, but this origin seems unlikely because the 43 and 55 peaks show significantly greater intensity than the surrounding peaks. In our experience (1, 2), all of the major peaks of a $C_nH_{2n\pm x}$ group are represented in LD spectra characterized by simple hydrocarbon fragments. The main C_4 peaks in the LD mass spectra of PE, PVC, and PP are at m/z 50 and are at least five times as intense as the peaks at 55; therefore, the intensity of the peak at m/z 55 cannot be explained as a hydrocarbon fragment.

The peak at m/z 106 corresponds to a Diels–Alder rearrangement fragment from the toluene-2,4-diisocyanate used to make the urethane bonds in Biomer. Other peaks in the Biomer positive ion spectrum that were assigned to fragments from urethane species include m/z 51, 65, 77, and 91 as well as those at m/z 132, 145, 159, 173, and 175. The first peaks correspond to $C_4H_3^+$, $C_5H_5^+$, $C_6H_5^+$, and $C_7H_7^+$, respectively, and would be expected for a molecule containing the toluene function. The peaks at m/z 175, 173, 159, 145, and 132 can be interpreted as the $(M + H)^+$ peak of toluene-2,4-diisocyanate (175) and some of its fragment ions: $C_9H_5N_2O_2^+$, $C_8H_3N_2O_2^+$, $C_8H_5N_2O^+$, and $C_8H_6NO^+$, respectively.

The remaining peaks in the positive ion spectrum form a series of clusters that have centroids 14 or 16 mass units apart. These cluster peaks were assigned to fragments of the polyether. The polyether of Biomer is polytetramethylene oxide (PTMO), and the molecular weight of the polyether before being segmented is approximately 1000. Because of the high molecular weight of the polyether relative to the diisocyanate and ethylenediamine (which is used as an extender in the polymerization process), the polyether is considered to be the backbone of Biomer. The large number of hydrogens in the polyether leads to multiple fragment ions clustered around the centroid masses. Table VIII shows the m/z values of the centroids of the polyether peak clusters and the probable ion composition for each centroid. The empirical formula for each centroid is $C_nH_{2n-x}O_m^+$, for

Table VIII. Peak Assignments for Biomer Positive Ion Spectrum

$$-[CH_2-CH_2-O-CH_2-CH_2-CH_2-CH_2-O-CH_2CH_2-]_x$$

m/z	Probable Ion Composition	Empirical Formula
43	$C_2H_3O^+$	
57	$C_3H_5O^+$	$C_nH_{2n-1}O^+$
71	$C_4H_7O^+$	
85	$C_5H_9O^+$	
115	$C_6H_{11}O_2^+$	
129	$C_7H_{13}O_2^+$	$C_nH_{2n-1}O_2^+$
143	$C_8H_{15}O_2^+$	
155	$C_9H_{15}O_2^+$	
169	$C_{10}H_{17}O_2^+$	$C_nH_{2n-3}O_2^+$
183	$C_{11}H_{19}O_2^+$	
197	$C_{12}H_{21}O_2^+$	
197	$C_{11}H_{17}O_3^+$	
211	$C_{12}H_{19}O_3^+$	
225	$C_{13}H_{21}O_3^+$	$C_nH_{2n-5}O_3^+$
239	$C_{14}H_{23}O_3^+$	
253	$C_{15}H_{25}O_3^+$	

the series $m = 1, x = 1; m = 2, x = 1,3;$ and $m = 3, x = 5$. Structures consistent with this composition require several double bonds and/or cyclization. For instance, the peak at *m/z* 115 could be represented by either Structure **1** or **2**, both of which have the formula $C_6H_{11}O_2^+$. Stability of cyclic organic structures was established for traditional positive ion MS (*27*), making structures like **1** seem more probable. A rational mass progression is evident from one cluster to the next that can be explained by subsequent addition of CH_2 or oxygen to each cluster, as necessary, to accommodate fragmentation of the polyether. Each cluster can be explained by cleavage α or β to oxygen, which is in agreement with traditional mass spectral fragmentation patterns (*24*). As the fragment ion chains of Biomer get longer, polycyclization may occur, which would explain the increased loss of hydrogen in the higher mass fragments.

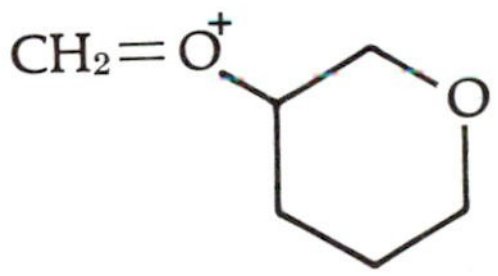

$$CH_2=\overset{+}{O}-CH_2-CH=CH-CH_2-O-CH_3$$

1

2

In the negative ion LD spectrum of Biomer (Figure 17 and Table IX), the major low mass peaks are at m/z 26, 42, 50, and 90, which correspond to CN^-, CNO^-, C_3N^-, and $C_7H_6^-$. The ions CN^-, CNO^-, and C_3N^- are formed by reactions in the plasma and are typical fragments seen in the negative ion spectra of nitrogen- and oxygen-containing organics. The $C_7H_6^-$ ion is probably a fragment ion from toluene-2,4-diisocyanate. The $(M - H)^-$ quasimolecular ion peak for toluene-2,4-diisocyanate at m/z 173 also is present.

The negative ion LD mass spectrum of Biomer contains clusters of peaks at various mass intervals that can be explained similarly to the positive ion spectrum. Many hydrogens are present in the backbone polyether so that several fragments that differ by only one or two mass units are possible for any given combination of carbon and oxygen atoms. The centroid of each cluster was assigned in Table IX as was done for the positive ion spectrum. All of the clusters were assigned to fragments from the polyether, and these fragments range from the tetramethylene oxide monomer at m/z 71 to a peak at m/z 345, corre-

Table IX. Peak Assignments for Biomer Negative Ion Spectrum

$-[CH_2-CH_2-O-CH_2-CH_2-CH_2-CH_2-O-CH_2-CH_2]_x-$

m/z	Probable Ion Composition	Empirical Formula
71	$C_4H_7O^-$	
85	$C_5H_9O^-$	$C_nH_{2n-1}O^-$
99	$C_6H_{11}O^-$	
113	$C_7H_{13}O^-$	
129	$C_7H_{13}O_2^-$	
143	$C_8H_{15}O_2^-$	$C_nH_{2n-1}O_2^-$
157	$C_9H_{17}O_2^-$	
171	$C_{10}H_{19}O_2^-$	
187	$C_{10}H_{19}O_3^-$	
201	$C_{11}H_{21}O_3^-$	
215	$C_{12}H_{23}O_3^-$	$C_nH_{2n-1}O_3^-$
229	$C_{13}H_{25}O_3^-$	
243	$C_{14}H_{27}O_3^-$	
259	$C_{14}H_{27}O_4^-$	
273	$C_{15}H_{29}O_4^-$	$C_nH_{2n-1}O_4^-$
287	$C_{16}H_{31}O_4^-$	
301	$C_{17}H_{33}O_4^-$	
303	$C_{16}H_{31}O_5^-$	
317	$C_{17}H_{33}O_5^-$	$C_nH_{2n-1}O_5^-$
331	$C_{18}H_{35}O_5^-$	
345	$C_{19}H_{37}O_5^-$	

sponding to $C_{19}H_{37}O_5^-$. All clusters are consistent with fragments corresponding to the formula $C_nH_{2n-1}O_m^-$, where m varies from 1 to 5 (*see* Table IX).

In the negative ion spectrum, a rational progression also is evident from one mass cluster to the next that can be explained by the addition of CH_2 or oxygen. For instance, a progression of polyether fragments at m/z 71, 85, 99, and 113 corresponds to subsequent additions of CH_2 to $C_4H_7O^-$, the proposed structure for m/z 71. Addition of CH_2 to the fragment for the peak at m/z 113 would result in a cluster with a centroid at m/z 127. However, the next centroid is at m/z 129, which involves the addition of oxygen to $C_7H_{13}O^-$, the proposed structure for the peak at m/z 113. On the basis of linear structures, the fragment ion at m/z 127 would be expected to be stable. However, a cyclic structure for m/z 127 would involve the formation of a nine-membered ring, although the structure at m/z 129 with the formula $C_7H_{13}O_2^-$ would require an eight-membered ring. This behavior suggests that the negative ion polyether structures probably are cyclic.

The negative ion fragments from PTMO apparently do not involve much hydrogen stripping, which makes them different from the structures proposed for the negative ion LD spectrum of PE. In the negative ion spectrum of PE, the ions are stripped of almost all hydrogen to yield polyacetylene-like structures that have the general formulas C_n^- and C_nH^-. If PTMO underwent similar hydrogen stripping, peaks at m/z 65, 79, 89, and 103 would result and would have empirical formulas C_4HO^-, $C_5H_3O^-$, C_6HO^-, and $C_7H_3O^-$, respectively. However, these peaks are not present in the negative ion spectrum of Biomer (Figure 17). The lack of hydrogen stripping in the polyether is probably due to many effects. One possibility is that oxygen blocks conjugation necessary for polyacetylene-like structures.

Avcothane. Avcothane is a more complex polymer than Biomer, because it contains not only polyether and urethane units but also a PDMS moiety (*see* Figure 16). In the positive LD spectrum of Avcothane (Figure 18 and Table X), the peaks at 23 and 39 amu are attributed to Na^+ and K^+ contamination. A series of peaks also occurs corresponding to CF^+, C_2F^+, CF_2^+, and CF_3^+ at 31, 43, 50, and 69 amu. These peaks are similar to those observed for Teflon; the presence of Teflon on the surface of aortic balloon pumps made from Avcothane was observed previously (*28*).

Toluene-2,4-diisocyanate is used to form the urethane segment of Avcothane. Fragments from it include the $(M + H)^+$ peak at 175 amu and other fragments at m/z 51, 65, 77, 91, 106, 132, 145, 159, and 173 as were seen for Biomer. Peaks in the positive ion spectrum of Av-

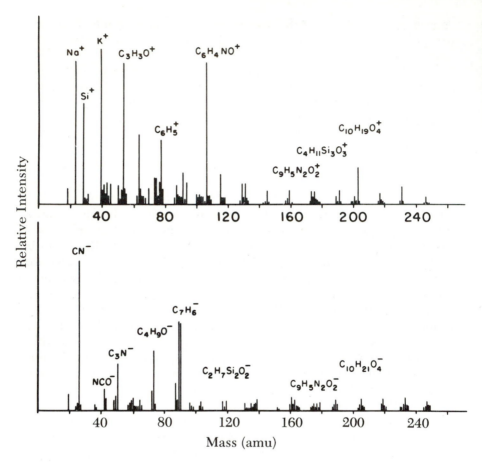

Figure 18. *Positive (top) and negative (bottom) ion LAMMA spectra of Avcothane.*

cothane that were assigned to PDMS fragments include those at 28, 45, 115, 175, and 191 amu. The peaks at m/z 28 and 45 correspond to Si^+ and $SiOH^+$, while the higher mass peaks are probably due to cyclic silicon-containing fragments such as $C_3H_7Si_2O^+$, $C_3H_7Si_3O_3^+$, and $C_4H_{11}Si_3O_3^+$. Such cyclic fragments of PDMS were reported for field desorption MS (*11*) and also were mentioned earlier in the chapter.

Polypropylene glycol (PPG) is the polyether of Avcothane and has a molecular weight of approximately 1000. Below m/z 87, only two peaks are assigned to the polyether: the one at m/z 43 is probably similar to the m/z 43 peak for Biomer; and the other peak at m/z 53 is probably a linear structure corresponding to $HC{\equiv}C{-}C{\equiv}O^+$. Clusters

Table X. Peak Assignments for Avcothane Positive Ion Spectrum

$$CH_3 \qquad CH_3 \qquad CH_3 \qquad CH_3 \qquad CH_3$$
$$-[CH_2CH-CH_2CH-O-CH_2CH-O-CH_2CH-O-CH_2CH-O]_x-$$

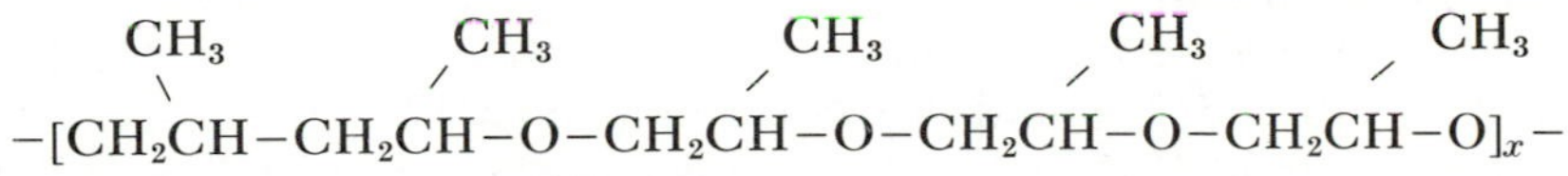

m/z	Probable Ion Composition	Empirical Formula
43	$C_2H_3O^+$	$C_nH_{2n-1}O^+$
87	$C_4H_7O_2^+$	
101	$C_5H_9O_2^+$	$C_nH_{2n-1}O_2^+$
115	$C_6H_{11}O_2^+$	
129	$C_7H_{13}O_2^+$	
145	$C_7H_{13}O_3^+$	
159	$C_8H_{15}O_3^+$	
173	$C_9H_{17}O_3^+$	$C_nH_{2n-1}O_3^+$
187	$C_{10}H_{19}O_3^+$	
201	$C_{11}H_{21}O_3^+$	
217	$C_{11}H_{21}O_4^+$	
231	$C_{12}H_{23}O_4^+$	$C_nH_{2n-1}O_4^+$
245	$C_{13}H_{25}O_4^+$	

of masses ranging from m/z 87 to 245 were observed for the polyether. The empirical formula that fits the data for the positive ion fragments from the polyether is $C_nH_{2n-1}O_m^+$; this formula requires that structures contain either two double bonds or one double bond and cyclization. For example, mass 129 Structures 3 and 4 represent possibilities for the $C_7H_{13}O_2^+$ fragment; again, the cyclic structure seems more likely. As with Biomer, the polyether clusters show a rational progression corresponding to addition of CH_2 or oxygen from one mass cluster to the next.

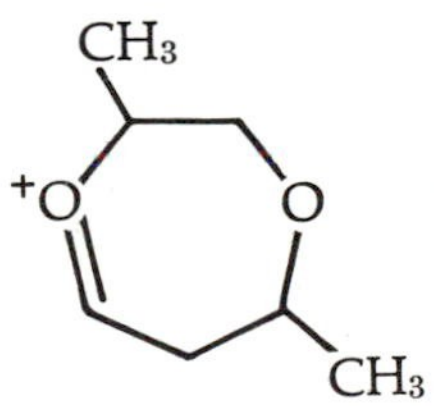

3

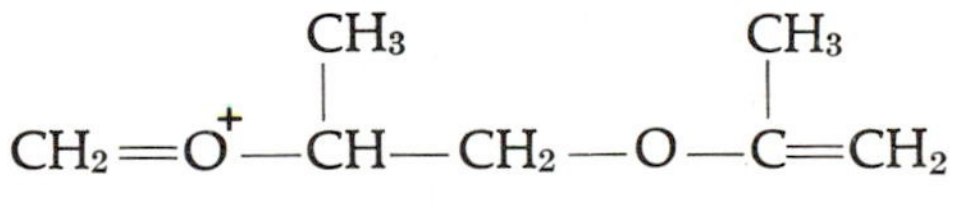

$$CH_2{=}\overset{+}{O}{-}\overset{\underset{\textstyle |}{CH_3}}{CH}{-}CH_2{-}O{-}\overset{\underset{\textstyle |}{CH_3}}{C}{=}CH_2$$

4

The negative ion LD spectrum of Avcothane contains clusters of peaks at various mass intervals that can be explained similarly to the positive ion spectrum. Teflon present on the polymer surface is re-

sponsible for a small peak at m/z 19. The peaks from toluenediisocya-
nate are at m/z 90 and 173, corresponding to $C_7H_6^-$ and $(M - H)^-$;
peaks at 26 and 42 correspond to CN^- and CNO^-. The PDMS fragments at 119, 135, 163, and 179 correspond to $C_2H_7Si_2O_2^-$, $C_2H_7Si_2O_3^-$,
$C_2H_7Si_3O_3^-$, and $C_2H_7Si_3O_4^-$. The peaks attributed to PDMS in both
the positive and negative LD spectra of Avcothane agree with LD
spectra of PDMS.

In addition to the peaks from Teflon, PDMS, and toluenediisocy-
anate, a series of clusters that can be assigned to fragmentation of the
polyether is evident. As in the positive-ion spectrum, a rational progression from one mass cluster to the next occurs, which can be explained by the subsequent addition of CH_2 or oxygen. Each cluster
has been assigned as a fragment corresponding to $C_nH_{2n+1}O_m^-$, where
n varies from 4 to 13 and m varies from 1 to 4 (*see* Table XI).

Conclusions

This chapter has addressed the application of a novel type of solid
state MS to the analysis of various polymer systems. We showed that
such analysis is feasible and that important structural information is
directly available from analysis of the spectra. Variation of operational
parameters was explored; the minimization of laser power density was
seen to be important in the generation of structure-related fragments.

Table XI. Peak Assignments for Avcothane Negative Ion Spectrum

$$
\begin{array}{ccccc}
\overset{CH_3}{|} & \overset{CH_3}{|} & \overset{CH_3}{|} & \overset{CH_3}{|} & \overset{CH_3}{|}
\end{array}
$$

$$-[CH_2CH-O-CH_2CH-O-CH_2CH-O-CH_2CH-O-CH_2CH-O]_x-$$

m/z	Probable Ion Composition	Empirical Formula
59	$C_3H_7O^-$	$C_nH_{2n+1}O^-$
73	$C_4H_9O^-$	
89	$C_4H_9O_2^-$	
103	$C_5H_{11}O_2^-$	$C_nH_{2n+1}O_2^-$
117	$C_6H_{13}O_2^-$	
131	$C_7H_{15}O_2^-$	
147	$C_7H_{15}O_3^-$	
161	$C_8H_{17}O_3^-$	$C_nH_{2n+1}O_3^-$
175	$C_9H_{19}O_3^-$	
189	$C_{10}H_{21}O_3^-$	
205	$C_{10}H_{21}O_4^-$	
219	$C_{11}H_{23}O_4^-$	$C_nH_{2n+1}O_4^-$
233	$C_{12}H_{25}O_4^-$	
247	$C_{13}H_{27}O_4^-$	

The backbone/side chain model of polymer structure proved valid for interpretation of a series of polymers with different backbones and different side chains. The method proved to be effective in differentiating among each of the methacrylate polymers, even though only small changes in side chain functionality are present throughout the series.

Additionally, this study showed the value of laser MS for studying complex biopolymer systems. The Biomer spectra show peaks corresponding to polyether and polyurethane components. The majority of clusters in both the positive and negative ion LD spectra are assigned to the backbone polyether. The most stable polyether peaks probably result from cyclization.

The Avcothane positive and negative ion LD spectra are complex and have peaks that can be assigned to Teflon contamination as well as to polyurethane, polyether, and PDMS. The majority of Avcothane peaks was assigned to the backbone polyether. The most stable polyether peaks in the positive ion spectrum are probably due to cyclic fragments, while in the negative ion spectrum the fragments appear to be linear. The most stable PDMS fragments were also those that produce cyclic structures and are comparable to those observed in FD.

The structures presented here are mostly only preliminary structures that explain the progression of peaks fairly well and do not require recombination reactions. However, these structures are not the only ones possible nor are they more likely than other possible structures; they merely help to demonstrate the rationality of the progressions in each of the spectra.

Because mass spectra up to and beyond the monomer unit of a polymer are possible using LDMS, the technique provides fragmentation patterns not readily available by other methods, particularly in the negative-ion spectrum.

Literature Cited

1. Gardella, J. A., Jr.; Hercules, D. M.; Heinen, H. J. *Spectrosc. Lett.* **1980,** *13,* 347–360.
2. Gardella, J. A., Jr.; Hercules, D. M. *Z. Anal. Chem.* **1981,** *308,* 297.
3. Gardella, J. A., Jr.; Hercules, D. M., presented at the "Workshop on Ion Formation from Organic Solids," University of Münster, Physikalisches Institut, October 6–8, 1980.
4. Graham, S. W.; Hercules, D. M. *Spectrosc. Lett.* **1982,** *15,* 1–19.
5. Graham, S. W., Ph.D. Thesis, Univ. of Pittsburgh, 1981.
6. Unsold, E.; Renner, G.; Hillenkamp, F.; Nitsche, R. *Adv. Mass Spectrom.* **1978,** *7,* 1423.
7. Hummel, D. O. In "Polymer Spectroscopy"; Hummel, D. O., Ed.; Verlag Chem: Weinheim, Berlin, 1974.

8. David, C. In "Degradation of Polymers," Vol. 14 of "Comprehensive Chemical Kinetics"; Bamford, C. H.; Tipper, C. F. H., Eds.; Elsevier: Amsterdam, 1975.
9. Bradt, P.; Dibeler, V. H.; Mohler, F. L. *J. Res. Natl. Bur. Stand. (U.S.)* **1953**, *50*, 201.
10. Udseth, H. R.; Friedman, L. *Anal. Chem.* **1981**, *53*, 29.
11. Kleinert, J. C.; Weschler, C. J. *Anal. Chem.* **1980**, *43*, 1245.
12. Matsuo, T.; Matsuda, H.; Katekuse, I. *Anal. Chem.* **1979**, *51*, 1331.
13. Neumann, G. N.; Cullis, P. G.; Derrick, P. J. *Z. Naturforsch. A* **1980**, *35*, 1090.
14. Kistemaker, P. G.; Boerboom, A. J. H.; Meuzelaar, H. L. C. In "Dynamic Mass Spectrometry"; Price, D.; Todd, J. F. J., Eds.; Heyden: London, 1976; Vol. 4, Chap. 9, pp. 139–52.
15. Coloff, S. G.; Vanderborgh, N. E. *Anal. Chem.* **1973**, *45*, 1507.
16. Macfarlane, R. D.; McNeal, C. J.; Hunt, J. E., presented at the 8th International Mass Spectrometry Conference, Oslo, Norway, August 12–18, 1979.
17. Gardella, J. A., Jr.; Hercules, D. M. *Anal. Chem.* **1980**, *52*, 226.
18. Tantsyrev, G. D.; Kleimenov, N. A. *Dokl. Akad. Nauk SSSR* **1973**, *213*, 649.
19. Barber, M.; Bordoli, R. S.; Sedgwick, R. D.; Tyler, A. J. *J. Chem. Soc. Chem. Commun.* **1981**, *1981*, 325.
20. Kaufmann, R.; Hillenkamp, F.; Wechsung, R. *Eur. Spectrosc. News* **1978**, *20*, 41.
21. Krueger, F. R. *Z. Naturforsch. A* **1977**, *32*, 1084.
22. Kaufmann, R.; Hillenkamp, F.; Wechsung, R. *Med. Prog. Technol.* **1979**, *6*, 109.
23. Kaufmann, R.; Hillenkamp, F.; Wechsung, R. *Eur. Spectrosc. News* **1978**, *20*, 41.
24. Heresch, F.; Schmid, E. R.; Huber, J. F. K. *Anal. Chem.* **1980**, *52*, 1803.
25. Wechsung, R.; Heinen, H. J.; Vogt, H. *LAMMA Applied to Quantitative Analysis*, Leybold-Heraeus GmbH, Research and Development Report 24, 1978.
26. Day, R. J., Univ. of Pittsburgh, unpublished data, 1981.
27. Biemann, K. "Mass Spectrometry Organic-Chemical Applications"; McGraw-Hill: New York, 1962; p. 116.
28. Graham, S. W.; Hercules, D. M. *J. Biomed. Mater. Res.* **1981**, *15*, 349.

RECEIVED for review October 14, 1981. ACCEPTED February 24, 1982.

38

Liquefaction Reactivity Correlations Using Pyrolysis/Mass Spectrometry/Pattern Recognition Procedures

KENT J. VOORHEES, STEVEN L. DURFEE, and
ROBERT M. BALDWIN

Colorado School of Mines, Departments of Chemistry and Chemical and
Petroleum Refining Engineering, Golden, CO 80401

Three suites of coal, each liquefied using a different process, were pyrolyzed directly into a mass spectrometer. Relationships between the mass spectra and the observed conversion yields were determined by computerized pattern recognition. For each of the three liquefaction processes (rocking bomb, stirred tank reactor, and stirred tank continuous reactor) the coals could be categorized according to their known conversion yields from the pyrolysis data. The results of the pyrolysis/mass spectrometric conversion yield correlations provided information about the effect of the homologous ion series that was produced, and the effect of the overall organic structure of the coal on the reactivity.

NUMEROUS TESTS HAVE BEEN DESIGNED [American Society for Testing and Materials (ASTM)] for the characterization of the composition of coal (1). Most of these procedures were formulated to define feedstocks for coking operations or for combustion. Attempts have been made to extend the use of these ASTM procedures to classify coal with respect to reactivity in liquefaction or gasification processes. In most cases, it is possible to observe drastic differences in reactivity and in the chemical composition of the liquefaction products from two coals of the same ASTM rank, obtained from different geographical locations. Because of the problems involved in designing and operating liquefaction reactors, such severe reactivity changes cannot be tolerated.

The failures of the standard ASTM tests for these applications can be rationalized easily because they provide minimal information con-

0065−2393/83/0203−0677$06.00/0

cerning the actual chemical structure of the coal. The structural information, and possibly the mineral composition of the feedstock, must be known to predict the reactivity and end-product composition.

Various insoluble or nonvolatile materials, such as synthetic polymers, kerogen, meteorite carbonaceous material, shales, coals, proteins, and bacterial cell walls have been investigated by several degradation methods, followed by an analysis of the degradation products. The usefulness of degradation processes for characterization is dependent on the specificity and reproducibility of these processes. For lysis procedures that show these characteristics, a correlation between the degradation products and the original structure can be made. Under properly controlled conditions, pyrolysis is a highly specific and reproducible procedure that, when combined with mass spectrometry, can be a useful method for providing structural information for the macromolecules in coal (2, 3).

This chapter addresses basic problems of these processes by defining a new analytical approach that uses pyrolysis/mass spectrometry which can be run efficiently and routinely for correlating the chemical structure of coals with reactivity, and with the eventual conversion products. In addition, this chapter identifies chemical structural features that affect the reactivity of coal toward direct hydrogenation.

Experimental

Pyrolysis Sample Preparation.   Coals obtained primarily from the Pennsylvania State University Coal Bank were manually ground such that a stable suspension in methanol (~5 mg of coal in 1 mL of methanol) could be prepared. Approximately 10 μL of the suspension was applied to a rotating ferromagnetic wire (4). Following evaporation of the solvent, the wire was transferred to the Curie-point pyrolyzer.

Pyrolysis/Mass Spectrometry/Pattern Recognition.   The pyrolysis was conducted using a Fisher Curie-point pyrolyzer (1.5-kW, 1.1-MHz power supply) in conjunction with an Extranuclear Spectrel mass spectrometer system (5). Curie-point wires composed of Fe, Ni, and Co (510 °C final temperature) were used throughout the study and low energy (14-eV) electron impact was used for ionization. A scan rate of 1000 amu/s was employed for all analyses. Mass spectral data were collected as summed spectra on a Hewlett Packard 2100 S computer. Data from this computer were transferred via magnetic tape to the PDP System 10 computer system for statistical calculations.

Statistical Calculations.   The data analysis involved normalization of the individual peaks to the total ion current (sum of all peak intensities) followed by multivariate analyses (6) using the SCALE, WEIGHT, and DISTANCE options of the ARTHUR[1] program. The SCALE routine assigns the intensities for each m/z value a mean of zero and unit standard deviation. WEIGHT gives weightings to the various features (m/z values) depending on the reproducibility of the relative intensities at a given m/z value and the differences be-

---

[1] ARTHUR is available from Infometrix, Inc., Seattle, WA 98125.

tween *m/z* intensities for the various samples (7). A high weighting value usually means that a particular feature is important in distinguishing the various samples from each other. DISTANCE establishes a distance matrix in multidimensional space where the Euclidian distance is used as a measure of the similarity or dissimilarity of various mass spectra. The dimensionality of the space is determined by the number of *m/z* values in the mass spectra. In this case the space for most of the data was defined by the scanned mass range (*m/z* 34–200) as 167. The matrix generated from DISTANCE gives all interspectra distances for replicate pyrolyses from the same coal sample, as well as the distance between spectra from different samples. A condensed form of the distance matrix can be constructed by computing an average distance between the members of two categories. A value on the diagonal is an average distance between spectra within a given category. When compared to the other values, the diagonal gives an idea of experimental reproducibility.

Although the distance matrix is useful for establishing degrees of similarity and dissimilarity between analyses of various samples, it is difficult to visualize totally. Two methods, nonlinear mapping (the NLM option) (8) and hierarchical clustering (the HIER option), have been used in this study for visualizing this relationship. The nonlinear mapping produces, through an iterative process, the best two-dimensional representation of the mass spectra in multidimensional space. The hierarchical clustering produces a dendrogram based on the similarity coefficients between the various spectra.

Liquefaction Conversion Data.   Liquefaction results were obtained from literature data generated from three different reactor systems. Extensive data have been reported (9, 10) for coals liquefied in both tubing bomb batch and continuous flow stirred reactors. Coals used in the Pennsylvania State University studies were highly characterized and represented a wide variety of geographical locations as well as chemical and physical properties.

For the rocking bomb experiments, reactivity was defined by the yield of benzene solubles after approximately 1 h of reaction time. Coals for this study were selected from the Colorado School of Mines collection. The continuous flow experiments utilized slightly different reaction conditions with conversion yields being defined as solubility in ethyl acetate after a 1-h residence time. Reactivity data (*11*) for the third suite of coals were based on tetrahydrofuran (THF) solubility after a 60-min residence time in a quick charge batch stirred autoclave. A majority of the coals for these three data sets were obtained from the Pennsylvania State University Coal Sample Bank. In the discussion that follows, the term reactivity has been defined as conversion to solvent soluble products, with reaction conditions dependent on the reactor system employed.

Results and Discussion

Three suites of coal, each liquefied using a different laboratory process, were investigated by pyrolysis/mass spectrometry (Py/MS). Table I summarizes the characterization properties for the suite of coals liquefied in the rocking bomb reactor. Typical Py/MS spectra for two of the coals are shown in Figure 1. The distance table, based on three replicate pyrolyses and calculated using ARTHUR, is illustrated in Table II. As previously mentioned, it is difficult to visualize totally the distance table. The nonlinear map of the data in Figure 2 allows for better comprehension. Each coal was run in triplicate and forms a

Table I. Summary of Coal Samples Studied in Rocking Bomb Reactor

Mine	Observed Liquefaction Reactivity (%)[a]	Sulfur Content	Rank
Island Creek, KY	>80	3.85	HVAB
Consolidated, WV	>80	3.50	HVAB
Island Creek-2, KY	>80	3.85	HVAB
University of Kentucky, KY	>80	3.46	HVAB
P & M, KY	>80	3.81	HVAB
Ireland Mine, WV	>80	4.45	HVAB
Eagle, CO	<50–65	0.32	LVB

Note: Complete analysis of coals is still being performed.

[a] Reactivity measured as solubility in benzene after reaction for approximately 60 min at 2000 psi (H_2), 400 °C, with tetralin as solvent.

triangular shaped data set on the nonlinear map. Also, the size of the area enclosed by the connected data points gives an idea of the reproducibility of the data for a selected coal.

The points on the nonlinear map clearly cluster into three groups. Comparison of the nonlinear map with the data in Table I shows that the clustering results from coals that represent three different geographical locations. The conversion data in Table I indicate that the eastern coals have better conversion characteristics than the single Colorado coal. Based solely on the implications of the geographical differences, the data suggest the possibility of detecting variations in coal reactivity by Py/MS.

The correlation of the various groups represented in the nonlinear map can also be shown by the hierarchical clustering dendrogram in Figure 3. Based on the similarity coefficients, the coals are clearly separated into three groups. The spectra in Figure 1 show the actual mass spectral differences between the best and the worst coal.

The most important observation that results from these data is the fact that the Py/MS/pattern recognition approach is capable of detecting and differentiating organic structural differences in coals. However, based on the limited number of poor reacting coals, it is not possible to state categorically that reactivity prediction has been demonstrated in this suite.

Because of the large number of coals studied by Given, Suite 2 could be selected more carefully to allow for a wider range of rank and characterization properties as well as reactivity differences. Table III summarizes pertinent data for the selected coals. The distance table and nonlinear map for the pyrolysis data are shown in Table IV and Figure 4, respectively. The reactivity for each coal is included, for reference, on the nonlinear map.

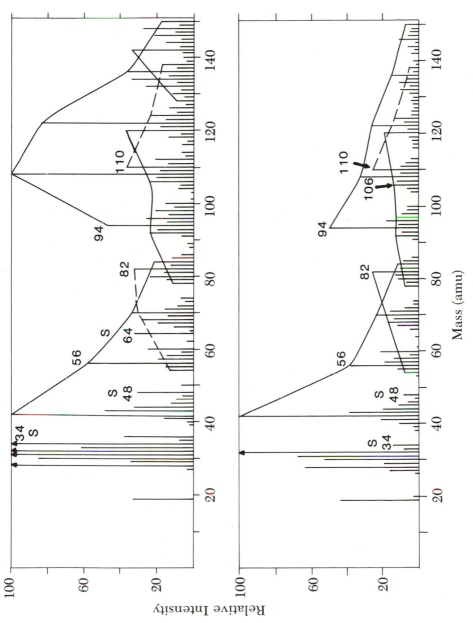

Figure 1. Py-MS of two selected coals. Key: top, Island Park, KY; and bottom, Eagle, CO.

Table II. Distance Table for Various Coals Studied in Tubing Bomb Reactor

	IC	Con	IC 2	UK	PM	IM	EC
Island Creek, KY (IC)	7.5						
Consolidated, WV (Con)	46.4	15.7					
Island Creek-2, KY (IC 2)	14.2	45.2	5.3				
University of Kentucky, KY (UK)	17.2	37.1	15.2	8.1			
P & M, KY (PM)	29.8	53.0	30.4	29.1	7.2		
Ireland Mine, WV (IM)	48.5	23.8	47.1	39.8	50.9	11.1	
Eagle, CO (EC)	49.6	53.7	47.1	41.8	45.2	58.4	6.1

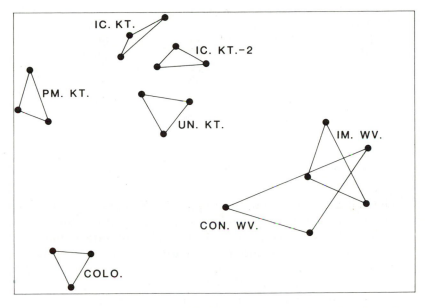

Figure 2. Nonlinear map of Py-MS data from rocking bomb coals.

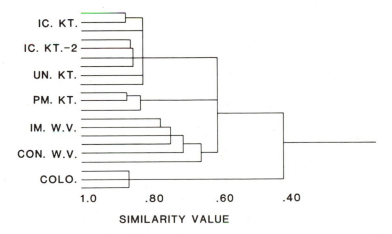

Figure 3. Hierarchical clustering dendrogram of Py-MS data from rocking bomb coals.

Table III. Characterization Data for Gulf Continuous Flow Reactor Coals

PSOC Number	Rank	Conversion (%)	%C	%S (dry)
265	HVA	44	86.7	0.6
271	HVA	49	86.5	—
276	HVA	72	83.5	3.5
278	HVA	83	80.5	5.6
302	HVA	29	88.9	0.71
305	HVB	77	82.3	4.1
307	HVA	77[a]	82.3	2.7
308	HVB	74	80.3	4.3
320	MV	15	90.4	1.2

[a] Measured in a tubing bomb reactor.

The points on the nonlinear map are clustered into two main groups. A comparison between the relative positions and the liquefaction conversion data suggests that a hypothetical line can be constructed through the center of the map to divide the coals into two major conversion classes. The coals above the dividing line exhibited conversion yields less than 50%, while those below the dividing line had yields greater than 50%. In addition, a very striking feature of the map is the fact that Coal 320 had the lowest conversion reactivity, while Coal 278 had the best liquefaction properties. The positions of these coals represent the extremes on the nonlinear map. The same type of clustering, based on 50% reactivity, is shown in the hierarchical clustering dendrogram in Figure 5.

The catalytic effect of mineral matter is an important factor in controlling liquefaction properties. Often, the contribution of the organic structure on liquefaction conversion has been ignored. We feel the Py/MS results of the Gulf continuous flow reactor coals present evidence that more emphasis should be placed on the organic portion.

The final group of coals investigated represented a suite that had been studied at Colorado School of Mines in a stirred batch reactor. Table V lists the characterization properties of these coals. The data analysis results for the Py/MS data are summarized in Table VI and Figure 6. The points on the nonlinear map (Figure 6) basically cluster into one major group with two outlying sets of points. Comparing the position of the data points on the nonlinear map to the reactivity data in Table V presents a picture consistent with that derived from the two previous coal suites. In general, the major cluster comprises coals that show conversion yields greater than 75%. The first set of outlying points (Fies Mine) above the major group shows good reactivity (87.5%). However, a review of the history of the sample indicated the possibility of weathering effects. No special precautions had been

Table IV. Distance Table for the Gulf Continuous Flow Reactor Coals

PSOC	265	271	276	278	302	305	307	308	320
265	28.3	102.0	158.3	158.9	119.3	128.9	146.4	179.1	115.5
271		41.4	117.7	146.9	95.0	105.4	110.1	139.2	122.4
276			27.4	126.3	179.8	69.8	52.4	75.7	179.6
278				43.8	157.7	96.6	125.5	144.1	176.4
302					28.6	164.9	174.0	201.6	117.6
305						25.2	48.4	73.6	167.9
307							36.5	64.7	170.9
308								22.8	209.5
320									29.2

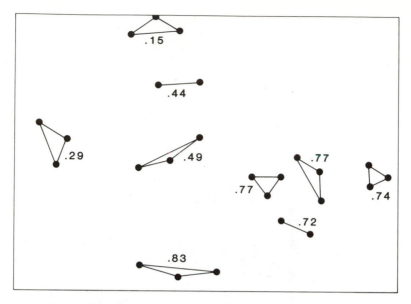

Figure 4. Nonlinear map of Pennsylvania State University coals. Numbers indicate percent conversion yields.

PSOC

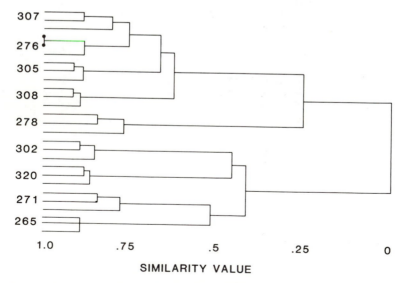

SIMILARITY VALUE

Figure 5. Hierarchical clustering dendrogram of Py-MS data from Pennsylvania State University coals.

Table V. Coals Used in Stirred Batch Reactor

PSOC Number	Rank	Conversion[a] (%)	%C	%S (dry)
071	HVCB	83.7	78.0	0.58
107	HVBB	81.0	82.0	0.53
130	MVB	<60.0	91.4	0.56
151	HVCB	80.0	78.4	0.47
370	HVAB	75.2	84.0	0.67
437	HVAB	85.0	80.3	0.51
444	HVBB	85.8	78.9	0.46
Fies (KY 9)	HVCB	87.5	66.9	3.85

[a] Conversion defined by product solubility in THF after 1 h reaction time at 400 °C, 2000 psi (H_2). Liquefaction solvent was tetralin.

taken for storage of this sample; therefore, it had been in contact with air for several years. Examination of the data in Table V shows an approximate conversion yield was measured for the PSOC 130 coal. This coal, when subjected to the standard conditions of the stirred batch liquefaction procedure, agglomerated to such an extent that reproducible data of the quality of the other measured reactivities in the suite could not be obtained. The Py/MS results suggest significant structural differences in this particular coal.

Information concerning the factors that affect the differences between samples within the three suites can be obtained by comparing the magnitude of the Fisher ratios within a conversion group. Tables VII and VIII list the 13 ions (plus possible compound class) that exhibit the highest ratios. The Fisher ratio reflects partially the intensity differences between ions. Therefore, each peak with a high Fisher ratio can be assessed as to its positive or negative influence, within the suite, with regard to conversion yields. Because of the similarity in the conversion yields in Suite 3, the Fisher ratio data for

Table VI. Distance Table for the Coals Studied in the Stirred Tank Reactor

PSOC	071	437	444	151	370	Fies	130	107
071	26.2	72.6	79.1	62.9	84.7	139.9	177.9	83.8
437		34.3	53.5	42.7	66.4	187.3	163.5	70.4
444			33.5	56.2	43.5	176.2	150.5	40.9
151				20.5	70.9	168.4	167.3	68.8
370					30.3	175.8	127.0	31.9
Fies						15.5	239.3	175.8
130							20.0	137.1
107								22.0

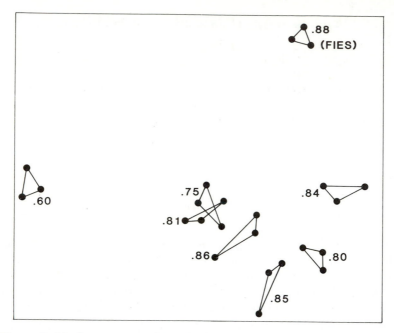

Figure 6. Nonlinear map of Py-MS data of CSM coals. Numbers indicate percent conversion yields.

Table VII. Fisher Weights for Tubing Bomb Coals

Mass	Fisher Weights	Probable Compound Class	Contribution to Conversion
36	57.6	Sulfur compounds	?
156	53.6	C_2-Naphthalenes	+
60	39.4	Carboxylic acid	?
142	32.9	C_1-Naphthalenes	+
97	28.0	Alkene	+
128	27.0	Naphthalenes	+
146	25.6	?	+
48	24.4	Sulfur compounds	+
106	23.2	Benzenes	?
157	16.5	—	
64	14.0	S_2-Sulfur compounds	+
57	12.8	Alkanes	?

Table VIII. Fisher Weights for Gulf Continuous Flow Reactor

Mass	Fisher Weights	Probable Compound Class	Contribution to Conversion
176	213.1	?	+
92	144.2	Benzenes	−
36	143.8	Sulfur compounds	?
85	139.1	Alkanes	−
41	115.9	Alkenes	?
64	84.4	S_2-Sulfur compounds	+
124	83.2	Lignin	?
83	79.7	Alkenes	−
128	73.0	Naphthalenes	?
207	72.3	?	?
156	61.3	C_2-Naphthalenes	+
162	55.3	?	?

the stirred batch reactor were not included. For the remaining two suites, there are only a few entries within the highest 13 Fisher ratio ions common to both suites. The group of ions m/z 128, 142, and 156 are probably alkylnaphthalenes and seem to have a positive correlation with reactivity in both groups. The ions that appear to correspond to the sulfur species—SO_2 or S_2 (m/z, 64) and CH_3SH (m/z, 48)—are produced most strongly in the coals with the best conversion performance. The total sulfur analysis for this suite (Table III) showed a weak relationship between sulfur content and conversion yields, with the highest sulfur content coal exhibiting the highest reactivity.

Fisher ratios can be used within a single suite of coals to determine positive or negative effects of homologous ion series on conversion yields. For the coals liquefied using the Gulf continuous flow reactor, a series of bivariate plots using homologous series that were identified in the top 50% of the Fisher ratios were constructed. For example, a plot showing the intensity of phenol vs. methylphenol, m/z 94 vs. 108, is shown in Figure 7. In this case, the highest liquefaction reactivity was observed in the coals with highest intensity for these homologs. This observation is consistent with the reported variation of liquefaction reactivity with total oxygen content of coal (*11*). Other series such as the sulfur species—m/z 34, 48, 64, and 76—and the series at 148, 162, and 176—have been plotted to give similar trends. In addition, a bivariate plot can be made from members of two different homologous series. Figure 8 illustrates a plot of m/z 83 vs. m/z 85 that represents the intensities of $C_6H_{11}^+$ (alkene fragment ion) vs. $C_6H_{13}^+$ (alkane fragment ion). From this plot, coals with the highest concentration of these ions showed the poorest liquefaction perfor-

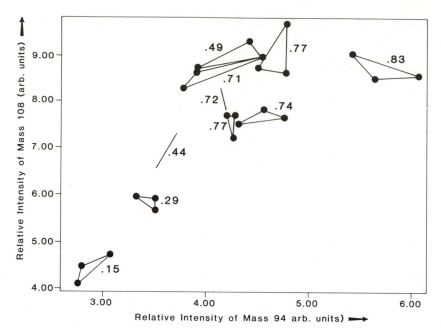

Figure 7. Bivariate plot of m/z *94 vs. 108.*

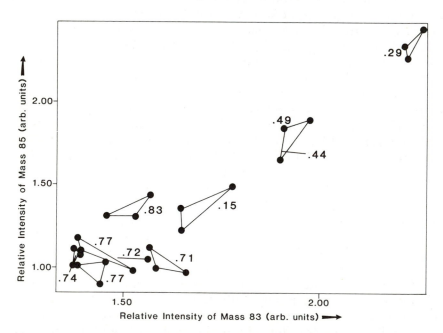

Figure 8. Bivariate plot of m/z *83 vs. 85.*

mance. The good correlation of the alkene:alkane signals is somewhat surprising because a broad spectrum of alkane yields had been reported for the liquefaction of various ranked coals (*13*). Once the trends have been established by a multivariate statistical calculation, the complexity of the data analysis and conversion correlations often can be reduced by the use of a bivariate statistical approach. However, bivariate techniques are no substitute for multivariate analysis.

The coals used for this study represent suites that have been liquefied by three microscale techniques. The data from the Py/MS results strongly correlate with the conversion parameters from the reactor studies. The Py/MS/pattern recognition results also generate information that can correlate structural features with the conversion data. We believe the Py/MS technique has a bright future in predicting liquefication reactivities and, when combined with new techniques such as MS/MS (*14*) will offer a new dimension in understanding the effects of coal structure on liquefaction reactivity.

Acknowledgments

Appreciation is given to the Mining and Mineral Resources Research Center for a fellowship to S.L.D. Gratitude is also given to the CSM Foundation for partial support of this work. A special thanks is extended to H. L. C. Meuzelaar for his cooperation during this study.

Literature Cited

1. *Annu. Book ASTM Stand.* Part 26, 1976.
2. Maters, W. L.; van de Meent, D.; Schuyl, P. J. W.; de Leeuw, J. W.; Schenck, P. A.; Meuzelaar, H. L. C. In "Analytical Pyrolysis"; Jones, C. E. R.; Cramers, C. A., Eds.; Elsevier: Amsterdam, 1977; pp. 203–16.
3. Van Graas, G.; de Leeuw, J. W.; Schenck, P. A. *J. Anal. Appl. Pyrolysis* **1980**, *2*, 265–76.
4. Meuzelaar, H. L. C.; Hileman, F. D. In "Pyrolysis Mass Spectrometry of Biomaterials"; Elsevier: Amsterdam, in press.
5. Kurzweg, L.; Thomas, E.; Meuzelaar, H. L. C. In "A Modular Pyrolysis Mass Spectrometer," National ACS Meeting, Las Vegas, NV, August 1980.
6. Jurs, P. C.; Isenhour, T. L. In "Chemical Applications of Pattern Recognition"; Wiley: New York, 1975.
7. Fisher, R. A. *Ann. Eugen.* **1936**, *7*, 179.
8. Sammon, J. A. *IEEE Trans. Comput.* **1968**, *C18*, 401.
9. Given, P. H.; Spackman, W.; Davis, A.; Walker, P. L., Jr.; Lovell, H. L.; Coleman, M.; Painter, P. C. In "The Relationship of Coal Characteristics to Coal Liquefaction Behavior," Pennsylvania State Univ. Quarterly Technical Progress Report, January–June 1978.
10. Yarzab, R. F.; Given, P. H.; Spackman, W.; Davis, A. *Fuel* **1980**, *59*, 81.
11. Furlong, M., Ph.D. Thesis, Colorado School of Mines, Golden, CO, 1981.
12. Yarzab, R. F.; Abdel-Baset, Z.; Given, P. H. *Geochim. Cosmochim. Acta* **1979**, *43*, 281.
13. Whitehurst, D. D.; Mitchell, T. O.; Farcasiu, M. In "Coal Liquefaction, The Chemistry and Technology of Thermal Process"; Academic: New York, 1980, p. 178.
14. Yost, R. A.; Enke, C. G. *Anal. Chem.* **1979**, *51*, 1251A.

RECEIVED for review October 14, 1981. ACCEPTED August 16, 1982.

Pyrolysis Gas Chromatographic Characterization, Differentiation, and Identification of Biopolymers— An Overview

FORREST L. BAYER

The Coca-Cola Company, Corporate Quality Assurance Department, Atlanta, GA 30301

Pyrolysis gas chromatographic (PGC) characterization, differentiation, and identification of biopolymers will be discussed in terms of repeatability, specificity, and sensitivity. An application overview will be presented for a wide range of biopolymers and their subunits including the PGC of amino acids, proteins, enzymes, and hemoglobins. Representative data displaying the degree of repeatability in terms of peak heights and retention times for complex macromolecular systems such as proteins will be shown. Retention time relative standard deviations in the order of 0.1% in 80-min PGC runs were achieved. Long-term repeatability of a PGC system will be demonstrated by the virtually superimposable pyrochromatograms of a compound taken 9 months apart.

T HE TECHNIQUE OF PYROLYSIS GAS LIQUID CHROMATOGRAPHY (PGLC) has become one of the most powerful tools in the analyst's arsenal of analytical equipment when dealing with intractable samples. This technique affords the analyst a means of obtaining characterization or identification information on complex, microgram-size samples in a relatively short time frame for a modest cost. With many samples, no preparation is required prior to PGLC analysis. Due to these attributes, this technique is becoming more popular for use in the analysis of biopolymers.

This chapter will give a brief overview of the literature with respect to the PGLC of biopolymers, including a review of amino acids, the basic building blocks of biopolymers, peptides, and proteins. In addition, inter-laboratory reproducibility will be discussed.

0065–2393/83/0203–0693$06.00/0

Amino Acids

Unraveling the complexities of biochemical/biopolymer systems inherently requires a firm understanding of their basic building blocks. For biochemical/biopolymer systems such as peptides and enzymes, the basic building blocks involve their structural units, amino acids. Over the past two decades, several investigators have studied the pyrolysis of amino acids (1-17). These investigations, although conducted under a variety of conditions, indicate that amino acids exhibit unique pyrochromatograms. Merritt and Robertson (8) pyrolyzed seventeen amino acids, and analyzed their pyrolysis products by mass spectrometry. They found that each amino acid pyrolyzed yielded a unique or most abundant primary product among those identified. The resultant products of amino acid pyrolysis, however, have been an area of controversy among investigators. For example, several investigators investigated the pyrolysis of proline and hydroxyproline. Giacobbo and Simon (16) initially reported that the pyrolysis of proline yielded pyrrolidine. Comparison of pyrograms in a later paper indicated that the pyrolysis of proline and hydroxyproline, yielded pyrrole (2). Vollmin (1) confirmed Simon's work demonstrating that proline yielded pyrrole. Smith et al. (10) studied the pyrolysis of proline and hydroxyproline more recently. In their review of the pyrolysis of proline and hydroxyproline, they stated that Vollmin found that hydroxyproline as well as proline yielded pyrrole. A closer inspection of Vollmin's work indicated that the hydroxyproline was not pyrolyzed in that particular study, however, pyrroline was shown to yield pyrrole. Studies by Merritt and Robertson (8) indicated that proline and hydroxyproline yielded pyrrole and methylpyrrole, respectively. Smith et al. (10) reported that the pyrolysis of proline and hydroxyproline yielded a tentative identification of 1-pyrroline as a major product of proline and pyrrole as the major product of hydroxyproline. Discrepancies of the pyrolysis products as noted earlier are a function of the total system and operating parameters. Therefore, if interlaboratory reproducibility is to be achieved, the analytical parameters of pyrolysis and chromatography must be standardized. Parameters that are important in achieving interlaboratory reproducibility are the temperature of pyrolysis (calibration of the pyrolyzer), the temperature·rise time of the pyrolyzer, the type of probe, the sample size and associated handling techniques, the type of carrier gas, the flow rate, the type of GC column, the percent loading of liquid phase, the coating and packing techniques, and the GC temperature program.

The potential for achieving interlaboratory reproducibility is demonstrated by looking at the repeatability data shown in Figure 1 and Table I for the pyrolysis of d,l-tyrosine. The two pyrochromatograms in Figure 1 were achieved on two different filament pyrolysis

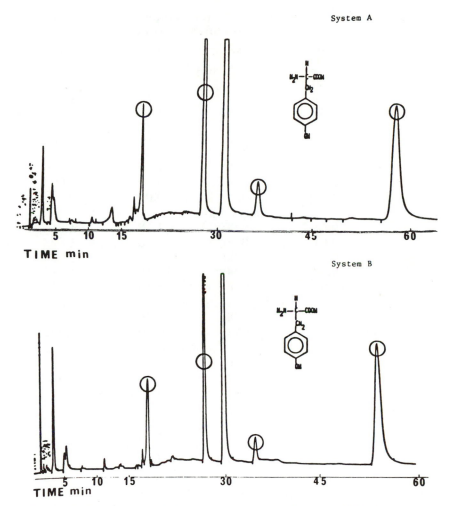

Figure 1. Pyrograms of d,l-tyrosine produced on two PGLC systems. Conditions: System A, CDS Pyroprobe 100 interfaced to a HP5840 GC with 3.65 m × 2 mm ID glass column packed with 5% Carbowax 20 M TPA on Anakrom ABS 110/120 mesh; and System B, CDS Pyroprobe 120 interfaced to a HP5830 GC with 3.65 m × 2 mm ID glass column packed with 5% Carbowax 20 M TPA on 110/120 Anakrom ABS.

systems. These systems consisted of a Chemical Data System (CDS) Pyroprobe 100 interfaced to a Hewlett Packard (HP) 5830A GC and a CDS Pyroprobe 120 interfaced to a HP5840 GC. The columns in both instances were 3.65 × 2 mm ID × 6.5 mm OD columns packed with 5% Carbowax 20 M TPA on 110/120 mesh Anakrom ABS. The amino acid was pyrolyzed at 900 °C, and the subsequent conditions of chromatography were identical on both instruments. All of the key features are

Table I. Statistics for Peak Retention Times for *d,l*-Tyrosine Produced on Two Different Systems

	Peak 1	Peak 2	Peak 3	Peak 4
		Data 5840A		
	0.60	0.90	1.17	1.87
	0.59	0.90	1.17	1.86
	0.59	0.90	1.17	1.86
	0.60	0.90	1.17	1.86
	0.59	0.90	1.17	1.88
Mean	0.594	0.90	1.17	1.866
SD	0.005	0.00	0.00	0.008
RSD	0.922	0.00	0.00	0.479
		Data 5830A		
	0.61	0.90	1.16	1.81
	0.61	0.90	1.16	1.82
	0.60	0.90	1.16	1.82
	0.60	0.90	1.16	1.81
	0.60	0.90	1.16	1.81
Mean	0.606	0.90	1.16	1.813
SD	0.009	0.00	0.00	0.005
RSD	1.47	0.00	0.00	0.301
		Data 5830A and 5840A		
Mean	0.60	0.90	1.165	1.84
SD	0.009	0.00	0.005	0.028
RSD	1.57	0.00	0.452	1.54

exhibited in both pyrochromatograms. Table I shows a comparison of the retention time (R.T.) relative to the most abundant peak at a retention time of 30.5−31.7 min. The percent relative standard deviation is less than 1.6% between the two systems for the circled peaks (Peak 1, R.T. 18.5 min, Peak 2, R.T. 28.45 min, Peak 3, R.T. 31.2 min, Peak 4, R.T. 55 min). The potential for the long-term stability of a pyrolysis system is shown in Figure 2, by comparison of the virtually superimposable pyrograms of the steroid 3-β-hydroxyandrostane-17-one taken 9 months apart. By using a light box, one is able to view the superimposition of the two pyrochromatograms one upon the other.

Peptides

Pyrolysis investigations of peptides have received only limited studies, or limited attention (*1, 2, 5, 8, 10, 18, 19*). Investigators present two opposing views regarding the pyrolysis products of peptides. One view is that the pyrolysis of peptides yields pyrograms that are linear combinations of the individual constituent amino acids (*1, 2,*

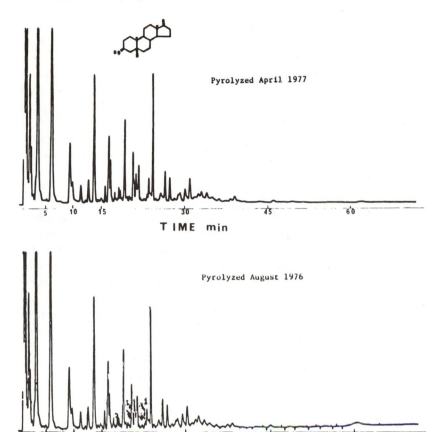

Figure 2. *Pyrograms of 3-β-hydroxyandrostane-17-one produced on the same PGLC system 9 months apart.*

18); yet others showed that the pyrolysis of peptides is influenced by the particular amino acid linkage or sequence (*8, 10, 19*). Most literature data tend to support the latter premise.

Merritt and Robertson (*8*) pyrolyzed the dipeptide pairs Gly–Ala, Ala–Gly, Gly–Leu, and Leu–Gly and discovered that the amino acid sequence affected the pyrolysis product. The dipeptides Ala–Gly yielded acetone, acetaldehyde, and ammonia as primary pyrolysis products, whereas, the dipeptide Gly–Ala yielded acetone and 2-methylpyrrole. Pyrolysis of Gly–Leu, where the leucine is carboxy terminal, yielded cyclopentane as the primary pyrolysis product; whereas pyrolysis of Leu–Gly, where the glycine carboxyl is terminal, yielded acetic acid as a unique pyrolysis product. Smith et al. (*10*) extended their investigations of the pyrolysis products of proline and hydroxyproline by looking at dipeptides and tripeptides containing these amino acids. More specifically, they pyrolyzed Pro−Gly, Gly−Pro, Gly−Hyp, Pro−Ala, Ala−Pro, and the tripeptide Gly−Pro−Ala using a curie point pyrolyzer at 770 °C with the samples at pH 2 and 9. Their studies of the individual amino acids indicated that the product was 1-pyrroline. Pyrolysis of the dipeptides and tripeptides at pH 2 and 9 yielded some interesting results. When proline's amino functionality was bound in an amide linkage as in Gly−Pro, Ala−Pro, and Gly−Pro−Ala, low yields of 1-pyrroline were found; this result is contrasted with proline being amino terminal, where high yields of 1-pyrroline were found. In addition, the yield of 1-pyrroline could be enhanced by pyrolysis at low pH.

Two studies were reported in which key structural information of a compound containing a peptide side chain for a peptide were established by the use of pyrolysis (18, 19). Mauger (18) worked out the structure of four actinomycins (D, C_3, II, and PIP2) that vary only in the second or third amino acid of the two pentapeptide side chains connected to the phenoxazone ring. The investigator assumed that the resulting diketopiperazines (DKPs) would reflect the neighboring pair of amino acids. Thus, actinomycin D with Val−Pro, actinomycin C_3 with D-allo−Ileu−Pro, actinomycin II with Sar−Sar, and actinomycin PIP2 with Val−Pip should all give the unique diketopiperazine arising from the cyclization of these neighboring pairs of amino acids. These compounds were confirmed by matching the retention times with authentic DKPs.

Schmid et al. (19) pyrolyzed the octapeptides lypressin and felypressin using a curie point capillary system. The differences in these two octapeptides exist in the cyclic portion of the peptide. Lypressin has a Glu−Phe−Tyr−Cys sequence, whereas felypressin has a Glu−Phe−Phe−Cys sequence. Phenol and cresol were identified in the pyrolysis of lypressin. If present in felypressin, they are in reduced concentration. The greater abundance of phenol and cresol is presumably due to the tyrosine found in lypressin but absent in felypressin.

Winters and Albro (5) pyrolyzed 30-mg samples of 18 different amino acids using a homemade filament pyrolyzer and attempted to identify various amines generated as pyrolysis products. These investigators then tried to apply this technique to the pyrolysis of proteins. Specifically, they pyrolyzed bovine serum albumin and crystalline egg albumin and then attempted to relate the profiles of the amino acid content of the proteins. To derive the amino acid composition of the particular proteins, the researchers initially calculated the amount of the various amines generated in the pyrolysis of protein and then factored in the contribution of the individual amino acids generating these particular amines. Analysis of the crystalline egg albumin pyrograms showed fair agreement with calculated values for several of the peaks. In their pyrolysis GC/MS work on amino acids and peptides, Merritt and Robertson (8) pyrolyzed crystalline bovine insulin. They had established that many of the amino acids yielded unique or most abundant products when pyrolyzed. Pyrolysis of the crystalline bovine insulin yielded pyrolysis products that could be attributed to amino acids known to be present in the insulin.

Investigations employing a specific or selective nitrogen–phosphorus detector first were accomplished by Myers and Smith (20) in their investigation of the low molecular weight fragments yielded from the pyrolysis of such substrates as proteins, carbohydrates, and protein-containing materials, e.g., beef collagen, bovine hemoglobin, cellulose, glucose, horse myoglobin, DNA, RNA, and various seeds and nuts. As one might expect in the pyrolysis of similar substances, these investigators found many of the pyrograms to be qualitatively similar, but found also that they varied in the range of their relative peak heights and peak height ratios. The pyrograms of hemoglobin and myoglobin were quite similar but distinguishable. This finding would be expected considering the similarity of their amino acid composition.

Other investigators studied the pyrolysis of hemoglobins (5, 21, 22). Bayer initially differentiated adult (Hgb-A) and fetal (Hgb-F) hemoglobins. These two hemoglobins are both composed of four polypeptide tetramers; both contain two α-chains consisting of 141 amino acids, but differ in the other two tetramers. The adult hemoglobin has two β-chains, whereas the fetal hemoglobin has two γ-chains. Each of these chains consists of 146 amino acids, but the amino acids in the β-chain differ from those in the γ-chain in 39 loci. In addition, the fetal hemoglobin has four residues of isoleucine not present in the adult hemoglobin. The discernable differences observed in the pyrograms for these two hemoglobins are illustrated in Figures 3 and 4. The use of capillary columns would allow for greater resolution of complex pyrolyzate matrices such as these and would thus aid in the analysis of such biopolymers.

Proteins

PGLC studies of proteins and proteinaceous substances covering a diverse range of macromolecular samples, such as albumins (5), enzymes (21–24), hemoglobins (5, 21, 22), plant seeds (20, 25), cockroaches (26), and toxic mushrooms (27) have been the subject of pyrolysis GC investigations. No reported investigation has systematically looked at a series of well-characterized proteins on the basis of their subunit amino acids and identified the resulting pyrolysis prod-

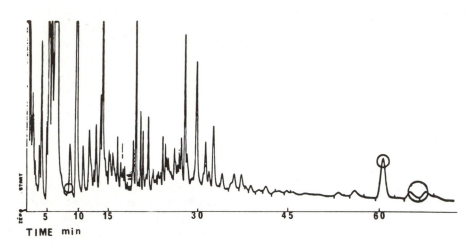

Figure 3. Pyrograms of adult (Hgb-1A) hemoglobin. Conditions: CDS Pyroprobe 120 at 900 °C for 5 s, ramp off, interface 180 °C HP5830 GC; column, 3.65 m × 2 mm ID glass packed with 5% Carbowax 20 M TPA on 110/120 Anakrom ABS. Program: 60 °C hold 4 min, 40 °C/ min to 165 °C; flow N_2 at 30 mL/min, FID at 250 °C.

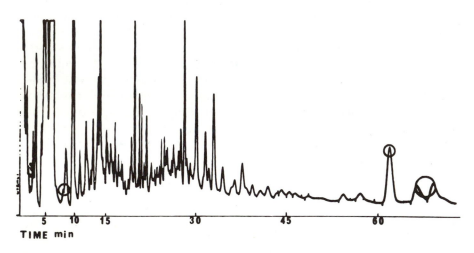

Figure 4. Pyrogram of fetal (Hgb-1F) hemoglobin. Conditions are the same as in Figure 3.

ucts. This observation is not surprising because the less complex peptides and their pyrolysis products have undergone only limited investigations of this nature.

Pyrolysis investigations of proteins and proteinaceous substances may be grouped loosely into one of three categories: (1) differentiation of proteins one from another; (2) studies involved in establishing the presence of a particular protein in a substance; and (3) studies involved in looking at the amino acid content of the protein for the purpose of identifying the protein or establishing the presence of an amino acid residue.

Bayer (21) investigated the PGLC of hemoglobinopathies, the most common of which is sickle cell anemia (Hgb-S). This particular hemoglobinopathy is characterized by having a valine substituted for a glutamic acid in the sixth position of the β-chains and afflicts approximately 10% of the Negro population. Preliminary studies on a limited number of normal adults vs. sickle cell patients indicated that PGLC, in fact, may be sensitive enough to pick out this subtle amino acid variation between the hemoglobins. This work requires further investigation.

Five vertebrate lyophilized hemoglobin samples also were differentiated (21). These samples were commercially obtained bovine, horse, rabbit, and dog hemoglobin and clinically obtained human hemoglobin. These hemoglobins all contained the same number of amino acids in the α- and β-chains with the exception of the bovine hemoglobin, which contains only 145 amino acid residues in its β-chain as opposed to 146 for the other species. The difference between the various hemoglobins again lies in the sequence of the individual amino acid residues. Variations in the sequencing between the various hemoglobins is from 10 to 20 amino acid residues per polypeptide chain. Based on the work of Merritt (8) and others involving the sequencing of dipeptides, we know that the sequence linkage is important with respect to the final product. Therefore, Bayer was able to find subtle differences in the peak height ratios for the five vertebrate hemoglobins, which allowed differentiation of the samples.

Enzymes

A limited number of investigations on the PGLC enzymes have appeared in the literature (8, 21 −24). Merritt (8) and Oyama (28) simultaneously published the first reports on the pyrolysis of enzymes. In both instances, the substrate was bovine insulin. The first report concerning the differentiation of enzymes via PGLC was conducted by Bayer (21). In this particular study, three commercially available enzymes, α-chymotrypsin, trypsin, and trypsinogen were pyrolyzed using a filament discharge system. Both α-chymotrypsin and trypsin are digestive enzymes containing 245 and 223 amino acid residues, respectively. Trypsinogen, on the other hand, is the inactive precursor enzyme containing 229 amino acid residues.

Cleavage of the sixth terminal amino acid residue of trypsinogen yields the active trypsin molecule. These two molecules therefore differ by less than 3% in their primary structure. Comparison of the three pyrochromatograms allowed for differentiation of the two enzymes and one proenzyme (22). Repeatability retention time for trypsin pyrolyzed at 900 °C is shown in Table II. Maximum relative standard deviations of less than 0.1% were achieved.

Danielson et al. (23) extended the work on enzyme differentiation by studying a series of six enzymes ranging in molecular weight from 22,600 to 483,000. The enzymes investigated included α-chymotrypsin, creatine kinase, lactate dehydrogenase (LDH), catalase, and acetylcholinesterase. In addition, their studies included the LDH enzymes from various sources as these enzymes vary in structure depending on the particular source. They also pyrolyzed some LDH isoenzymes.

Pyrolysis of the enzymes at 800 °C using a filament discharge-type pyrolyzer yielded five prominent peaks whose peak height ratios could be used to distinguish the enzymes α-chymotrypsin, jack bean urase, creatine kinase, lactate dehydrogenase, catalase, and acetylcholinesterase. They identified four out of five of the major peaks as phenol, p-cresol, indole, and skatole. Differentiation was achieved for all of the isoenzymes except the pig H_4 and M_4 enzymes. These authors also employed an alkali flame ionization detector and found significant differences for a few of the peaks, but overall the variations were not as striking as expected.

One of the more practical applications for the use of PGLC in the study of enzymes and proteins is that of Danielson and Rogers (24). They previously reported that the pyrolysis of tryptophan yielded the unique pyrolysis product skatole (23). Skatole first was reported as a pyrolysis product of tryptophan by Shulman and Simmonds (11). Danielson et al. calibrated their PGLC system to the height of the skatole peak vs. micrograms of pyrolyzed tryptophan. They then were able to determine the number of tryptophan residues present in sev-

Table II. Repeatability Retention Time for Trypsin Pyrolyzed at 900 °C

Peak Retention Time (s)

	Peak 1	Peak 2	Peak 3	Peak 4
Mean	1734.8	1945.2	4022.8	4533.6
SD	0.692	0.599	2.42	3.71
RSD	0.039	0.030	0.06	0.081

eral enzymes based on the concentration of skatole liberated during pyrolysis. They compared the results from the pyrolysis technique with those of the more standard Spies and Chambers method and obtained good agreement.

Investigations involving dental research by Stack (29) demonstrated the potential of distinguishing normal from abnormal (chalky) enamel in patients. In addition to pyrolyzing teeth enamel and dentin, this researcher pyrolyzed collagen, gelatin, crystalline bovine albumin, horse α-globulin, and bovine γ-globulin; all the various proteins except collagen and gelatin were differentiated.

Hair has been the subject of pyrolysis investigations from a potential forensic standpoint (30, 31). Hair often is found at the scene of a crime, and owing to the belief that everyone has unique hair just as they have unique fingerprints, it is considered a highly valuable piece of evidence relating to a crime. Thus, differentiation of hair by PGLC would add a very strong piece of instrumentation to the array of techniques used by the forensic chemist. Initial PGLC work by Kirk (30) indicated that considerable discrimination of samples from 50 individuals could be found. More extensive work by Deforest (31) and Kirk (30) was not quite as gratifying. They found that one of the limiting factors was lack of resolution of the constituents. Keratinaceous substances such as fingernail gave similar pyrograms to hair. Pyrolysis of fingernail casein and gelatin at 400 °C also gave similar but distinguishable pyrograms.

PGLC/MS has been utilized in the search for extraterrestrial life. The Viking mission to Mars had a highly compact system that allowed for the 500 °C pyrolysis of samples followed by separation and mass spectral detection of up to 200 amu. The ground work for this project was primarily due to the efforts of Simmonds and coworkers (11 –14). They investigated the PGLC/MS of a number of bioorganic terrestrial substances such as proteins, carbohydrates, nucleic acids, lipids, and porphyrins. Simmonds and coworkers characterized the predominant molecules liberated from the pyrolysis of the just mentioned terrestrial bioorganic materials. These data coupled with a comprehensive literature survey resulted in a list of substances that would be obtained on pyrolysis of various types of terrestrial bioorganic materials. Thus, pyrolysis of an unknown sample should yield characteristic fragments that then could be attributed to a particular type of bioorganic matter.

Chemotaxonomic applications to proteinaceous matrices using pyrolysis were investigated by Hall and Bennett (26). These investigators pyrolyzed 2-mg samples of cockroaches spiked with 10 μg of C_{18} plus 30 μg of C_{32} n-alkane standards at 600 °C using a curie point pyrolyzer. The authors were able to identify specimens at the species level with 100% accuracy in those instances where reference pyrograms were available. In the pyrolysis of the 52 coded samples, specimens for which no reference pyrograms were available were placed easily into groups of unknowns.

The pyrolysis of toxic mushrooms has been investigated (27). Pyrolysis studies were conducted on the ethanol extracts of toxic mushrooms using a homemade temperature-regulated filament pyrolyzer. Chromatographic resolution of the pyrolyzed entities was not very gratifying; however, the authors indicated that they could differentiate 16 different mushroom extracts.

Literature Cited

1. Vollmin, J.; Kriemler, P.; Omura, I.; Seibl, J.; Simon, W. *Microchem. J.* **1966**, *11*, 73.
2. Simon, W.; Giacobbo, H. *Angew. Chem. Int. Ed.* **1965**, *4*, 938.
3. Kojima, T.; Morishita, F. *J. Chromatogr. Sci.* **1970**, *8*, 471.
4. Merritt, C., Jr.; DiPietro, C.; Robertson, D. H.; Levy, E. J. *J. Chromatogr. Sci.* **1974**, *12*, 668.
5. Winter, L. N.; Albro, P. W. *J. Gas Chromatogr.* **1964**, *2*, 1.
6. Stack, M. V. *Biochem. J.* **1965**, *96*, 56.
7. Kanomata, K.; Mashiko, Y. *Nippon Kagaku Zasshi* **1966**, *87*, 57.
8. Merritt, C., Jr.; Robertson, D. H. *J. Gas Chromatogr.* **1967**, *5*, 96.
9. Simon, W.; Kriemler, P.; Vollmin, J. A.; Steiner, H. *J. Gas Chromatogr.* **1967**, *5*, 53.
10. Smith, R. M.; Shawkat, S. A.; Haynes, W. P. *Analyst* **1980**, *105*, 1176.
11. Shulman, P. G.; Simmonds, P. G. *Chem. Commun.* **1968**, 1040.
12. Simmonds, P. G.; Shulman, G. P.; Stembridge, G. H. *J. Chromatogr. Sci.* **1969**, *7*, 36.
13. Simmonds, P. G.; Medley, E. E.; Ratcliff, M. A.; Shulman, G. P. *Anal. Chem.* **1972**, *44*, 2060.
14. Ratcliff, M. A.; Medley, E. E.; Simmonds, P. G. *J. Org. Chem.* **1980**, *39*, 1481.
15. Smith, R. M.; Solabi, G. A.; Hays, W. P.; Stretton, R. J. *J. Anal. Appl. Pyrolysis* **1980**, *1*, 197.
16. Giacobbo, H.; Simon, W. *Pharm. Acta Helv.* **1964**, *39*, 162.
17. Means, J. C.; Perkins, E. G. In "Analytical Pyrolysis"; Jones, C. E. R.; Cramers, C., Eds.; Elsevier: Amsterdam, 1977; p. 249.
18. Mauger, A. B. *Chem. Commun.* **1971**, 39.
19. Schmid, J. P.; Schmid, P. P.; Simon, W. In "Analytical Pyrolysis"; Jones, C. E. R.; Cramers, C., Eds.; Elsevier: Amsterdam, 1977; p. 99.
20. Myers, A.; Smith, R. N. L. *Chromatographia* **1972**, *5*, 521.
21. Bayer, F. L., Ph.D. Thesis, Emory Univ., Atlanta, GA, 1976.
22. Bayer, F. L.; Hopkins, J. J.; Menger, F. M. In "Analytical Pyrolysis"; Jones, C. E. R.; Cramers, C. A., Eds.; Elsevier: Amsterdam, 1977; p. 217.
23. Danielson, N. D.; Glajch, J. L.; Rogers, L. B. *J. Chromatogr. Sci.* **1978**, *16*, 455.
24. Danielson, N. D.; Rogers, L. B. *Anal. Chem.* **1978**, *50*, 1680.
25. Reiner, E. *J. Gas Chromatogr.* **1967**, *5*, 65.
26. Hall, R. C.; Bennett, G. W. *J. Chromatogr. Sci.* **1973**, *11*, 439.
27. Benoit-Guyod, J. L.; Benkheder, K.; Seigle-Murandi, F.; Duc, C. L. *J. Agric. Food Chem.* **1980**, *28*, 1317.
28. Oyama, V. I.; Carle, G. C. *J. Gas Chromatogr.* **1967**, *5*, 151.
29. Stack, M. V. *J. Gas Chromatogr.* **1967**, *5*, 22.
30. Kirk, P. L. *J. Gas Chromatogr.* **1967**, *5*, 11.
31. Deforest, P. R.; Kirk, P. L. *Criminologist* **1973**, *8*, 5.

RECEIVED for review February 12, 1982. ACCEPTED June 21, 1982.

Pyrolysis/Gas Chromatography/Mass Spectrometry of Biological Macromolecules

E. REINER[1] and T. F. MORAN

Georgia Institute of Technology, School of Chemistry, Atlanta, GA 30332

In pyrolysis/gas chromatographic studies of biological macromolecules, more definitive information is attained by adding a mass spectrometer and data system to the combined instrument. Key peaks that impart specificity or uniqueness to the pyrogram (fingerprints) can be identified by mass spectrometry. Consequently, microbial or mammalian cells can be characterized with greater confidence through their pyrolysis fragments. Highlights of studies on pathogenic microorganisms and mammalian cells are discussed. The latter include primary normal human tissue cells, cultured cells reflecting genetic disease, e.g., cystic fibrosis, and cells involved with malignant disease processes. Also included is a discussion of the possible use of computers in disease diagnosis.

FROM OUR INITIAL WORK with pyrolysis/gas chromatography (Py/GC) techniques in 1964, we suspected that the small molecular weight series of peaks that comprise the pyrogram were undoubtedly fragments split off from macromolecular structures. Several lines of evidence pointed in this direction. The peaks that proved to be unique or specific for a strain of pathogenic microorganisms were also closely associated with their antigenic structure. (*See* Table I.) Considerable antigenic activity resides in cell wall biopolymers, and the association has been confirmed by numerous structural studies on the chemistry of cell walls (*1*).

More evidence emerged from comparison of pyrograms derived from pure bacterial cell walls with those derived from whole cells.

---

[1] Current address: Emory University, Department of Chemistry, Atlanta, GA 30322.

0065−2393/83/0203−0705$06.00/0

Table I. Comparative Serology and Chemistry of Salmonella Species

Species Name	Somatic (O) Antigen (6)	H Antigen (6)		Kauffmann (6) White Group	Chemotype (7)	Py/GC Characteristic Peaks[a]
		Phase 1	Phase 2			
Paratyphi A	2, 12	a	—	A	XV	6½
var. Durazzo	1, 4, 5, 12	r	1, 2	B	XIV	2, 4
Heidelberg						
Chester	4, 5, 12	e, h	e, m, z	B	XIV	2, 4
Derby	1, 4, 5, 12	f, g	—	B	XIV	2
Bredeney	1, 4, 12, 27	1, v	1, 7	B	XIV	2
Paratyphi B	1, 4, 5, 12	b	1, 2	B	XIV	2, 4
Java	1, 4, 5, 12	b	(1, 2)	B	XIV	2, 4
var. Odense	1, 4, 12	b	1, 2	B	XIV	2
Typhimurium	1, 4, 5, 12	i	1, 2	B	XIV	2, 4
var. Copenhagen	1, 4, 12	i	1, 2	B	XIV	2
St. Paul	1, 4, 5, 12	e, h	1, 2	B	XIV	2, 4
Choleraesuis	6, 7	c	1, 5	C₁	III	—
var. Kunzendorf						
Bareilly	6, 7	y	1, 5	C₁	III	—
Montevideo	6, 7	g, m, s	—	C₁	III	—
Braenderup	6, 7	e, h	e, m, z	C₁	III	—
Tennessee	6, 7	z_{29}	—	C₁	III	—
Oranienberg	6, 7	m, t	—	C₁	III	—

Infanteis	6, 7	r	1, 5	C_1	III	—
Thompson	6, 7	k	1, 5	C_1	III	2
Newport	6, 8	e, h	1, 2	C_2	XIV	2
Manhattan	6, 8	d	1, 5	C_2	XIV	2
Blockley	6, 8	k	1, 5	C_2	XIV	2
Kentuckey	(8), 20	i	z_6	C_2	XIV	2
Muenchen	6, 8	d	1, 2	C_2	XIV	2
Typhi H901W	9, 12	d	—	D	XVI	3
Typhi 0901W	9, 12	—	—	D	XVI	3
Typhi 2v	9, 12(Vi)	d	—	D	XVI	3
Typhi Watson	9, 12(Vi)	d	—	D	XVI	3
Typhi GS, Rough	(Vi)	(d)	—	D	XVI	3
Typhi Vi I, Rough	(9), (Vi)	(d)	—	D	XVI	3
Typhi R_2, Rough	—	d	—	D	XVI	3
Enteritidis	1, 9, 12	g, m	—	D	XVI	3
Gallinarum	1, 9, 12	—	—	D	XVI	3
Pullorum	9, 12	—	—	D	XVI	3
Panama	1, 9, 12	l, v	1, 5	D	XVI	3
Typhi Lact$^+$	9, 12, (Vi)	d	—	D	XVI	3
Javiana	1, 9, 12	l, z_{28}	1, 5	D	XVI	3
Typhi T_2	9, (12), (Vi)	d	—	D	XVI	3
Anatum	3, 10	e, h	1, 6	E	XIII	—
Senftenberg	1, 3, 19	g, s, t	—	E_4	XIII	—
Illinois	(3), (15), 34	z_{10}	1, 5	E_3	XIII	—
Newington	3, 15	e, h	1, 6	E_2	XIII	—
Give	3, 10	e, v	1, 7	E_1	XIII	—
Cubana	1, 13, 23	z_{29}	—	G	VI	—
Worthington	1, 13, 23	z	1, w	G	VI	—
Poona	13, 22	z	1, 6	G	VI	—

[a] See Figure 1.

The pyrograms were very similar, but the former proved more definitive in their fingerprint characteristics (2). However, in a study of
Salmonella organisms an opposite conclusion was reached (3).

The most important line of evidence, of course, came directly
from mass spectrometric (MS) identification of pyrolysis fragments.
Simmonds (4) had deduced the probable macromolecular origin of
fragments from the pyrolysis of two microorganisms (Table II). Other
studies (5, 6) have enabled investigators to catalog the identified fragments from protein thermal decomposition and other sources of biological macromolecules. There is considerable agreement on the identity of these macromolecule fragments.

Before undertaking MS identification of pyrolysis fragments,
certain aspects of the technique had to be evaluated. These included
variability in cell cultures and gas chromatograph (GC) instrumental
parameters.

Experimental

Cultured cells were washed three times in a small volume of distilled
water. The heavy cell suspension was lyophilized and samples ranging from
40 to 120 μg were transferred to a tared quartz boat and weighed to the nearest
microgram on an electrobalance. After completion of a pyrolysis run, the residual coke was weighed to determine the quantity of gaseous degradation
products analyzed from each sample. Py/GC conditions have evolved over the
years. A typical recent set of parameters is listed for a Chemical Data Systems
Pyroprobe 100 Unit linked to a Varian 3700 GC. The GC was in turn coupled
to a Varian MAT 112-S MS.

The temperatures were as follows: pyrolyzer interface, 200 °C; pyrolysis
sample, 800 °C for 10.0 s. The temperature program began at 65 °C and was
held for 4 min. It was then raised 6 °C/min to a final temperature of 165 °C,
which was maintained isothermally.

Recent work with a Carbowax 20M SCOT capillary column (43 m × 0.5
mm ID) had a resolution capability of 37,600 effective plates. Helium carrier
gas velocity was measured at 38.5 cm/s. GC inlet and flame ionization detector
(FID) temperatures were maintained at 250 °C. Electrometer amplifier was
operated at a setting of 10^{-12} amps full scale.

The GC was connected to the MS by means of an open-split coupling.
Interface and ion source temperatures were 250 °C. The electron beam was 80
eV and 1.5 mA. Ion accelerating voltage was 850 eV, resolution was 700, scanning range was 1 s per mass decade. Spectra were acquired and stored by
means of a Varian MAT SS 200 data system. In addition, exact mass measurements of significant ions were obtained as well as chemical ionization
spectra.

Since undertaking these studies, we have examined several thousand
samples of microbial and mammalian cells. From their pyrograms, we came to
a number of general conclusions, some of which will be discussed here.

Py/GC: A Simple, Rapid Technique

Samples are prepared by washing and lyophilizing the cells. Lyophilized cells store well with no discernable change when examined over yearly intervals (7). Oxborrow cultured cells on moistened filter paper are placed over the medium. Inoculating several different strains, e.g., *Pseudomonas*, on one paper membrane is feasible. This method favors the inclusion of metabolic products as well as structural elements. Cells completely free of contaminating media were removed with a spatula and aseptically dried in 7 min in a stream of warm nitrogen (8, 9). The technique must be rigorously standardized. Py/GC analysis time requires about 1 h to achieve a classification or identification. Ordinarily, minimum time from harvesting cells to end of analysis is about 2 h. By traditional methodology, such identification would require several days.

Pyrolysis Fragments Are Immutable

The same structures were recognized in the pyrogram of microorganisms even when operating parameters, column, or different culture of the same strain of bacteria were used. These phenomena have been observed on numerous occasions. The avian and Battey strains of mycobacteria (now collectively called *M. intracellare*) gave characteristic triple end peaks that differed only in amplitude. Over a period of years (10–13), these characteristic end peaks were observed consistently and remained essentially unchanged.

In another case, the anaerobic organism causing botulism can produce spores that can elaborate simple globular protein toxins. Py/GC enabled us to differentiate toxin types, and the toxic (proteolytic) strains proved to be different from nonproteolytic strains by the increasing amplitude of three high boiling compounds (a staircase effect) of the proteolytic strains. In two laboratories, with different instrumentation (capillary vs. packed column), and analyses performed 3 years apart, the staircase as a marker for proteolytic activity persisted (9, 14).

Reproducibility of Py/GC and Py/GC/MS Techniques

Pyrolysis techniques are so highly reproducible (15) that profiles can be superimposed to give essentially a single line. If operating conditions were closely controlled, relative standard deviations of retention times were usually in the 0.2% range even after a 1-h recording (14), and area measurements were in the 8–20% range. Good sample preparation, uniform, carefully controlled procedures, and

Table II. Assignment of Pyrolysis Fragments Found in Both *B. subtilis* and *M. luteus* to Biological Classes[a]

Protein	Carbohydrate	Nucleic acid	Lipid	Porphyrin
Ethanenitrile (9)[b]	Acrolein (1)[c]	Acrylonitrile (9)	Acrolein (1)[c]	Pyrrole (25)
Acrylonitrile (9)	Acetone (1)[c]	Ethanenitrile (9)	Ethene (1)[c]	Methylpyrroles (26, 27)
Propanenitrile (10)	Butanone (4)	Propanenitrile (10)	Propene (1)[c]	Dimethylpyrroles (29, 31)
Butanenitrile (12)	Pentanone (8)	Butanenitrile (12)	Butene (1)[c]	C_3 alkylpyrrole (34)
2-Methylpropanenitrile (13)	Propanal (1)[c]	Pyridine (18)		C_4 alkylpyrroles (36, 37)
Methylbutanenitrile (15)	Methylpropanal (2)	Methylpyridine		
Methylpentanenitrile (20)	Methylbutanal (6)			
Benzonitrile (30)	Furan (1)[c]			
Phenylacetonitrile (39)	2-Methylfuran (3)			
Tolunitrile (40)	Dimethylfuran (7)			
Phenol (42)	Furfural (24)			
o-Cresol (41)	Methylfurfural (28)			
p-Cresol (43)	Furfuryl alcohol (33)			
Ethylphenol (44)	Methylbutadiene			
Xylenols (45, 46)				
Indole (47)				
Methylindole (48)				
Pyrrole (25)				

(5)
Benzene (7)

Methylpyrroles (26, 29)
Methanethiol (1)[c]
Methane (1)[c]
Ethene (1)[c]
Propene (1)[c]
Butene (1)[c]
Methylpropene (1)[c]
Methylbutene (1)[c]
Benzene (7)
Toluene (11)
Styrene (21)
Ethylbenzene (14)
m-Xylene, p-xylene (15b)
o-Xylene (17)
Propylbenzene (16)
C_3 alkylbenzene (19)
Pyridine (18)
Methylpyridine (22)
Dimethylpyridine (23)
Ethylene oxide (1)[c]

[a] Acetamide (35), propionamide (38), and acetophenone (32) could not be readily assigned to the classes listed above.
[b] Numbers in parentheses refer to peak number on chromatogram (Ref. 4, p. 568).
[c] Mass spectra were recorded every 2 s during elution of the initial chromatographic peak (numbered 1) which permitted identification of individual compounds in the mixture.

microprocessor instrumentation all contribute to this extraordinary reproducibility. These precise results have inevitably led several investigators to consider applying computer methodology to the characterization of biological samples.

Computer Applications: Cell Characterization or Identification

Menger and coworkers (16) devised a simple program to differentiate pathogenic bacteria. The algorithm used was based solely on retention time and the area of a few key peaks of the profile. Similar attempts have been undertaken by others (17–21).

A scheme for automatic identification or classification of cells from their recorded mass pyrograms has been conceived (22). Briefly, the scheme calls for establishing a reference pattern for each cell type. Unknown samples would be compared systematically to appropriate sets of mass pyrograms. Finally, statistical techniques would be invoked to arrive at a diagnosis. Schafer also describes a time-warping function which permits alignment of mass pyrograms recorded under different operating conditions (22).

Microbial Studies

Heretofore, cell characterization has been accomplished by simple visual comparison of test and reference pyrograms. It is perhaps fortuitous that simple diagnostic information can be obtained from pyrolysis of an extremely complex matrix.

In contrast to nonpyrolytic chromatographic studies recorded in the literature, all of our investigations except those utilizing primary mammalian cells have been completed with coded samples. One notable study involved double blind samples of human tuberculosis and other mycobacterial cells. This was the extent of the information provided to the analysts, who were given no clues with regard to which of the 50 samples were of the same or different strain, how they were grown, how inactivated, etc. The cells were subjected to Py/GC on two different instrumental systems and each sample was analyzed at least twice. Except for some uniform variation in retention time, the two sets of profiles were remarkably similar. By laborious visual comparisons of over 200 pyrograms, we were able to select certain key peaks in the profiles that we considered to be significant. Pyrograms were sorted into five groups of 10 samples each. The criteria for differentiation, the key peaks, proved to be a simple process once a pattern was recognized. For example, lyophilized cells of *Mycobacterium tuberculosis* always yielded a pyrogram having a short peak flanked by two peaks of greater amplitude in a specific retention time band. The relationship always held.

When the decoding took place, it was learned that all 48 cultures had been classified correctly—or identified if reference profiles had been available. (We had indicated to the bacteriologist that two of the cultures had given bizarre profiles; on examination, it was learned that these two came from tubes contaminated by a mold.)

The work on mycobacterial identification has great practical significance. For those suspected of harboring the tuberculosis bacillus, a diagnosis can take from 3 to 6 weeks or more. (The organism is slow growing.) In only 2 weeks' subculture, Py/GC enabled the analyst to make a definitive diagnosis. Moreover, essentially the same patterns were produced when cultured in three different media. This phenomenon appears to be unusual because most workers report that each medium gives it own pyrogram pattern.

The mycobacteria, which include the leprosy bacillus, are unique in their cell wall structure, some 40% or more being of complex lipid. Some of these, the mycolic acids, have chains or rings numbering 88 carbon atoms. Structural studies are still incomplete but excellent Py/GC/MS investigations on mycolic acids have been carried out (23).

Another study of pathogenic bacteria proved to be interesting. This study involved the enteric organisms, Salmonellae, which commonly cause food poisoning. Fifty-four of the most commonly encountered (coded) strains were subjected to Py/GC. They were correctly sorted and classified. Results are summarized in Table I (24). In most studies, Py/GC differences are attributed to changes in relative peak height. In this study, however, some of the strains were characterized by the presence or absence of peaks.

Of special interest was the close correlation between antigenic structures and unique pyrolysis peaks. The pyrolysis peaks also showed a close relationship with the so-called chemotypes. The chemotypes are unusual dideoxy hexoses such as paratose, abequose, and tyvelose sugars that are found in their cell walls (25). These sugars are considered to be prominent markers.

The pyrolysis technique, when applied to biological macromolecules, can bring out small yet significant marker peaks. In a different context, the neurotransmitter, acetylcholine, has been isolated from neural tissue (26). Dipicolinic acid has served as a marker for the bursting forth or release of bacterial spores (8). In a most interesting study dealing with pyrolysis of enzymes, four major peaks were tentatively identified as compounds derived from tyrosine and tryptophan (27).

Mammalian Cells

Several investigations were undertaken on both cultured and primary cells. Py/GC profiles were recorded for mouse and rabbit kidney and red blood cells, and for normal and leukemic white blood cells. In addition, carcinogenic experiments were performed on Chinese and Syrian hamster cells (28).[2] Reliable fingerprinting, i.e., differentiation, was achieved in those studies and lends some encouragement to the proposition that chemical differences between normal and cancer cells do exist.

During this period, another line of investigation was followed. We received a batch of cultured cells derived from human patients afflicted with various inherited biochemical disorders. These cells (fibroblasts) were examined by Py/GC and their profiles showed distinct differences, although the GC columns used at that time were quite inefficient by present day standards (29).

We reported on a study of cultured fibroblasts from patients with the inherited disease, cystic fibrosis (CF) (30). CF is an autosomal genetic disease which afflicts one out of every 2000 live births among Caucasians. The cause of the disease is unknown. If heterozygotes (gene carriers, but the individuals themselves are not afflicted) could be designated, such designation would constitute a major advance in diagnosis.

The pyrograms did indeed show slight but repeatable differences between a heterozygous father and his two CF sons. The pyrograms were compared to that of a normal female subject. All of the CF cultured cells appeared to have the same profile characteristics. Results on a limited number of samples (nine) were encouraging, and future work in this area is contemplated utilizing the combined Py/GC/MS system.

The first report on Py/GC/MS of normal human tissue cells appeared in 1979 (22). In this work, frozen, unfixed samples of liver, brain, spleen, and kidney taken from recent accident victims were converted into analytical samples. The conversion consisted merely of grinding small portions of tissue in distilled water and lyophilizing the resultant suspension. Results of this investigation brought some interesting facts to light. The four tissues could be clearly differentiated and, quite unexpectedly, the same tissue obtained from different individuals gave very similar profiles. Molecular pyrolysis products, 44 in number, were identified by MS. Not surprisingly, many of the identified fragments have been found in microorganisms and even in geological samples.

Single and multiple ion mass chromatographic techniques were also used to exploit the characteristic differences between tissue specimens. Conceivably, in the future, baseline information such as that recorded in this study could be used to discriminate normal and pathological cells.

[2] The caption of Figure 3, Reference 28, should read "Ch. Hamster—Normal Tumor Cell Line."

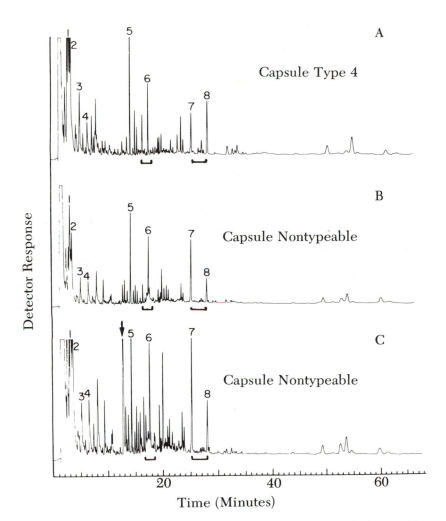

Figure 1. *Py/GC chromatograms of capsular typable and nontypable strains of* K. pneumoniae. *Key: A, C2; B, C6; and C, C1. Key to peak identification: 1, acetonitrile; 2, toluene; 3, ethylbenzene; 4, pyridine; 5, pyrrole; 6, furfuryl alcohol; 7, phenol; and 8, cresol.*

Epidemiological Studies

Py/GC/MS can be used in a practical sense for solving epidemiological problems (*31*). A number of samples of *Klebsiella pneumoniae* isolated from hospital-acquired infections proved to be untypeable by the usual serological methods. However, nine coded duplicate strains of capsular nontypeable *K. pneumoniae* were analyzed and correctly matched by Py/GC/MS. Pyrograms of three of the cultures are shown in Figure 1.

The large peak that appears in Figure 1C at a retention time of 12 min, and precedes the prominent pyrrole peak (4), has been identified as acetic acid. This latter peak serves as an important marker for differentiation of several *Klebsiella* strains.

Assignment of the eight compounds to their probable macromolecular origin can be made by referring to Table I.

Conclusions

Pyrolysis techniques have great potential as a means of obtaining important information on biological macromolecules. This potential extends to both pure research and practical, nonresearch applications. In the future it is conceivable that pyrolysis techniques could form a new system of taxonomy—a chemotaxonomy. Such a scheme could be a valuable adjunct to classical methods of classifying flora, fauna, and geological samples.

Acknowledgments

The authors would like to thank Edgar Ribi of the NIH Rocky Mountain Laboratory who kindly supplied the pure cell wall samples for this study.

Literature Cited

1. Salton, M. R. J. In "The Bacterial Cell Wall"; Elsevier: Amsterdam, 1964; pp. 133–85.
2. Reiner, E., unpublished data.
3. Emswiler, B. S.; Kotula, A. W. *Appl. Environ. Microbiol.* **1978**, *35*, 97.
4. Simmonds, P. G. *Appl. Microbiol.* **1970**, *20*, 567.
5. Merritt, C., Jr.; Robertson, D. H. *J. Gas Chromatogr.* **1967**, *5*, 96.
6. Stack, M. V. *J. Gas Chromatogr.* **1967**, *5*, 22.
7. Reiner, E.; Ewing, W. H. *Nature* **1968**, *217*, 194.
8. Oxborrow, G. S.; Fields, N. D.; Puleo, J. R. In "Analytical Pyrolysis"; Elsevier: Amsterdam, 1977; p. 69.
9. Reiner, E. In "Analytical Pyrolysis"; Elsevier: Amsterdam, 1977; p. 49.
10. Reiner, E. *Nature* **1965**, *206*, 1272.
11. Reiner, E. *J. Gas Chromatogr.* **1967**, *5*, 65.
12. Reiner, E.; Kubica, G. P. *Am. Rev. Respir. Dis.* **1969**, *99*, 42.
13. Reiner, E.; Beam, R. E.; Kubica, G. P. *Am. Rev. Respir. Dis.* **1969**, *99*, 750.
14. Reiner, E.; Bayer, F. L. *J. Chromatogr. Sci.* **1976**, *16*, 623.
15. Irwin, W. J.; Slack, J. A. *The Analyst* **1978**, *103*, 673.
16. Menger, F. M.; Epstein, G. A.; Goldberg, D. A.; Reiner, E. *Anal. Chem.* **1972**, *44*, 423.
17. Carmichael, J. W.; Sekhon, A. S.; Sigler, L. *Can. J. Microbiol.* **1973**, *19*, 403.
18. Macfie, H. J. H.; Gutteridge, C. S.; Norris, J. R. *J. Gen. Microbiol.* **1978**, *104*, 67.
19. Eshuis, W.; Kistemaker, P. C.; Meuzelaar, H. L. C. In "Analytical Pyrolysis"; Elsevier: Amsterdam, 1977; p. 151.
20. Schulten, H. R.; Beckey, H. D.; Meuzelaar, H. L. C.; Boerboom, A. J. H. *Anal. Chem.* **1973**, *45*, 191.
21. Kistemaker, P. G.; Meuzelaar, H. L. C.; Posthumus, M. In "New Ap-

proaches to the Identification of Microorganisms"; Heden, C.; Illeni, T., Eds.; Wiley: New York, 1975; p. 179.
22. Reiner, E.; Abbey, L. E.; Moran, T. F.; Papamichalis, P.; Schafer, R. W. *Biomed. Mass Spectrom.* **1979,** *6,* 491.
23. Etemadi, A. H.; Miquel, A.-M.; Lederer; Barber, M. *Bull. Soc. Chim. Fr.* **1964,** 3274.
24. Reiner, E.; Hicks, J. J.; Ball, M. M.; Martin, W. J. *Anal. Chem.* **1972,** *44,* 1058.
25. Luderitz, O.; Staub, A. M.; Westphal, O. *Bacteriol. Rev.* **1966,** *30,* 192.
26. Schmidt, D. E.; Szilagyi, P. I. A.; Alkon, D. L.; Green, P. *Science* **1969,** *165,* 1370.
27. Danielson, N. D.; Glsjch, J. L.; Rogers, L. B. *J. Chromatogr. Sci.* **1978,** *16,* 455.
28. Reiner, E.; Hicks, J. J. *Chromatographia* **1972,** *5,* 525.
29. Reiner, E.; Hicks, J. J. *Chromatographia* **1972,** *5,* 529.
30. Reiner. E. J.; Moran, T. F. *J. Chromatogr. (Biomed. Appl.)* **1980,** *221,* 371.
31. Abbey, L. E.; Highsmith, A. K.; Moran, T. F.; Reiner, E. J. *J. Clin. Microbiol.* **1981,** *13,* 313.

RECEIVED for review October 14, 1981. ACCEPTED March 2, 1982.

SPECIAL TOPICS

Recent Advances in the Use of Scattering for the Study of Solid Polymers

R. S. STEIN and G. P. HADZIIOANNOU[1]

University of Massachusetts, Polymer Research Institute and Materials
Research Laboratory, Amherst, MA 01003

*Polymers in the solid state may be characterized by
scattering of light, x-rays, or neutrons. These three
techniques may give information about the size, the
shape, and the distribution in the space of the domains
(or structural elements). Light scattering is useful for
studying larger species having dimensions comparable
with the wavelength of the radiation used. Rapid data
acquisition techniques were developed that permit following the time dependence of scattering accompanying
phase transformation or sample deformation.*

General Principles of Scattering

Origin of Scattering. When a beam of radiation impinges on a
sample, a portion of the radiation is scattered. This scattered fraction
depends on both the nature of the radiation and the composition of the
sample.

The scattering of visible light is related to the fluctuation of the
refractive index or polarizability of the sample and its anisotropy. This
polarizability is related to the mobility of the electrons, which is affected by the molecular structure of the sample. The scattering of
x-rays, however, is related only to the fluctuation of the electron density and is not affected by the electron strength of binding to the
nuclei. Neutron scattering power, on the other hand, is a nuclear
property, that varies with the composition of the nucleus and is independent of the chemical nature of the species within which these nuclei
reside. Thus, it may be affected by isotopic substitution, which proves
to be a useful labeling technique as will be discussed.

[1] Current address: IBM Research Laboratories, San Jose, CA 95193.

0065−2393/83/0203−0721$10.00/0

Scattering Theory. AMPLITUDE OF THE SCATTERING POWER. In all cases, the amplitude of the scattering may be described by an equation of the type:

$$E_s(\mathbf{q}) = \sum_j p_j \exp\left[i(\mathbf{q} \cdot \mathbf{r}_j)\right] \tag{1}$$

where p_j is the scattering power of the jth scattering element of the system located at position $\mathbf{r}_j$, $\mathbf{q}$ is the scattering vector of the system defined by

$$\mathbf{q} = \mathbf{q}_1 - \mathbf{q}_0 \tag{2}$$

where $\mathbf{q}_1 = (2\pi/\lambda)\mathbf{s}_1$ and $\mathbf{q}_0 = (2\pi/\lambda)\mathbf{s}_0$ where $\mathbf{s}_0$ and $\mathbf{s}_1$ are unit vectors along the incident and scattered rays, and λ is the wavelength of radiation within the scattering medium (Figure 1). The magnitude of $\mathbf{q}$ is

$$q = |\mathbf{q}| = (4\pi/\lambda) \sin (\theta/2) \tag{3}$$

where θ is the scattering angle between the incident and scattered rays measured within the medium. (The parameter q often is designated elsewhere by Q, h, K, or μ. For x-ray scattering, the angle normally is designated by the Bragg angle, θ_B, equal to half of θ.)

For light (1), x-ray (2), and neutron (3) scattering, p_j is proportional to the polarizability (α_j), the electron density (ρ_j), and to the neutron scattering length (b_j) at position j.

For light scattering, the light interacts with the valence electrons, so the scattering power may be dependent on the molecular orientation. In such cases, the p_j will be a tensor quantity given by:

$$p_j = K_L(\mathbf{M}_j \cdot \mathbf{O}) \tag{4}$$

where K_L is a constant, $\mathbf{M}_j$ is the induced dipole at position j given by

$$\mathbf{M}_j = |\boldsymbol{\alpha}_j|\mathbf{E} \tag{5}$$

where $\boldsymbol{\alpha}_j$ is the polarizability tensor at position j, and $\mathbf{E}$ is the electrical field of the incident light wave within the medium acting on the scattering element. For uniaxially symmetrical scattering elements with principal polarizabilities $\boldsymbol{\alpha}_{1j}$ and $\boldsymbol{\alpha}_{2j}$, $\mathbf{M}_j$ proves to be

$$\mathbf{M}_j = \delta_j (\mathbf{E} \cdot a_j) \boldsymbol{\alpha}_j + \alpha_{2j}\mathbf{E} \tag{6}$$

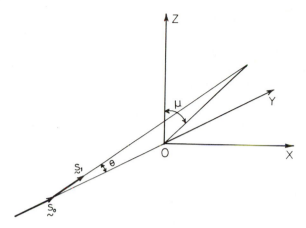

Figure 1. Geometrical configuration of a scattering experiment.

where $\delta_j = \alpha_{1j} - \alpha_{2j}$ and is the anisotropy at position j, and a_j is a unit vector along the principal polarizability axis at position j. The vector **O** is a unit vector along the polarization direction of an analyzer in the scattered light beam. A consequence of anisotropy is that the light scattering from an anisotropic system is dependent on the polarization of the incident and scattered rays.

For neutron scattering, the neutrons interact with the matter through nuclear or magnetic interaction (3, 70–72). The neutron scattering cross-section, which is the ratio between the scattered neutron flux to the incident one, is given by the relation $\sigma = 4\pi b^2$ and contains two components, one coherent (σ_{coh}) and one incoherent (σ_{inc}) cross-section. Only part of the scattered neutrons take part in the interference process, which is described by the coherent scattering cross-section ($\sigma_{coh} = 4\pi b^2_{coh}$). The incoherent cross-section is related to the neutron or nuclear spin. This incoherent cross-section does not exist for zero-spin nucleus, such as ^{12}C or ^{16}O but is very important for hydrogen (101).

SCATTERING INTENSITY IN A DISCRETE APPROACH.   The intensity of scattering, energy per unit area per unit time, is obtained from

$$I_s(\mathbf{q}) = K[E_s(\mathbf{q}) \cdot E_s{}^*(\mathbf{q})] \qquad (7)$$

where the constant K depends on the nature of the radiation, and $E_s{}^*(\mathbf{q})$ is the complex conjugate of $E_s(\mathbf{q})$.

The calculation of the scattering from any system involves the summation of Equation 1 over all scattering elements. This step requires a knowledge of p_j for

$$E_s(\mathbf{q}) = \int p(\mathbf{r}) \exp\left[i(\mathbf{q} \cdot \mathbf{r})\right] d^3\mathbf{r} \tag{8}$$

The integral is three-dimensional over all orientations and lengths of the vector $\mathbf{r}$. For spherically symmetrical systems where $p(\mathbf{r}) = p(r)$, the integral becomes one-dimensional as

$$E_s(q) = 4\pi \int_{r=0}^{\infty} p(r) \frac{\sin(qr)}{qr} r^2 dr \tag{9}$$

and depends only on the variation of $p(r)$ with r. This expression may be evaluated readily for particular model systems, for example, for isolated spheres of radius R_s and scattering power p_s imbedded in a medium of scattering power p_0, leading to (2, 5)

$$E_s(q) = V_s (p_s - p_0) (3/U^3) (\sin U - U \cos U) \tag{10}$$

where $V_s = (4/3)\pi R_s^3$ is the volume of the sphere and $U = qR_s$. This treatment results in a scattering intensity that oscillates with U (and as a consequence with angle θ) but generally decreases with θ at a rate that depends on (R_s/λ). The scattering falls off more rapidly with large (R_s/λ).

[This result implies the Rayleigh–Gans–Debye approximation, that is, that the direction of the incident field $\mathbf{E}$ is unmodified in crossing the boundary of the scattering particle. This approximation is good for x-rays and neutrons and applies for light provided that too great a polarizability difference does not exist between the particle and its surroundings and that R_s/λ is not too large. More exact theories have been advanced by Mie and others (1, 5).]

Equivalent expressions have been found for other shape particles (6–9). In general, this observation of the rate of fall off of scattered intensity with scattering angle provides a measure of the particle size.

For isotropic spherically symmetrical particles, the scattered intensity is circularly symmetrical around the incident beam and is independent of the azimuthal angle μ (Figure 1). This observation also will be true for anisotropically shaped particles such as rods that are randomly oriented. However, for oriented anisotropically shaped particles, the variation of the intensity with θ depends on μ. This relationship provides a means for the measure of the dimensions of particles in different directions.

For polarized light scattering from isotropic particles, the scattered light is polarized in the same direction as the incident light. Thus, intensity is observed for V_v polarization (vertically polarized incident and scattered light where vertical is the direction perpendicular to the plane in which θ is measured), but intensity is zero for H_v polarization (scattered light viewed through a horizontal analyzer).

SCATTERING FROM SPHERULITIC-TYPE MORPHOLOGY.   Crystalline polymers often exhibit spherulitic morphology (*10*). These spherulites are spherically symmetrical aggregates of crystalline and amorphous regions (Figure 2). They are anisotropic and exhibit different polarizabilities in the radial, α_r, and tangential, α_t, directions. Their scattering can be approximated quite well by that from an anisotropic sphere leading to the equations (*4, 11*)

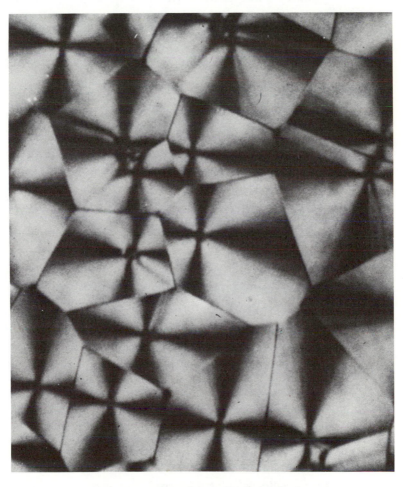

Figure 2. Spherulites of polyethylene.

$$I_{H_v} = KV^2 \left\{ (3/U^3) \, (\alpha_t - \alpha_r) \cos^2 (\theta/2) \sin \mu \cos \mu \right.$$
$$\left. \cdot \, (4 \sin U - U \cos U - 3 \, SiU) \right\}^2 \tag{11}$$

$$I_{V_v} = KV^2 \left\{ (3/U^3) \, (\alpha_t - \alpha_s) \, (2 \sin U - U \cos U - SiU) \right.$$
$$+ (\alpha_r - \alpha_s) \, (SiU - \sin U) - (\alpha_t - \alpha_r) \cos^2 (\theta/2) \tag{12}$$
$$\left. \cdot \, \cos^2\mu \, (4 \sin U - U \cos U - 3 \, SiU) \right\}^2$$

where V is the volume of the spherulite, $U = qR$, R is the radius of the spherulite, $SiU = \int_0^U (\sin x/x)dx$, and μ is the azimuthal scattering angle. In this case, the polarized scattering patterns depend on μ being fourfold symmetric for H_v (Figure 3) and twofold for V_v. The H_v intensity is finite and depends on the anisotropy $(\alpha_r - \alpha_t)$ of the spherulite. It exhibits a maximum with θ at a value of $U = 4.08$ such that

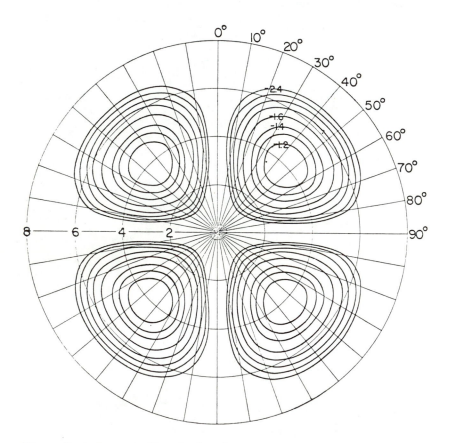

Figure 3. A theoretically calculated H$_v$ *scattering pattern from an isolated ideal spherulite (100).*

$$4.08 = (4\pi R_s/\lambda) \sin (\theta_m/2) \tag{13}$$

This equation provides a means for the determination of spherulite size, as discussed later. The intensity of scattering varies with $(\alpha_r - \alpha_t)^2$, which depends on the degree of crystallinity of the spherulite and may be used for following its change with temperature or time.

Two of the terms in the V_v expression (which are independent of μ) depend on a_0, the polarizability of the surrounding of the spherulite. These surroundings may be amorphous phase with a polarizability α_a or other spherulites with an average polarizability $(\alpha_r + 2\alpha_t)/3$. Results indicated (*12*) that

$$\alpha_s = \phi_s(a_r + 2\alpha_t)/3 + (1 - \phi_s) \alpha_a \tag{14}$$

where ϕ_s is the volume fraction of spherulites. Thus, the shape of the V_v scattering pattern will change with ϕ_s as the spherulites grow, as was observed (*13*).

When spherulitic samples are deformed, the spherulites change from spherical to ellipsoidal objects. This change may be modeled to yield deformed scattering patterns as shown in Figure 4 (*14*). Their study during deformation permits the analysis of deformation mechanisms.

SCATTERING INTENSITY IN A DEBYE–BUECHE CONTINUUM APPROACH FOR ISOTROPIC SYSTEMS.   Samples with randomly arranged structures can best be treated in terms of a statistical description involving the fluctuation of their scattering power from its average value $\bar{p}$

$$n_j = p_j - \bar{p} \tag{15}$$

Debye and Bueche (*15*) showed that for isotropic systems

$$I_s(q) = K\overline{\eta^2} \int \gamma(\mathbf{r}) \exp \left[i(\mathbf{q} \cdot \mathbf{r})\right]d^3\mathbf{r} \tag{16}$$

where $\gamma(\mathbf{r})$ is a spacial correlation function defined by

$$\gamma(\mathbf{r}) = \frac{\langle \eta_j \eta_\ell \rangle_{\mathbf{r}}}{\overline{\eta^2}} \tag{17}$$

where the symbol $\langle \ \rangle_r$ designates an average over all pairs of scattering elements separated by r. The quantity r varies from unity at $r = 0$ to zero at $r = \infty$ in a manner dependent on the structure of the system. For spherically symmetrical systems, Equation 16 reduces to

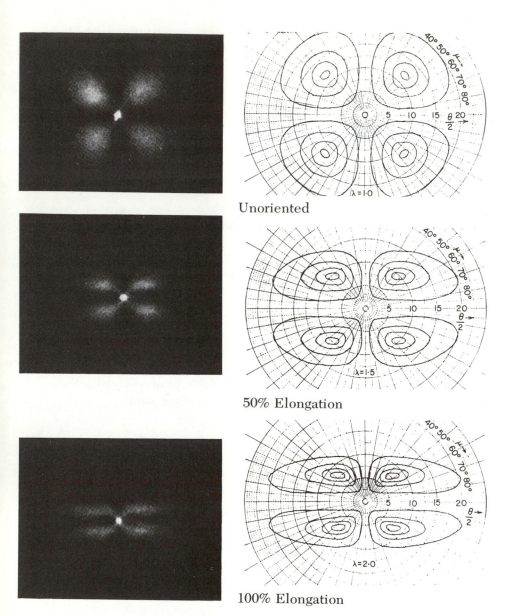

Unoriented

50% Elongation

100% Elongation

Figure 4. Theoretically predicted and experimentally determined changes in the H_v small angle light scattering pattern of isotactic polypropylene film with elongation. The film stretch direction and the polarization of the incident beam are vertical while the scattered beam is viewed with a horizontal analyzer (14).

$$I_s(q) = 4\pi \, K\overline{\eta^2} \int_{r=0}^{\infty} \gamma(r) \frac{\sin(qr)}{qr} r^2 dr \tag{18}$$

In this case, $\gamma(r)$ may be obtained by Fourier inversion to give

$$\gamma(r) = \frac{K'}{\overline{\eta^2}} \int_{q=0}^{\infty} I_s(q) \frac{\sin(qr)}{qr} q^2 dq \tag{19}$$

For randomly dispersed two-phase systems, $\gamma(r)$ usually can be approximated by an exponential function (*15, 16*)

$$\gamma(r) = \exp(-r/a_c) \tag{20}$$

where a_c is a correlation distance characterizing the scale of the heterogeneity.

Kratky and Porod (*17*) showed that for a two-phase system containing volume fractions ϕ_1 and ϕ_2 of the two phases, the average chord lengths through the phases are given by

$$\overline{\ell_1} = a_c/\phi_2 \text{ and } \overline{\ell_2} = a_c/\phi_1 \tag{21}$$

The substitution of Equation 20 in Equation 18 leads to the result (*16*)

$$I_s(q) = K''\overline{\eta^2} \frac{a_c^3}{(1 + q^2 a_c^2)^2} \tag{22}$$

This equation predicts an intensity that increases with a_c^3 at small a_c and decreases as $1/a_c$ at large a_c. It passes through a maximum at $a_c = \sqrt{3}/q$. At large q, this expression predicts that the intensity varies as $1/q^4$. This relationship is known as Porod's law (*18, 19*) and is characteristic of three-dimensional systems with sharp boundaries. Deviations from this dependence may be used to characterize larger boundary thicknesses for systems with diffuse boundaries (*20—22*). Equation 22 indicates that a plot of $I^{-1/2}$ vs. q^2 should be linear with a slope/intercept equal to a_c^2. Such a Debye–Bueche plot (*16*) may be used to test the validity of the exponential correlation function and serves as a convenient means for determining a_c.

Scattering Invariant. Because when $r = 0$ then $\gamma(r) = 1$ and $\sin(qr)/qr = 1$, Equation 19 may be rearranged to give

$$\overline{\eta^2} = K' \int_0^{\infty} I_s(q) \, q^2 dq \tag{23}$$

This integral is known as the total integrated intensity or the scattering invariant (2, 17–19) and serves as a means for experimentally evaluating $\overline{\eta^2}$. For a two-phase system

$$\overline{\eta^2} = \phi_1\phi_2 \, (p_1 - p_2)^2 \tag{24}$$

while for a multiphase system

$$\overline{\eta^2} = \sum_j \sum_\ell \phi_j\phi_\ell \, (p_j - p_\ell)^2 \tag{25}$$

Thus, for the two-phase system, if ϕ_1 and ϕ_2 are known, an experimental determination of $\overline{\eta^2}$ serves to characterize $(p_1 - p_2)$, which depends on the composition difference between the phases.

SCATTERING INTENSITY IN A DEBYE–BUECHE CONTINUUM APPROACH FOR RANDOMLY ANISOTROPIC SYSTEMS. The Debye–Bueche formulation suffices for isotropic systems. For randomly anisotropic systems, it has been generalized to give approximately for spherically symmetrical systems (23)

$$I_{H_v}(q) = \frac{4\pi K\overline{\delta^2}}{15} \int_{r=0}^{\infty} f(r)\frac{\sin{(qr)}}{qr} r^2 dr \tag{26}$$

and

$$I_{V_v}(q) = 4\pi K \left\{ \overline{\eta^2} \int_{r=0}^{\infty} \gamma(r)\frac{\sin{(qr)}}{qr} r^2 dr + 4/45 \, \overline{\delta^2} \int_{r=0}^{\infty} f(r)\frac{\sin{(qr)}}{qr} r^2 dr \right\} \tag{27}$$

(A randomly anisotropic system is one where the correlation of orientation of optic axes depends only on their separation, r, and not on the angle that the optic axis vector a_j makes with r. The domains of correlation possess spherical symmetry.)

These equations reduce to the Debye–Bueche case when the mean-squared anisotropy of the anisotropic scattering element $\overline{\delta^2}$ is equal to zero, in which case $I_{H_v}(q) = 0$ and $I_{V_v}(q) = I_s(q)$. The correlation function, $f(r)$, is an optic axis orientation correlation function defined as

$$f(r) = [3\langle\cos^2\theta_{j\ell}\rangle_r - 1]/2 \tag{28}$$

where $\theta_{j\ell}$ is the angle between the optic axis of the jth and ℓ^{th} scattering elements. This function, like $\gamma(r)$, varies from unity at $r = 0$ (perfect correlation when $\theta_{j\ell} = 0°$) toward zero as angular orientation becomes uncorrelated ($\langle\cos^2\theta_{j\ell}\rangle = 1/3$) as r becomes large.

Thus, $I_{H_v}(q)$ depends on orientation correlations and its Fourier transform yields $f(r)$ characterizing the size of the region of correlated orientation. The term $I_{V_v}(q)$ depends on both orientation and average polarizability (density) correlations. Thus, $\gamma(r)$ may be obtained from a Fourier transform of $I_{V_v}(q) - (4/3)I_{H_v}(q)$.

Density correlations predominate for phase-separated polymer blends or for crystalline polymers at early stages of crystallization when crystalline structures are not volume filling, whereas orientation correlations predominate with highly crystalline systems and with liquid crystals.

This theory applies to random orientation correlations and leads to circularly symmetrical scattering patterns (independent of μ). It was generalized (24, 25) to cover more complex correlations in which the orientation correlation function must be expanded in spherical harmonics and involves angular variables as well as r.

SCATTERING INTENSITY AT LOW SCATTERING ANGLES, GUINIER REGION.   For scattering at low angles (low q), the $\sin(qr)$ term in Equation 18 may be expanded in a power series leading to

$$I_s(q) = K\overline{\eta^2}\int_0^\infty \gamma(r)\,[1 - (qr)^2/3! + (qr)^4/5!\ldots]\,r^2dr$$

$$(29)$$

$$= K\overline{\eta^2}\int_0^\infty \gamma(r)r^2dr\left\{1 - \frac{q^2}{3!}\frac{\int_0^\infty\gamma(r)r^4dr}{\int_0^\infty\gamma(r)r^2dr} + \ldots\right\}$$

When we have low concentrations of scattering particles, $\gamma(r) = Cp_{22}(r)$ where $p_{22}(r)$ is the probability that two volume elements separated by distance r both reside within the particle. In this case, the ratio of the integrals in Equation 29 equals twice the average square radius of gyration ($\overline{R_g^2}$) of the particle so that

$$P(q) = I_s(q)/I_s(0) = \left[1 - \frac{\overline{Rg^2}}{3}q^2 + \ldots\right]\quad(30)$$

where $P(q)$ is known as the single particle interference function. Thus, at low q, a plot of $P(q)$ vs. q^2 has an initial slope of $\overline{Rg^2}/3$ leading to a means for determination of $\overline{Rg^2}$ independent of the nature of the particle (26).

An alternate approach is to notice that the first two terms of Equation 30 are identical with what is obtained by expanding the exponential of

$$P(q) = \exp\left[-\overline{Rg^2}q^2/3\right] \tag{31}$$

so that $Rg^2/3$ may be obtained from the initial slope of a plot of $\ln P(q)$ vs. q^2, known as a Guinier plot (2). Such plots do not require a knowledge of the absolute intensities. However, the absolute intensity is relevant to the determination of the molecular weight of the scattering particle, as discussed below.

SCATTERING INTENSITY BY A MOLECULAR APPROACH AT LOW AND WIDE SCATTERING ANGLES.   For particular models of particles, $\gamma(r)$ may be determined. For example, for spheres of radius R_s, it proves to be

$$\gamma(r) = 1 - \frac{3}{4}\left(\frac{r}{R_s}\right) + \frac{1}{16}\left(\frac{r}{R_s}\right)^3 \tag{32}$$

which on substitution in Equation 18 leads to the following scattering expression

$$I_s(q) = I_s(0)KV_s^2\,\overline{\eta^2}\left[\frac{3}{U^3}(\sin U - U\cos U)\right]^2 \tag{33}$$

where V_s is the sphere volume and $U = qR_s$.

For polymer coils, $\gamma(r)$ can be evaluated from the Gaussian statistics of polymer chains and leads, after integration in Equation 17, to the Debye (27) result

$$P(q) = \frac{2}{v^2}\left[v - 1 + \exp(-v)\right] \tag{34}$$

where $v = q^2Rg^2$. At high v (high q or θ), the equation approaches

$$P(q) \rightarrow \frac{2}{v} = \frac{2}{q^2Rg^2} \tag{35}$$

Thus, a plot of $q^2P(q)$ vs. q should approach an asymptotic value given by Equation 35 at large q as shown in Figure 5. Neutron scattering results verify this prediction but show deviations at very large q characteristic of the failure of Gaussian statistics at small separation of scattering elements.

MOLECULAR DIMENSIONS AND THERMODYNAMIC PARAMETERS DETERMINED FROM THE SCATTERING INTENSITY OBTAINED BY POLYMERIC SOLUTIONS SYSTEMS.   If condensed systems were perfectly uniform, their scattering should vanish because of destructive interference by all pairs of scattering elements. However, real systems scatter because of fluctuations in the scattering power of volume elements arising

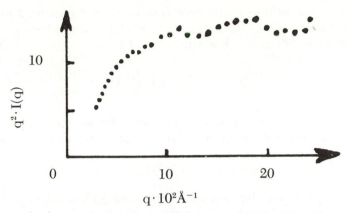

Figure 5. A Debye plot of $I(q)q^2$ *vs.* q *for a polymer coil.*

from fluctuations in density, orientation, and concentration (for multicomponent systems). For a dilute solution of a scatterer in a solvent, the density fluctuation contribution may be approximately subtracted out by subtracting the scattered intensity of the solvent from that of the solution. The residual scattering primarily arises from concentration fluctuations that may be related to the opposing force of osmotic pressure (and hence molecular weight) through the Einstein fluctuation theory (28). For ideal solutions, the result is that the scattered intensity may be expressed in terms of the Rayleigh factor defined as

$$R(q) = I_s(q)d^2/I_0V \qquad (36)$$

where d is the sample to detector distance, I_0 is the incident intensity, and V is the scattering volume. (The Rayleigh ratio conventionally used in light scattering is equivalent to the differential scattering cross section used by neutron scattering.) The Rayleigh factor is proportional to the molecular weight, M, and the concentration, c, of the solute for an ideal dilute solution (29–33)

$$R(q) = K_iMcP(q) \qquad (37)$$

The single particle interference factor, $P(q)$, accounts for intraparticle interference and approaches one as q approaches zero. The constant, K_i, depends on the nature of the radiation, thus for light (31)

$$K_i = K_L = \frac{2\pi^2 2n^2(\delta n/\delta c)^2}{\mathcal{N} \; \lambda_0^4} \qquad (38)$$

where n is the solution refractive index, $\mathcal{N}$ is Avogadro's number, λ_0 is the wavelength of the radiation in vacuo, and c is the concentration of the solute (g/cm³). For x-rays (2, 34)

$$K_i = K_x = \mathcal{N} i_e \left(\frac{\delta\rho}{\delta c} \right)^2 \tag{39}$$

where i_e is the Thomson scattering factor for an electron equal to $(mc^2/e^2)^2$, and ρ is the electron density. Finally, for neutrons (35, 36)

$$K_i = K_N = (\mathcal{N}/m^2)\,(a_1 - a_2)^2 \tag{40}$$

where a_1 and a_2 are the density of the coherent neutron scattering lengths of the two components, and m is the molecular weight of the repeat unit. Generally, K_i represents the contrast and depends on the difference in scattering power of the components.

For a dilute nonideal solution, Equation 37 is modified to give

$$R(q) = \frac{K_i c}{\dfrac{1}{MP(q)} + 2A_2 c} \tag{41}$$

where A_2 is the second virial coefficient characterizing solute–solvent interaction. (Note that the solvent may be a large molecule, e.g., a polymer.)

This equation may be rearranged to give (1, 26, 31)

$$\frac{K_i \cdot c}{R(q)} = \frac{1}{MP(q)} + 2A_2 c \tag{42}$$

So, as proposed by Zimm (26, 31), an extrapolation of $K_i c/R(q)$ to zero q and zero c should lead to a value of $1/M$. The variation of $K_i c/R(q)$ with q at zero c leads to a characterization of $1/P(q)$ and thus a measure of Rg^2, and its variation with c at zero q provides a value of A_2. A geometric procedure, the Zimm plot (26), was proposed for accomplishing this double extrapolation. Although the procedure originated for the determination of molecular weight and size for polymer molecules in a low molecular weight solvent using light scattering, it was extended to x-ray and neutron scattering (37). The latter two techniques, although experimentally more complex, have the advantage of being able to measure a smaller value of the radius of gyration (Rg) because of the shorter wavelength of the radiation. The solvent can be a high rather than low molecular weight compound, permitting studies of polymer blends (36, 38, 39). Neutron scattering has the additional advantage of permitting the enhancement of K_N through isotopic substitution. This substitution is principally done by replacing hydrogen in one of the species by deuterium, which has a considerably different scattering length. In fact, the molecular dimensions of a polymer in a bulk state can be studied by dissolving a deuterated polymer in a hydrogenous matrix of the identical polymer (or vice versa) (*40*). Although earlier studies of this sort utilized dilute solutions of deuterated polymer in a hydrogenous matrix, the technique has been extended to concentrated solutions in which it has been shown that to the ideal dilute solution approximation, Equation 37 may be replaced by (*41–44*)

$$R(q) = K_N Mc(1 - c) P(q) \qquad (43)$$

This method has the advantage that considerably greater scattered intensities occur at higher concentrations, resulting in greater precision of measurement or shorter data acquisition times.

Studies may be extended to oriented systems in which case the scattering becomes dependent on the azimuthal angle μ. The directional dependence of $\overline{Rg^2}$ then may be dependent from the studies of the variation of $P(q)$ with q at various μ (*45–48*).

Experimental

All scattering experiments have certain features in common. A radiation source provides a parallel monochromatic beam of radiation that passes through the sample. A detector measures the scattered intensity as a function of the angle θ and μ. Features unique to the various radiation types are described next.

Light Scattering. Although early light scattering experiments were performed using conventional light sources, these methods have been displaced almost entirely by lasers, which provide extremely intense parallel and monochromatic sources. Light from lasers may be polarized or not. In the former case a polarization rotator usually is desired, whereas in the latter a polarizer (usually a Polaroid film type) is used. The sample is best in the form of a film, normal to the light beam. To avoid contributions due to surface scattering, the film may be sandwiched between flat glass plates using a refractive index matching immersion fluid. The thickness of the film was such that it provided sufficient scattering for accurate measurement but not so much as to lead to appreciable multiple scattering. A film that will transmit 90–95% of incident intensity is usually desirable. Such films may be of the order of 100 μm thick for typical crystalline polymer films but need be thinner for highly scattering liquid crystal materials or thicker for amorphous homopolymers.

For quantitative intensity measurements, multiple scattering corrections should be made. For small amounts of multiple scattering, a correction for the reduction intensity due to scattering out of the incident and scattered beams suffices according to Reference 49.

$$I_{\text{meas}} = I_{\text{corr}} \exp (-\tau t) \tag{44}$$

where τ is the turbidity of the sample obtained from

$$\tau = (1/t) \ln (I_0/I_{\text{trans}}) \tag{45}$$

where t is the sample thickness, and I_0 and I_{trans} are the incident and trans-mitted intensities measured at normal incidence, respectively. For higher amounts of multiple scattering, a more complex correction including the effect of rescattering into the observed beam is required (50). For polarized light experiments, the effect of multiple scattering on the state of polarization must be included (51). For colored samples, a correction for intensity loss due to absorption analogous to Equation 44 should be employed.

Because the theoretical expressions require values of the scattering angle and wavelength within the sample, corrections for those dependent on the refractive index of the sample are required (49). Also, a correction for the reflection at interfaces should be employed (49).

For oriented samples that are birefringent, retardation may affect the phase and state of polarization of the incident and scattered beams and cor-rections need to be made (52).

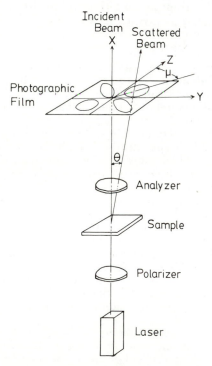

Figure 6. The photographic light scattering apparatus (82).

The simplest means of detection of the scattered radiation is photographic, as shown in Figure 6. Polaroid instant film is convenient. Sample-to-film distance may be adjusted to accommodate a range of scattering angles. Photographs permit the rapid assessment of the shape of patterns and the angles of scattering maxima. The angular scale may be calibrated readily by using a diffraction grating in place of the sample. Although intensity measurements may be made by densitometry of photographs, measurement of the intensities directly by photometry is more precise. Mechanical angular scanning of the intensity has been carried out but, at least for small angle scattering, an electronic scanning technique proves more convenient. This technique may be accomplished using a one- or two-dimensional multichannel analyzer (OMA) (53, 54) as shown in Figure 7. The scattering pattern is scanned by a vidicon camera that is connected to a small computer. Scattered intensities are digitalized, stored in computer memory or on floppy disks, and may be computer processed as desired. For example, an intensity contour plot as illustrated in Figure 8 may be obtained. With present techniques, such output can be achieved in time scales of the order of a second.

For absolute light scattering, intensity measurements calibration may be accomplished by converting data into Rayleigh factors by directly measuring I_s/I_0 and taking into account geometric factors. Because I_0 is usually very much greater than I_s, it is usually necessary to reduce it using calibrated neutral filters. Alternatively, calibration may be accomplished by referring I_s to the scattering of a reference sample of known Rayleigh factor. This sample could be a pure liquid or a polymer of known molecular weight in solution. However, such standards scatter much more weakly than the usual samples so that the accuracy of establishing the intensity ratio is not great. A working standard, which may be a calibrated polymer film having stable properties, may be employed conveniently. The scattering by the samples can be referred to the scattering of the working standard.

x-Ray Scattering.   x-Ray sources are usually conventional sealed anode or, for higher intensities, rotating anode tubes (55, 56). For time-dependent experiments, high intensity synchrotron sources have been employed (57, 58). For tube sources, rough monochromatization can be accomplished by filtration and pulse-height analysis, but use of a diffracting crystal monochromator is better. For the continuous radiation from a synchrotron, the latter method is necessary.

Collimation systems are of two types, slit and pinhole. The former has the advantage that intensities are much higher, but it requires slit-desmearing procedures to correct for the distortion of $I(q)$ by the slit geometry. A commonly used camera with slit collimation is that due to Kratky (34). For very high resolution, a multiple diffracting crystal collimator of Bonse–Hart (59) type has advantages.

Samples for x-rays are generally thicker than those for light scattering and often in the range of a few millimeters. Because of the usual small scattering cross section, multiple scattering is usually not a problem. Also, refraction and reflection corrections are usually not significant, although the change in the effective scattering volume of the sample with scattering angle should be considered (55, 56, 60).

Photographic film often is used for detection. A commonly used pinhole collimation camera employing film is that due to Statton, manufactured by Warhus. When necessary, intensities can be obtained by densitometry. Point-by-point counting employing proportional or scintillation counters can be done with the Kratky or Rigaku–Denki apparatus, but because of the

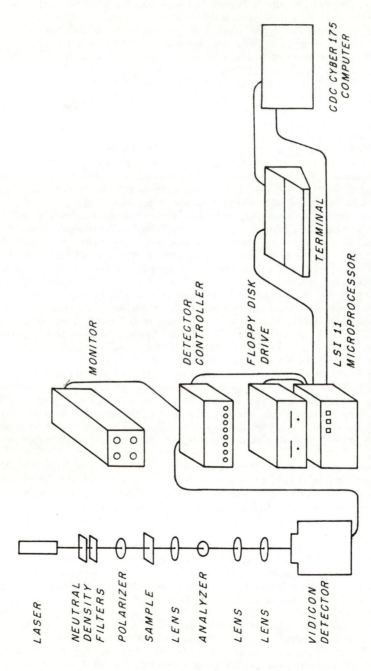

Figure 7. The optical multichannel analyzer light scattering apparatus (54).

Figure 8. An experimental H_v *light scattering contour diagram obtained with the OMA for a PET sample (54).*

relatively long data-gathering time at each point, the procedure is tedious and computer-controlled data collection is advantageous. Recently, position-sensitive detectors have become available *(61, 62)*. In their one-dimensional form, these detectors employ a counting wire several centimeters long. The position on the wire at which a photon arrives is encoded from the difference in time taken by the pulse to arrive at the two ends of the wire. In this way, data are acquired (through a multichannel analyzer) over many angles at once. This procedure reduces data gathering time and eliminates errors arising from incident beam intensity fluctuations. A two-dimensional position-sensitive detector employing a two-dimensional detector wire grid has become available at the National Center for Small Angle Scattering Research at the Oak Ridge National Laboratories, which allows intensity measurements with pinhole collimation with oriented samples exhibiting μ-dependent scattering *(63)*.

Photon counters are often not fast enough to accommodate high intensity synchrotron sources. For these, ionization chambers or fluorescent screens with image intensifiers and OMAs *(64)* have been used.

Reduction of air scattering is desirable, particularly in long path length cameras. This task is done by evacuation of the beam path or by filling with helium gas. Windows are made of x-ray transparent material such as beryllium.

Calibration most often is accomplished by referring intensities to that of a standard polyethylene sample that was calibrated and furnished by Kratky *(65)*.

Corrections for incoherent (Compton) scattering are more important for x-rays than for light (where the correction usually is neglected). This correction is done either by assuming an angularly independent background or else by fitting its angular variation to an empirical variation fitted to the scattering at higher angles above which coherent small angle scattering contributes but below which wide angle diffraction is encountered.

Corrections should be made for angular-dependent polarization, if data are to be interpreted in terms of a one-dimensional theory, a Lorentz correction should be employed. Data processing computer programs have been offered by Vonk (66) and Glatter (67–69).

Neutron Scattering. Neutron sources are commonly steady state nuclear reactors giving a black body distribution of neutron velocities (70–72). A cold source consisting of a liquid hydrogen or deuterium cooled finger within the reactor is often advantageous so as to provide a lower black body temperature and consequently a greater flux of longer wavelength neutrons. Such a cold source is employed in the ILL apparatus (73) at Grenoble but is not available at Oak Ridge. Because of the low energy of the neutrons, radiation damage to samples is negligible, in contrast to x-rays.

Monochromatization is accomplished either by a velocity selector (Grenoble) employing a rotating drum with helicoidal channels, or else by diffraction from a crystal such as graphite (Oak Ridge) (74). Collimation is achieved through pinholes and beam guides.

Samples for neutron scattering are often thicker than those used for x-ray scattering and of the order of several millimeters to a centimeter. Although surface quality is not important, it is important for samples to be free of microvoids. Their presence often can be detected from visual turbidity or from their small angle x-ray scattering. As with x-rays, it is desirable to have beam paths evacuated. Windows can be made of pure aluminum or from silica.

The scattered radiation usually is detected with a two-dimensional position-sensitive detector. Because of the necessity for converting the neutron beam into ionizing radiation by the detector gas, the spatial resolution of the detector is low, so that large sample-to-detector distances (5–30 m) are necessary to achieve good angular resolution.

Counting times are usually relatively long, leading to the desirability of computer-assisted acquisition and automatic sample changer.

Incoherent neutron scattering intensities are relatively high, often comparable with or greater than those for coherent scattering. Thus, subtraction of incoherent neutron scattering intensity is essential. It usually is evaluated from observation of high q scattering, assuming q independence of the incoherent component.

Intensity calibration of neutron scattering usually is accomplished three different ways (75). First, by measuring the intensity of the direct beam using calibrated attenuators, gold or cadmium. Second, by measuring the incoherent scattering from water or vanadium. This scattering is monotonic but for water it must be corrected for the high inelastic scattering contribution that is wavelength dependent. When vanadium is used, correction for multiple scattering due to the presence of strong absorption should be made. The most accurate method is the third, to use a standard polymer sample with known molecular weight measured by the conventional light scattering technique. For our experiments, we use a polystyrene of $M_w = 80,000$ and $M_w/M_n = 1.05$ (where M_w and M_n represent the weight-average and the number-average molecular weight, respectively) (90, 91).

Some Recent Applications of Scattering Measurements on Solid Polymers

We shall not attempt to review exhaustively applications of scattering methods, but shall select some specific examples illustrating particular features of the scattering technique. Because of familiarity, undue weight shall be placed on our own work.

With Light Scattering. Much of the work on light scattering from solid polymers has focused on studies of crystalline spherulitic polymers. A typical variation of the photographic H_v scattering pattern occurring during the crystallization of polyethylene terephthalate (PET) is shown in Figure 9 (76). The decrease in the angle of the scattering maximum with time reflects the growth in the spherulite radius, according to Equation 13. Photometric measurements indicate that the intensity of the scattering maxima increases with time, associated with the increase in number and anisotropy of the spherulites.

The change in intensity obtained with an OMA scan of the $\mu = 45°$ line of the H_v pattern during the melting of the PET is shown in Figure 10. Because the number and size (estimated from the position of the scattering maximum) remain constant with melting, the decrease in intensity is associated with the decrease in spherulite anisotropy ($\alpha_t - \alpha_r$) in Equation 11. This anisotropy primarily arises from the contribution of oriented crystals within the spherulites. This contribution leads to the expectation that as a first approximation, the intensity should drop with ϕ_c^2, where ϕ_c is the volume fraction crystallinity.

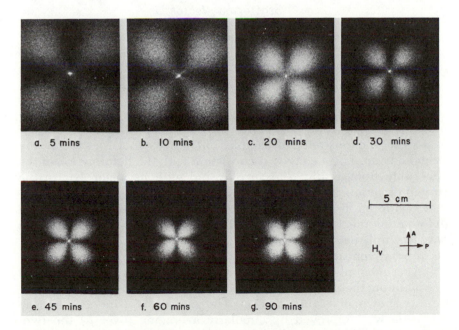

a. 5 mins   b. 10 mins   c. 20 mins   d. 30 mins

e. 45 mins   f. 60 mins   g. 90 mins

5 cm

H_v A / P

Figure 9. The variation in H_v light scattering patterns during the course of crystallization of PET (76).

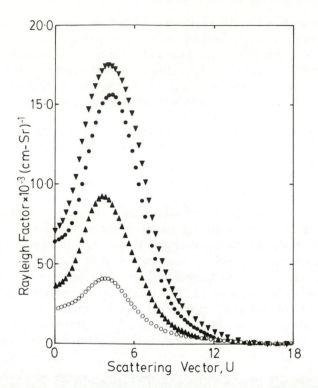

Figure 10. The variation of H_v *light scattering intensity with* θ *at* μ =
45 °C during the melting of PET (99). T (°C): $\bigcirc$, *40.0;* $\blacktriangle$, *200.0;* $\bullet$, *240.0;*
and $\blacktriangledown$, *245.0.*

A comparison of a quadrant of the H_v scattering pattern from unstretched and a 60% stretched polyethylene sample is shown in Figure 11 (54). In agreement with Equation 11, the maximum occurs at μ
= 45° for the unstretched sample but moves to larger μ for the
stretched one. This behavior is consistent with the deformation of the
spherulite from a sphere to an ellipsoid with its major axis in the
stretch direction. The detailed analysis of this pattern and its consistency with the theory provide a means for analysis of the spherulitic
deformation mechanism (77, 78). The analysis of rapid deformation
was accomplished previously using cinematography (79). This analysis is being repeated quantitatively using the OMA.

A recent application of light scattering from amorphous systems is
in examining the morphology of polymer blends during phase separation as shown in Figure 12 (54). The pattern is μ-independent so
precision was improved by circularly averaging (over μ) the two-
dimensional OMA data. The intensity maximum is associated with a
persistent spacing of the phase-separated domain, and it is found to

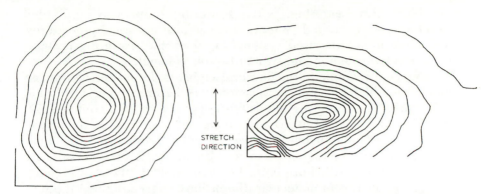

Figure 11. A quadrant of an H_v scattering pattern for unstretched high density polyethylene film (left) and stretched high density polyethylene film (right) (54).

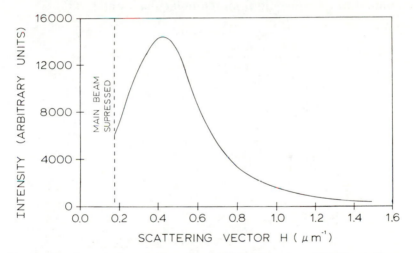

Figure 12. A typical variation of scattered intensity with θ for a phase-separated sample of a blend of 40% poly(o-chlorostyrene) (POCS) with 60% polystyrene (PS) (54).

shift to smaller angles and to grow in intensity with time as the domains grow and change in composition.

Phase-separated blends were studied by Yuen and Kinsinger (80) using the Debye-Bueche correlation function technique.

Anisotropic shock-cooled polypropylene samples were studied by Keijzers et al. (81). From measurements of H_v and V_v scattering, orientation and density correlation functions were obtained.

With x-Ray Scattering. x-Ray scattering studies may be divided into those of disordered systems, analyzed by statistical procedures, and those from structured systems in which the scattering is interpreted in terms of a model structure such as lamellar.

A recent example in the former category is the scattering from an amorphous blend of polycaprolactone (PCL) in polyvinyl chloride (PVC) (82). The scattering intensity increases with PCL content (Figure 13) (38). Debye–Bueche plots were obtained leading to correlation distances found using Equation 22 and were resolved into chord lengths through the two phases using Equation 21. Those results are plotted vs. composition in Figure 14. In dilute PCL, the chord length, $\bar{\ell}_{PCL}$, of the PCL is of molecular dimensions, suggesting that the scattering units are individual molecules. An alternative approach is to perform a Guinier plot using Equation 31, as shown in Figure 15. From the initial slope at low PCL concentration, a Rg of 10.0 nm was obtained (38, 39), which compares with a value of 6.7 nm estimated for this unperturbed molecule from its molecular weight (83).

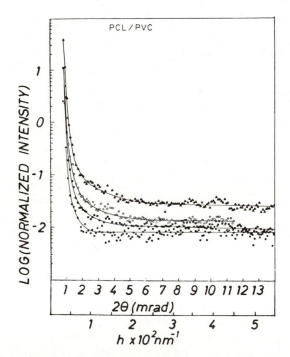

Figure 13. The variation of normalized small angle x-ray scattered intensity with θ for a blend of PCL with PVC for various PVC concentrations: $\bullet$, 0; $\blacksquare$, 1.05×10^{-2}; $\bigcirc$, 1.46×10^{-2}; *and* $\blacktriangle$, 6.16×10^{-2} g PCL/cm^3 PVC (38).

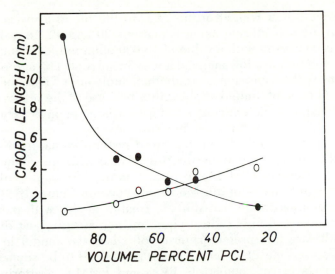

Figure 14. *Chord lengths of PCL (●) and PVC (○) phases as a function of concentration in their blend (39).*

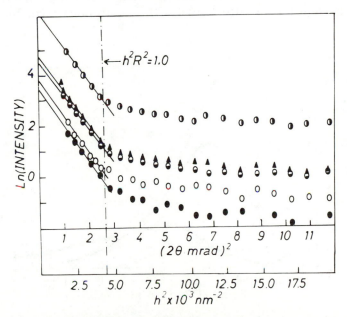

Figure 15. *A Guinier plot for the small angle x-ray scattering from the PCL/PVC blend at concentrations: ●, 1.05×10^{-2}; ▲, 1.46×10^{-2}; ○, 1.70×10^{-2}; ◑, 1.74×10^{-2}; and ◓, 6.16×10^{-2} g PCL/cm³ PVC (38).*

The molecular weight of the PCL in the blend also was determined by x-ray scattering using Equations 39 and 42. Its value was 50,000, as compared with a value of 43,000 obtained for this polymer in dilute solution in a low molecular weight solvent. The agreement is consistent with dispersion of individual molecules. This technique has proved to be of limited applicability because of the relatively few systems having sufficient contrast to give a large enough value of K_x (Equation 39) to permit precise measurement.

An example of a study on structured systems is that from a blend of crystallized polyvinylidene fluoride (PVF_2) with amorphous polymethyl methacrylate (PMMA). Typical Lorentz-corrected small angle x-ray scattering curves for this blend are shown in Figure 16 (84). The presence of a maximum that shifts to smaller angles with increasing PMMA concentration is indicative of an increasing repeat distance. The results are interpreted in terms of a lamellar model in which stacks of crystalline PVF_2 lamellae are postulated to be separated by amorphous layers of compatible PVF_2 and PMMA. Scattering from such systems can be interpreted using a one-dimensional correlation function as described by Vonk (85–87), or by a Hosemann (88) one-dimensional paracrystalline model involving as parameters the number of crystals in a stack, the thickness of the amorphous and crystalline layers, and the statistics of their distribution. Either approach leads to the conclusion that the thickness of the amorphous layer increases

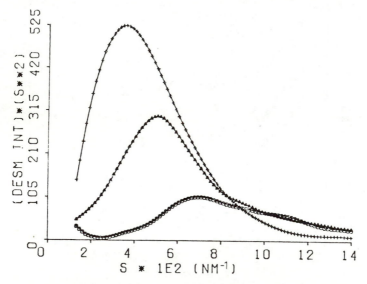

Figure 16. The variation in small angle x-ray scattered intensity with s [2 sin (θ/2)/λ] for PVF_2 and two of its blends with PMMA. Volume % PVF_2: ○ , 100.0; △, 85.8; and +, 50.1 (84).

with PMMA content, as shown in Figure 17 (*84*). The result is in quantitative agreement with the PMMA residing in these interlamellar layers.

Figure 16 shows that the x-ray scattered intensities increase with PMMA content. This evidence may be quantized by calculating the invariant using Equation 23 and leads to the obvious result that $\overline{\eta^2}$ increases with PMMA content. This phenomenon may be understood using a modification of Equation 24

$$\overline{\eta^2} = \phi_{PVF_2} \, \phi_B \, (\rho_{PVF_2} - \rho_B)^2 \tag{46}$$

where the subscript B refers to the amorphous blend of PVF_2 and PMMA, ϕ_{PVF_2} is the volume fraction of crystalline PVF_2 with electron density ρ_{PVF_2}, and $\phi_B = 1 - \phi_{PVF_2}$ is the volume fraction of the amorphous phase having electron density ρ_B. This relationship is given by

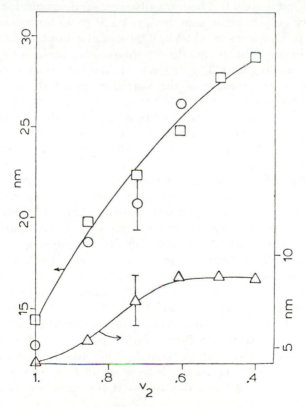

Figure 17. The variation of the correlation function peak, the long period, and the crystal thickness with composition for a PVF₂/PMMA blend. Key: ○, correlation function peak; □, long period; and △, crystalline thickness (84).

$$\rho_B = \phi_{a-\mathrm{PVF}_2} \cdot \rho_{a-\mathrm{PVF}_2} + \phi_{a-\mathrm{PMMA}} \cdot \rho_{a-\mathrm{PMMA}} \qquad (47)$$

where $\phi_{a-\mathrm{PVF}_2}$ and $\phi_{a-\mathrm{PMMA}}$ are the volume fractions of PVF$_2$ and PMMA in the amorphous phase, respectively. The electron densities of those amorphous polymers are $\rho_{a-\mathrm{PVF}_2}$ and $\rho_{a-\mathrm{PMMA}}$.

Because $\rho_{a-\mathrm{PMMA}} < \rho_{a-\mathrm{PVF}_2}$, an increase in $\phi_{a-\mathrm{PMMA}}$ leads to a decrease in ρ_B and hence an increase in the difference $(\rho_{\mathrm{PVF}_2} - \rho_B)$ and as a consequence an increase in $\overline{\eta^2}$. This calculation quantitatively supports the increase in observed $\overline{\eta^2}$ with PMMA content.

The opposite trend is seen for the blend of crystalline PCL with PVC. The parameter $\overline{\eta^2}$ is found to decrease with increasing PVC concentration. This difference occurs in this case because the electron density of the PVC is greater than that of the PCL (38, 39).

An illustration of the application of the pinhole collimated small angle x-ray apparatus at Oak Ridge to the study of oriented systems is shown in Figure 18 (89). The intensity contour diagrams for a series of low density polyethylene samples drawn at room temperature to the indicated percentages is shown. The circular contours for the unoriented sample indicate that the repeat distance for lamellar structure is equal in all directions. The distortion of the patterns for the oriented samples is a consequence of the variation in interlamellar spacing with lamellar orientation, as is consistent with the model of the spherulite deformation shown in Figure 19. The deformation proves not to be affine with the percentage increase in spacing for those lamellae oriented perpendicularly to the stretching direction being greater than affine.

With Neutron Scattering. An application of neutron scattering is in the study of the blend of amorphous isotactic polystyrene (IPS) with atactic polystyrene (APS) (90, 91). Various ratios of APS/IPS were taken, and the neutron scattering variation was observed as a function of concentration of added deuterated atactic polystyrene (APSD). The results were represented by a Zimm plot as shown in Figure 20. Radii of gyration and molecular weights of the APSD were obtained as indicated in Equation 42, and the results are summarized in Table I. The molecular weights and radii of gyration of the APSD in the blend are relatively independent of composition and are approximately equal to the values obtained in APS alone. Furthermore, the zero slope of the bottom line suggests a zero value of A_2, indicating a zero value of the Flory χ_{12} parameter. Thus, the indication is that APS and IPS are compatible and that a negligible interaction occurs between the components.

Similar results were obtained for the amorphous PVF$_2$/PMMA blend and the blend of APS with polyvinyl methyl ether (PVME).

An extensive study of the compatibility of APS with PVME by

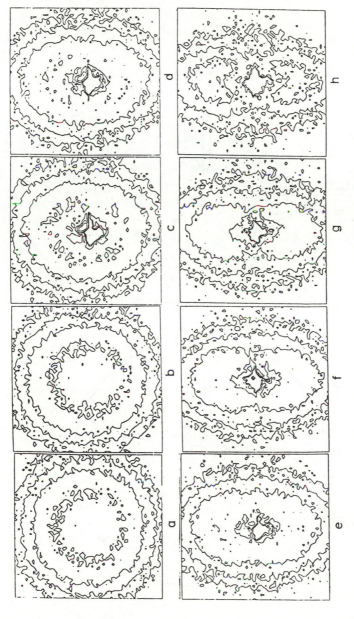

Figure 18. Small angle x-ray scattering contours for low density polyethylene samples stretched to various elongations: a, 0%; b, 10%; c, 20%; d, 30%; e, 40%; f, 50%; g, 60%; and h, 83%.

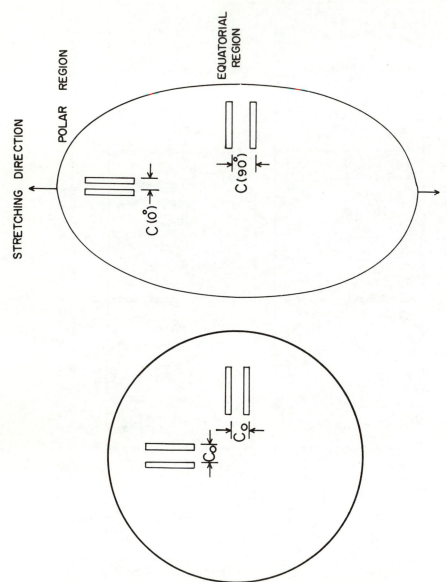

Figure 19. A model for lamellar separation in deformed spherulites (89).

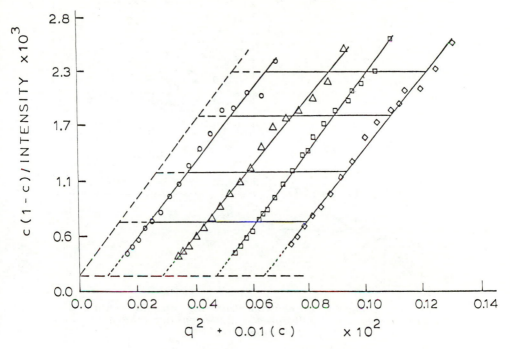

Figure 20. A Zimm plot for the blend of 25% IPS with 75% APS containing various amounts of APSD (91).

neutron scattering was carried out by Schmitt and Kirste (92), who obtained a Zimm plot as a function of temperature for this blend and showed that A_2 varies from a positive to a negative value with increasing temperature as the blend becomes more incompatible.

A study of orientation by the neutron scattering technique is illustrated in Figure 21, where scattering intensity contours are plotted for unoriented APS containing 5% APSD and that of a similar sample oriented to a draw ratio of 4.2 using Porter's extrusion orientation technique (93). The marked anisotropy of the scattering intensity is seen with the intensity falling off more rapidly with q in the draw

Table I. Small Angle Neutron Scattering Results for IPS/APS Blends

Composition APS/IPS	$(\overline{R_g})_w$, A°	$\overline{M}_w \times 10^{-5}$
Pure IPS	236 ± 43	4.59 ± 0.71
25/75	175 ± 14	5.14 ± 2.96
50/50	229 ± 27	7.61 ± 2.09
75/25	201 ± 19	6.09 ± 1.09
θ-Condition	203	5.47

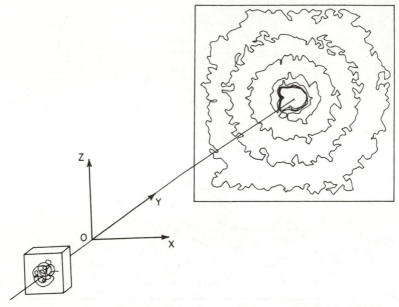

Figure 21a. A small angle neutron scattering intensity contour diagram for an unoriented polystyrene sample containing 5% deuterated
polystyrene (45).

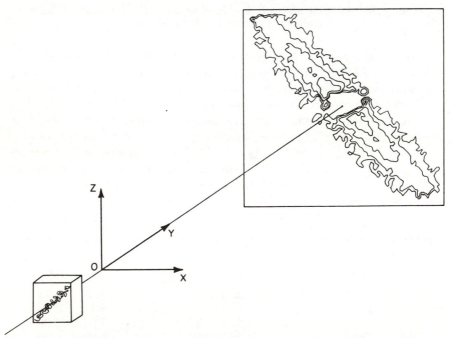

Figure 21b. A small angle neutron scattering intensity contour diagram for an extrusion-oriented (draw ratio = 4.2) polystyrene sample
containing 5% deuterated polystyrene (45).

direction than in the perpendicular direction. By performing Guinier plots in these two directions, we concluded that the radius of gyration parallel to draw $Rg_\parallel$ increases with elongation while that perpendicular to draw $Rg_\perp$ decreases, as shown in Figure 22. The changes are in approximate agreement with an affine deformation model (*45*).

The neutron scattering technique was applied to the study of the conformation of chains in crystalline polymer to define the regularity of chain folding in polymer crystals grown from solution (*94*) and from the melt (*95*). The earlier work on polyethylene recently was extended to polypropylene (*96*). Preliminary reports were made of the anisotropy of scattering from deformed crystalline polymer samples (*97*, *98*).

Conclusions

We illustrated applications of the light, x-ray, and neutron scattering techniques for studying a number of phenomena involving

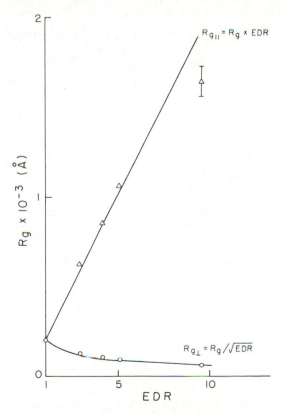

Figure 22. Radii of gyration for polystyrene molecules parallel and perpendicular to the draw direction as a function of draw ratio for extrusion-oriented APS samples (45).

solid polymers. These phenomena involve studies of crystallization kinetics, deformation, blend compatibility, and phase separation. Advances in experimental methods for such measurements permit their more convenient application to the elucidation of polymer structures and their changes.

Acknowledgments

This work was supported in part by grants from the National Science Foundation and the Army Research Office (Durham).

Literature Cited

1. Kerker, M. "The Scattering of Light and Other Electromagnetic Radiation"; Academic: New York, 1969.
2. Guinier, A.; Fournet, G. "Small Angle Scattering of X-Rays"; Wiley: New York, 1955.
3. Bacon, G. E. "Neutron Diffraction"; Oxford Univ. Press: New York, 1975.
4. Stein, R. S.; Rhodes, M. B. *J. Appl. Phys.* **1960,** *31,* 1873.
5. Van de Hulst, H. C. "Light Scattering by Small Particles"; Wiley: New York, 1957.
6. Kratky, O.; Porod, G. *Acta Phys. Austriaca* **1948,** *2,* 133.
7. Porod, G. *Acta Phys. Austriaca* **1948,** *2,* 255.
8. Mittelbach, P.; Porod, G. *Acta Phys. Austriaca* **1961,** *14,* 185.
9. Schmidt, W. P. *Acta Crystallogr.* **1955,** *8,* 722.
10. Geil, P. "Polymer Single Crystals"; Wiley: New York, 1963.
11. Stein, R. S. "Structure and Properties of Polymer Films"; Lenz, R. W.; Stein, R. S., Eds.; Plenum: New York, 1973; p. 1.
12. Yoon, D. Y.; Stein, R. S. *J. Polym. Sci., Polym. Phys. Ed.* **1974,** *12,* 735.
13. Clough, S.; Rhodes, M. B.; Stein, R. S. *J. Polym. Sci., Part C* **1967,** *18,* 1.
14. Samuels, R. J. "Structured Polymer Properties"; Wiley: New York, 1974.
15. Debye, P.; Bueche, A. *J. Appl. Phys.* **1949,** *20,* 518.
16. Debye, P.; Anderson, H. R.; Brumberger, H. *J. Appl. Phys.* **1957,** *28,* 679.
17. Kratky, O. *Pure Appl. Chem.* **1966,** *12,* 483.
18. Porod, G. *Kolloid-Z.* **1951,** *124,* 83.
19. Porod, G. *Kolloid-Z.* **1951,** *125,* 51.
20. Stribeck, N.; Ruland, W. *J. Appl. Crystallogr.* **1978,** *11,* 535.
21. Koberstein, J.; Morra, B.; Stein, R. S. *J. Appl. Crystallogr.* **1980,** *13,* 34.
22. Hashimoto, T.; Nagatoshi, K.; Toda, A.; Hasagawa, H.; Kawai, H. *Macromolecules* **1974,** *7,* 364.
23. Stein, R. S.; Wilson, P. R. *J. Appl. Phys.* **1962,** *33,* 1914.
24. Stein, R. S.; Erhardt, P.; Clough, S.; Adams, G. *J. Appl. Phys.* **1966,** *37,* 3980.
25. van Aartsen, J. J. In "Polymer Networks, Structure and Mechanical Properties"; Chompf, A. J.; Newman, S., Eds.; Plenum: New York, 1971; p. 307.
26. Zimm, B. H. In "Light Scattering from Dilute Polymer Solutions"; McIntyre, D.; Gornick, F., Eds.; Gordon & Breach: New York, 1964; pp. 149, 157.
27. Debye, P. "Topics in Chemical Physics"; Prock, A.; McConkey, G., Eds.; Elsevier: New York, 1962; p. 252.
28. Einstein, A. *Ann. Phys. (N.Y.)* **1908,** *25,* 205.
29. Debye, P. "Topics in Chemical Physics"; Prock, A.; McConkey, G., Eds.; Elsevier: New York, 1962; p. 164.
30. Debye, P. *J. Phys. Colloid Chem.* **1947,** *51,* 18.
31. Zimm, B. H.; Stein, R. S.; Doty, P. *Polym. Bull.* **1945,** *1,* 90.

32. Zimm, B. H. *J. Chem. Phys.* **1948**, *16*, 1093.
33. Zimm, B. H. *J. Chem. Phys.* **1948**, *16*, 1099.
34. Kratky, O. "Progress in Biophysics"; Pergamon: Elmsford, New York, 1953; Vol. 13, p. 105.
35. Kirste, R. G.; Kruse, W. A.; Ibel, K. *Polymer* **1975**, *16*, 120.
36. Walsh, D. J.; Higgins, J. S.; Doube, C. P.; McKeown, J. G. *Polymer* **1981**, *22*, 168.
37. Higgins, J. S.; Stein, R. S. *J. Appl. Crystallogr.* **1978**, *11*, 346.
38. Russell, T. P.; Stein, R. S. *J. Polym. Sci., Polym. Phys. Ed.* **1982**, *20*, 1593.
39. Russell, T. P., Stein, R. S. *J. Macromol. Sci., Part B* **1980**, *17*, 617.
40. Benoit, H.; Decker, D.; Higgins, J. S.; Picot, C.; Cotton, J. P.; Farnoux, B.; Jannink, G.; Ober, R. *Nature* **1973**, *245*, 13.
41. Williams, C.; Nierlich, M.; Cotton, J. P.; Jannink, G.; Boue, F.; Daoud, M.; Farnoux, B.; Picot, C.; deGennes, P. G.; Rinaudo, M.; Moan, M.; Wolff, C. *J. Polym. Sci., Polym. Lett. Ed.* **1979**, *17*, 379.
42. Akcasu, A. Z.; Summerfield, G. C.; Jahshan, S. N.; Han, C. C.; Kin, C. Y.; Yu, H. *J. Polym. Sci., Polym. Phys. Ed.* **1980**, *18*, 863.
43. Boue, F.; Nierlich, M.; Leibler, L., unpublished data.
44. Wignall, G. D.; Hendricks, R. W.; Koehler, W. C.; Lin, J. S.; Wai, M. P.; Thomas, E. L.; Stein, R. S. *Polymer* **1981**, *22*, 886.
45. Hadziioannou, G.; Wang, L. H.; Stein, R. S.; Porter, R. S. *Macromolecules* **1982**, *15*, 800.
46. Picot, C.; Duplessix, R.; Decker, D.; Benoit, H.; Cotton, J. P.; Daoud, M.; Farnoux, B.; Jannink, G.; Nierlich, M.; deVries, A. J.; Pincus, P. *Macromolecules* **1977**, *10*, 436.
47. Boue, F.; Jannink, G. *J. Phys., Orsay, Fr.* **1978**, *39(C2)*, 183.
48. Benoit, H.; Duplessix, R.; Ober, R.; Daoud, M.; Cotton, J. P.; Farnoux, B.; Jannink, G. *Macromolecules* **1975**, *8*, 451.
49. Stein, R. S.; Keane, J. J. *J. Polym. Sci.* **1955**, *17*, 21.
50. Prud'homme, R. E.; Bourland, L.; Natarajam, R. T.; Stein, R. S. *J. Polym. Sci., Polym. Phys. Ed.* **1974**, *12*, 1955.
51. Nataragam, R. J.; Prud'homme, R. E.; Bourland, L.; Stein, R. S. *J. Polym. Sci., Polym. Phys. Ed.* **1976**, *14*, 1541.
52. Stein, R. S.; Stidham, S. N. *J. Polym. Sci.* **1966**, *A2(4)*, 89.
53. Wasiak, A.; Peiffer, D.; Stein, R. S. *J. Polym. Sci., Polym. Lett. Ed.* **1976**, *14*, 381.
54. Tabar, R.; Stein, R. S. *J. Polym. Sci.*, in press.
55. Klug, H.; Alexander, L. E. "X-Ray Diffraction Procedures"; Wiley: New York, 1974.
56. Guinier, A. "Theorie et Technique de la Radiocristallographie"; Dunod: Paris, 1964.
57. "Synchrotron Radiation Research"; Winick, H.; Doniach, S., Eds.; Plenum: New York, 1980.
58. Elsner, G.; Zachmann, H. G. *Makromol. Chem.* **1981**, *182*, 657.
59. Bonse, U.; Hart, M. "Small Angle X-Ray Scattering"; Brumberger, H., Ed.; Gordon & Breach: New York, 1967.
60. Alexander, L. E. "X-Ray Diffraction Methods in Polymer Science"; Wiley: New York, 1969.
61. Borkowski, C. J.; Kopp, M. K. *IEEE Trans. Nucl. Sci.* **1970**, *NS-17*, 340; **1972**, *NS-19*, 161.
62. Hashimoto, T.; Suehiro, S.; Shibayama, M.; Saijo, K.; Kawai, H. *Polym. J.* **1981**, *13*, 501.
63. Hendricks, R. W. *J. Appl. Crystallogr.* **1978**, *11*, 15.
64. Reynolds, G. T.; Milch, J. R.; Gruner, S. M. *Rev. Sci. Instrum.* **1978**, *49*, 1241.
65. Kratky, O.; Pilz, I.; Schmitz, P. J. *J. Colloid Interface Sci.* **1966**, *21*, 24.
66. Vonk, G. C. *J. Appl. Crystallogr.* **1971**, *4*, 340.
67. Glatter, O. *J. Appl. Crystallogr.* **1974**, *7*, 147.

68. Glatter, O. *J. Appl. Crystallogr.* **1977**, *10*, 415.
69. Glatter, O. *Acta Phys. Austriaca* **1977**, *47*, 83.
70. Egelstaff, P. A. "Thermal Neutron Scattering"; Academic: New York, 1965.
71. Turchin, V. F. "Slow Neutrons"; Israel Prog. Sci. Trans. Ltd., 1968.
72. Gurevitch, I. I.; Tarasov, L. V. "Low Energy Neutron Physics"; North-Holland: Amsterdam, 1968.
73. Schmatz, W., Springer, T.; Schelten, J.; Ibel, K. *J. Appl. Crystallogr.* **1974**, *7*, 96.
74. Schelten, J.; Hendricks, R. B. *J. Appl. Crystallogr.* **1978**, *11*, 297.
75. Jacrot, B. *Rep. Prog. Phys.* **1976**, *39*, 911.
76. Yuasa, T., M.S. Thesis, Univ. of Massachusetts, Amherst, MA, 1975.
77. Stein, R. S.; Clough, S.; van Aartsen, J. J. *J. Appl. Phys.* **1962**, *36*, 3072.
78. van Aartsen, J. J.; Stein, R. S. *J. Polym. Sci.* **1971**, A2 9, 295.
79. Erhardt, P. F.; Stein, R. S. *Appl. Polym. Symp.* **1967**, *5*, 113.
80. Yuen, H. K.; Kinsinger, J. B. *Macromolecules* **1974**, *7*, 329.
81. Keijzers, A. E. M.; van Aartsen, J. J.; Prins, W. *J. Am. Chem. Soc.* **1968**, *90*, 3167.
82. Khambatta, F. B.; Warner, F. P.; Russell, T.; Stein, R. S. *J. Polym. Sci., Polym. Phys. Ed.* **1976**, *14*, 1391.
83. Koleske, J. V.; Lundberg, R. D. *J. Polym. Sci., Part B* **1969**, *7*, 897.
84. Morra, B., Ph.D. Thesis, Univ. of Massachusetts, Amherst, MA, 1980.
85. Vonk, C.; Kortleve, G. *Kolloid Z. Z. Polym.* **1967**, *220*, 19.
86. Vonk, C. G.; Kortleve, G. *Kolloid Z. Z. Polym.* **1968**, *225*, 124.
87. Vonk, C. G. *J. Appl. Crystallogr.* **1973**, *6*, 81.
88. Hosemann, R.; Bagghi, S. N. "Direct Analysis of Diffraction by Matter"; North-Holland: Amsterdam, 1962; Chap. XVII.
89. Baczek, S., Ph.D. Thesis, Univ. of Massachusetts, Amherst, MA, 1977.
90. Wai, M.; Hadziioannou, G.; Stein, R. S. *Bull. Am. Phys. Soc.* **1981**, *26*, 328.
91. Wai, M., Ph.D. Thesis, Univ. of Massachusetts, Amherst, MA, 1982.
92. Jelenic, J.; Kirste, R. G.; Schmitt, B. J.; private communication of results to be published in *Makromol. Chem.* **1982**, *183*.
93. Griswold, P. D.; Zachariades, A. E.; Porter, R. S. "Flow-Induced Crystallization in Polymer Systems"; Miller, R. L., Ed.; Gordon & Breach: New York, 1979; p. 205.
94. Sadler, D. M.; Keller, A. *Polymer* **1976**, *17*, 37.
95. Schelten, J.; Ballard, D. G. H.; Wignall, G. D.; Longman, G.; Schmatz, W. *Polymer* **1976**, *17*, 751.
96. Stamm, M.; Schelten, J.; Ballard, D. G. H. *Colloid Polym. Sci.* **1981**, *259*, 286.
97. Sadler, D. M.; Odell, J. A. *Polymer* **1980**, *21*, 479.
98. Ballard, D. G. H., private communication.
99. Leite-James, P., Ph.D. Thesis, Univ. of Massachusetts, Amherst, MA, 1979.
100. Stein, R. S. In "Rheology Theory and Applications"; Eirich, F. R., Ed.; Academic: New York, 1969; Vol. 5, p. 297.
101. Allen, G.; Higgins, J. S. *Rep. Prog. Phys.* **1973**, *36*, 1073.

RECEIVED for review December 21, 1981. ACCEPTED March 25, 1982.

Excimer Fluorescence as a Molecular Probe of Blend Miscibility

Comparison with Differential Scanning Calorimetry

STEVEN N. SEMERAK and CURTIS W. FRANK

Stanford University, Department of Chemical Engineering, Stanford, CA 94305

Results obtained using differential scanning calorimetry (DSC) on toluene-cast blends containing 35% poly(2-vinylnaphthalene) (P2VN) indicate miscibility with polystyrene (PS) and immiscibility with polymethyl methacrylate (PMMA) when the latter two polymers have molecular weights of 2000. The same conclusions are reached when excimer fluorescence of the P2VN components is used to analyze similar blends containing only 0.3% P2VN. As the molecular weight of the nonfluorescent host polymer is increased, the increase in the excimer to monomer intensity ratio of the guest P2VN suggests that the 0.3% blends move toward greater immiscibility. A fluorescence study of these blends during the solvent-casting process showed that no major differences exist in the drying behavior of the low and high molecular weight PS and PMMA matrices, and that phase separation occurs before the solvent concentration drops below 25%.

T HE CONSIDERABLE TECHNOLOGICAL SIGNIFICANCE of polymer blends has resulted in a continuing challenge to the polymer physicist to describe blend structure on the molecular level (*1–5*). A host of techniques, including differential scanning calorimetry (DSC), dynamic mechanical and dielectric spectroscopy, NMR spectroscopy, and light, x-ray, and neutron scattering have been employed to determine the degree of mixing in presumably miscible blends and the size of the

NOTE: This is part V in a series.

0065–2393/83/0203–0757$06.00/0
© 1983 American Chemical Society

domain structure in partially miscible or immiscible blends. In general, however, existing methods are limited to systems in which the minor component must be at a concentration of at least 1%, or even as high as 10%, for its presence to be detected. Although some low-concentration studies have been performed (6–8), new approaches must be developed to elucidate the details of the component interaction on the segmental level.

One class of methods that shows considerable promise for study of dilute blends is based on the emission behavior. Fluorescence spectroscopy can provide both high sensitivity at low concentrations and experimental flexibility with respect to a variety of parameters. The most common techniques are fluorescence and phosphorescence quenching (9) and depolarization (10–15). In addition, a recently developed approach utilizes nonradiative energy transfer (16–18). These methods have in common the use of specific chemical moieties, usually aromatic, that are either free dopants in the polymer system or chemically bound to one or more polymer chains.

Before the information derived from a procedure that employs such probes to infer the details of molecular structure can be accepted, however, the influence of the probe itself on the phenomenon under study must be understood. For example, the use of small-molecule probes may be hindered by difficulties in determining the precise location of the molecules in the blends owing to aggregation effects. This difficulty may be avoided by attaching the luminescent tag or label to the polymer molecule by copolymerization or grafting. The latter approach is particularly appealing for development of fluorescence methods as general purpose analytical tools. However, the number of such labels on any given chain must be small if the tagged polymer is to behave chemically like an untagged species. Consequently, the overall concentration of the tagged polymer must be increased generally beyond the dilute region of interest for studies on the early stages of phase separation.

The low-concentration advantage is regained when the fluorescent probe polymer contains a large number of chromophores, as is the case for the aromatic vinyl polymers. The formation of excited dimers, or excimers, is well known in these polymers (19, 20). In solution, an excimer may form between an electronically excited aromatic ring and an adjacent ground-state ring, which move into a coplanar sandwich arrangement because of backbone motions occurring during the lifetime of the excited state. In rigid media, on the other hand, segmental motion is restricted so that suitable excimer-forming sites (EFS) occur between two aromatic rings nominally situated in a preexisting sandwich arrangement. Although the number of such EFS is relatively small, their influence is enhanced by the occurrence of energy migration, which funnels a photon absorbed at an isolated ring along the polymer chain into the EFS. This phenomenon is common in the aromatic vinyl polymers (*21*). The occurrence of phase separation of a fluorescent polymer from a nonfluorescent host causes additional intermolecular EFS and may increase the efficiency of the energy migration process as well. Both factors increase the observed excimer fluorescence.

The measurement of excimer fluorescence is experimentally quite straightforward. The broad, structureless fluorescence of the excimer is distinctly different from the structured vibrational progression of an isolated aromatic ring, or monomer. The ratio of excimer to monomer emission intensities, I_D/I_M, provides a convenient, albeit relative, experimental measure of the degree of phase separation in a blend containing an aromatic vinyl polymer. Quantitative determination of the domain size will require transient measurements of the excimer and monomer emission as well as a detailed understanding of the energy migration process. Although these points are presently under study, recent research has demonstrated conclusively that the technique has the potential of providing the molecular level information of interest. These works include studies of the effects of the molecular weight of poly(2-vinylnaphthalene) (P2VN) in a second polymer matrix (*22, 23*), the solubility parameter of the host polymer (*24, 25*), and the concentration of the P2VN guest (*23, 25, 26*) on the blend fluorescence.

The objective of the present work was to relate the classical methods of optical clarity and DSC to excimer fluorescence from the guest in a study of the effect of host molecular weight on blend miscibility. The host polymers used are polystyrene (PS) and polymethyl methacrylate (PMMA) having molecular weights ranging between 1100 and 390,000. In addition, excimer fluorescence from the blends was monitored during the solvent-casting process to determine the point of phase separation and the moment at which the amorphous blends become glassy.

Experimental

The P2VN guest polymer sample, prepared as described previously (*24*), had a viscosity-average molecular weight of 70,000. The 1,3-bis(2-naphthyl)propane ($\beta\beta$DNP) was prepared using a published procedure (*27*). The PS host samples were obtained from the Pressure Chemical Co. with M_n (the number-average molecular weight) = 2200, 4000, 9000, 17,500, 35,000, 100,000, 233,000, and 390,000. They all had M_w/M_n values less than 1.1 (where M_w represents the weight-average molecular weight). The PMMA host samples with M_n = 54,000, 79,000, 92,000, 180,000, and 350,000 also were obtained from the Pressure Chemical Co., and had M_w/M_n values less than 1.15. In addition, a broad molecular weight distribution PMMA sample having M_w = 12,000 was obtained from Scientific Polymer Products. Finally, a second broad-distribution PMMA sample with $M_w = 20,000$ was received from Rohm and Haas Co., as were PMMA samples having $M_w = 1100$ and 2500 with M_w/M_n values less than 1.2. The P2VN and PS samples were repeatedly precipitated from toluene into methanol to remove residual fluorescent impurities, but the PMMA samples were clean enough to be used as received.

Low-concentration blends containing 0.3 wt % P2VN were prepared by solvent casting from toluene onto sapphire disks at room temperature to produce film thicknesses of 25 μm. The films subsequently were examined using backface illumination with a spectrofluorometer that has been described previously (24). An alternate procedure utilized glass microscope slides for the casting substrate. In this case, frontface illumination was required to obtain fluorescence spectra. Both methods gave identical results, within experimental error. Films were dried in air at room temperature for 24 h, at which point they typically contained 5–8% residual solvent.

Results and Discussion

Effect of Molecular Weight of the Host Matrix on the P2VN Guest Fluorescence.   Typical fluorescence spectra of P2VN in blends with PS (2200) and PS (233,000) are shown in Figure 1 as Curves A and B, respectively. Each spectrum consists of a broad structureless band, corresponding to the excimer emission, superimposed on a structured high-energy (monomer) emission similar to that of 2-ethylnaphthalene. For comparison, the spectrum of 2-ethylnaphthalene in PS is given by

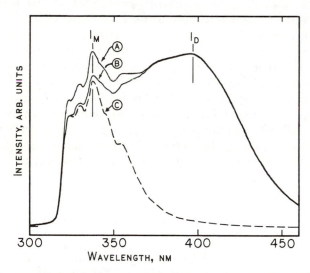

Figure 1. Uncorrected fluorescence spectra of 0.3 wt % P2VN (70,000) and 2-ethylnaphthalene in PS. Excitation wavelength was 290 nm. Curve A, P2VN in PS (2200); Curve B, P2VN in PS (233,000); and Curve C, 2-ethylnaphthalene in PS (158,000). D and M refer to dimer and monomer, respectively. Curve B is normalized to Curve A at 398 nm; Curve C is normalized to Curve B at 323 nm. (Reproduced from Ref. 23. Copyright 1981, American Chemical Society.)

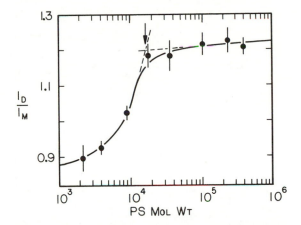

Figure 2. Fluorescence ratio, I_D/I_M, of 0.3 wt % P2VN (70,000) in PS vs. PS mol wt. The "knee" of the curve is indicated by the arrow. (Reproduced from Ref. 23. Copyright 1981, American Chemical Society.)

the dashed line, Curve C. The intensities of the monomer emission, I_M, and the excimer emission, I_D, were measured at 337 and 398 nm, respectively.

The effects on the fluorescence caused by increasing the molecular weight of the host matrix for the PS and PMMA host polymers are presented in Figures 2 and 3, respectively. Two observations are significant. First, I_D/I_M for the P2VN guest appears to level off at host molecular weights below 4000. An upper limit of the ratio is obtained for both host polymers at molecular weights above 20,000, and sigmoidal curves describe the ratio at intermediate molecular weights. Second, the fluorescence ratio is consistently higher for the PMMA host compared with PS, being 100% higher at the low molecular weight limit and 180% higher at the high molecular weight limit.

In previous work (26), the I_D/I_M ratio was shown to depend on both the number of EFS and the efficiency with which these sites are sampled by the randomly migrating exciton. At present, it is not possible to distinguish unambiguously between an increase in the number of intermolecular EFS or an increase in the likelihood of exciton hopping across loops of the same chain or between different chains as sources of explanation for an increase in I_D/I_M. This point is of little importance for this study, however, because both phenomena could result from aggregation of guest polymer chains in a thermodynamically incompatible host polymer matrix. Because an increase in host molecular weight will lead to a reduction in the combinatorial entropy of mixing, phase separation of an initially miscible blend could result for modest molecular weights if the binary interaction parameter is small and positive (28, 29). In earlier work on polystyrene hosts, part

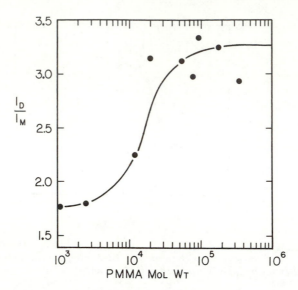

Figure 3.   Fluorescence ratio, I_D/I_M, of 0.3 wt % P2VN (70,000) in PMMA vs. PMMA mol wt.

of which is reproduced in Figure 2, the high molecular weight break point in I_D/I_M could be explained satisfactorily from binodals calculated using classical Flory–Huggins theory with a binary interaction parameter based on the solubility parameters of Hildebrand regular solution theory (*23*). Corresponding calculations for the P2VN/PMMA system will be reported separately (*30*).

In earlier studies on low- and moderate-concentration blends of P2VN with PS and PMMA, the fluorescence behavior and optical quality of the blends led to the conclusion that the P2VN/PS blends were apparently more thermodynamically compatible than the P2VN/PMMA blends (*24, 25*). Although there is essentially no difference in the high molecular weight break points in Figures 2 and 3, the larger values of I_D/I_M for P2VN/PMMA blends are consistent with a higher degree of P2VN segmental clustering in the PMMA host.

One of the main theses of the present research effort is that the excimer fluorescence method shows considerable potential for elucidation of molecular structure in the amorphous solid state. To establish credibility for any new method, however, correlating the results with more established classical techniques is necessary. Two such methods relevant to the study of immiscible polymer blends are the use of optical clarity to observe visible phase separation and DSC to detect multiple glass transition temperatures. Both approaches are pursued in this work.

In consideration of the first method, only those PMMA blends having molecular weights greater than 12,000 show visible signs of immiscibility with a faint bluish tint, whereas the remaining PMMA blends and all PS blends are optically clear. In the visibly cloudy blends, phase separation probably has occurred. On the other hand, the interpretation of optically clear films is ambiguous because in fact a two-phase system may still exist with domains that are either very small or arranged in a bilayer structure. Also, the refractive index difference between blend polymers may be small enough for virtual elimination of scattered light, independent of domain size or amount. Nevertheless, the use of visual appearance is widespread, for example, in the generation of equilibrium cloud point curves to represent the system binodal. Thus, correlation of the fluorescence results with the optical quality is of interest.

In previous work on blends with P2VN in a series of poly(alkyl methacrylates) and in PS, films with I_D/I_M ratios less than 3.0 ± 0.5 were optically clear although those with larger values were cloudy (26). The present results are consistent with these observations. A detailed model of the significance of the absolute value of I_D/I_M in the blends is under study (31). The important point for this work is that, although all of the P2VN/PS blends are optically clear, the same qualitative increase in I_D/I_M with increasing host molecular weight is observed in both host systems. Apparently, phase separation is occurring in both cases. Thus, excimer fluorescence is sensitive to phase separation, which is not apparent from the optical quality of the 0.3% P2VN/PS blend.

To obtain a better understanding of the morphological state of the blends with the low molecular weight hosts that exhibit clear films and low values of I_D/I_M, DSC was employed to search for multiple glass transitions. Figure 4 presents DSC thermograms of the pure polymers PS (2200), PMMA (2500), and P2VN (70,000). The glass transition temperatures, as determined from the midpoint between the arrows designating deviation from the low- and high-temperature heat capacity slopes, are 60 °C for PMMA (2500), 64 °C for PS (2200), and 152 °C for P2VN (70,000). Unfortunately, as noted earlier, DSC is totally insensitive to the existence of multiple phases in the low-concentration regime of interest for the fluorescence work. As a result, it was necessary to increase the amount of P2VN in the blends to 35 wt % to detect two separate phases, should they exist. Because these blends are two orders of magnitude more concentrated in P2VN than those examined in the fluorescence work, the experiment presents a stringent test of miscibility of the low molecular weight host systems.

Thermograms for 35% blends of P2VN (70,000) with PMMA (2500) and PS (2200) are shown in Figure 5. The lower curve indicates that the P2VN (70,000)/PS (2200) blend is miscible because a single,

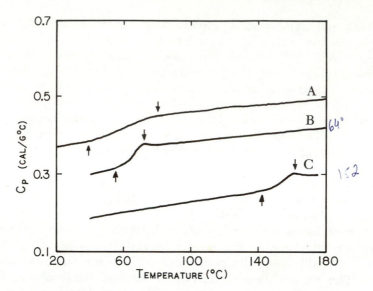

Figure 4. Heat capacity, C_p, of unblended polymers vs. temperature for samples heated at 20 °C/min. Curve A, PMMA (2500); Curve B, PS (2200); and Curve C, P2VN (70,000). Curve C is true; the other curves have been displaced 0.1 cal/g °C from the curves immediately beneath them. (See text for scheme used to locate the arrows.)

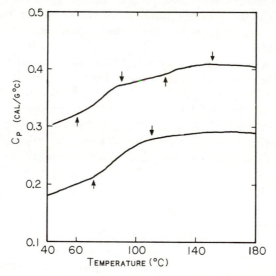

Figure 5. Heat capacity, C_p, of solvent-cast blends containing 35 wt % P2VN (70,000) vs. T for samples heated at 20 °C/min. The second component of the blend is given for each curve. The upper curve [PMMA (2500)] has been displaced 0.1 cal/g °C from the lower curve [PS (2200)]. The upper temperature arrow for the P2VN/PS blend is found by extrapolating the data between 80 and 95 °C and the data above 145 °C. Other transition arrows are located using the same scheme as for Figure 4.

broad transition having T_g = 91 °C, which is intermediate between the transitions of the unblended polymers, is shown. In addition, this blend was optically clear. On the other hand, the blend containing low molecular weight PMMA proved to be immiscible, as shown in the upper curve of Figure 5. Transitions at 75 and 125 °C were observed in the DSC trace. Although this blend was hazy and white, the fact that the transition temperatures of PMMA and P2VN in the blend are shifted toward each other indicates a small degree of mixing. This solvent-cast blend was originally clear before it was heated in preparation for the DSC experiments.

The relevance of the DSC results for 35% P2VN blends to the miscibility of 0.3% P2VN blends requires extrapolation of temperature and concentration. The DSC results are interpreted to indicate that the 0.3% P2VN/PS (2200) blend is miscible, and the corresponding blend with PMMA (2500) is nearly miscible. The deviation from miscibility in the latter case presumably is reflected in the increase in I_D/I_M at low host molecular weight.

The major difference between the fluorescence technique and the use of optical clarity should now be apparent: P2VN/PS (>15,000) blends appear immiscible using excimer fluorescence, and appear miscible (clear) under visible light. Furthermore, for identical molecular weight, PMMA blends are less miscible than PS blends. A small domain size in the P2VN/PS blends probably is responsible for the insensitivity of the optical technique, although a small refractive index difference between P2VN and PS may also be a factor. Data for the refractive index of P2VN were unavailable. Nevertheless, the fluorescence technique will be more sensitive to phase separation than a light scattering method in such cases. Although light scattering is still the most flexible technique for studying low-concentration blends because neither polymer is required to be fluorescent, this study provides additional evidence for the proposal that excimer fluorescence can signal immiscibility before conventional techniques are able to detect such a change.

Correlation of P2VN Fluorescence and Solvent Content During Casting. In the previous section, the blends were treated as binary systems, in spite of the fact that residual solvent is likely to remain in the solvent-cast films. Because the presence of residual solvent has a strong plasticizing effect on the T_g of the blend and, in addition, the T_g values of the host polymers vary over a 60 °C range, it is quite possible that there may be kinetically caused differences in the glassy blends that are cast at the same temperature. The objective of this section is to examine the solvent-casting process in more detail, with particular attention to the solvent concentration at which the phase structure of the resulting blend is fixed. The fluorescence behavior and the solvent evaporation were monitored continuously, in separate experiments, during the casting of 0.6 wt % blends of P2VN with PS (2200), PS (233,000), PMMA (2500), and PMMA (180,000).

In the first set of experiments, glass slides used as film substrates were mounted horizontally in the spectrofluorometer sample chamber and a casting solution prepared in the usual manner was spread onto the slide at time zero. Although the basic spectrofluorometer has been described previously (24), the sample illumination optics had to be modified to allow front face excitation in the horizontal plane. A spectral scan, lasting 2 min, was begun immediately. All spectra were taken with exciting light at 290 nm and were not corrected for instrumental response. Repeat scans were made at intervals of 10 min or more, with the sample unilluminated between scans. No attempts were made to exclude oxygen from the casting solution or sample chamber.

In these experiments, the fluorescence ratio I_D/I_M is expected to be initially large because, in fluid solution, backbone rotations in P2VN facilitate rapid sampling of conformations (both excimer-forming and non-excimer-forming). This process effectively moves the small number of EFS to the excited aromatic rings. When backbone rotations slow down relative to the lifetime of the excited ring, the fluorescence ratio should decrease. Energy migration, which is less important in fluid solution, then becomes the sole mechanism for transferring electronic energy to the EFS. Consequently, the fluorescence ratio decreases as long as no phase separation occurs. If phase separation does occur, however, the ratio should increase owing to an increased number of excimer-forming sites. These opposing effects will control the fluorescence ratio of an immiscible blend during the casting process.

The fluorescence results for P2VN blends with low and high molecular weight PS and PMMA are presented in Figure 6. As expected, all blends exhibit a sharp drop in I_D/I_M between 10 and 20 min of drying time. This behavior also is shown by the double-ring model compound for P2VN, which is $\beta\beta$DNP. The fluorescence behavior of $\beta\beta$DNP in PS (2200) was identical to the behavior observed in the other three polymer hosts. After the fluorescence freeze in point, small, noticeable decreases in the ratio for all P2VN blends were observed up to 100 min of drying time. Little change was observed in the fluorescence behavior between 100 min and 24 h except for the PMMA (2500) sample, which showed a modest decrease in the ratio and for the PS (233,000) sample, which showed a slight increase in the ratio. After 24 h, all blends were optically clear except the PMMA (180,000) host, which had a bluish tinge.

In the second experiment, a weight-loss study was performed to

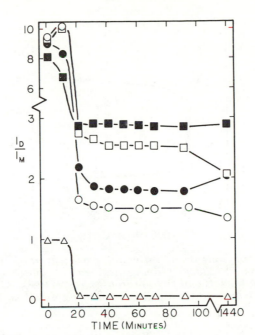

Figure 6.   Fluorescence ratio, I_D/I_M, of solvent-cast blends containing 0.6 wt % P2VN (70,000) or ββDNP vs. drying time. P2VN blends with PS (2200), ○; PMMA (2500), □; PS (233,000), ●; PMMA (180,000), ■. ββDNP blends with any of the preceding polymers, △. All casting solutions initially contain 8 g/dL polymer in toluene. (See Figure 7 for more details.)

estimate the solvent concentration as a function of time in the fluorescence casting experiment. To do so, a sample slide was placed on the pan of a Sartorius model 2432 balance, and a casting solution containing PS (233,000) was spread onto the slide in a manner identical to the fluorescence study. The weight of the solvent-cast film subsequently was recorded vs. time. When other host polymers were used, virtually no change in the drying behavior was observed. During the first 20 min of drying, a constant rate of solvent evaporation (in mass per time unit) was recorded. If the balance chamber were airtight, then the amount of toluene vapor contained in the chamber after one film had been cast and completely dried would be only one-sixth of the amount contained in the chamber when filled with toluene-saturated air. Thus, the solvent evaporates into essentially toluene-free air, as reflected by the constant drying rate. Although the sample chamber of the fluorescence spectrometer is only one-half as large as the balance chamber, constant-rate drying also is expected for the fluorescence samples.

The concentration of solvent in the blends during the fluores-

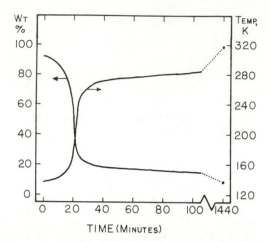

Figure 7. Solvent content and estimated glass transition temperature for a solvent-cast PS (233,000) film vs. drying time. Solvent content and T_g are given on the left and right ordinates, respectively. Glass transition temperatures are computed using Equation 1.

cence study of the casting process can be estimated from these weight-loss measurements, which are given in Figure 7. Also shown are estimates of the glass transition temperature for the PS–toluene solution during drying. These temperatures were calculated from Equation 1, which is an empirical Kelly–Bueche equation drawn from the data of Adachi et al. (32):

$$T_g = 373 \left\{ \frac{1 + 0.13w}{1 + 2.22w} \right\} \tag{1}$$

in which T_g is the glass transition temperature in K and w is the weight fraction of solvent. Although these temperatures are liable to have an error of ±50 °C, they should provide an estimate of the effective cooling rate, that is, how rapidly the T_g of the PS–toluene solution increases with time. As seen in Figure 7, the drying film changes rapidly between 15 and 30 min elapsed drying time, going from 80% to 20% solvent and showing an increase in T_g of 110 °C. The effective cooling rate is about 10 °C/min.

Analysis of the dynamics of the guest–host coil interaction as the solution becomes more concentrated on solvent evaporation will not be dealt with in this chapter. Rather, the important point of interest is to characterize the point at which phase separation occurs in the ternary system. Comparison of the weight loss and I_D/I_M measurements shows that the decrease in I_D/I_M coincides closely with the loss of solvent and subsequent increase in T_g. The high rate of decrease in both parameters and the slight ambiguities associated with use of two different, albeit similar, drying environments make it difficult to state precisely the solvent content in the film at the time that I_D/I_M becomes fixed. A realistic lower limit would be that present after 25 min, about 25 wt %. The upper limit is more difficult to discern, but it could be as large as 60%. Although a more detailed study of the nonequilibrium aspects of the casting process is necessary, analysis of the thermodynamics of the resulting solvent cast blend must explicitly include the residual solvent. A first attempt at this analysis was made recently (33), and the effects of other casting solvents now are being studied (34).

Summary

Solvent-cast blends containing 0.3% P2VN in both PS and PMMA hosts were examined using the techniques of optical clarity, DSC, and excimer fluorescence of the P2VN component. The results of the different methods are compared in Table I. Lowering the molecular weight of the host polymer appears to encourage blend miscibility, but the effect is relatively minor compared to differences in miscibility caused by chemically different host polymers. The results obtained using the fluorescence technique are consistent with previous studies and add support to the interpretation that changes in the fluorescence ratio precede visible signs of phase separation. The fluorescence study of the solvent-casting process showed that most phase separation occurs before the solvent concentration drops below 25%, at which point the backbone motions of the P2VN guest become infrequent on a nanosecond timescale.

Table I. Miscibility of 0.3% P2VN Blends Cast from Toluene

Host (Molecular Weight)	Technique		
	Optical	DSC[a]	Fluorescence
PS(<15,000)	clear[b]	miscible	miscible
PS(>15,000)	clear[b]	c	immiscible
PMMA(<15,000)	clear	immiscible	partially miscible
PMMA(>15,000)	cloudy	c	very immiscible

[a] P2VN is not detectible at 0.3%; DSC blends are annealed at 150 °C.
[b] Refractive index difference may be small enough to feign miscibility.
[c] Glass transition temperatures of blend polymers are not sufficiently different.

Acknowledgments

This work was supported by the Polymers Program of the National Science Foundation under grants DMR 77–09372 and DMR 79–16477.

1,3-Bis(2-naphthyl)propane was prepared by R. Trujillo of Sandia Laboratories, Albuquerque, NM. Various PMMA samples were kindly provided by D. J. McDonald of Rohm and Haas Co., Bristol, PA.

Literature Cited

1. Platzer, N. A. J. In "Copolymers, Polyblends and Composites"; ADVANCES IN CHEMISTRY SERIES No. 142; ACS: Washington, D.C., 1975.
2. Klempner, D.; Frisch, K. C. In "Polymer Alloys—Blends, Blocks, Grafts and Interpenetrating Networks"; Polymer Science and Technology, Vol. 10; Plenum: New York, 1977.
3. Sperling, L. H. In "Recent Advances in Polymer Blends, Grafts and Blocks"; Plenum: New York, 1974.
4. Paul, D. R.; Newman, S. In "Polymer Blends"; Academic: New York, 1978; Vol. 1.
5. Olabisi, O.; Robeson, L. M.; Shaw, M. T. In "Polymer–Polymer Miscibility"; Academic: New York, 1979.
6. Yuen, H. K.; Kinsinger, J. B. *Macromolecules* **1974**, *7*, 329.
7. Anavi, S; Shaw, M. T.; Johnson, J. F. *Macromolecules* **1979**, *12*, 1227.
8. Roe, R.-J.; Zin, W.-C. *Macromolecules* **1980**, *13*, 1221.
9. Somersall, A. C.; Dan, E.; Guillet, J. E. *Macromolecules* **1974**, *7*, 233.
10. Nishijima, Y. *Ber. Bunsenges Phys. Chem.* **1970**, *74*, 778.
11. Nishijima, Y. *J. Macromol. Sci. Phys.* **1973**, *8*, 389.
12. Miller, L. J.; North, A. M. *J. Chem. Soc. Faraday 2* **1975**, *71*, 1233.
13. Rutherford, H.; Soutar, I. *J. Polym. Sci., Polym. Phys. Ed.* **1977**, *15*, 2213.
14. Jarry, J. P.; Monnerie, L. *J. Polym, Sci. Polym. Phys. Ed.* **1978**, *16*, 443.
15. Chapoy, L. L.; Dupré, D. B. *Methods Exp. Phys.* **1980**, *16(A)*, 404.
16. Morawetz, H.; Amrani, F. *Macromolecules* **1978**, *11*, 281.
17. Amrani, F.; Hung, J. M.; Morawetz, H. *Macromolecules* **1980**, *13*, 649.
18. Mikês, F.; Morawetz, H.; Dennis, K. S. *Macromolecules* **1980**, *13*, 969.
19. Birks, J. B. In "Photophysics of Aromatic Molecules"; Wiley-Interscience: New York, 1970; Chap. 7.
20. Klöpffer, W. In "Organic Molecular Photophysics"; Birks, J. B., Ed.; Wiley: New York, 1973; Vol. 1, Chap. 7.
21. Klöpffer, W. *Spectrosc. Lett.* **1978**, *11*, 863.
22. Ito, S.; Yamamoto, M.; Nishijima, Y. *Rep. Prog. Polym. Phys. Jpn.* **1978**, *21*, 393.
23. Semerak, S. N.; Frank, C. W. *Macromolecules* **1981**, *14*, 443.
24. Frank, C. W.; Gashgari, M. A. *Macromolecules* **1979**, *12*, 163.
25. Frank, C. W.; Gashgari, M. A.; Chutikamontham, P.; Haverly, V. J. In "Structure and Properties of Amorphous Polymers"; Walton, A. G., Ed.; Elsevier: New York, 1980; pp. 187–210.
26. Frank, C. W.; Gashgari, M. A. *Ann. N.Y. Acad. Sci.* **1981**, *366*, 387.
27. Chandross, E. A.; Dempster, C. J. *J. Am. Chem. Soc.* **1970**, *92*, 3586.
28. Flory, P. J. In "Principles of Polymer Chemistry"; Cornell Univ. Press: Ithaca, NY, 1953; Chaps. 12 and 13.
29. Krause, S. In "Polymer Blends"; Paul, D. R.; Newman, S., Eds.; Academic: New York, 1978; Vol. 1, Chap. 2.

30. Semerak, S. N.; Frank, C. W., in press.
31. Gelles, R.; Frank, C. W. *Macromolecules* **1982**, *15*, 747.
32. Adachi, K.; Fujihara, I.; Ishida, Y. *J. Polym. Sci., Polym. Phys. Ed.* **1975**, *13*, 2155.
33. Gashgari, M. A.; Frank, C. W. *Macromolecules* **1981**, *14*, 1558.
34. Semerak, S. N.; Frank, C. W., in press.

RECEIVED for review October 14, 1981. ACCEPTED December 9, 1981.

INDEX

Copy Editors: Deborah Corson and Katherine Mintel
Indexer: Deborah Corson
Production Editor: Paula M. Bérard
Managing Editor: Janet S. Dodd
Jacket Designer: Kathleen Schaner

Typesetting: Sheridan Press, Hanover, PA
and Service Composition Co., Baltimore, MD
Printing: Maple Press Company, York, PA